Handbook of
VALVES, PIPING AND PIPELINES

1st Edition

by R.H.Warring

TRADE & TECHNICAL PRESS LTD.,
MORDEN, SURREY, SM4 5EW,
ENGLAND.

ISBN. 85461 087 1

Printed in Great Britain by TTP (Printers) Ltd., East Molesey, Surrey.

PREFACE

The most intricate and reliable piping scheme ever created is embodied within the human frame. With relief valves, safety valves, non-return valves, temperature control valves, actuators, regulators, float controls, traps and drainers, integral pump, flexible piping, sealing, expansion and contraction joints, corrosion protection, fault diagnosis and monitoring, etc, it is as complete in content as the Handbook of Valves, Piping and Pipelines. Designed to work continuously under widely variable conditions for upwards of seventy or eighty years, it is the most perfect piping system ever designed. Leonardo da Vinci marvelled at the mechanics of the human circulatory system. Yet even this requires regular maintenance, occasional repair and even artificial substitution. A three minute arrest can destroy or irreparably damage the whole system. How much more then does a mere mortal engineer need to consider in order to design a reliable and efficient piping scheme? The known variables are numerous, the unknown innumerable. This First Edition contains all necessary and essential, plus much useful, information and data, in ready-reference format, on industrial valves, piping and pipelines. It is a vital contribution to modern industry and an indispensable part of a pumping engineer's bookshelf.

The Publishers

Acknowledgements

Aiton & Co Ltd
Burkert
Dyno-Rod Ltd
Fagersta
Fike Metal Products
S. Forster
Furmanite International Engineering
Gestra (UK) Ltd
Hopkinsons Ltd
D. L. Hore
IMI Bailey Valves Ltd
The International Meehanite Metal Co Ltd
Lancashire Fittings Ltd
Latty International
Neptune Glenfield Ltd
Plascoat International
PI Corrosion Engineers Ltd
The Reiss Engineering Co Ltd
Rotork Controls Ltd
Stanton & Staveley Ltd
Siemens Aktiengesellschaft
TA Controls Ltd
TAC Construction Materials Ltd
Tour and Anderson Ltd
Dr R. J. Wakelin
R. H. Warring
R. Watson

CONTENTS

SECTION 1

Introductory

Pipes and Pipelines - Definitions and Explanations

ACCORDING TO the Oxford dictionary a *pipe* is a tube, whereas a *tube* is a long, hollow cylinder. Neither is of any help in establishing true definitions, for there are recognized differences between pipes and tubes — but not those the dictionary gives. The more obvious distinction is that 'a pipe is a big tube, and a tube is a small pipe' — which is not far from the truth in application. But we are also concerned with differences in usage of terms in different industries — and different countries.

Taking the 'big tube/small pipe' premise as substantially correct, we can further comment that pipes which may run up to several feet in diameter are cast, spun, welded up or otherwise fabricated, depending on the materials and sizes involved. Nobody could logically visualize producing very small sizes of pipes — say under 1 inch diameter — by such time consuming methods. It is much quicker — and cheaper — to produce them by extrusion. Hence tubes are basically (but not exclusively) extruded products, involving reduction in size during manufacture in the case of metal tubes, and a moulding process in the case of plastic tubes.

Just to confuse the issue some tubes are produced by rolling to shape and seam welding or seam jointing; and large size plastic tubes, which then become pipes, are produced by the same methods as small plastic tubes. But ignore that for the moment. A main difference does emerge from the two different methods of manufacture. Inherently, tubes have a smooth bore as manufactured. Pipes will have a varying degree of bore roughness, depending both on the material involved and the actual fabrication method. Once you extend tube-manufacturing process to pipe production, then these pipes also have a smooth bore (*eg* plastic pipes). Pipes produced by pipe-manufacturing methods normally require specific after treatment to render them smooth bore.

With this difference (and there are exceptions to the rule), we can further differentiate between the two by size ranges, and terminology adopted by different industries. One of the main users of smooth bore small diameter tubes, for example, is the hydraulic industry where line sizes may range from 1/8 in bore up to ¼ in bore; or larger in low pressure hydraulic systems. And we have called them *lines* — not pipes *or* tubes. The industry itself may call them hydraulic pipes, hydraulic tubes or hydraulic lines. And larger (pipes sizes!) of hydraulic tubes are produced for cylinder tubes.

Industries and applications concerned with the conveyance of fluid products (almost) invariably refer to their tubular products as *pipes.* Again sizes may range down into 'tube' sizes (and even be

Circumferential winding of a cylinder type prestressed concrete pipe.

Polyethylene sleeving of a 600 mm dia. pipeline at Aylesbury by South West Water Authority. (Stanton & Staveley)

Manufacturing the reinforcement for centrifugally spun concrete pipes.

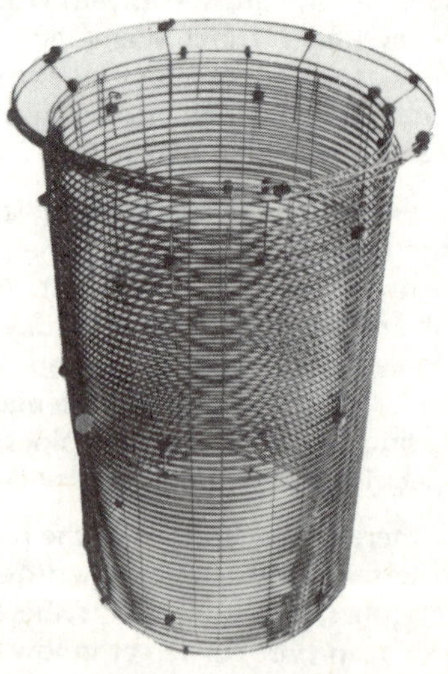

Typical double-cage reinforcement for spun concrete pipes.

drawn or extruded products or true tubes) – *eg* gas pipes and small bore water service pipes. But they are all still *pipes.* And the system they provide is a pipeline.

Hopefully this has established a satisfactory definition and explanation of why the title of this Handbook is specifically concerned with PIPES AND PIPELINES, for these are the user-fields mainly covered. And those user fields *call* their tubular products *pipes.* But *tubes* are mentioned and described where appropriate.

There remains one distinction between British and American practice to clarify. In the UK the handling and installation of pipes, performance calculations, *etc,* embracing the complete system are commonly referred to as *pipework, eg* pipework installations, pipework calculations, *etc.* In America the word 'pipework' does not appear to be accepted and is seldom, if ever, used. In the interest of rationalization this Handbook uses the single description *pipeline.* It means the same as 'pipework'.

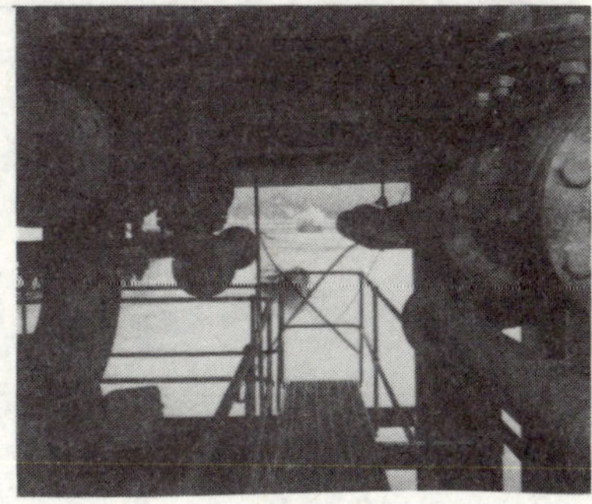

Pipework from chemical dosing tanks at Water Treatment Works by the Peterborough Division of the Anglian Water Authority
(Stanton & Staveley)

SUMMARY OF PIPE MATERIALS – METALLIC

Material	Manufacturing Process	Size Range	Typical Applications	Remarks
Aluminium	Drawing or rolling (seamless tube)		Cryogenic and chemical pipelines; lightweight hydraulic pipes	Low weight and good corrosion resistance
Copper	Drawing or rolling (seamless tubing)	Mainly small bore tubes	Marine applications. Hot water services (domestic).	Resistant to corrosion but costly.
Ductile iron	Spinning	Up to 24 inches (600 mm)	Gas and water distribution systems	Stronger than cast iron.
Grey cast iron	Casting	Up to 48 inches (1 200 mm)	Gas, water and drainage systems.	Brittle material.
Malleable iron	Heat treated casting		Mainly used for small fittings.	Less brittle than cast iron.
Steel	Various	Up to 160 inches (4000 mm)	Gas and oil pipelines	Available in a wide range of tensile strengths.
Stainless steel	Various		Cryogenic and chemical pipelines. Stainless steel tubing for domestic water supplies, plumbing and heating.	Corrosion resistant, but high cost
Tungum	Extrusion	Mainly small bore tubes	Marine applications. Specialized hydraulic systems	Corrosion resistant, non-sparking material.

SUMMARY OF PIPE MATERIALS – NON-METALLIC

Material	Size Range	Corrosion Resistance	Typical Applications	Remarks
Asbestos-Cement	2–42 inches (50–1050 mm)	Very good in most soils	Buried water pipelines and drainage systems	Brittle material
Clay		Very good	Drainage pipelines and ducts.	Brittle material. Normally salt-glazed.
Concrete	6–76 inches (150–1950 mm)	Very good in most soils	Drainage pipelines	Produced in unreinforced and reinforced forms.
Spun Concrete		Good resistance to sulphate attack and sewer gas	Sewerage,drainage, etc.	Smooth, concentric bore. Smooth external finish. High density, steel reinforced.
Prestressed concrete	Up to 120 inches (3000 mm)	Very good in most soils	Large water and drainage pipelines	Suitable for very large diameters
Pitch/Fibre	2–9 inches (50–225 mm)	Very good in most soils	Small drainage pipelines.	

PLASTIC PIPES: ABS	½–6 inches (12–150 mm)	Corrosion free but lower chemical resistance than PVC	Alternative to PVC where better mechanical pipelines required.	Suitable for solvent jointing.
GRP	Up to 190 inches (4800 mm)	Corrosion free	Large water and drainage pipelines.	Thermoset material. Also available in other reinforced plastic matrix (RPM) constructions. Disadvantage: high cost
Polyvinyl chloride UPVC	Up to 42 inches (1050 mm)	Corrosion free	General purpose pipelines suitable for a wide range of exterior and interior applications.	Unplasticized PVC. Suitable for solvent welding. Widely available.
Polyvinyl chloride CPVC	Up to 14 inches (360 mm)	Corrosion free	Cold and hot water services, domestic plumbing, etc.	Rigid PVC
Polypropylene (PP)	Up to 40 inches (UK sizes up to 16 inches) (1000 mm)	Similar to PE, but superior for resistance to detergents	Applications required good combined temperature/pressure pipelines, eg effluent, pulp mills etc.	Subject to embrittlement at low temperatures
Polypropylene (CO-PP)	Ur to 12 inches ([illegible] mm)			Copolymer of PP with better resistance.
Polypropylene (PVDF)		High chemical resistance including acids, alkalis and hydrocarbons	Specialized applications; higher service temperatures than possible with other thermoplastic pipes	Fusion jointed
Polybutylene (PB)	Up to 24 inches (600 mm)		Hot water applications – suitable for temperatures up to 110°C (230°F).	Cheaper than PEX, better abrasion resistance than PEH, (particularly at elevated temperatures). Cannot be solvent welded.
Polythene (PE) (PEL)	Up to 8 inches (200 mm)	Corrosion free	Agriculture and irrigation	Low density polythene: pressures to 6 bar
Polythene (PEM)	Up to 20 inches (500 mm)	Corrosion free	Gas distribution. General purpose pipelines for exterior and interior applications.	Medium density polythene: fused jointed or mechanical joints.
Polythene (PEH)	Up to 70 inches (1800 mm)	Corrosion free	Water distribution, sewage, industrial effluent, etc. Gas distribution.	High density polythene: fusion jointed (socket, butt or saddle); also mechanical forms.
Polythene HMW–PEH	Up to 48 inches (1200 mm)	Corrosion free		High molecular weight PEH – limited availability in pipe forms and expensive
PEX			Hot water applications	Cross-linked PE
Fluorocarbon (FEP, PFA, PTFE)		Outstanding	Used as liners bonded to GRP or metallic pipes for complete corrosion resistance. Limited availablity in tube form (FEP and PFA)	

750 mm diameter RPM pipelines for transporting water at Stallingborough, Lincolnshire.

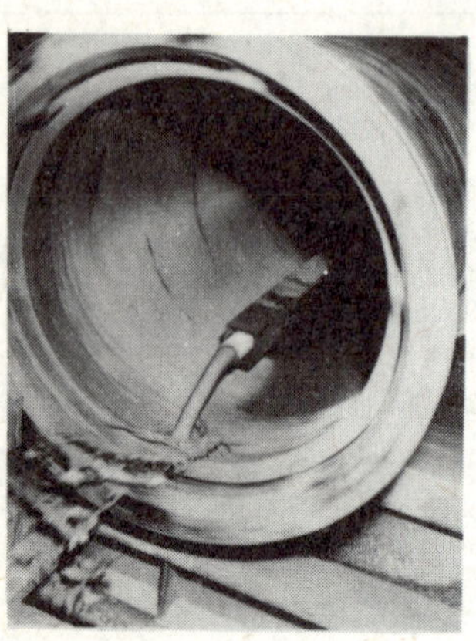

The cement mortar lining process being applied to ductile iron pipes at Stanton & Stavely.

Application of sand into the pipe wall during the manufacturing process of RPM in pipes at Stanton & Staveley.

It is to be regretted that similar rationalization is not possible between English (and American) and metric units and standards. This leads to differences in values of 'flow loss' coefficients for pipe bends, valves, *etc* – the English/American coefficient being based on 1 US gallon per minute at 1 lb/in^2 pressure loss; and the metric coefficient being based on $metres^3$/hour at 1 bar pressure loss.

Equally, pipe sizes are standard in both inch and millimetre sizes, together with matching fittings and valves. There are no exact equivalents. You work in standard manufactured sizes – either in inches *or* millimetres. To give 'equivalent' sizes in tabular data for either would be meaningless. With rare exceptions, the exact equivalent size just is not obtainable.

That is a problem, too, which complicates the presentation of working formulas. We have attempted, within reason, to cover most possibilities in the case of the main formulas for flow performance calculation in alternative forms embracing all the units most likely to be used – both in English and metric units. Here, in fact, English units are often less rational than their metric equivalents – with volumes expressed in cubic inches, US gallons, Imperial gallons or barrels, for example, depending on the industry or application involved.

In other more specialized cases, solutions and formulas are presented in one set of units only, being those most generally used, or in which the original solutions were derived. In that case conversion tables will be necessary if you want to use these with different units entered. As a final comment here, do remember that 'g' or gravitational acceleration *is* the same in English or metric units – 32.2 ft/sec = 9.81 m/sec = g.

PIPE SPECIFICATIONS : AMERICAN AND BRITISH STANDARDS

The following Table lists ASTM (American) pipe specifications and grades with British Standard equivalents and basic material descriptions.

ASTM	Material	BS Equivalent
A120	Carbon steel	1387
A53 Gr.A	Carbon steel	3601/23
A53 Gr.B	Carbon steel	3601/27
A106 Gr.A	Carbon steel	3602/23
API 5L Gr.A	Carbon steel	3602/27
A106 Gr.B	Carbon steel	2602/27
API 5L Gr.B	Carbon steel	3602/27
A333 Gr.1	Killed carbon steel	3063/LT50
A333 Gr.3	3.5% nickel	3603/503LT100
A335 Gr.P1	½% molybdenum	3604/240
A335 Gr.P12	1% Cr ½% Mo	3604/620
A335 Gr.P11	1¼% Cr ½% Mo	3604/621
A335 Gr.P22	2¼% Cr 1% Mo	3604/622
A335 Gr.P5	5% Cr ½% Mo	3604/625
A335 Gr.P7	7% Cr ½% Mo	3604/627
A335 Gr.P9	9% Cr 1% Mo	3604/629
A312 Gr.Tp304	Austenitic chromium nickel	3605/304 S18 (EN58E)
A312 Gr.Tp304L	Austenitic chromium nickel (extra low carbon)	3605/304 S14
A312 Gr.Tp316	Austenitic chromium nickel molybdenum bearing	3605/316 S18 (EN58J)
A312 Gr.Tp316L	Austenitic chromium nickel molybdenum bearing (extra low carbon)	3605/316 S14
A312 Gr.Tp321	Austenitic chromium nickel titanium stabilized	3605/321 S18 (EN58B)
A312 Gr.Tp347	Austenitic chromium nickel niobium stabilized	3605/347 S18 (EN58G)

EQUIVALENT SPECIFICATIONS FOR STAINLESS AND HIGH-RESISTANT STEELS

				SWEDISH				FRENCH		GERMAN
Description	BS970 EN No	AISI Type	Avesta	Fagersta	Nyby	Sandviken	Uddeholm	Government	Ugine	Krupps
12/14% Chromium Low carbon	331S42 (56A)	410	393	R.R.J.10	1410	2.C.27	S/S.1	Z.12.C.13	FIA/FIB	V13F
12/14% Chromium 15% Carbon	420S29 (56B)	410	393H	R.R.J.11	1415	4.C.27	S/S.31		FIU12	V5M
12/14% Chromium 35% Carbon	(56D)	420	739H	R.R.S.72	1435	7.C.27	S/S.6	Z.36.C.13	S12	V3M
18/20% Chromium 2% Nickel	431S29 (57)	431	249EH			4N2C36	S/S.22	Z.15.CN 18–02		VIM
18% Chromium		430	249	R.R.M.20	1710	1.C.36 2.C.34	S/S.2		F17	V17F Extra
26% Chromium 5% Nickel			453E	R.R.V.62	27–4	I.R.8	S/S.45			V.2.A. Supra Special
26% Chromium 5% Nickel 1.5% Molybdenum			435S	R.R.V.64	27–5MO	I.R.10	S/S.44			V.8.A. Supra Special
12% Chromium 12% Nickel	(58D)		832P	R.R.N.J.39	14–12	2.R.I.	S/S.33		Inoxargent	V.12.A. Supra
18% Chromium 8% Nickel 0.08% Carbon	304S15 (58E)	304	832M	R.I.M.291	18–8EL	O.R.2	S/S.3.M.M.	Z.5.CN. 18–08	N.S.22.S.	V.2.A. Supra
18% Chromium 8% Nickel	320S25 (58A)	302	832	R.R.N.J.32	18–8	2.R.2	S/S.3	Z.10.CN. 18–08		V.2.A. Normal
18% Chromium 8% Nickel Free machining	303S21 (58M)	303	832C			2.R.2.A.	S/S.43			
18% Chromium 10% Nickel 1.5% Molybdenum	315S16 (58H)		832SV	R.R.N.J.41	18-8EMO	O.R.3	S/S.4.M.M.			
18% Chromium 8% Nickel 1.5% Molbydenum	315S16 (58H)		832S	R.R.N.J.40	18–81MO	2.R.3	S/S.4			V.8.A. Normal
18% Chromium 8/10% Nickel 1/2% Titanium	321S12 (58B)	321	832T	R.R.N.J.51	18–8T	I.R.4	S/S.53	Z.10.CNT. 18–08	N.S.20C	V.2.A. Extra
18% Chromium 10/12% Nickel 1% Niobium	347S17 (58G)	347	832T			I.R.41				
17% Chromium 20% Nickel			254	R.R.T.80	20–20	2.R.6.	S/S.15			
25% Chromium 20% Nickel		310	254E	R.R.T.83	25–20	3.R.9.	S/S.25	Z.20.CNS 25/20	N.S.30	N.C.T.3
18% Chromium 10% Nickel 2.5% Molybdenum	304S15 (58E)	316	832SK	R.R.N.J.44	18–20 Mo	O.R.11	S/S.24	Z.8.CND. 18–08	N.S.M.C.	V.4.A. Supra

BRITISH, AMERICAN AND GERMAN EQUIVALENT STEEL SPECIFICATIONS

En BS 70	Type of Steel	SAE	AISI	Werkstoff	DIN
3A	'20' C steel (hot rolled or normalized)	1020	C1020	0402	C22
5	'30' C steel	1030	C1030	0501	C35
6A	Bright C steel	1035	C1035	0503	C45
8	'40' C steel	1040	C1040	0503	C45
9	'55' C steel	1055	C1055	0601	C60
11	'60' C-Cr steel	5160	5160	8161	58 Cr-V4
14B	C-Mn steel	1027 1330	C1027 1330	5066	30 Mn 4
18	1% Cr steel	5140	5140	7035	41 Cr 4
18B	1% Cr steel	5132	5132	7035	34 Cr 4
18C	1% Cr steel	5135	5135	7034	37 Cr 4
18D	1% Cr steel	5140	5140	7035	41 Cr 4
19	1% Cr-Mo steel	4140	4140	7220	34 Cr MO 4
19A	Cr-Mo steel	4140	4140	7225	42 Cr-Mo 4
19C	1% Cr-Mo steel	4140	4140	7225	42 Cr-Mo 4
23	3% Ni-Cr steel			5755	22(31)Ni-Cr 14
24	1.5% Ni-Cr-Mo steel	4340	4340	6582	34 Cr-Ni-Mo 6
56A	Cr-rust-resisting steel	51410	410	4006	x 10 Cr 13
56B	Cr-rust-resisting steel	51410	410	4021	x 20 Cr 13
56C	Cr-rust-resisting steel	51420	420	4021	C 20 Cr 13
56D	Cr-rust-resisting steel	51420	420	4034	x 40 Cr 13
56AM	Cr-rust-resisting steel	51416	416	4024	x 20 Cr 13
		51416 Se (S1416Se)	416 Se		
110	Low Ni-Cr-Mo steel			6582	34Cr-Ni-Mo 6

COLOUR CODES FOR PIPE LINE IDENTIFICATION

ORIGINALLY PIPES or sections of pipes were painted in colours for identification. Identification colours are now more commonly applied with bands of self-adhesive tapes, with colour-fast resistance to washing down, heat, *etc.*

Colour coding employed in UK practice is based on BS1710–1960, BS1710–1971 and BS1710–1975.

Pipe Contents	Ground Colour	Colour Band
BS1710 – 1960		
Water		
Cooling (primary)	Sea Green, 217	
Boilerfeed	Strong Blue, 107 (Light French Blue, formerly 175)	
Condensate	Sky Blue, 101	
Drinking	Aircraft Blue, 108	
Treated	Aircraft Grey Green, 283	
Central heating below 140 °F	French Blue, 166	
Central heating 140 °F–212 °F	French Blue, 166	Post Office Red, 538
Central heating above 212 °F	Crimson, 540	French Blue, 166
Cold water down service from storage tanks	Brilliant Green, 221	
Domestic hot water supply	Eau-de-nil, 216	
Hydraulic power	Mid Brunswick Green, 226	
Sea, river, untreated	Grass Green, 218	
Air		
Compressed, up to 200 lb/in²	White	
Compressed, over 200 lb/in²	White	Post Office Red, 538
Vacuum	White	Black
Steam	Aluminium or Crimson, 540	
Drainage	Black	
Electrical Services	Light Orange, 557	
Town Gas	Canary Yellow, 309	
Oils		
Diesel fuel	Light Brown, 410	
Furnace fuel	Dark Brown, 412	
Lubricating	Salmon Pink, 443	
Hydraulic power	Salmon Pink, 443	Sea Green, 217
Transformer	Salmon Pink, 443	Light Orange, 557
Fire Installations	Signal Red, 537	
Chemicals	Dark Grey, 632	
Industrial Gases	Dark Grey, 632	
Medical Gases	Dark Grey, 632	
Hazards		
Lightly Radio Active	Jasmin Yellow, 397	Black diagonal crossed stripes
Heavy Radio Active	Light Orange, 557	Black diagonal crossed stripes
Other Hazards	Golden Yellow 356 and Black	Diagonal and of equal size

cont...

Pipe Contents	Basic Colour		Colour Code Banding	
BS1710 – 1975				
Water				
Drinking	Green 12-D-45	White	Blue 18-E-53	White
Cooling (primary)	Green 12-D-45	White	White	White
Boiler Feed	Green 12-D-45	Crimson 04-D-45	White	Crimson 04-D-45
Condensate	Green 12-D-45	Crimson 04-D-45	Emerald Green 14-E-53	Crimson 04-D-45
Chilled	Green 12-D-45	White	Emerald Green 14-E-53	White
Central heating below 100°C	Green 12-D-45	Blue 18-E-53	Crimson 04-D-45	Blue 18-E-53
Central heating above 100°C	Green 12-D-45	Crimson 04-D-45	Blue 18-E-53	Crimson 04-D-45
Cold down service	Green 12-D-45	White	Blue 18-E-53	White
Hot water supply	Green 12-D-45	White	Crimson 04-D-45	White
Hydraulic power	Green 12-D-45	White	Salmon Pink 04-C-33	White
Sea, river, untreated	Green 12-D-45	White	Green 12-D-45	White
Fire extinguishing	Green 12-D-45	White	Safety Red 04-E-53	White
Air				
Compressed	Light Blue 20-E-51	White	Light Blue 20-E-51	White
Vacuum	Light Blye 20-E-51	White	White	White
Steam	Silver Grey 10-A-03	White	Silver Grey 10-A-03	White
Other Fluids				
Drainage	Black	White	Black	White
Electrical				
Electrical conduits	Orange 06-E-51	White	Orange 06-E-51	White
Gas				
Manufactured	Yellow Ochre 08-C-35	White	Emerald Green 14-E-45	White
Natural	Yellow Ochre 08-C-35	White	Canary Yellow 10-E-53	White
Oils				
Diesel fuel	Brown 06-C-39	White	White	White
Furnace fuel	Brown 06-C-39	White	Brown 06-C-39	White
Lubricating	Brown 06-C-39	White	Emerald Green 14-E-53	White
Hydralulic power	Brown 06-C-39	White	Salmon Pink 04-C-33	White
Transformer	Brown 06-C-39	White	Crimson 04-D-45	White
Acid and Alkalis	Violet 22-C-37	White	Violet 22-C-37	White

SAFETY COLOUR CODES

These are used as bands on the basic colours, usually 100 mm wide and could be used on a smaller pipe.

Fire Fighting	Safety Red 04-E-53
Warning	Golden Yellow 356 (08-E-51) with black diagonal lines for danger
Warning	Golden Yellow 356 (01-E-51) with black trefoil for ionizing radiation
Fresh water	Auxiliary Blue 18-E-53. Used as a central band on Basic Green 12-D-45

Pipe Contents	Background Colour	BS Colour to 381 or 2660	Optional additional colour banding codes. To be used in conjunction with the basic background colour when deemed necessary.		
BS1710 – 1971					
Water					
Drinking	Green	5-065		Strong Blue 107	
Cooling (primary)	Green	5-065		White	
Boiler feed	Green	5-065	Crimson 540	White	Crimson 540
Condensate	Green	5-065	Crimson 540	Emerald Green 228	Crimson 540
Chilled	Green	5-065	White	Emerald Green 228	White
Central heating below 100°C	Green	5-065	Strong Blue 107	Crimson 540	Strong Blue 107
Central heating above 100°C	Green	5-065	Crimson 540	Strong Blue 107	Crimson 540
Cold down service	Green	5-065	White	Strong Blue 107	White
Hot water supply	Green	5-065	White	Crimson 540	White
Hydraulic power	Green	5-065		Salmon Pink 447	
Sea, river, untreated	Green	5-065		Green 5-065	
Fire extinguishing	Green	5-065		Signal Red 537	
Steam	Silver Grey	9-099		Silver Grey 10-A-03	
Oils					
Diesel fuel	Brown	2-032		White	
Furnace fuel	Brown	2-032		Brown 2032	
Lubricating	Brown	2-032		Emerald Green 228	
Hydraulic power	Brown	2-032		Salmon Pink 447	
Transformer	Brown	2-032		Crimson 540	
Gas					
Natural	Yellow Ochre	358		Yellow Ochre 358	
Manufactured	Yellow Ochre	358		Emerald Green 228	
Oxygen	Yellow Ochre	358		White 9-102	
Nitrous oxide	Yellow Ochre	358		French Blue 166	
Nitrous oxide/oxygen – mixed	Yellow Ochre	358		French Blue 166	
Carbon dioxide	Yellow Ochre	358		French Grey 630	
Spare medical gas	Yellow Ochre	358		Yellow Ochre 358	with SMG written on
Acids and Alkalis	Violet	797		Violet 797	
Often additional information needed here showing chemical gas symbols			Diagonal stripes for danger needed in middle sometimes		
Air					
Medical	Light Blue	8-088	White		Black
Medical vacuum	Light Blue	8-088		Primrose 310	
Pathological	Light Blue 8	8-088	Terracotta 444		Primrose 310
Compressed	Light Blue	8-088		Light Blue 8-088	
Vacuum	Light Blue	8-088		White	
Drainage	Black	9-103		Black	
Electrical conduit & ducts	Light Orange	557		Light Orange	
Danger Signs					
General danger	As required		Golden Yellow	Black with diagonal stripes	
Ionizing radiation			Golden Yellow with trefoil		

Standard Service Codes

Letter symbols are also used to identify pipes and pipe lines, fittings, *etc*. The following summarizes British practice.

Water (various)	
Cooling water	CLW
Hot (domestic) water	HWS
Steam	S
Treated water	TW
Waste water	WW
Boiler feed water	BFW
Brine	B
Cold water	
Mains	MWS
Down service	CWS
Drinking	DWS
Flushing	FWS
Pressurized	PWS
Cold water down supply	CWDS
Chilled water	CHW
Fire fighting	
Fire extinguisher	FE
Fire hydrant	FH
Gases	
Town	G
Oxygen	O_2
Nitrous Oxide	N_2O
Heating	
Low pressure water	LPHW
Medium pressure water	MPHW
High pressure water	HPHW
Valves	
Air release	ARV
Air	AV
Auto air	AAV
Ball	BV
Gate	GV
Lockshield	LSV
Non-return	NRV
Pressure reduction	PRV
Safety	SV
Sluice	SV
Wheel	WV

Sewers	
Foul water	FWS
Surface water	SWS
Drains	
Foul water	FWD
Surface water	SWD
Pipes	
Discharge pipe	DP
Rain water pipe	RWP
Vent pipe	VP
Fittings	
Bath	b
Bidet	bt
Wash basin	wb
Shower	sh
Urinal	u
Flushing cistern	fc
Sink	s
Drinking fountain	df
Water closet	wc
Manholes, etc	
Back drop	BD
Invert	INV
Inspection chamber	IC
Manhole	MH
Fresh air inlet	FAI
Position	
High level	HL
Low level	LL
From below	FB
To below	TB
From above	FA
To above	TA
Flow	F
Return	R
Effluents	
Foul water	FW
Radio Active Water	RAW
Rain water	RW
Surface water	SW

Gullies	
Access	AG
Back inlet	BIG
Grease trap	GT
Road	RG
Sealed	SG
Yard	YG

Miscellaneous	
Half round channel	HRC
Rain water head	RWH
Condensate	C
Fuel	F
Vacuum	V
Cold feed	CF
Feed and expansion	F & E
Plug cock	PC

Access Points	
Access cover	A/C
Cleaning eye	CE

Dry weather flow	DWF
Fire Hydrant	FH
Compressed air	CA
Refrigerants (identified by symbol for particular gas)	R_o
Draw-off point	DO
Open vent	OV
Stop cock	SC

Computer Aided Design

COMPUTER AIDED DESIGN (CAD) as opposed to conventional methods of data processing, is relatively new in the pipeline industry. The first prominent system (ISOPEDAC*) initiated in 1965, concentrated on the problems of estimating, detailing, procurement, manufacture and erection of pipelines for the process industries where thousands of components may have to be selected, fabricated, stored, tested, erected, *etc.* Initially it would handle only flanged pipelines and produce bills of materials, but by 1970 the system (shown in chart 1) was extensively developed to provide the following facilities:-

(i) Estimate, using statistical techniques, material quantities and cost of pipework required.

(ii) Monitor orders for pipework components.

(iii) Calculate actual material quantities required, based on a detailed description of each pipeline, and produce costs which fabricators will accept on invoices.

(iv) Forecast discrepancies between materials ordered and materials actually required for material control purposes.

(v) Select automatically from the data-bank the pipework fitting required in accordance with the pipework specification, *eg* BS, DIN, ANSI *etc.*

(vi) Handle Imperial or SI units or even mixed units, *eg* nominal pipe bore in inches and length in metres.

(vii) Choose spool sizes (*ie* flange or field weld positions) to meet constraints on the maximum spool size that can be transported to the construction site, or accept cast iron or rubber lined pipe where the lining thickness must be considered at each point.

(viii) Draw, using ISOPLOT, an isometric plan which a fabricator will accept without retouching, and which includes a bill of materials and cut lengths of pipe, automatically deciding how many isometric drawings each pipeline requires.

(ix) Analyze in detail the materials and costs at the end of a construction project.

*ISOPEDAC = an Integrated System of Pipework Estimating, Detailing and Control (Imperial Chemical Industries PLC).

CHART I – Schematic Diagram of ISOPEDAC System

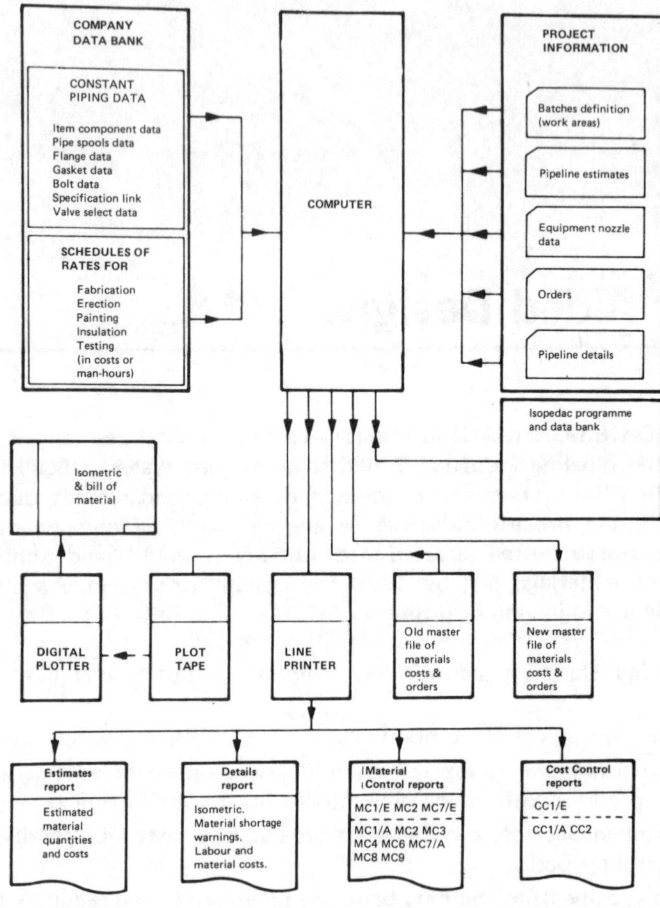

ISOPEDAC is written in FORTRAN and PL/1, and will accept free-format data from punched cards, visual display unit keyboards, word processors, magnetic tape or disc storage. All the necessary data is stored and results output to line printers, visual display units, typewriters, and a digital plotter. ISOPEDAC has been implemented on IBM/360/370 and 303X computers, and in addition ISOPLOT has been implemented on UNIVAC, PRIME and DEC computers.

Estimating, detailing material control and costing is the most labour-intensive part of pipework design, now well covered by ISOPEDAC and alternative CAD systems. These continue to be developed to cover all the major activities in pipeline design – *eg* see chart 2. The immediate aim in future CAD systems is to avoid the need for re-input data separately for each activity, *ie*

(i) with an integrated system covering all activities

or

(ii) automatic interfaces between all the separate systems for each activity.

CHART II – Pipeline Design Flowchart

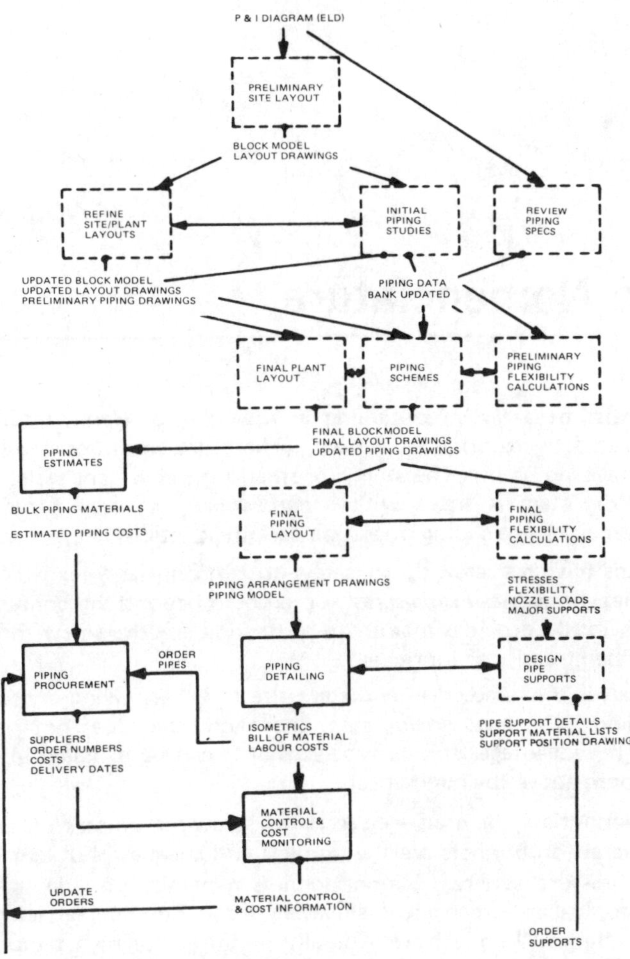

Pipeline Simulation

A further equally important field of pipeline design is the static and dynamic analysis of pipeline systems. Here the particular problem of applying CAD is that traditional design is a mixture of art and science and subjective judgements have to be reduced to a form of logic that can be accepted by a computer. Original software programmes were largely restricted to relatively simple calculations and thus limited in scope. This state of affairs was changed dramatically in the early 1970s with the development of pipeline simulation models for batch computing.

Today it is possible to find CAD programmes to analyze almost any pipeline design problem via simulation models. Further, when the pipeline is laid and in operation, dynamic performance – *ie* flows, pressures and temperatures, can be predicted from the model which is kept on the computer. Leaks can be predicted if the actual data differs from the model and the effect of other potential failures predicted. The development of such technology is rapid and innovative.

Basic Valve Nomenclature

MOST VALVES consist of a *body* containing a *flow control element* (discs, plug, gate, *etc*) attached to and operated by rotation of a *stem.* (There are exceptions; swing check and pinch valves for example, have no stem). The stem, together with any stem seals, is enclosed within a *bonnet.* The top of the stem is fitted with a *handwheel* (or lever) for rotation of the stem (although some stems may have a sliding operation for quick action).

With threaded stems (giving a screw-down, screw-up motion) the threaded portion may be fully enclosed by the bonnet, known as *inside-screw;* or exposed beyond the bonnet, known as *outside-screw.* The former obviously provides maximum protection for the screw thread. Outside screws have the advantage of being easier to lubricate.

With *rising-stem* valves the handwheel and stem rise together, giving a visual indication of the degree of valve opening. With a *non-rising stem* the handwheel does not rise (or fall) with the turning movement. The advantage of this type is that it can be installed in situations providing only minimum headroom above the handwheel.

Various types of bonnet may be used — *eg* screw-in, screw-on, union *style and bolted or flanged bonnet.* Screw-in or screw-on bonnets are the simplest and cheapest, but largely limited to smaller valves used on low pressure services. Union bonnets generally provide tighter sealing and are particularly suitable for valves which are dismantled frequently for servicing. Plain (flat) flange and male and female flanged bonnets are generally preferred for high temperature or high pressure valves, and also larger sizes of valves. An alternative type for high pressure and/or high temperature services is the breech-lock bonnet.

Valve Trim

Trim is the term used to describe the parts of a valve which are replaceable, *ie* normally those parts likely to be subject to wear or degradation. The following parts are considered as trim:

Gate valves — stem, seat ring, wedge, back seat bushing

Globe and angle valves — stem, seat ring, disc, disc nut, back seat bushing

Disc valves — disc, disc nut, back seat bushing

Swing check valves — disc, disc holder, disc nut, side plug, carrier pin, disc holder pin, disc nut pin, seat rings

Lift check valves — disc, disc guide, seat rings

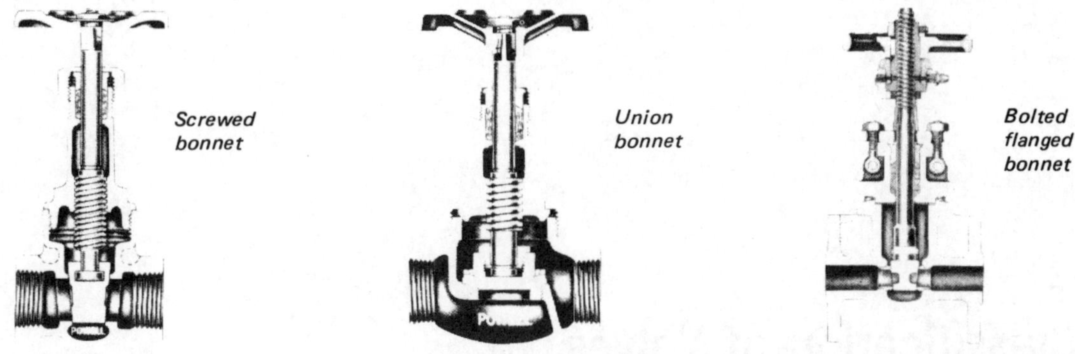

Screwed bonnet

Union bonnet

Bolted flanged bonnet

Stem seals and other internal seals (where fitted) are arguably included under the definition of trim, but are not normally used in describing trim materials.

Standard Abbreviations

The following abbreviations are used to describe or designate valve parts, features, *etc*

All iron — all parts of iron construction
All bronze — all parts of bronze construction
BB — bolted (flanged) bonnet
CWP — cold working pressure
DD — double disc
DW — double wedge
FE — flanged end (connection)
FF — flat flange
IBBM — iron body bronze mounted
IPS — iron pipe size
ISNRS — inside screw non-rising stem
ISRS — inside screw rising-stem
NRS — non-rising stem
RF — raised flange
RS — rising stem
SIB — screwed bonnet
SW — solid wedge
S int, or int S — internal seat
S ren, or ren S — renewable seat
OS&Y — outside screw and yoke
WOG — water, oil, gas pressure rating

Nomenclature covering the individual parts of various different types of valves is included in the individual chapters in Section 2.

Classification of Valves

VALVES MAY be classified in a number of ways, *eg* by category (general type), specific type, purpose or name; or by flow characteristics (*eg* straight-through, full flow or throttled flow). Descriptions can also differ slightly in different countries although the main type names are established internationally (with some exceptions).

Classification of valves by *category* is given in Table I. This follows British Standards and general practice adopted by British manufacturers; but is also generally applicable to American practice. One major difference in this respect is that the important class of *ball valves* is considered as a type of plug valve in the tabular summary, whereas American practice would favour regarding it as a separate category. The *ball valve* is, in fact, a major type in its own right.

Valves are classified and described by specific *type* in Section 2 chapters, which also include a number of individual designs best categorized as 'miscellaneous'. Some other valve types are given in Table II.

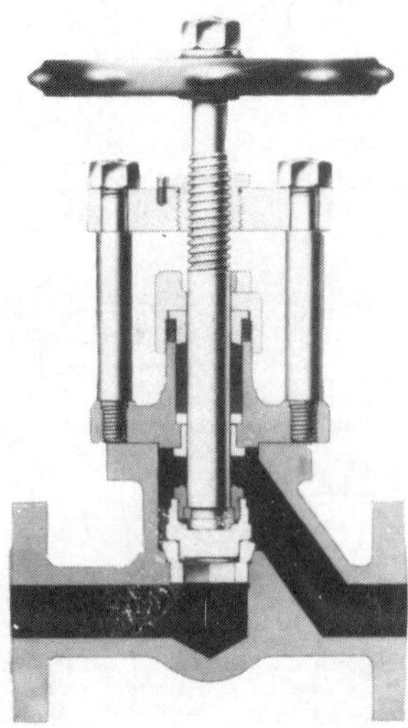

Screw-down stop valve. (Hopkinson Ltd).

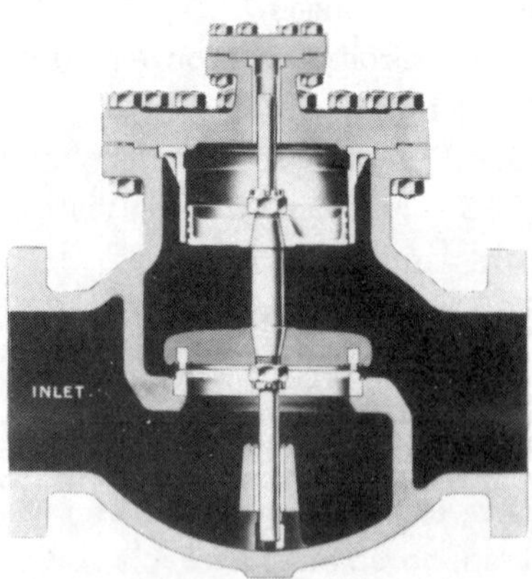

Damped left-check valve. (Hopkinson Ltd).

MANNESMANN
DEMAG

Metallverformung
MEER Armaturen
Postfach 365 · Ohlerkirchweg 66
D-4050 Mönchengladbach 1
Tel. (2161) 350-1 · Telex 852525

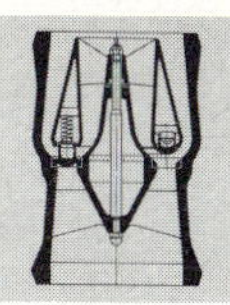

DRV-B type during manufacture for one of the largest water transmission systems
DN: 56″
PN: ANSI Class 600

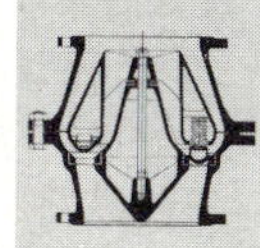

DRVg – type in a nuclear power plant cooling water system
DN: 20″
PN: ANSI Class 900
Operating temp.: 360°F (180°C)

COUPON

Please send us a copy of your Nozzle Check Valves brochure

Name/Company ____________________

Address ____________________

____

4706e

Fabricated gate valve, through-conduit, bevel gear operated, weld end. (Grove Valve & Regulator Co).

World's largest gate valve 120 inches — for diversion tunnel flow control at dam on American river. (Grove Valve & Regulator Co).

Descriptions of various valve types may also differ, and here Table III lists some alternative descriptions, standard terminology in this case being based on that adopted for Table I. This is by no means complete, but is offered as a general guide.

Classification of valves by *function* yields the following general list where any individual type of valve may be capable of performing one or more of these functions. Excluded from this list are specific functions or specialized services for which special designs of valves are normally employed.

(i) on-off service
(ii) throttling or flow control
(iii) preventing of reverse flow
(iv) pressure control
(v) directional flow control
(vi) sampling
(vii) flow limiting

Valves classified by duty or the service they are intended to perform are described in Section 4 chapters. Necessarily these embrace types already described under specific types and the relevant chapters can be studied together where appropriate. A further source of reference and information in this respect is the chapter on *Valve Selection Guides.*

TABLE I – CLASSIFICATION OF VALVES

Category	Patterns	Types of Construction	Remarks
Cock	(i) Plug (ii) Gland (iii) Packed cock (iv) Compound gland	Tapered plug Plug retained by gland or packing. Packing between plug face and body seat Stuffing box in cover	Also parallel plug
Plug valve	(i) Plain (ii) Lubricated	(i) Taper plug (ii) Parallel plug (iii) Ball plug	Passage through port in rotatable plug supported or mounted to reduce friction
Screw-down stop valve	(i) Inside screw (ii) Outside screw	(i) Globe valve (ii) Angle (iii) Oblique (iv) Others	Spherical body Spherical body with ends at right angles Spherical body, stem axis oblique Usually described by type (eg needle valve) or geometry of body (eg tee valve)
Gate valve (wedge gate valve) Gate valve (slide valve)	(i) Inside screw (ii) Outside screw (iii) Lever (a) sliding stem (b) rotary stem	(i) Wedge (gate) (ii) Sluice (valve) (iii) Double disc	Closure effected by wedge action (a) solid wedge or (b) split wedge Solid wedge gate valve Gate composed of parallel sliding discs or slides
Check valve	(i) Horizontal (ii) Vertical (iii) Angle	(i) Swing (check) (ii) Lift (a) disc (b) piston (c) ball (c) ball (iii) Foot (valve)	Hinged flap check mechanism Disc check mechanism Disc plus piston check mechanism Ball check Check valve fitted to bottom of a suction pipe
Butterfly valve	(i) Double flanged (ii) Water (a) single flange (b) flangeless	Each flange and individually bolted. Primarily designed for insertion between pipe flanges	Based on rotatable disc valve

Diaphragm valve		Flexible diaphragm mounted over a weir	
Pinch valve	(i) Mechanical (ii) Pneumatic	Flexible tube capable of being flattened and released	
Ball (float valve)	(i) Single beat (ii) Double beat	(i) Direct (lever) operated (ii) Pressure operated (iii) Droptight (iv) Non-droptight	Single beat – flow through single seating ring Double beat – flow through two seating rings
Safety valve	(i) Direct spring loaded (ii) Direct weight loaded (iii) Lever and spring loaded (iv) Lever and weight loaded (v) Tension spring loaded (vi) Torsion bar		Also designated by: (i) High lift valve (ii) Full lift valve (iii) Pilot operated valve (iv) Electrically assisted valve
Relief valve	(i) Direct spring loaded (ii) Direct weight loaded		Also designated by: (i) Full lift (relief) valve (ii) Pilot operated (relief) valve
Pressure control valve	(i) Self-contained (ii) Spring loaded (iii) Weight loaded (iv) Pressure loaded (v) Externally piped (vi) Tight closing (vii) Non-tight closing (viii) Relay operated	(i) Pressure reducing (ii) Pressure retaining (iii) Indirect	
Air relief valve	(i) Single orifice LP (ii) Single orifice HP (iii) Single orifice with integral isolating valve (iv) Double orifice with integral isolating valve		
Turbine valve	(i) Regulating (ii) Quick-closing (iii) Starting (iv) Exhaust (v) Guarding		
Free discharge valve	(i) Needle type (ii) Hollow jet type (iii) Sleeve type		

TABLE II – SOME OTHER VALVE TYPES

Category	Description
Flow regulating valve	For controlling rate of flow in a system.
Temperature regulating valve	For controlling fluid temperature level in a system.
Automatic process control valve	For controlling rate of flow relative to value of a command system.
Anti-vacuum valve	An automatic type of air valve for the prevention of the formation of vacuum or the release of vacuum in large bore pipelines.
Blow down valve	A valve which is used for cleaning sludge and other foreign matter from a boiler.
Bulkhead valve	A gate valve.
Free ball valve	A valve in which a ball, free to rotate in any direction, is moved at 90° to the flow stream from a position removed from the flow stream until finally rolling into a circular orifice for shut off.
Fusible link or fire valve	A fire prevention valve which has a weighted lever held open by a wire and fusible link which melts at an increase in room temperature.
Hydraulic valve	A control valve for either water, oil, or hydraulic systems.
Jet dispersal valve	A valve incorporating an element by virtue of which the energy within the emitting jet is dissipated.
Penstock	A single faced type of valve consisting of an open frame and door, and used in terminal positions only; usually located in tanks or channels as a means of controlling flow into a pipe.
Plate valve	A gate valve incorporating a sluicing effect.
Radiator valve	A valve for controlling the flow of water through a radiator.
Rotary slide valve	A valve in which rotation of internal parts regulates flow by opening or closing a series of segmental ports.
Rotary valve	A spherical plug valve, in which the plug, which rotates through 90°, is provided with a circular waterway to match the body and ports.
Solenoid valve	A valve operated by an electrical solenoid.
Spectacle-eye valve	A type of parallel slide valve in which the 'spectacle gate' has one 'lens' of circular waterway and the other of solid section.
Thermostatic mixing valve	A valve which combines temperature selection and flow control in the same body.
Throttle valve	A non-tight closing butterfly valve with a centrally hinged flap which can be locked in any desired position.

TABLE III

General or 'Popular' Description	Standard Terminology
Back-pressure valve	Check valve
Block and bleed valve	Gate valve
Clack valve	Check valve
Conduit valve	Gate valve with full bore aperture
Controllable check valve	Screw-down stop and check valve
Controllable non-return valve	Screw-down stop and check valve
Dashpot valve	Check valve (piston check disc type)
Excess flow valve	Flow regulating valve
Excess or minimum pressure valve	Flow regulating valve
Flap valve	Check valve (swing type)
Follower-ring valve	Gate valve
Fullway valve	Gate valve
Governor valve	Pressure control valve
Non-return self-closing valve	Check valve
Parallel gate valve	Gate valve (double disc type)
Pop safety valve	Safety valve (direct spring loaded)
Proportional flow valve	Flow regulating valve
Reflux valve	Check valve
Retention valve	Check valve
Screw-down non-return and flood valve	Screw-down stop and check valve
Wheel valve	Screw-down stop valve
Y-type valve	Oblique valve

Valve Connections

Valves are normally designed to take either threaded pipe ends, or with flanges for flanged connection. Threaded connections are simpler and cheaper to produce and more easily installed. However, it can prove difficult to remove valves so mounted without dismantling a considerable portion of the piping unless a number of extra fittings, such as unions, are incorporated.

Flanged ends make a stronger, tighter, and more leak-proof connection. Where heavy viscous media are to be controlled, as in refineries, process and chemical plants, *etc,* flanged end valves are normally used. The initial cost is higher, not only because of more metal but because the flanges must be carefully and accurately machined. Also the installation cost is greater, because companion flanges, to which the valve end flanges are bolted, as well as gaskets, bolts and nuts must be provided.

All flat faces are commonly termed plain faces. Bronze and iron flat faces can have a machined finish. Cast iron raised faces may be smooth finished or have a serrated finish (preferably with no less than 16 serrations per inch) which may be spiral or concentric. Steel flat faces and raised faces should have a serrated finish of approximately 32 serrations per inch. The serrations may be either spiral or concentric.

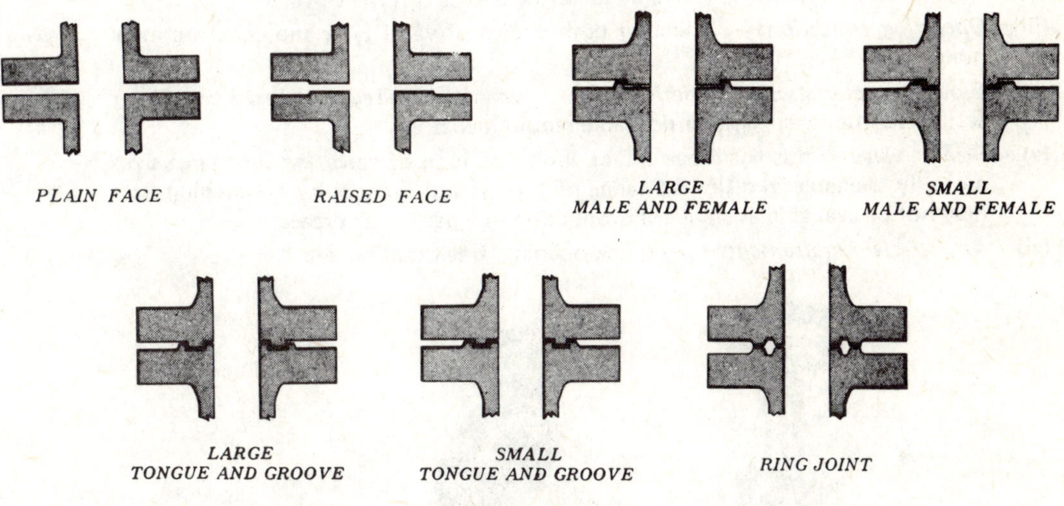

Fig 1 Flanged ends.

Steel male and female and tongue and grooved faces should have a smooth finish. Steel ring joint faces should have smooth finished grooves. If spiral wound gaskets are used on flange faces, the flanges should have a smooth finish. Examples of flanged ends are shown in Fig 1.

Socket or butt-welded ends are used on all-welded pipeline systems. For specific services valves are also produced to permit connection to pipes by soldering or brazing. In the latter case the valve may be supplied with integral preformed brazing material inserts.

Valve Selection Guides

THE MAIN parameters concerned in selecting a valve or valves for a typical general service are:-

(i) *Fluid to be handled* – this will affect both type of valve and material choice for valve construction.

(ii) *Functional requirements* – mainly affecting choice of type of valve.

(iii) *Operating conditions* – affecting both choice of valve type and constructional materials.

(iv) *Flow characteristics and frictional loss* – where not already covered by (ii), or setting additional specific or desirable requirements.

(v) *Size of valve* – this again can affect choice of type of valve (*eg* very large sizes are only available in a limited range of types); and availability (matching sizes may not be available as standard production in a particular type).

(vi) *Any special requirements* – *eg* quick-opening, free draining, *etc.*

Gunmetal gate, globe and check valves from Peglers.

TABLE I – APPLICATIONS OF VALVE TYPES

Valve Category	General Application(s)	Actuation	Remarks
Screw-down stop valve	Shut-off or regulation of flow of liquids and gases (eg steam)	(i) Handwheel (ii) Electric motor (iii) Pneumatic actuator (iv) Hydraulic actuator (v) Air motor	(a) Limited application for low pressure/low volume systems because of relatively high cost (b) Limited suitability for handling viscous or contaminated fluids
Cock	Low pressure service on clean, cold fluids (eg water, oils, etc)	Usually manual	Limited application for steam services
Check valve	Providing flow in one direction	Automatic	(a) Swing check valves used in larger pipelines (b) Lift check valves used in smaller pipelines and in high pressure systems
Gate valve	Normally used either fully open or fully closed for on-off regulation on water, oil, gas, steam and other fluid services	(i) Handwheel (ii) Electric motor (iii) Pneumatic actuator (iv) Hydraulic actuator (v) Air motor	(a) Not recommended for use as throttling valves (b) Solid wedge gate is free from 'chatter' and jamming
Parallel slide valve	Regulation of flow, particularly in main services in process industries and steam power plant		(a) Offers unrestricted bore at full opening (b) Can incorporate venturi bore to reduce operating torque
Butterfly valve	Shut-off and regulation in large pipelines in waterworks, process industries, petrochemical industries, hydroelectric power stations and thermal power stations	(i) Handwheel (ii) Electric motor (iii) Pneumatic actuator (iv) Hydraulic actuator (v) Air motor	(a) Relatively simple construction (b) Readily produced in very large sizes (eg up to 18 ft or more)
Diaphragm valve	Wide range of applications in all services for flow regulation	(i) Handwheel (ii) Electric motor (iii) Pneumatic actuator (iv) Hydraulic actuator (v) Air motor	(a) Can handle all types of fluids, including slurries, sludges, etc, and contaminated fluids (b) Limited for steam services by temperature and pressure rating of diaphragm
Ball valve	Wide range of applications in all sizes, including very large sizes in oil pipelines, etc.	(i) Handwheel (ii) Electric motor (iii) Pneumatic actuator (iv) Hydraulic actuator	(a) Unrestricted bore at full opening (b) Can handle all types of fluids (c) Low operating torque (d) Not normally used as a throttling valve
Pinch valve	Particularly suitable for handling corrosive media, solids in suspension, slurries, etc.	(i) Mechanical (ii) Electric motor (iii) Pneumatic actuator (iv) Hydraulic actuator (v) Fluid pressure (modified design)	(a) Unrestricted bore at full opening (b) Can handle all types of fluids (c) Simple servicing (d) Limited maximum pressure rating
Automatic process control valve	Designed to meet particular service conditions	To meet particular service conditions	Most commonly of single or double beat globe valve configuration
Air relief valve	Used in water works, etc, to release entrapped air and prevent formation of vacuum 'pockets'	Automatic – responding to changes in flow pressure	
Turbine valves	Designed to meet requirements of steam and water turbines in industrial, marine and power generation services	To meet particular service conditions	Provide guaranteed control over maximum and minimum turbine speeds and power in association with other valves

TABLE II – VALVE TYPES FOR SPECIFIC SERVICES

Service	Main	Secondary
Gases	Butterfly valves Check valves Diaphragm valves Lubricated plug valves Screw-down stop valves	Pressure control valves Pressure-relief valves Pressure-reducing valves Safety valves Relief valves
Liquids, clear up to sludges and sewage	Butterfly valves Screw-down stop valves Gate valves Lubricated plug valves Diaphragm valves Pinch valves	
Slurries and liquids heavily contaminated with solids	Butterfly valves Pinch valves Gate valves Screw-down stop valves Lubricated plug valves	
Steam	Butterfly valves Gate valves Screw-down stop valves Turbine valves	Check valves Pressure control valves Presuperheated valves Safety and relief valves

TABLE III – VALVE TYPE SUITABILITY

Valve Type	SERVICE OR FUNCTION										
	On–Off	Throttling	Diverting	No Reverse Flow	Pressure Control	Flow Control	Pressure Relief	Quick Opening	Free Draining	Low Pressure Drop	Handling Solids in Suspension
Ball	S	M	S	–	–	–	–	S	–	S	LS
Butterfly	S	S	–	–	–	S	–	S	S	S	S
Diaphragm	S	M	–	–	–	–	–	M	M	–	S
Gate	S	–	–	–	–	–	–	S	S	S	–
Globe	S	M	–	–	–	M	–	–	–	–	–
Plug	S	M	S	–	–	M	–	S	S	S	LS
Oblique (Y)	S	M	–	–	–	M	–	–	–	–	–
Pinch	S	S	–	–	–	S	–	–	S	S	S
Slide	–	M	–	–	–	M	–	M	S	S	S
Swing check	–	–	–	S	–	–	–	–	–	S	–
Tilting disc	–	–	–	S	–	–	–	–	–	S	–
Lift check	–	–	–	S	–	–	–	–	–	–	–
Piston check	–	–	–	S	–	–	–	–	–	–	–
Butterfly check	–	–	–	S	–	–	–	–	–	–	–
Pressure relief	S	–	–	–	–	–	S	–	–	–	–
Pressure reducing	–	–	–	–	S	–	–	–	–	–	–
Sampling	S	–	–	–	–	–	–	–	–	–	–
Needle	–	S	–	–	–	–	–	–	–	–	–

S = Suitable choice
LS = Limited suitability
M = May be suitable in modified form

In the case of specific services, choice of valve type may be somewhat simplified, *eg* by following established practice or selecting from valves specifically produced for that particular service.

On a broad basis, Table I summarizes the applications of the main types of general purpose valves. It has only limited use as a selection guide – *ie* can be regarded as a starting point. Table II carries general selection a stage further in listing valve types normally used for specific services. Table III is a particularly useful expansion of the same theme relating the suitability of different valve types to specific functional requirements.

Normally, for general services (and for many specific services), several valve types may appear as a possible choice. These may then need evaluating individually, and comparing on the basis of the flow characteristics they offer. Even more important, calculations may be necessary to establish a suitable size of valve to meet a specific performance requirement – *eg* a maximum acceptance pressure drop or head loss through the valve.

Joucomatic motorized valves
(electric motor actuators)

Valve Coefficients and Flow Values

The *valve coefficient* is a convenient method of relating flow rates to pressure drop through valves and, in fact, is sometimes called a *flow value.* This coefficient can only be determined empirically for a specific type of valve as it will be influenced by detail design and construction. It will also vary with the physical *size* of the valve and the degree of *opening* of the valve. Valve coefficient values are normally quoted for 100% opening (full open), with individual values for each size.

Some confusion can arise from the fact that the coefficient quoted for a valve can have three different values, depending on the basic units on which it was computed. Normally these are apparent from the designation of the valve coefficient, viz:

C_V in units of US gal/min, lb/in^2

K_V or k_V in units of litres/min, bar

f in Imperial units of Imp gal/min, lb/in^2

The following conversions apply:

	K_V	C_V	f
K_V	–	14.28	17.09
C_V	0.07	–	1.1966
f	0.0585	0.8357	–

Flow Characteristics of Valves

Where the flow characteristics through the valve are of significance, the following notes can be useful.

Plug valves (Fig 1) offer a straightway passage through the ports with a minimum of turbulence. Flow can be in either direction and a quarter-turn will fully open or fully close the valve. Similar comment applies to *ball valves.*

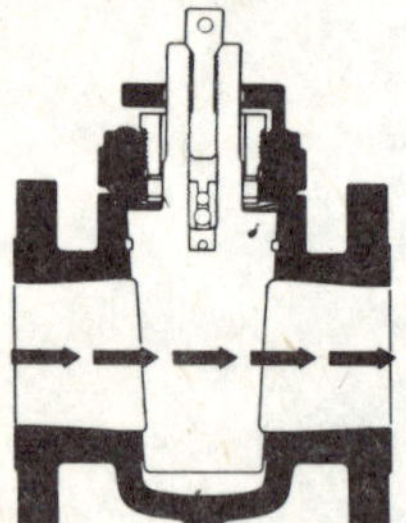

Fig 1 Lubricated plug valve.

Firesafe (BS5146) one piece integrally flanged ball valves available in carbon steel, stainless steel and S.G. iron, with ANSI 150 (or 300) flanges. (Worcester Controls Ltd).

UPVC plastic ball valve.

Inlet and outlet manifolds using Cameron ball valves.

Gate valves (Fig 2) present a substantially straightway flow through the ports in the full open position since the wedge or 'gate' is lifted clear of the flow passage. Turbulence and pressure drop is low. Again flow can be in either direction.

Globe valves (Fig 3) are normally installed so that pressure is under the disc, assisting operation and eliminating a certain amount of erosive action. Turbulence and pressure drop is higher than with straightway valves.

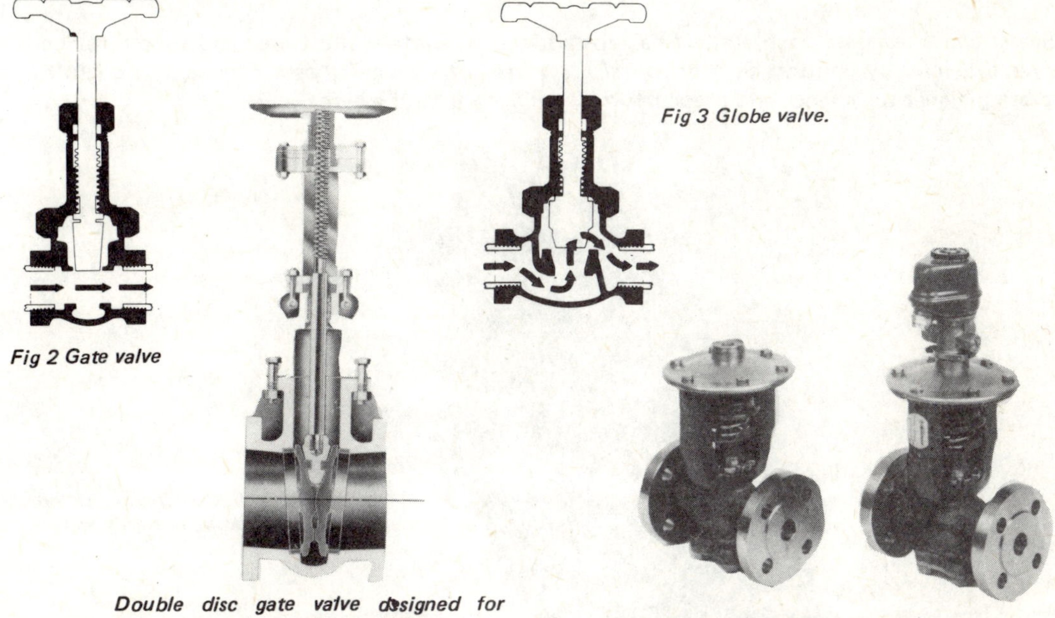

Fig 2 Gate valve

Fig 3 Globe valve.

Double disc gate valve designed for sulphuric acid services. (Chas. S. Lewis & Co).

Joucomatic diaphragm globe valve for solenoid-pilot operation.

Angle valves (Fig 4) have similar characteristics to globe valves, with flow directed through 90 degrees. Again flow is normally directed under the disc. Reverse flow may be used in the case of high temperature steam. Ball, globe and angle valves are suitable for throttling.

Oblique or Y-valves (Fig 5) combine the advantages of the gate and globe valve with a nominally straightway flow and good throttling characteristics.

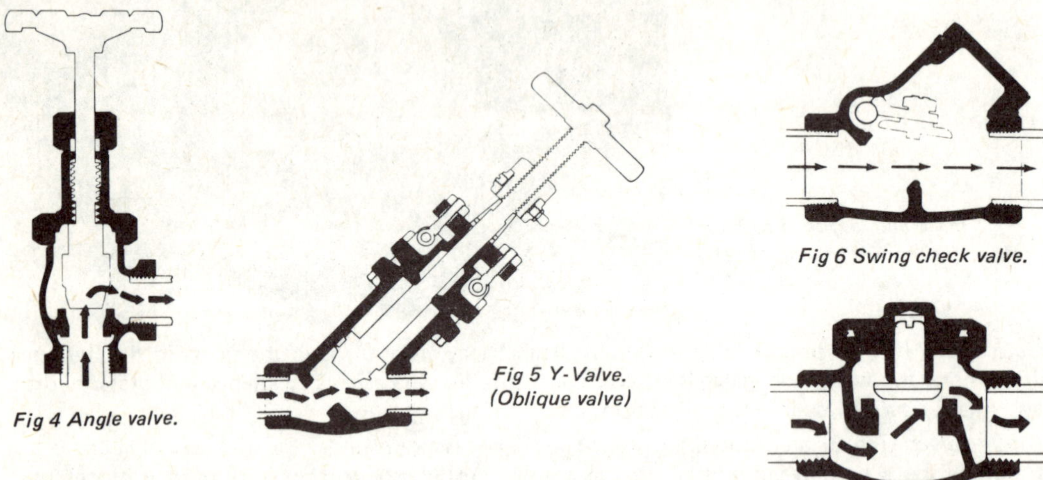

Fig 4 Angle valve.

Fig 5 Y-Valve. (Oblique valve)

Fig 6 Swing check valve.

Fig 7 Horizontal lift check valve.

Swing check valves (Fig 6) provide a substantially straightway flow area and thus turbulence generated is low. By contrast the *horizontal lift check valve* (Fig 7) has a deflected throughflow, generating higher turbulence and pressure drop similar to that of a globe valve.

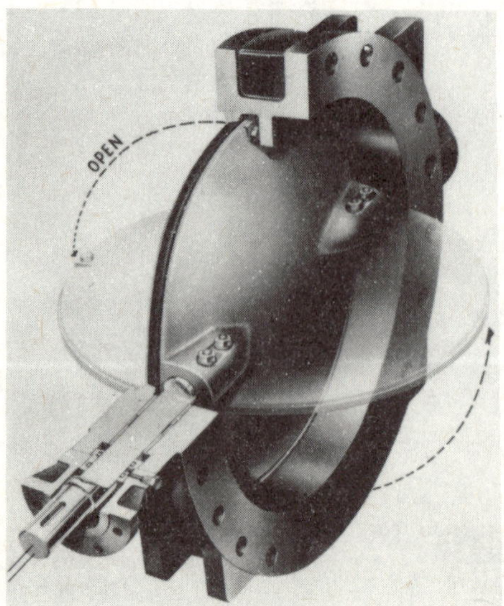

Blakeborough 'Conesphere' pattern rubber seal butterfly valve.

Flow Values

Examples of flow values (k_V values) for five different types of valves in a range of different sizes are given in Fig 8. Flow factor for any given (percentage) opening can be read from the appropriate graph. It should be emphasized that these are 'typical average' values. Individual designs of the same valve types may exhibit different values. More specific data are given in later chapters in Section 2 for different valve types. (See also Table IV).

The value of charts of this type is in calculating likely losses and/or valve sizes required.

Nominal Size	k_v100
⅜	41
½	95
¾	180
1	327
1¼	484
1½	725
2	1130
2½	1700
3	–

Main seat

Nominal Size	k_v100
⅜	49
½	77
¾	146
1	260
1¼	437

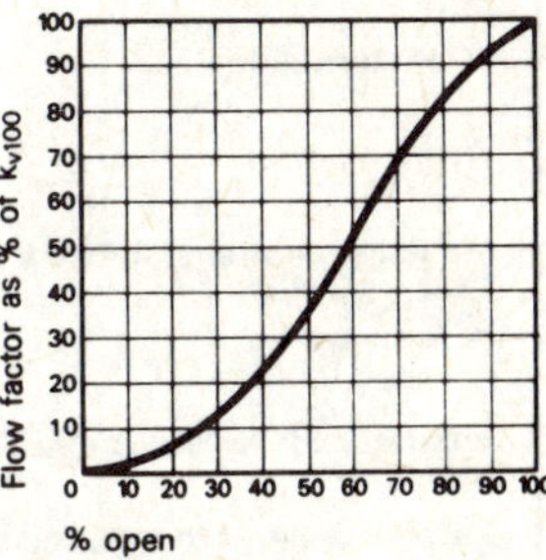

L valve

Nominal Size	k_v100
⅛	22
¼	32
⅜	71
½	185
¾	350
1	700
1¼	1000
1½	1600
2	3100
3	6500
4	11000

Ball

Nominal Size	k_v100
½	63
¾	121
1	187
1¼	332
1½	416
2	704
3	1700
4	2700

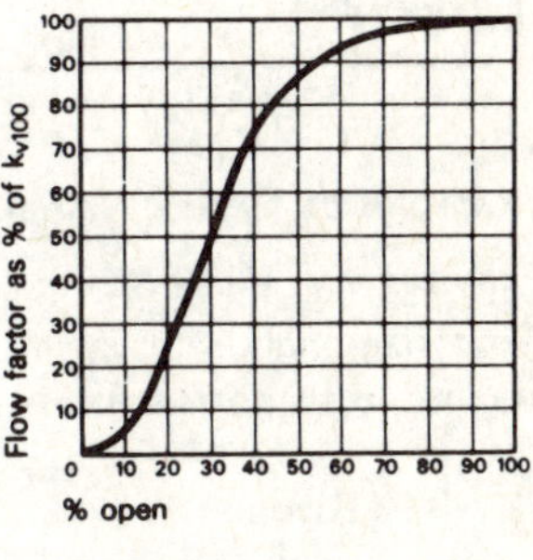

Diaphragm

Nominal Size	k_v100
2½	1500
3	3000
4	5000
5	8000
6	12000
8	17000

Fig 8

TABLE IV – TYPICAL 'K' VALUES AND PRESSURE DROPS FOR VARIOUS 150 mm (6 inch) BORE VALVES*

Valve Type	K Value	PRESSURE DROP†	
		bar	lb/in^2
Globe	5.0	0.59	8.5
Swing check	3.5	0.40	5.9
Y pattern	2.9	0.34	4.9
Angle (globe)	2.2	0.25	3.7
Venturi parallel slide (with eyepiece)	1.1	0.13	1.9
Butterfly	1.0	0.12	1.7
Parallel slide without eyepiece	0.15	0.021	0.3
Parallel slide with eyepiece	0.05	0.007	0.1
Ball (full bore)	0.05	0.007	0.1
Straight pipe (the length of an average 6 inch bore valve)	0.045	0.005	0.075

† Flow 40 m/s (140 ft/s) at 24 bar (350 lb/in^2) sat steam *Hopkinsons Ltd

Examples of Typical Calculations (using Fig 8 data)

Example 1 – Valve Sizing

What size angle seat valve is required to handle 250 litres per minute of water, assuming that the maximum acceptable pressure drop is given as 0.3 kg/cm^2?

Given

Flow rate Q = 250 litres/min

Specific gravity γ = 1 kg/dm^3

Pressure drop ΔP = 0.3 kg/cm^2

Flow value k_{V100} = ?

$$K_{V100} = Q \sqrt{\frac{\gamma}{\Delta P}}$$

$$= 250\sqrt{\frac{1}{0.3}}$$

$$\therefore K_{V100} = 456$$

TABLE V – TYPICAL SIZE AND OPERATING RANGES OF VALVES

Valve	Size Minimum inches (mm)	Size Maximum inches (mm)	Pressure Range Minimum lb/in² (bar)	Pressure Range Maximum lb/in² (bar)	Temperature Range Minimum °F (°C)	Temperature Range Maximum °F (°C)
Ball	¼ (6)	48 (1220)	Atmospheric	7500 (525)	–65 (–55)	575 (300)
Butterfly	2 (50)	72 (1830)	Vacuum	1220 (84)	–20 (–30)	1000 (538)
Butterfly check	1 (25)	72 (1830)	Atmospheric	1200 (84)	0 (–18)	500 (260)
Gate	1/8 (3)	48 (1220)	Vacuum	10000 (700)	–455 (–277)	1250 (675)
Globe	1/8 (3)	30 (760)	Vacuum	10000 (700)	–455 (–272)	1000 (540)
Plug lubricated	¼ (6)	30 (760)	Atmospheric	5000 (350)	–40 (–40)	600 (315)
Plug non-lubricated	¼ (6)	16 (406)	Atmospheric	3000 (210)	–100 (–75)	425 (220)
Swing check	¼ (6)	24 (610)	Atmospheric	2500 (175)	0 (–18)	1200 (540)
Swing check Y-type	¼ (6)	6 (150)	Atmospheric	2500 (175)	0 (–18)	1200 (540)
Lift check	¼ (6)	10 (250)	Atmospheric	2500 (175)	0 (–18)	1200 (540)
Titling disc	2 (50)	30 (760)	Atmospheric	1200 (84)	–450 (–260)	1100 (590)
Diaphragm	1/8 (3)	24 (610)	Vacuum	300 (21)	–60 (–50)	450 (230)
Y (Oblique)	1/8 (3)	30 (760)	Vacuum	2500 (175)	455 (–272)	1000 (540)
Slide	2 (50)	75 (1900)	Atmospheric	400 (28)	0 (–18)	1200 (650)
Pinch	1 (25)	12 (305)	Vacuum	300 (21)	–100 (–75)	300 (260)
Needle	1/8 (3)	1 (25)	Vacuum	10000 (700)	–100 (–78)	500 (260)

The k_V table for angle seat valves gives a k_{V100} factor of 327 for size 1 inch, 484 for size 1¼ inch and 725 for size 1½ inch.

In this example the correct size to use is 1¼ inch. (See also Table V).

Example 2 – (Figs 9 and 10)

(i) What is the k_V factor for a 1¼ inch water pipeline with a flow of 300 litres/min, an inlet pressure of 0.5 bar and an outlet pressure of 0 bar?

(ii) If a valve has to be fitted and the minimum acceptable flow rate in the pipeline is 250 litres/min, which type of valve should be used?

Solution to (i) (Fig 9)

Calculate the k_v for the pipeline (k_{vp})

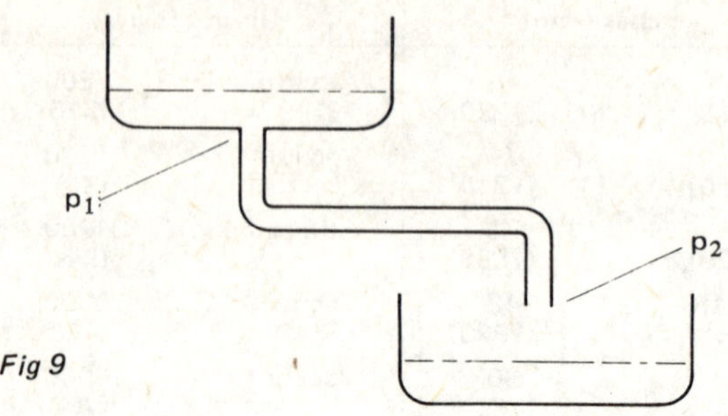

Fig 9

Given

$Q = 300$ litres/min

$\gamma = 1$ kg/dm^3

$\Delta P = P_1 - P_2 = 0.5 - 0 = 0.5$ bar

then

$$k_{vp} = Q\sqrt{\frac{\gamma}{\Delta P}}$$

$$= 300\sqrt{\frac{1}{0.5}}$$

$$k_{vp} = 424$$

Solution to (ii) (Fig 10)

First it is necessary to calculate the k_v factor for the *total* system (k_{vt}).

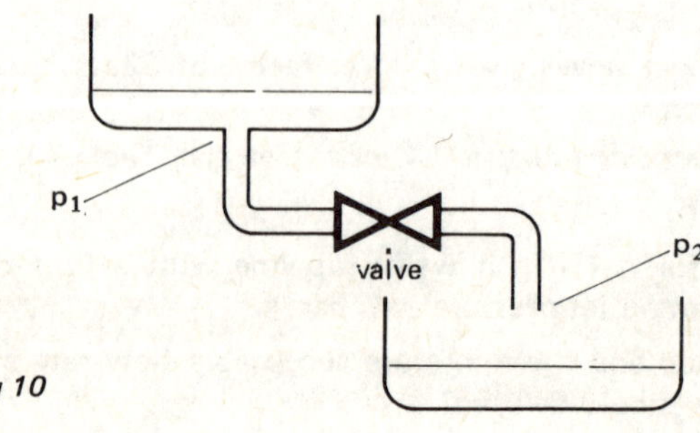

Fig 10

Given

$$Q = 250 \text{ litres/min}$$
$$\gamma = 1 \text{ kg/dm}^3$$
$$\Delta P = P_1 - P_2 = 0.5 - 0 = 0.5 \text{ kg/cm}^2$$

Then

$$k_{vt} = Q\sqrt{\frac{1}{\Delta P}}$$
$$= 250\sqrt{\frac{1}{0.5}}$$
$$k_{vt} = 354$$

The k_v factor for the valve (k_{vv}) can now be established by subtracting the k_v factor for the pipeline (k_{vp}) from the k_v factor for the total system (k_{vt}). For this purpose, the formula for calculating flow factors in series should be used, which is

$$\frac{1}{K_v^2{}_x} = \frac{1}{k_v^2{}_1} + \frac{1}{k_v^2{}_2} + \ldots\ldots \frac{1}{k_v^2{}_n}$$

thus

$$\frac{1}{k_v^2{}_t} = \frac{1}{k_v^2{}_v} + \frac{1}{k_v^2{}_p}$$

$$\therefore \quad \frac{1}{k_v^2{}_v} = \frac{1}{k_v^2{}_t} - \frac{1}{k_v^2{}_p}$$

$$k_v^2{}_v = \frac{1}{354^2} - \frac{1}{424^2}$$
$$= 7.98 \times 10^{-6} - 5.56 \times 10^{-6}$$
$$= 2.42 \times 10^{-6}$$

$$\therefore \quad k_{vv} = \sqrt{\frac{1}{2.42 \times 10^{-6}}}$$
$$= 643$$

The calculation shows that the valve used must be one with a minimum k_{v100} factor of 640.

From the k_v tables it can be seen that a 1¼ inch ball valve has a k_{v100} factor of 484 and a 1¼ inch diaphragm valve a k_{v100} factor of 332.

Therefore only the 1¼ inch *ball* valve can be used.

SECTION 2

Valves (Types, Design, Construction)

Cocks (Plug Valves)

THE DESCRIPTION *cock* (in the UK) or *plug valve* (USA) is given to the simplest form of valve comprising a body with a tapered, or less frequently a parallel, seating into which a plug fits. The plug is formed with a through-port, the relative position of the port controlling the amount of opening through the valve – Fig 1. A 90 degree rotation of the plug fully opens or closes the fluid flow.

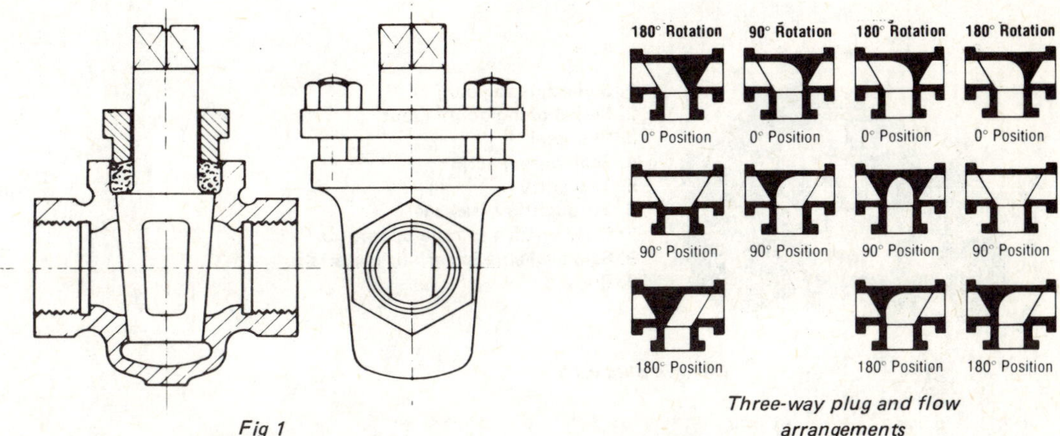

Fig 1

Three-way plug and flow arrangements

The simple cock is generally suitable for low pressure, low temperature applications, and can be made in quite large sizes; 10–12 in (250–300 mm) bore is quite common in some applications. Its main limitation is that if wide variations in fluid temperature are involved differential expansion is inevitable, leading either to undue stiffness of operation or loss of pressure-tightness.

This can be overcome to some extent by employing a packed gland on which the plug rides – Fig 2. The packing is commonly shredded asbestos. In the smaller range, the sleeve-packed cock represents a distinct step forward in cock design (Fig 3). Not only does this have a perfectly cylindrical plug, more economical to produce than a tapered one, but the resilience afforded by the asbestos fabric sleeve longitudinally compressed by the two plugs screwed into both top and bottom of the body provides for temperature variations, and thereby prevents binding.

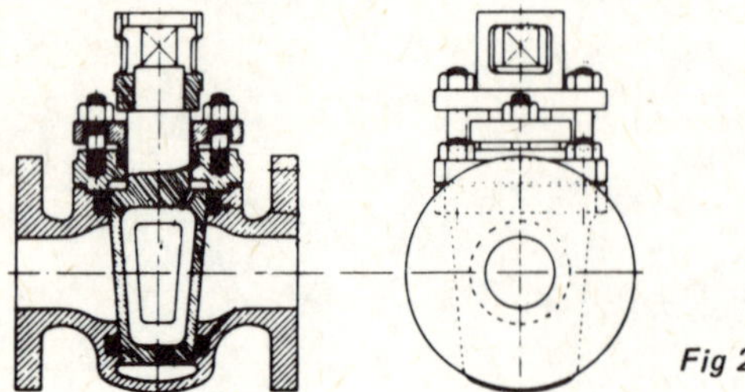

Fig 2

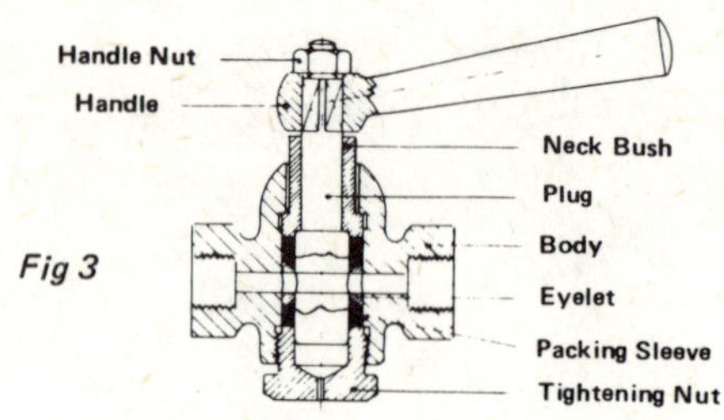

Fig 3

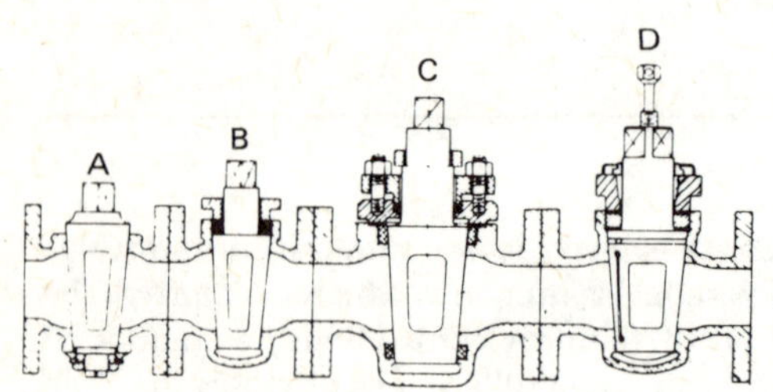

Fig 4

A – Ground plug cock with nut and washer base.
B – Ground plug cock gland packed.
C – Asbestos groove-packed plug cock with gland and holding down plate.
D – Lubricated plug cock gland packed.

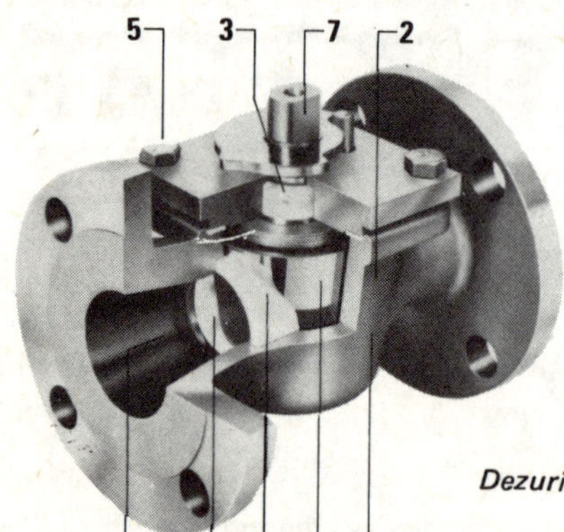

1. Self-adjusting plug
2. Self-aligning double seat
3. Skin seal
4. Seat wiping action
5. Top entry
6. Throughflow passage
7. Quarter-turn spindle operation
9. Seat available in various materials
10. Body

Dezurik plug valve

3-way 2-port

3-way 3-port

4-way 4-port

Transflo plug

Examples of multi-port arrangements

Rectangular port, round port and diamond port plug valves. (W.K.M.)

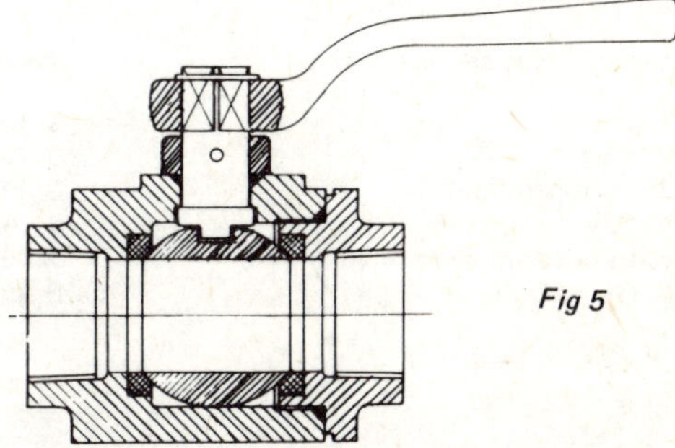

Fig 5

In the UK the description *plug valve* is specifically given to a cock which incorporates special design features to reduce the friction between the plug face and the body seat. The plug itself may be tapered or parallel, and the movement plain or lubricated (Fig 4). There is also a further variation known as a ball plug valve, where the plug element is spherical, with circular ports rotating between circular seats of concave section (Fig 5).

Plug valves may be further categorized by pattern:-

(i) round opening — with full bore round ports in both plug and body.

(ii) rectangular (rectangular opening) with rectangular or similar shaped ports of substantially full bore section.

(iii) standard opening — where the area through the valve is less than the area of standard pipe.

(iv) diamond port — where the opening through the valve is diamond shaped. Such valves are also normally of venturi design.

(v) multipoint — with three or more pipe connections, used mainly for transfer or diverting services.

(vi) venturi design — with reduced area porting (down to 40%) and featuring venturi flow through the body.

(vii) short — with reduced area ports and/or reduced face-to-face dimensions.

(viii) vertical — with reduced area seating ports and the plug passages reduced in section to form a throat.

Nomenclature:

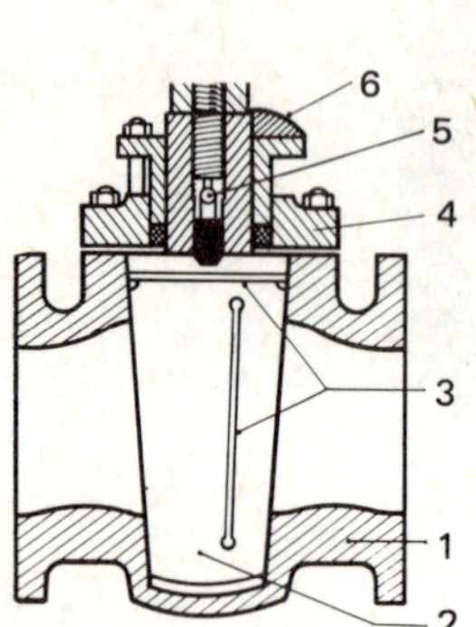

Taper plug valve (lubricated)

1. Body
2. Plug
3. Lubricant grains
4. Cover
5. Lubricant check valve
6. Gland follower

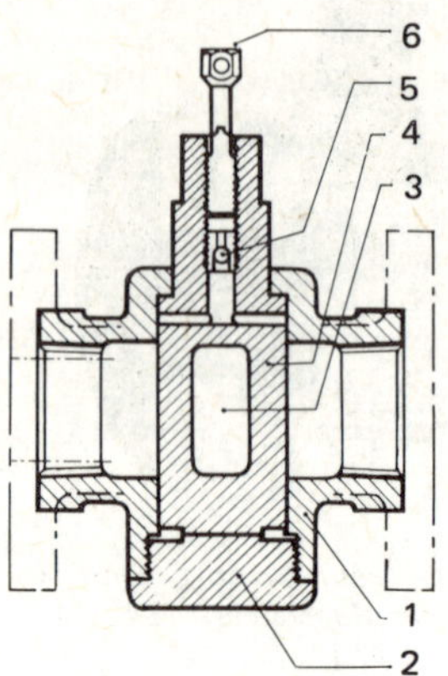

Parallel plug valve

1. Body
2. Bottom cover
3. Plug port
4. Plug
5. Lubricant grains
6. Lubricant screw

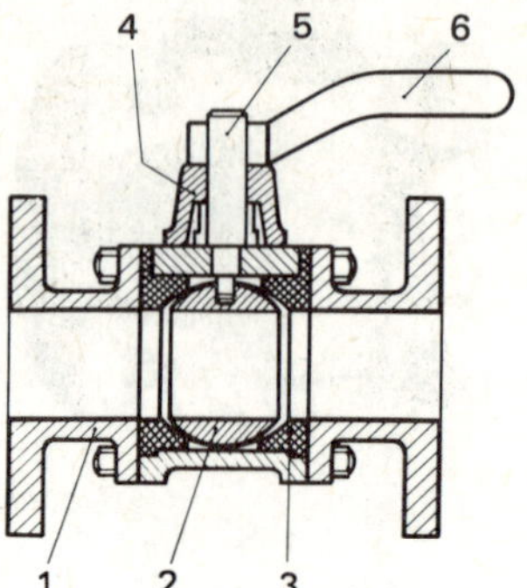

Ball plug valve

1. Body
2. Ball
3. Seal
4. Bonnet
5. Spindle
6. Handle

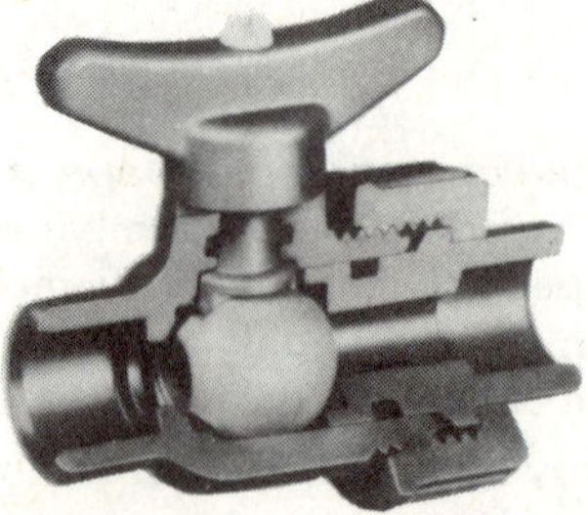

Laboratory ball cock valve in UPVC

Materials

Cocks and plug valves are produced in a variety of metals and plastics, and also include lined types. Metals most commonly used are brass, bronze, steel and stainless steel.

Basic Design Proportions

A rectangular or trapezoid section port is commonly preferred as this can be accommodated in a plug of smaller diameter than that required for a circular port of the same area. The width of the port is then often made less than half of the bore to provide an effective *positive lap* for sealing. The length of port is then given by $\pi d/2$, where d is the pipe diameter. In practice a small addition is usually made to this length to allow for radiusing the corners of the opening.

In the case of multiport cocks or plug valves, *negative lap* may be called for to ensure that there is no complete shut-off during the transition of ports. This applies particularly when connected to a positive displacement pump (*ie* to prevent the pump pumping against a closed outlet).

See also *Ball Valves.*

Ball Valves

BALL VALVES are amongst the least expensive but most widely used of all valve types, as well as being available in an extremely wide range of sizes. Basic geometry involves a spherical ball located by two resilient sealing rings in a simple body form (Fig 1). The ball has a hole through one axis, connecting inlet to outlet with full bore flow when aligned with the axis of the valve. Rotating the ball through 90 degrees completely closes the flow passage with positive sealing via the sealing rings. Sealing is equally effective in both directions.

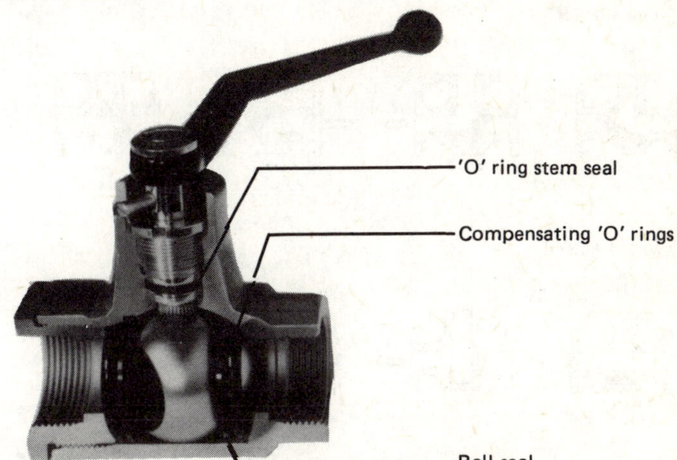

Fig 1 Legris ball valve with 'O'-ring seals.

Body forms and matching ball hole may provide straight through (full bore parallel), reduced flow, or venturi flow.

The ball itself may be free floating, in which case the squared off or splined end of the stem fits into a matching recess in the top of the ball. On larger valves the ball may be trunnion mounted. Trunnion mounting reduces operating torque to about two-thirds that of a floating ball (Fig 2).

Ball valves are produced in top entry and split body forms for assembly and for renewal of the seals and ball. They are also produced in multi-port configurations, thus normally requiring a larger size of ball to accommodate multi-port drillings. These ports can be proportioned to give positive lap or negative lap as required. (See also Fig 3).

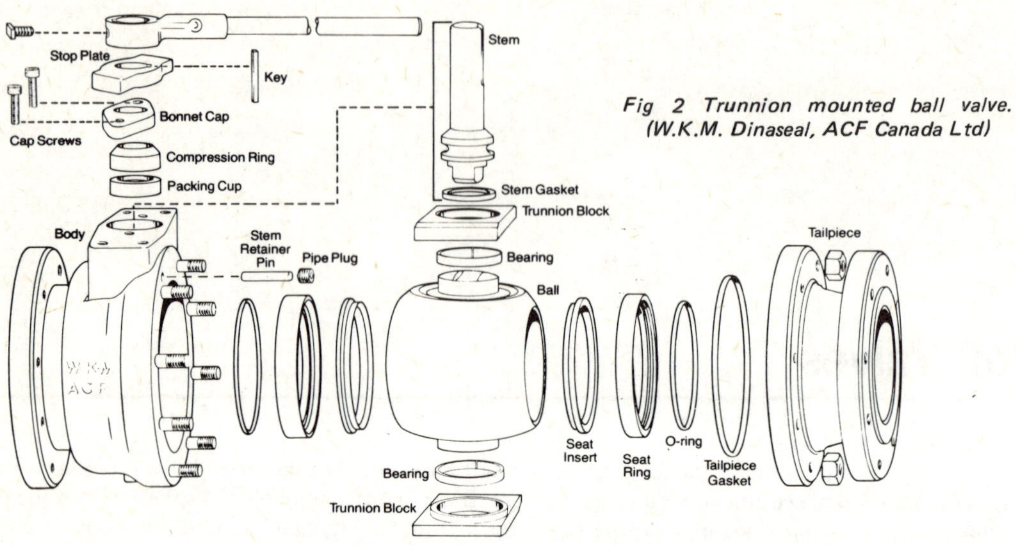

Fig 2 Trunnion mounted ball valve. (W.K.M. Dinaseal, ACF Canada Ltd)

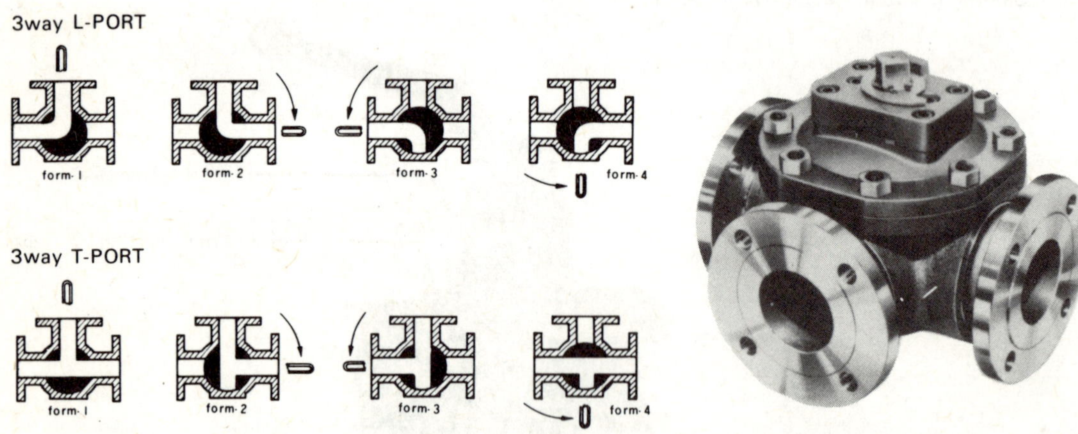

Fig 3 KTM 3-way ball valve.

Full operating movement is 90 degrees rotation of the ball. Steps may be incorporated to limit movement of the operating lever, or continuous rotation may be possible. In either case the lever position is in line with the axis of the valve in the open position and at right angles to it in the closed position. Larger ball valves may be operated by handwheels through reduction gearing, or by powered actuators. In all cases opening/closing torque is low since the only friction forces involved are those of the ball rotating against its seals and the friction offered by the stem gland. The latter can range from O-rings to glands fitted with die-formed packing rings.

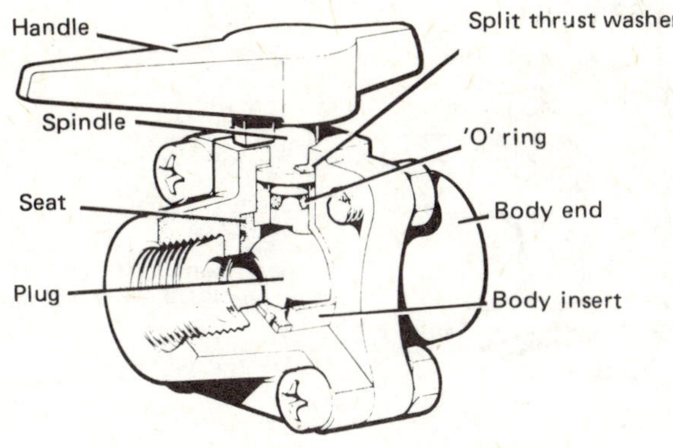

Saunders 'MC' type lightweight ball valve.

Trufio flanged top entry ball valves. Sizes ½–12 inch nominal bore manufactured in carbon and stainless steels.

Argus flanged ball valve in steel or stainless steel. (Powerite Limited)

Compression Test: A 24 inch ANSI 600 class Cameron ball valve was tested to confirm that it will function properly when loaded with an axial compressive load in excess of the load required to yield line pipe. The compressive force caused the pipe to yield at 1 800 000 pounds. The force reached a maximum of 2 000 000 pounds. The valve held drop tight during the entire test and the operating torque did not increase

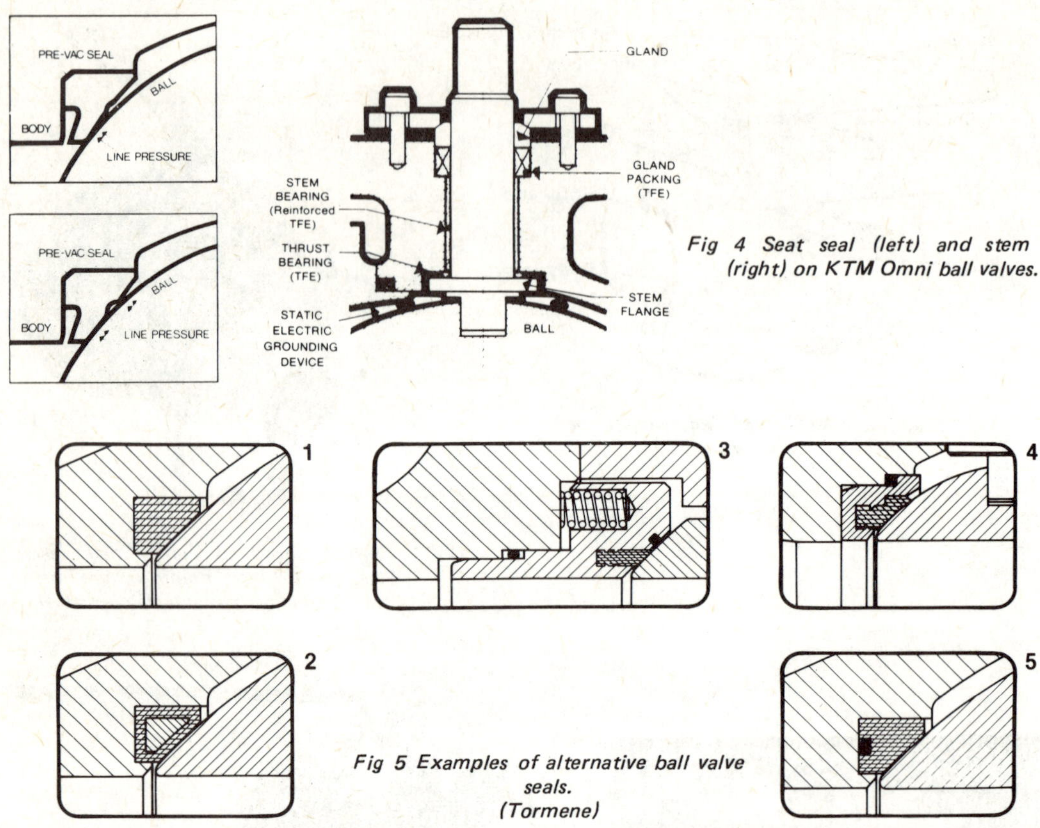

Fig 4 Seat seal (left) and stem seal (right) on KTM Omni ball valves.

Fig 5 Examples of alternative ball valve seals. (Tormene)

Ball seals are commonly simple rings of elastomeric or plastomeric material — *eg* rubber, neoprene, nylon, virgin TFE and filled PTFE. Choice is largely dictated by fluid compatibility and maximum service temperature requirements. More complex sealing arrangements may be used on larger valves, or valves designed for specialized applications, including multiple rings, spring loaded rings for automatic adjustment, metal seats with inserted seals, and sealing rings mounted on rotating seats to distribute wear evenly and increase seal life. With fixed seals wear tends to be highest at the point subject to the high velocity initial flow when the valve is opened. (See also Figs 4 and 5).

Ball valves are made in a very wide range of materials. Materials commonly used are carbon steel for the body, with the ball in low alloy steel either nickel or chrome plated, or stainless steel. Seat rings normally made in carbon steel are cadmium or nickel plated. Plastics may be used for corrosive duties in smaller sizes of valves, *eg* plastic ball and/or body. Glass fibre/vinyl ester composite construction has also been used successfully in place of stainless steel with a cost saving of some 30%. Graphite based composites offer even better wear surfaces for the ball.

Flow Characteristics

Typical flow characteristics for a ball valve are shown in Fig 6. Equal percentage flow characteristics are achieved at about 60% opening. Conventional ball valves are not usually chosen as throttling valves, however, since under partial flow conditions high velocity flow impinges against a

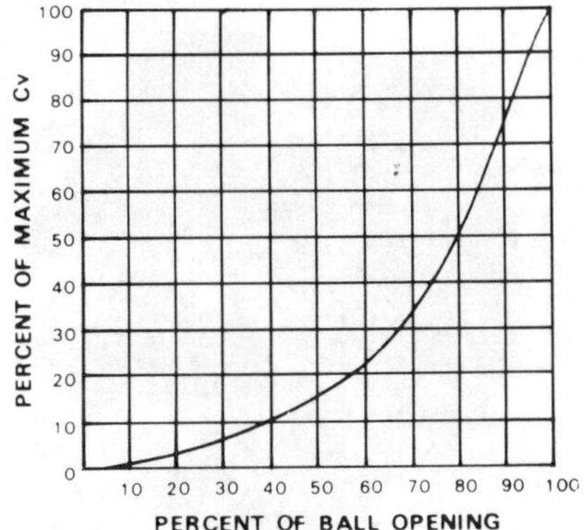

Fig 6 Flow characteristics of a typical ball valve.

Nil-Cor 310 fibreglass reinforced ball valve — body and insert moulded from glass fibre reinforced single ester. Designed as an economic alternative to stainless steel ball valves. (Babcock & Wilcox Ltd)

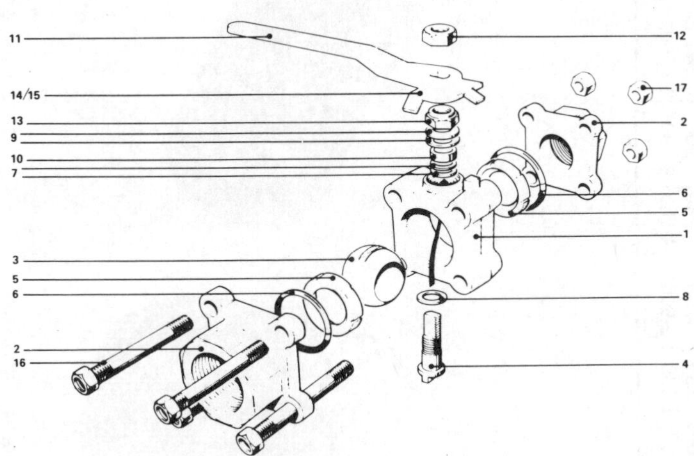

Worcester aluminium and brass series 44 ball valve.

1. Body
2. Body connector
3. Ball
   8–25 mm valve
   32–50 mm valve
4. Stem
5. Seat ring
6. Body connector seal
7. Gland packing
8. Stem thrust seal
9. Disc spring
10. Gland
11. Wrench
12. Wrench nut
    8–20 mm
    25–50 mm
13. Gland nut
14. Identification plate
15. Wrench sleeve
16. Body connector bolt
17. Body connector nut

localized area of the ball and seals. This particular limitation can largely be overcome by detail changes making such types of ball valves suitable for control applications over a wide range of flows.

Typical flow coefficients for ball valves are given in Table I. See also chapter on *Water Services.*

TABLE I – TYPICAL VALVE COEFFICIENTS FOR BALL VALVES

FULL BORE VALVES																
Valve Size																
in	½	¾	1	1½	2	2½	3	4	6	8	10	12	14	16	18	20
mm	12	19	25	37.5	50	62.5	75	100	150	200	250	300	350	400	450	500
Cv	26	50	94	260	480	750	1,300	2,300	5,400	10,000	16,000	24,000	31,400	43,000	57,000	73,000
REDUCED BORE VALVES																
Valve Size																
in	3	4	6	8	10	12	14	16	18	20						
mm	75	100	150	200	250	300	350	400	450	500						
Cv	420	770	1,800	2,500	4,500	8,000	12,000	14,000	18,000	22,000						

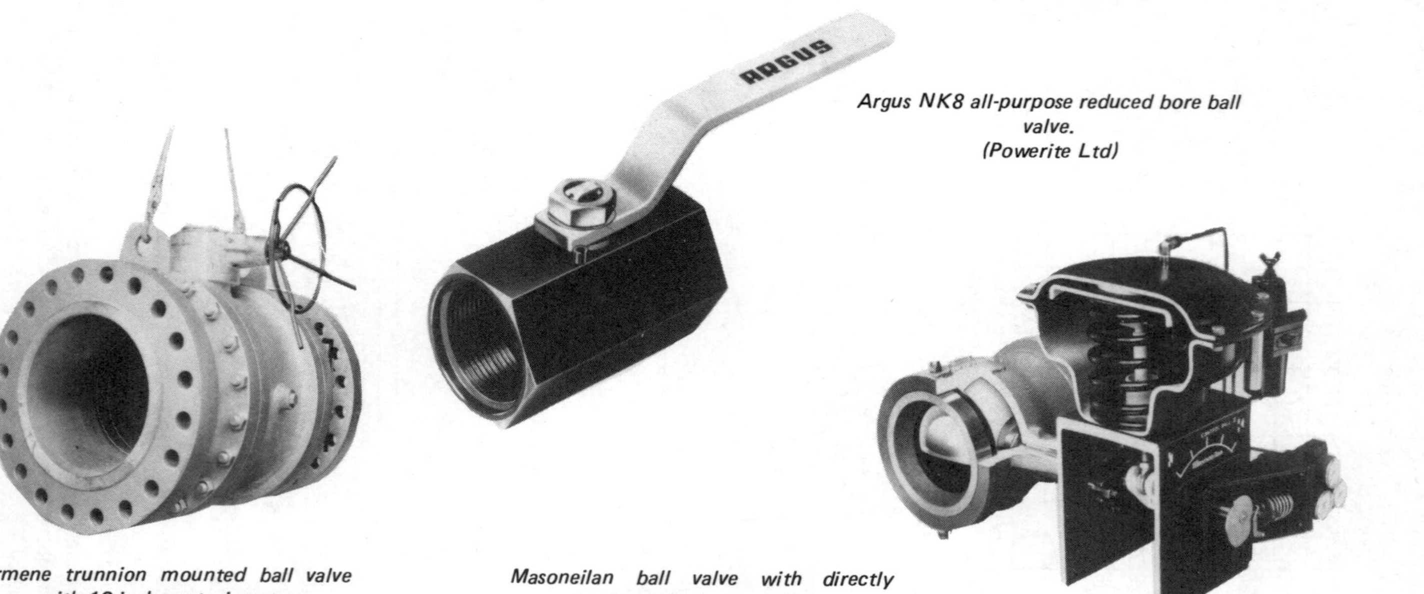

Argus NK8 all-purpose reduced bore ball valve. (Powerite Ltd)

Tormene trunnion mounted ball valve with 12 inch venturi port.

Masoneilan ball valve with directly connected valve positioner.

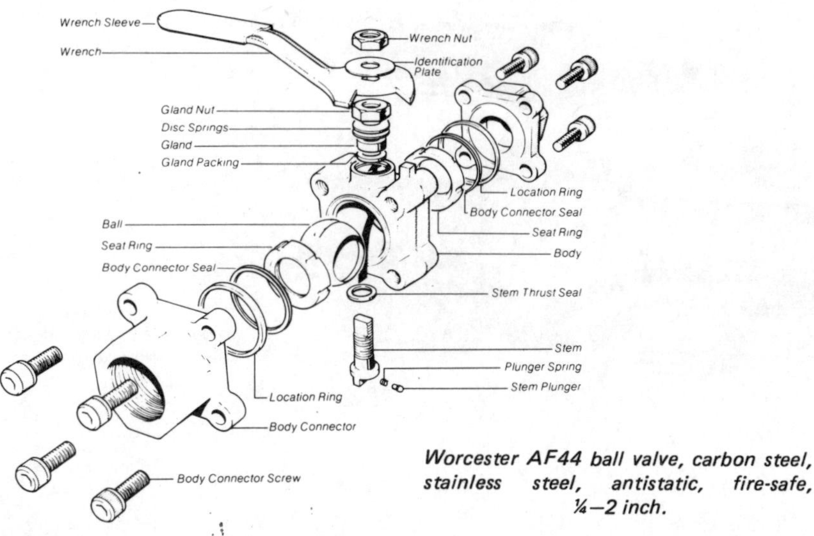

Worcester AF44 ball valve, carbon steel, stainless steel, antistatic, fire-safe, ¼–2 inch.

Fire-safe Ball Valves

Extensive use is made of ball valves in gas and oil pipelines, where 'fire-safe' characteristics are a necessary feature. 'Fire-safe' requires that the valves remain operable with nil or specified minimal leakage (depending on test specification adopted) after being burnt. Burning temperatures will destroy soft seats and seals and various methods may be used to overcome this effect. One approach is to use sacrifical stem bearings which burn away in heat allowing the ball to 'float' into the downstream seal. This can result in the valve being very difficult to operate. Others provide secondary metal seats to maintain contact between ball and a sealing surface in the event of the non-metallic seat being destroyed by heat.

See also chapter on *Fire Safe Valves.*

Bend Test: A 20 inch ANSI 600 class Cameron ball valve is exposed to a bending moment of 1.33 million foot pounds. This bending moment is greater than the pipe that is normally used would withstand. The valve operated and sealed satisfactorily during all parts of the test.

Powerite ball valve in UPVC and polypropylene

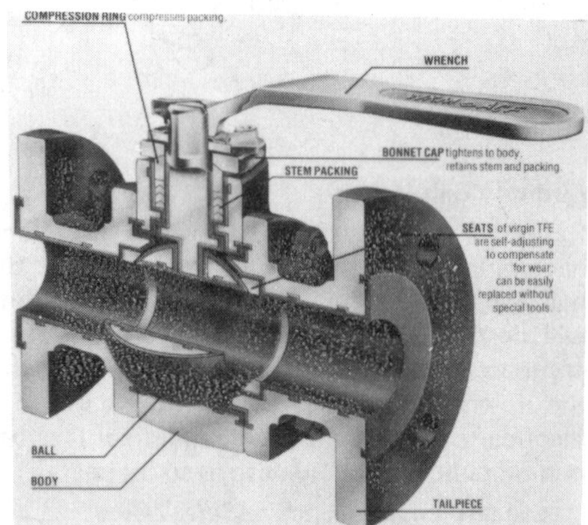

Ball valve with virgin PTFE seal. (W.K.M. Dinaseal, ACF Canada Ltd)

Ball Valves in Hydroelectric Power Plants

It has been not quite twenty years that the ball valve, in applications both above and below water, took the position of pre-eminence in hydroelectric power plants in the use with turbines, pumps and turbo-pumps, replacing the sluice valve and the rotary piston valve. This was after extensive testing of the hydraulic behaviour of the ball valve with regard to its action while partially closed, which turned out to be positive, and after appropriate redesigning of the valve in order to eliminate unwanted vibratory action. Such changes resulted in the development of a ball valve having the smallest reliable size as well.

At the same time as these practical tests were being carried out, theoretical studies were also being worked on to determine the requirements for ball valves in applications, and for evaluating the equivalence of various models of valves. This has meant that today we have available representative values, which, for example, can be used for determining the range of cavitation or calculating the operating moments of the valve. The types of ball valves used in hydroelectric power plants have ring seats and can be opened or closed using either water or oil pressure, and have given these types of shut-offs their name, although from the point of view of design a valve is actually meant.

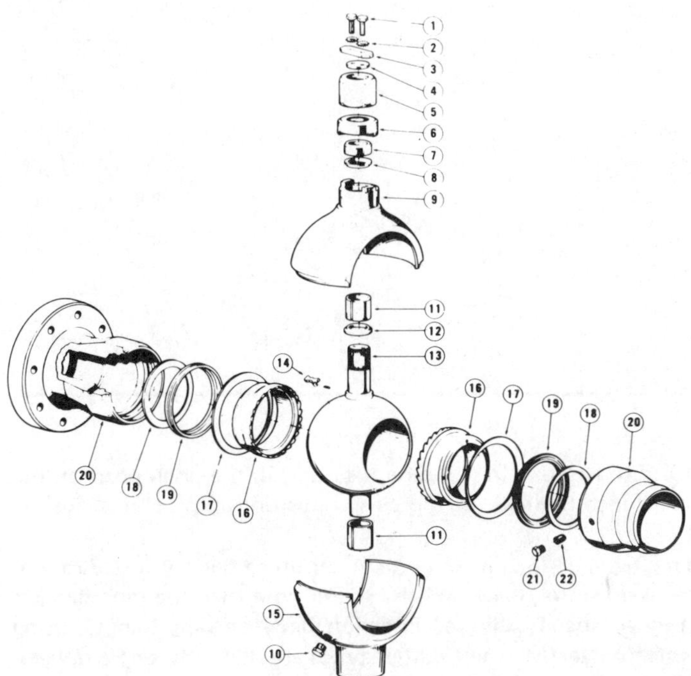

Nomenclature: Ball valve.

1. Screws, cap
2. Washers, lock
3. Indicator
4. Plate, loading
5. Nut, operating
6. Ring, stem stop
7. Gland, stem seal
8. Seal, stem, outer
9. Body, upper
10. Plug, pipe
11. Bearing
12. Seal, stem, inner
13. Assembly, ball
14. Dog
15. Body, lower
16. Ring, seat
17. Sediment barrier
18. Spring, seat loading
19. Seal, lip
20. Connection, end
21. Plug, pipe
22. Check valve

Cameron all-welded ball valve

The ball valves in such applications are mostly equipped with an alternative opening which is actuated separately from the main seal, and are predominantly actuated from a system using oil pressure.

The type of actuating and regulating system used here requires a high degree of maintenance and supervision; in hydroelectric power plants, the maintenance and supervision, which is achieved by other mechanical or electronic devices, is generally under the control of the operating personnel. Also, in hydroelectric power plants, due to their relatively high annual amount of hours of maintenance and operation, it is possible to make use of a greater amount of technology and still be economically competitive in terms of the aggregate of equipment that would be possible in the case of drinking water supply systems.

Butterfly Valves

A BUTTERFLY VALVE consists of a disc rotating in trunnion bearings. In the open position the disc is edge-on, offering minimum resistance to flow. In the closed position it is rotated against a seat.

Butterfly valves are available in sizes from an inch or so in diameter up to twenty feet, (25 mm to 6 m). They are particularly attractive since they take up little more room than the pipe flanges. In fact some butterfly valves are designed specifically for insertion between pipe flanges, using through bolts. These are known as *wafer butterfly valves.* Other types are normally *single flanged* or *double flanged,* known also as lug body valves — Fig 1.

Fig 1 Lug body (left) and wafer body (right) butterfly valves. (De Zurik)

Butterfly valve movement is simple and straightforward, requiring only 90° rotation of the butterfly for full movement (or somewhat less in most designs). The main disadvantage is that a simple butterfly design is difficult to render completely 'tight' when closed, due to the absence of any wedging effect in this plane. This can be alleviated to a certain extent by seating the butterfly on a resilient ring on closure, and/or offsetting the axis of rotation slightly so that, as well as rotating, the blade has a movement in a plane at right angles to the spindle axis.

from

mono valve engineering ltd.

17 CHALFORD INDL. ESTATE, CHALFORD, STROUD, GLOS.
Tel: Brimscombe (0453) 883866
Telex: 43587

The true facts behind the most advanced range of automatic process control valves in the world.

Camflex II
The only all purpose automatic control valve with totally enclosed moving parts which can handle about 80% of all control valve applications in a typical process plant.

Sigma F
The versatile Sigma F system – another Masoneilan step towards standardisation. It combines a choice of only three unique side mounted actuators with three types of valve body developed from earlier ranges.

MiniTork II
The second generation of butterfly valves designed for automatic throttling control offers an unparalleled performance in either direction but with all external linkages now eliminated.

Masoneilan

Northern Office: 19 Vine Street, Hazel Grove. Stockport, Cheshire SK7 4JS. Tel: (061) 483 0819 Telex: 666461
Southern Office: 49 Church Street, Maidstone, Kent ME14 1DS. Tel: Maidstone (0622) 59861 Telex: 965668

Gearbox operated 8 in lugged stainless steel butterfly valve in chemical plant. (Charles Winn (Valves) Ltd).

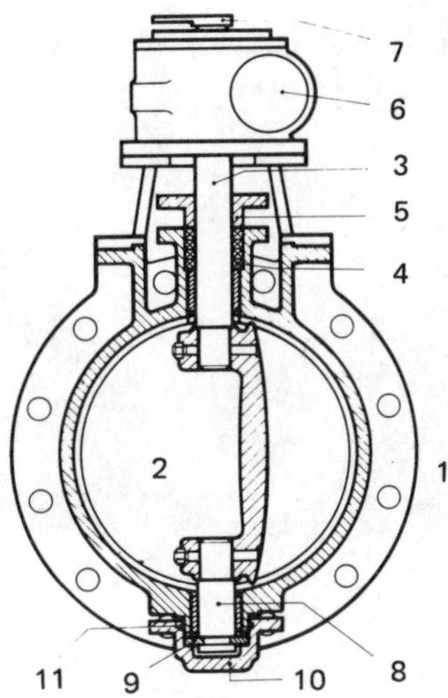

1. Body
2. Disc (butterfly)
3. Shaft
4. Shaft gland
5. Gland receiver
6. Manual operator or actuator
7. Position indicator
8. Stub shaft and lower bearing
9. Thrust collar
10. Cover
11. Stub shaft seal

Nomenclature: Butterfly valve.

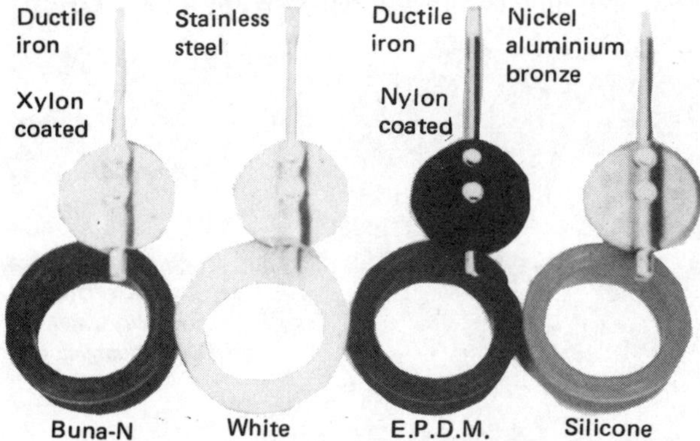

Alternative disc and seal materials for general purpose butterfly valves. (Keystone Valve (UK) Ltd).

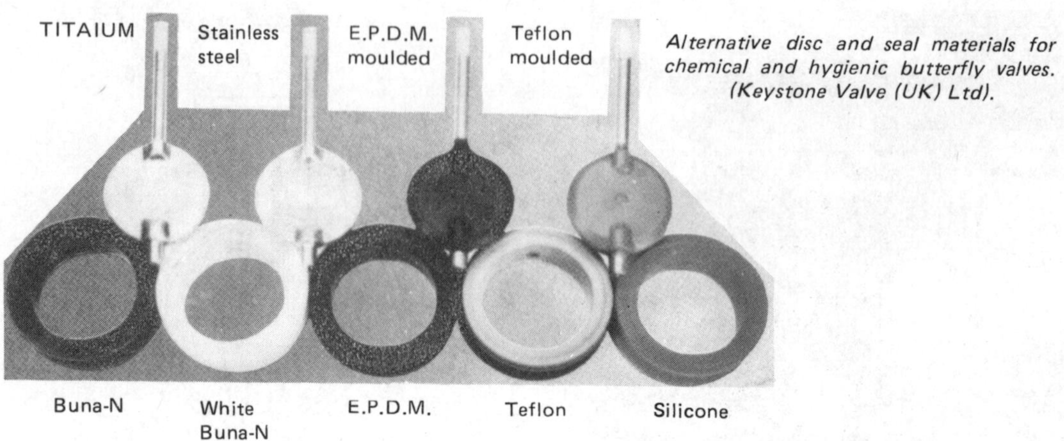

Alternative disc and seal materials for chemical and hygienic butterfly valves. (Keystone Valve (UK) Ltd).

One lesser drawback of the butterfly is that in the fully open position the butterfly, (now edge-on), offers some resistance to flow, and could collect fibrous or stringy materials when used with a contaminated fluid. It is thus more suited for use with clean fluids, where its basic advantages and low actuating power requirements can make it an attractive choice in all sizes.

Body construction is normally cast iron, although various other materials may be used depending on size and application. Welded materials (*eg* steel, stainless steel and titanium) may be used for certain valves for the chemical industry, and in particular where percolation of gases through cast components is to be prevented.

Discs are also usually of cast iron, although again alternative materials may be specified for particular services. Profile shapes vary, most having some form of convex streamline shape to minimize head loss.

Seal design can vary considerably. A basic arrangement is a corrosion-resistant metal seat (*eg* bronze or stainless steel), into which a continuous rubber ring seal is fitted. Alternatives include resilient plastic seal rings (for chemical duties) and flexible metal seal rings (for high temperature services) Fig 2. Detail design may provide automatic adjustment to any eccentric motion of the disc, and/or automatic compensation for seal wear. Drop-tight closure of any butterfly valve is normal and can be retained for a considerable time before seal replacement is necessary.

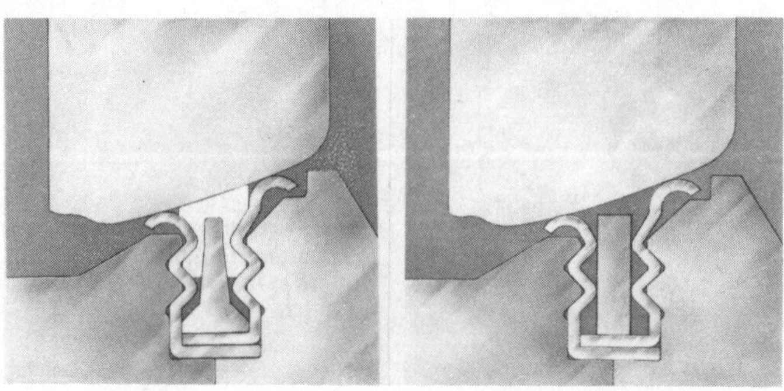

Fig 2 Metallic seats, alternative to soft seats on 'DynaCentric' butterfly valves. (ACF Industries Inc)

Blakeborough type 'SL' butterfly valve with electric actuator.

TABLE I – SELECTION TABLE FOR AUTOMATIC CLOSING BUTTERFLY VALVES

			NO		YES				
	is power available ?								
	... during emergency closing ?								
	... any voluntary remote control ?								
Control Device	Electrically motorized valve								●
Control Device	Hydraulic jack		●	●	●	●	●	●	
Control Device	Counterweight		●	●	●	●			
Control Device	Oleopneumatic accumulator						●		
Control Device	Mechanical locking	manual resetting	●	●					
Control Device	Mechanical locking	automatic resetting			●				
Control Device	Automatic leakage recovery					●	●		
Control Device	Motorized pump			●	●	●	●	●	
Control Device	Hand pump		●						
Control Signal	Defect transmission	mechanical	●						
Control Signal	Defect transmission	electrical						●	●
Control Signal	Defect transmission	optional		●	●	●	●		
Control Signal	Electrical remote control				●	●	●	●	●

Actuation

All types of control mechanisms can be used to operate butterfly valves – manual by lever or handwheel with reduction gear, electrical by actuator or reduction gear, hydraulic actuator or pneumatic actuator. Choice largely depends on the size of the valve and the specific application. Special control systems can also be used for automatic closing – *eg* see Table I.

Hydraulic Performance

Typical flow characteristics of a butterfly valve relative to percentage opening are shown in Fig 3 and Table II. Here it will be seen that linear characteristics are normally present after about 60% opening. The flow coefficient (C_V) follows a similar course. Fig 4 gives flow coefficients for a series of butterfly valves of similar design but different size, those being representative of good design.

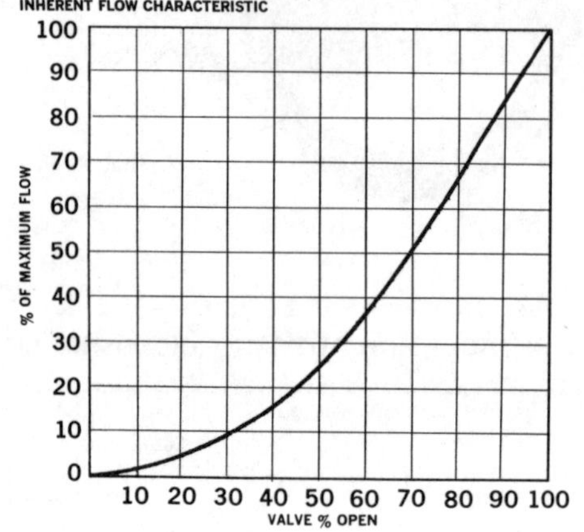

Fig 3

TABLE II – TYPICAL VALVE COEFFICIENTS (C_v) FOR BUTTERFLY VALVES (De Zurik Valves)

Valve Size		Disc Angle – Degrees Open				Maximum Flow Velocity (ft/sec)	
in	mm	90° Wide Open	70°	50°	30°	ANSI 150	ANSI 300
2	50	85	65	35	15	90	90
2½	65	160	120	65	29	80	90
3	80	260	195	104	47	80	80
4	100	475	356	190	86	80	80
5	125	770	577	308	139	80	80
6	150	1125	844	450	203	75	80
8	200	2110	1583	844	380	70	75
10	250	3350	2513	1340	603	60	70
12	300	4800	3600	1920	864	50	60
14	350	6900	5175	2760	1224	50	–
16	400	9000	6750	3600	1620	50	–
18	450	11800	8850	4720	2124	50	–
20	500	14300	10725	5720	2574	50	–

Flow in gal/min of water at I lb/in^2 pressure drop

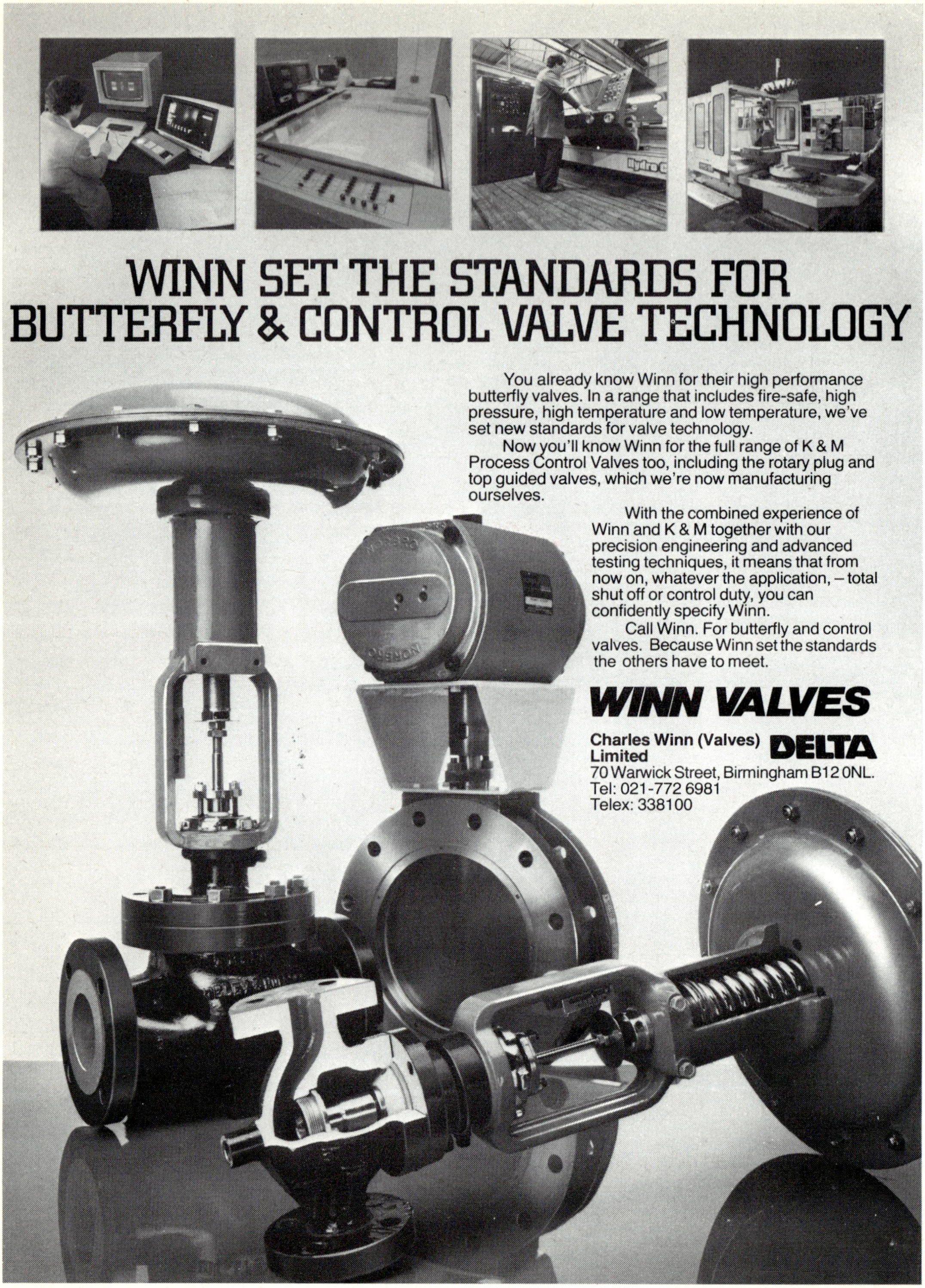
WINN SET THE STANDARDS FOR
BUTTERFLY & CONTROL VALVE TECHNOLOGY
You already know Winn for their high performance butterfly valves. In a range that includes fire-safe, high pressure, high temperature and low temperature, we've set new standards for valve technology.
Now you'll know Winn for the full range of K & M Process Control Valves too, including the rotary plug and top guided valves, which we're now manufacturing ourselves.
With the combined experience of Winn and K & M together with our precision engineering and advanced testing techniques, it means that from now on, whatever the application, – total shut off or control duty, you can confidently specify Winn.
Call Winn. For butterfly and control valves. Because Winn set the standards the others have to meet.
WINN VALVES
Charles Winn (Valves) Limited
DELTA
70 Warwick Street, Birmingham B12 0NL.
Tel: 021-772 6981
Telex: 338100

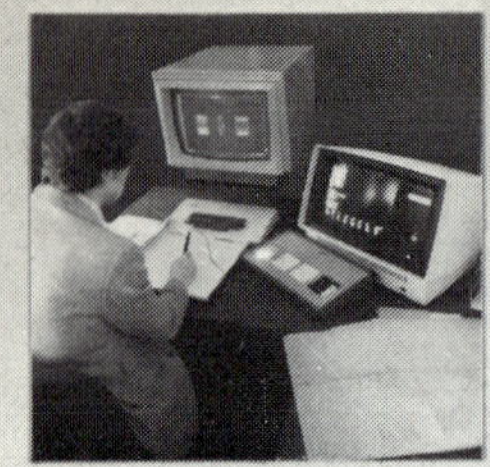

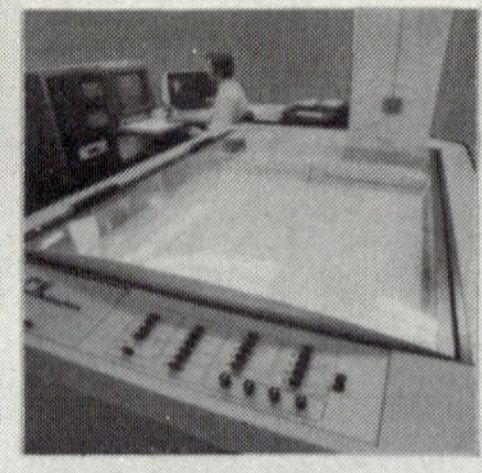

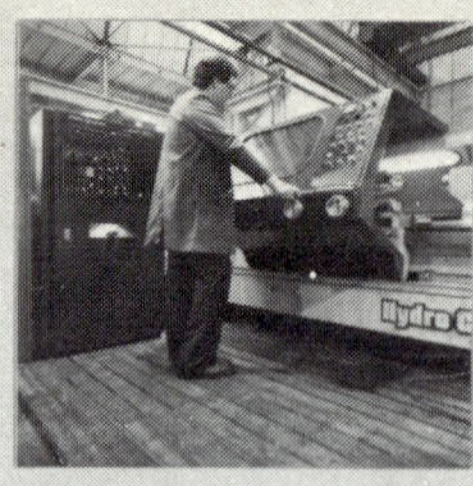

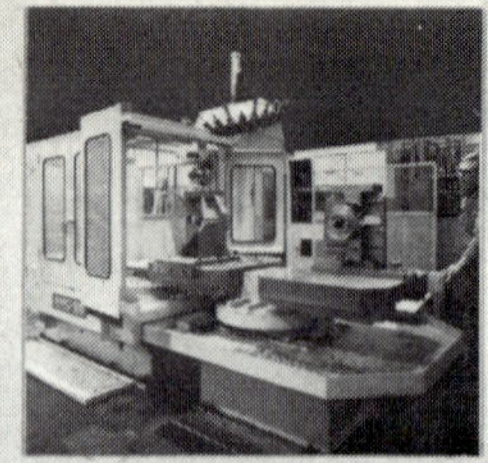

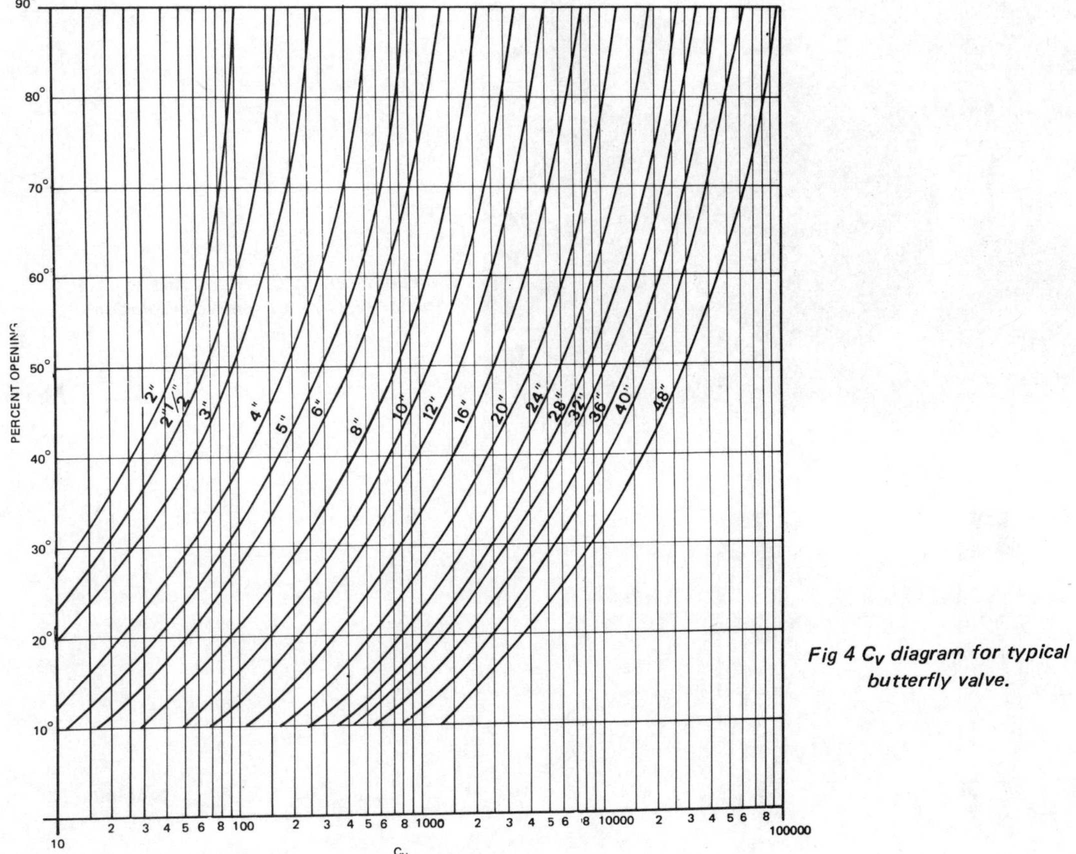

Fig 4 C_V diagram for typical butterfly valve.

Head loss (or pressure drop) is proportional to the square of the flow velocity (V) and to a coefficient K_α which is a function of the angular position of the disc, *ie*

$$\text{Head loss} = K_\alpha \frac{V^2}{2g}$$

$$\text{Pressure drop} = K_\alpha \rho \frac{V^2}{2g}$$

where ρ is the mass density of the fluid

Values of the coefficient K_α can only be determined empirically and are thus specific to the geometry of individual designs of butterfly valves. The values given in Fig 4 are, however, fairly typical of modern design practice.

Blakeborough 2200 mm rubber seal butterfly valve for reservoir isolation duty.

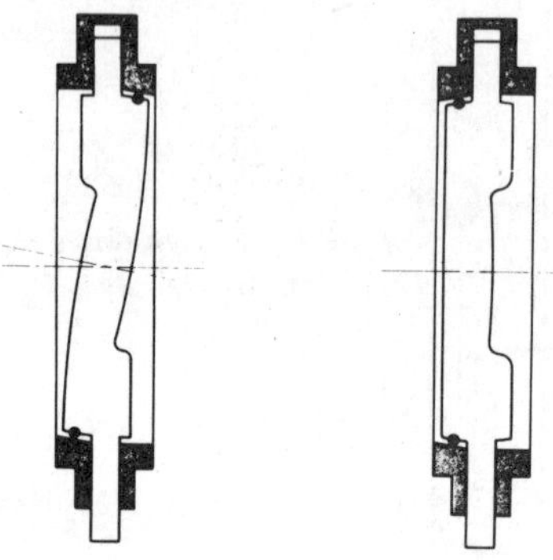

Oblique and straight butterfly discs with seals.
(Neyrtec)

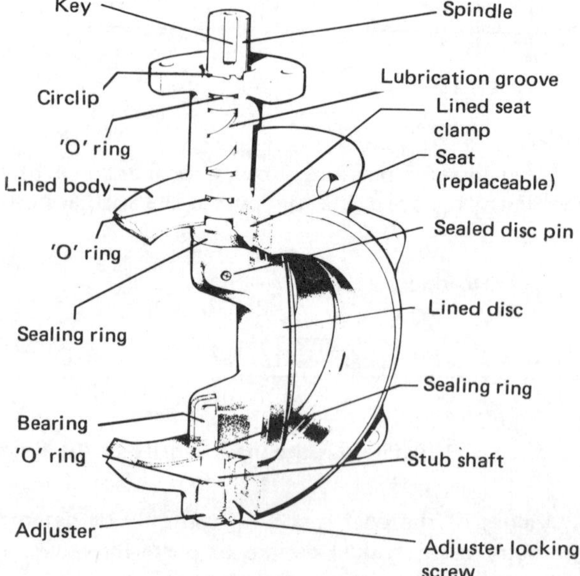

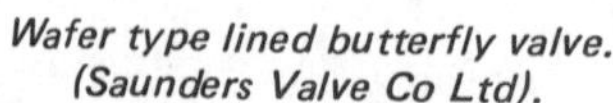

Wafer type lined butterfly valve.
(Saunders Valve Co Ltd).

Key:

1. Steel to BS970 Grade EN6
2. Integral mounting flange
3. Shaft bearing
   – glass filled PTFE
4. 'O' ring – nitrile rubber
5. Shaft bearing
   – glass filled PTFE
6. Disc/Shaft
   Stainless steel
   BS3100 Grade 316C16
   Electro polished
7. Upper body half — BS1452 Grade 14 Electroless nickel
8. Lower body half — Plated and white epoxy coated
9. Sleeve
   Ethylene propylene
   Rubber (black)
   or Nitrile (white)
10. Shaft bearing
    – glass filled PTFE
11. Body screw
    Stainless tsteel BS970
    Grade 302S25

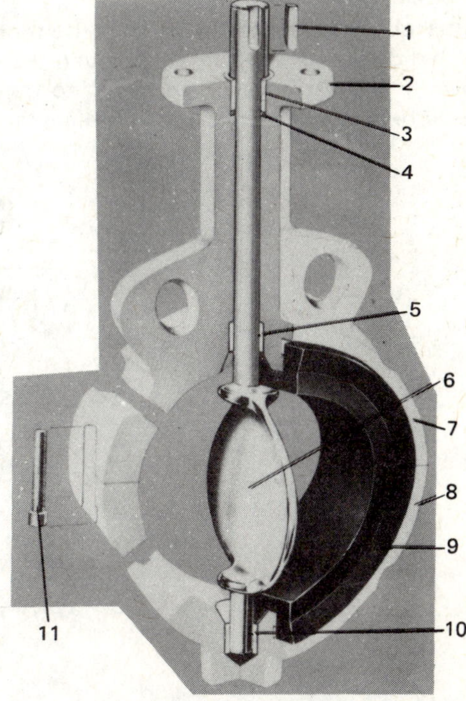

Hygienic butterfly valve.
(Saunders Valve Co Ltd).

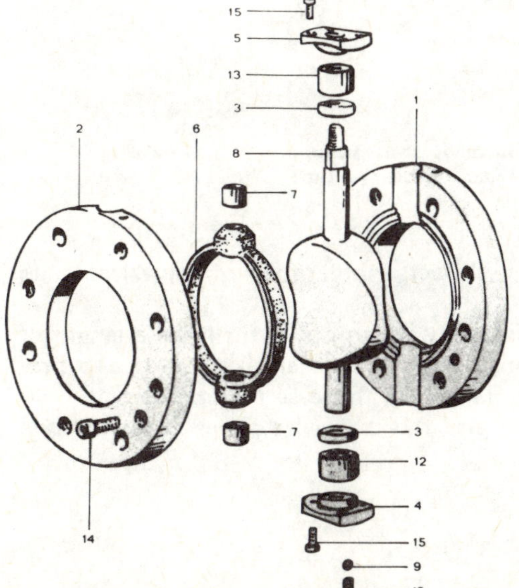

1. Grooved body ring
2. Tongued body ring
3. Adjustment ring
4. Lower cover plate/packing gland
5. Upper cover plate/packing gland
6. Sealing ring
7. Sliding ring.
8. Butterfly disc
9. Ball bearing
10. Adjustment screw
11. Nut
12. Roller bearing
13. Roller bearing
14. Set screw
15. Set screw

Heavy duty butterfly valve with double spherical section butterfly.
(Tormene S.U.C).

Butterfly Check Valves

Butterfly check valves differ from conventional butterfly valves in employing a hinged instead of a pivoted disc, with a sealing ring around the edge of the disc — Fig 5. With forward flow the two halves of the disc are swung together to trail downstream. With reverse flow the two halves open to approximately 45 degrees each, sealing the bore.

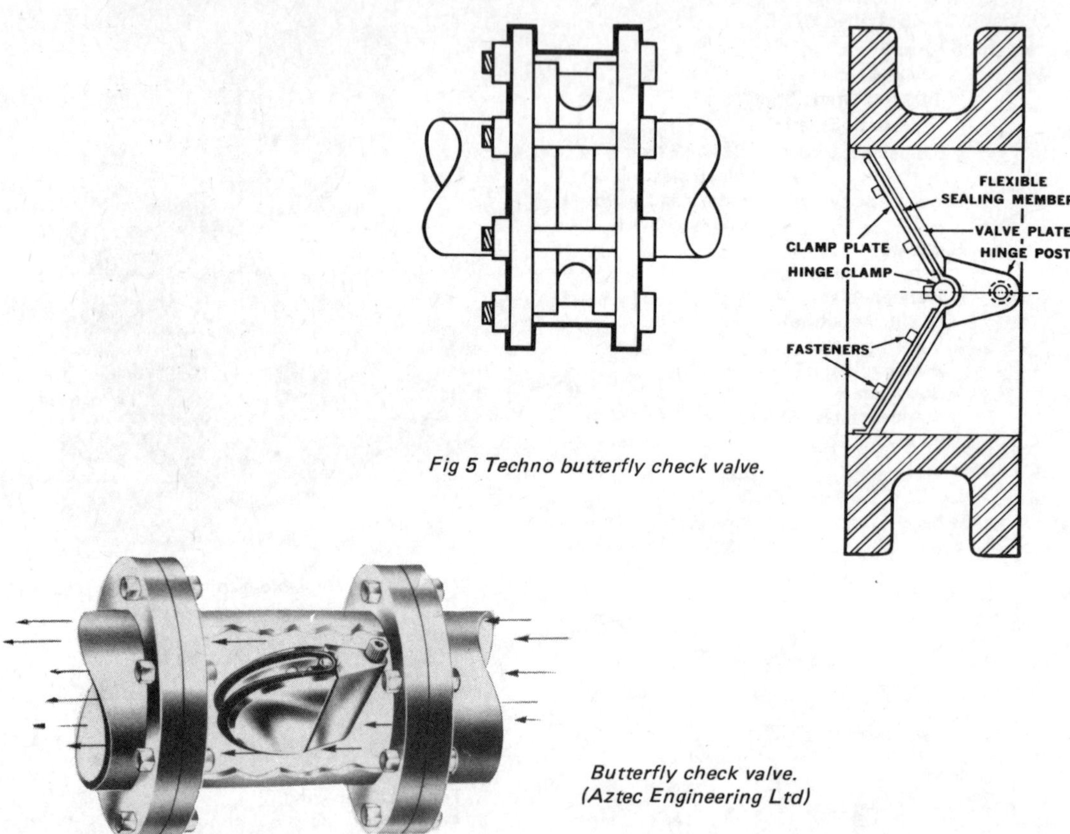

Fig 5 Techno butterfly check valve.

Butterfly check valve.
(Aztec Engineering Ltd)

With this mode of working a butterfly check valve is, in fact, another form of flap valve. It has the advantage of rapid action with a resilient seal.

Butterfly check valves require only a short length of body which can virtually be a length of standard pipe flanged at both ends (or threaded or plain ends in smaller sizes). It is also produced in wafer form for clamping between two pipe flanges. In this case the body length only needs to be sufficient to accommodate the hinge post and the two valve plates in their closed position.

Globe Valves

THE GLOBE valve is characterized by a baffle or partition separating the two halves of the body, with an interconnecting part at the centre opened and closed by a screw-down/screw-up disc or plug mounted at right angles to the body — Fig 1. The name derives from the fact that original body shapes were spherical, the more usual modern form being semi-spherical, or even substantially parallel sided.

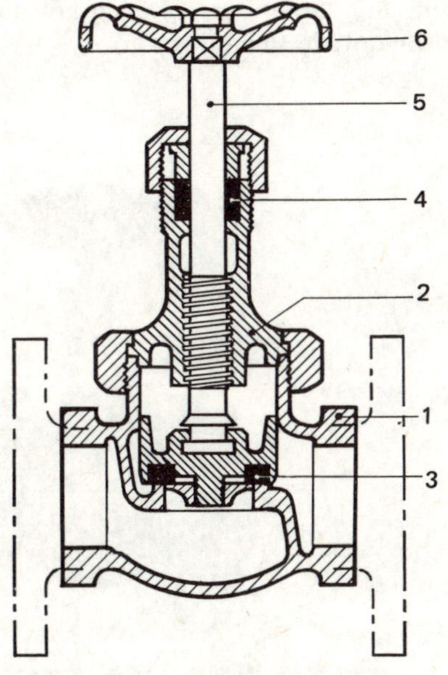

1. Body
2. Bonnet
3. Cover
4. Gland packing
5. Stem
6. Handwheel

Nomenclature: Screw-down valve

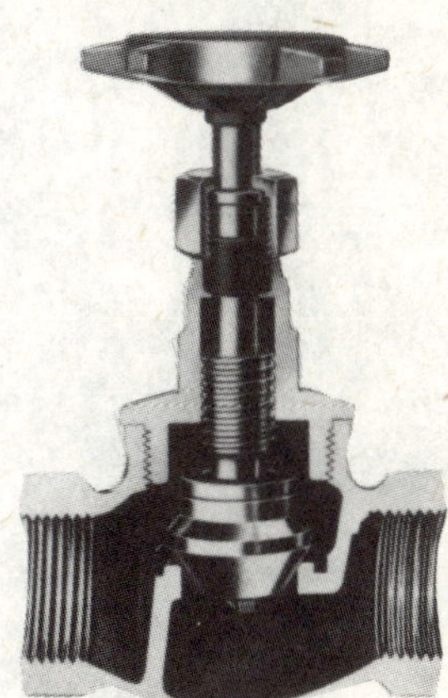

Fig 1 Sectional view of Peglers 1031 globe valve.

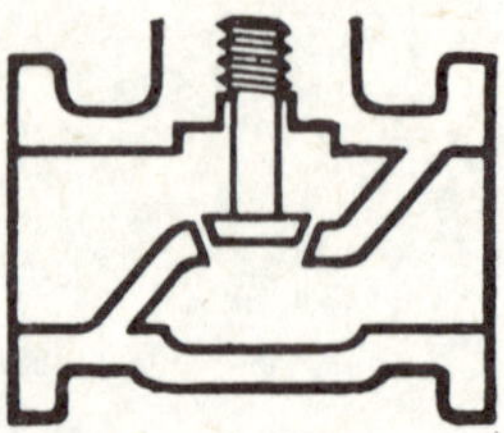

Fig 2

The globe valve offers good regulation characteristics but high resistance because of the tortuous flow path. This can be reduced to some extent by making the throat area equivalent to that of the pipe (calling for a more bulbous body), rounding the partition to smooth flow, or inclining it to the flow – Fig 2. Reduction of head loss by such treatment is generally minimal and so right angled partitions are commonly used.

Globe valves are produced with a variety of discs or plugs and seals. *Discs* provide line contact with the seat which can be broken by solid deposits forming on the seat. Discs are thus mainly suited only for clean fluids. Disc valves, too, are not as effective as plugs for throttling duties and so are normally used only on shut-off valves. Another limitation of the disc is that it does not provide positive shut-off for air and gases, unless made of a resilient material.

Plugs are used in a variety of configurations, ranging from needle shapes to semi-discs. The plug contour also governs the throttling characteristics, *eg* equal percentage plugs (percentage flow proportional to valve lift); linear plugs (linear flow proportional to valve lift); *etc.* Needle plugs provide the finest flow control. Globe valves fitted with this latter type of plug are normally referred to as *needle valves.* Needle valves are normally made only in small sizes.

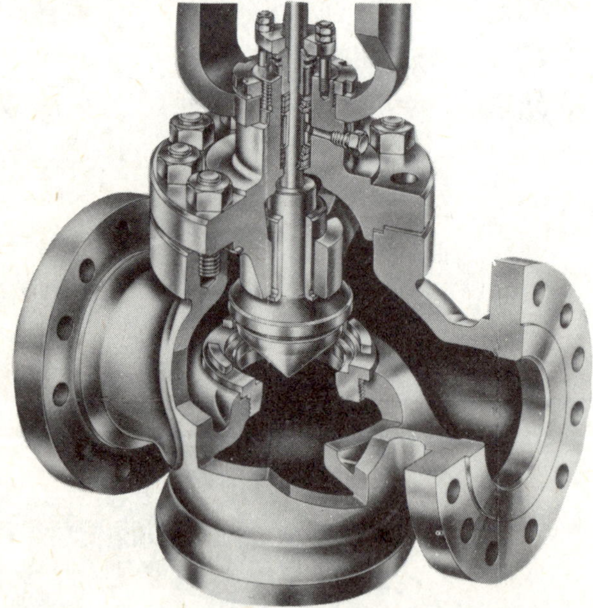

Introl 'Classic 10' globe valve. (Kent Process Control Ltd).

Peglers 1026 bronze globe valve.

Seats

Globe valve seats may be cast integral with the body, or take the form of screwed-in, pressed-in, or spot welded rings. A variety of materials may be used for seats, depending on the application, including coated seat rings or seat rings with plastic inserts. Normally only screwed-in seat rings are replaceable.

Stems

Possible stem assemblies include inside screw and outside screw (both rising stem), and sliding stem. Inside stem rising screw is usual. A stem seal (stem gland) is necessary to eliminate leakage. This is normally a gland-type packing or gland rings. A diaphragm bonnet seal or bellows bonnet seal may also be used on globe valves, the former isolating the working parts of the valve from the fluid as well as preventing leakage to atmosphere. Metallic bellows seals are often used on valves intended for high vacuum duties.

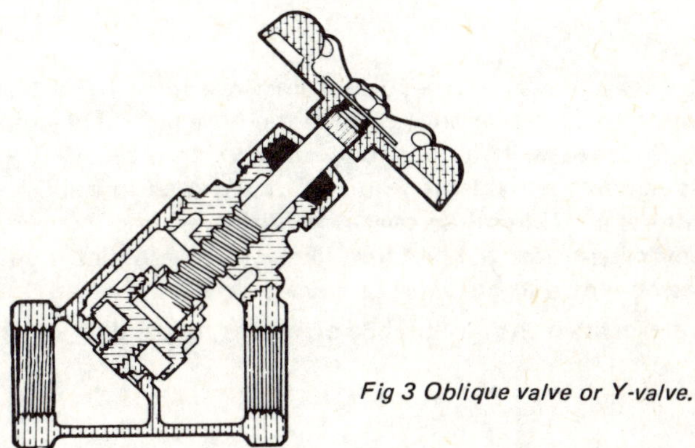

Fig 3 Oblique valve or Y-valve.

Oblique Valves

The *oblique valve* or *Y-valve* is a hybrid globe valve characterized by the stem being angled – Fig 3. As a consequence the flow path is less tortuous, with reduced pressure drop compared with a conventional globe valve. It retains the same good throttling characteristics as a globe valve and can be fitted with similar types of plugs.

Construction is basically the same as globe valves with the option of integral or fitted seat rings, stem options and stem seal gland treatment.

Gate Valves

GATE VALVES feature a disc or wedge sliding in a track or seat which can be lifted in a direction at right angles to the valve until clear of the flow path. The common feature of all gate valves is that they are intended to be either fully open, when they offer little resistance to flow, or fully closed. For this reason they are the principal valves used in bulk pumping practice and are available in sizes from ½ in (12 mm) to over 6 ft (2 metres) and in a very wide range of materials. It is possible that any pressure or temperature likely to be associated with rotodynamic pumps can be met by some member of the gate valve family.

Gate valves are divided into a number of classes, depending on the design of the 'gate' and its seating faces.

Ductile iron gate valve.
(Eurovalve Ltd)

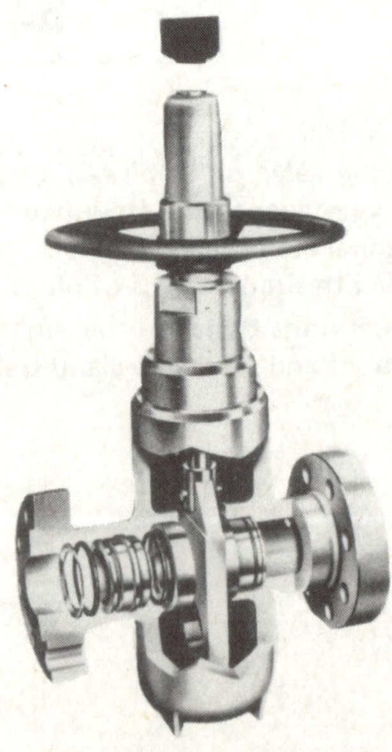

Grove G-3 gate valve.
(Grove Valve & Regulator Co)

Blakeborough wedge gate valve with by-pass valve and spur gearbox for manual operation.

Blakeborough wedge gate valve with bevel gearbox with rising indicator.

Blakeborough 900 mm bore wedge gate valve for water supply duty. Equipped for operation by spur gear box and electric actuator.

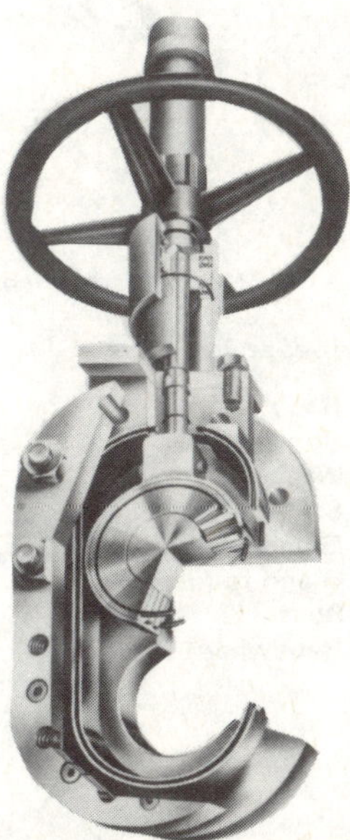

Grove G-5 compact full-opening gate valve for pipeline services 2–120 in.

Wedge Gate Valves

The gate is wedge-shaped and seats on corresponding faces in the valve body. Wedges and seats can be made of, or coated with, resistant material or faced with plastic such as PTFE. The plastic is often contained in a groove to prevent it spreading. Fig 1 shows a wedge gate valve with external screw and Fig 2 one with internal screw.

Flexible wedge valves and split wedge valves are similar to the aforementioned, but with provision for slight seat misalignment.

Wedge gate valves can be further described as inside screw or outside screw patterns — see Fig 3. They are widely used for oil, gas and air services, and also for handling slurries, *etc.*

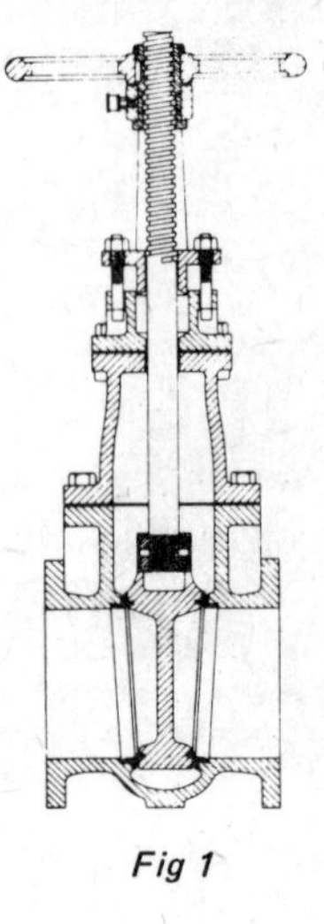

Fig 1

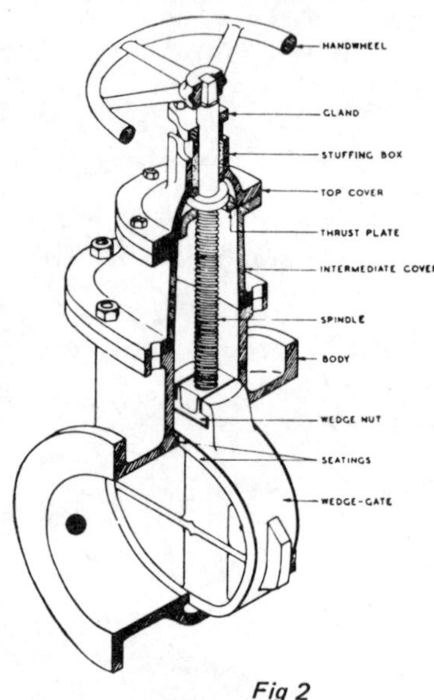

Fig 2

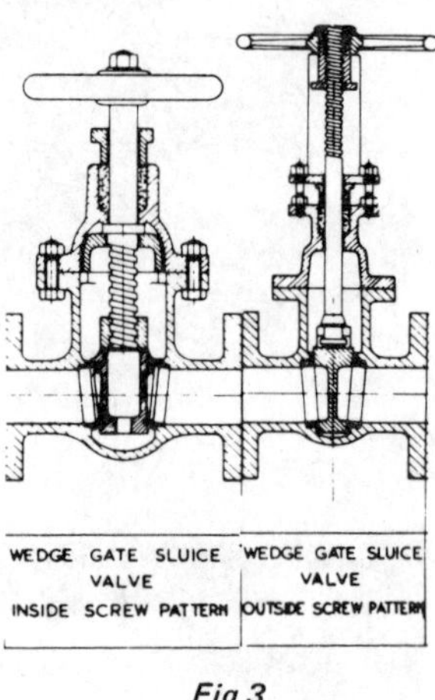

Fig 3

Nomenclature:

Solid wedge gate valve.

1 — Body
2 — Bonnet
3 — Wedge
4 — Seat ring
5 — Gland packing
6 — Gland follower
7 — Stem
8 — Handwheel

Parallel slide gate valve.

1 — Body
2 — Disc
3 — Spring
4 — Bonnet
5 — Gland
6 — Stem
7 — Yoke
8 — Handwheel

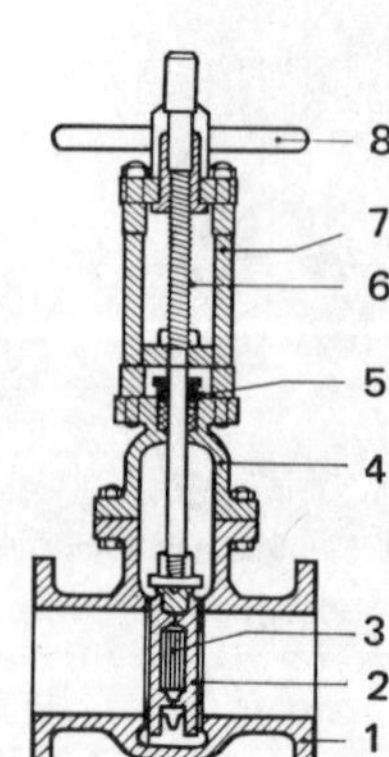

Double Disc Valves

In these valves the gate is in the form of two discs which are forced apart against parallel seats by a spring. This provides tight sealing without relying on fluid pressure, making this type of valve particularly suitable for steam duties as well as handling gases and light oils.

Parallel Slide Valves

Sealing of the valve relies on the upstream pressure acting on a flat parallel gate valve employed extensively in paper mills for the control of pulp of varying consistencies.

Sluice Valves

This is a name applied to solid wedge valves for waterworks.

Actuation

Manual actuation of gate valves is invariably by screw and handwheel. The screw mechanism may be exposed or protected and the screw 'rising' or 'non-rising'. A variety of materials for the working parts is offered by some makers.

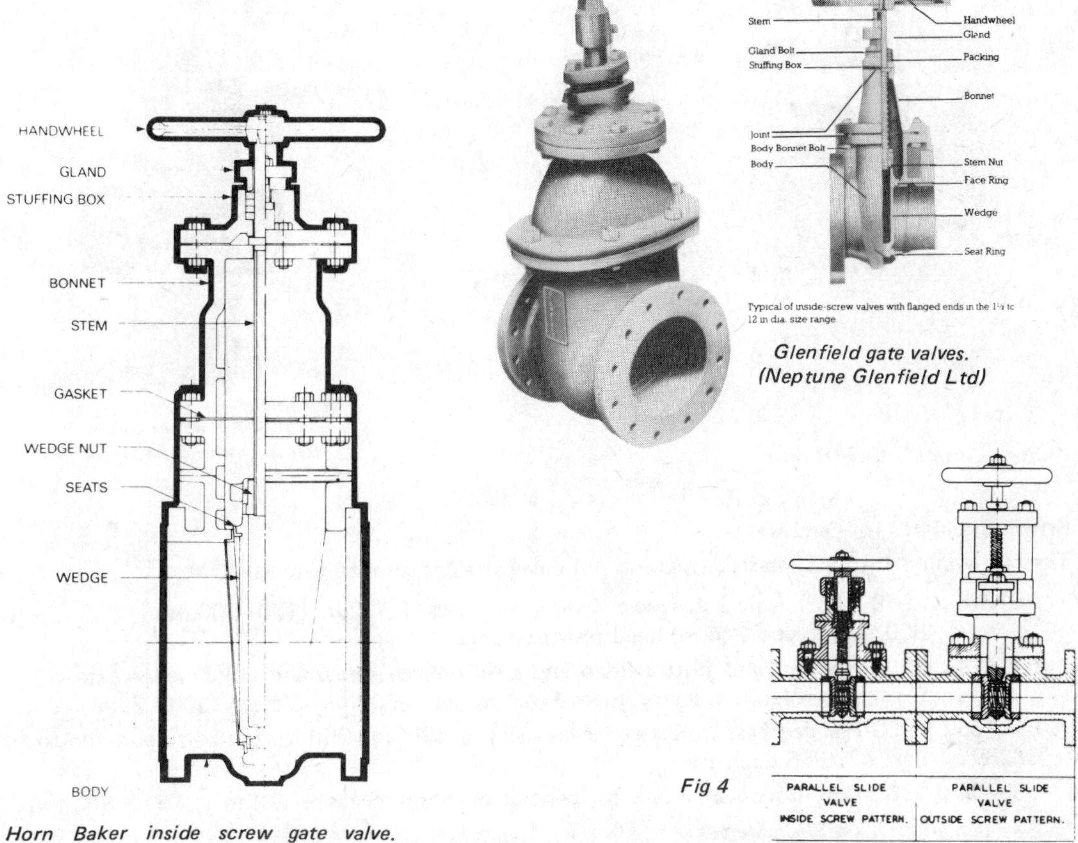

Glenfield gate valves. (Neptune Glenfield Ltd)

Fig 4

Horn Baker inside screw gate valve.

Power actuators are very often fitted, especially where valves are difficult of access and are operated frequently. Automation and semi-automation control schemes also make extensive use of actuators.

Parallel Slide Valves

This is a further type of gate valve, the simplest (and most effective) layout employing two discs as valve members, initially separated by a spring – Fig 4. The function of the spring is to prevent the discs from rattling and to encourage a wiping action on the downstream disc when under pressure and on both discs when there is virtually no pressure in the line. This is to avoid – as far as possible – grit or scale becoming trapped between the vulnerable seating faces which might impair their sealing properties.

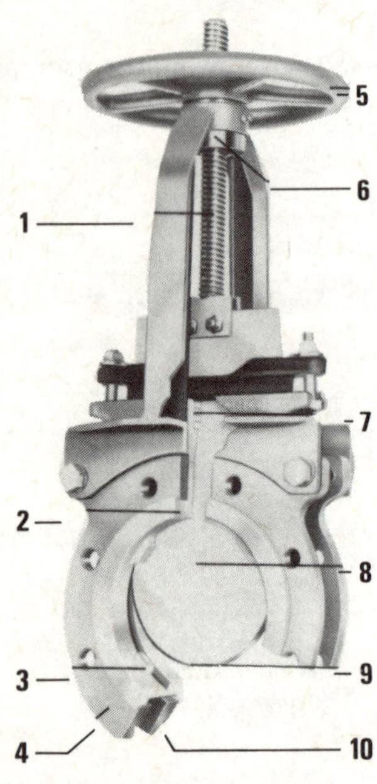

Dezurik knife-edge gate valve.

1 – Stainless steel stem
2 – Cast body
3 – Gate guides and jams
4 – Raised face flanges
5 – Handwheel
6 – Yoke
7 – Packing
8 – Full round port
9 – Stainless steel gate
10 – Welded on steel flanges

Through-conduit gate valve. (Grove Valve & Regulator Co)

British Standards for Gate Valves

The following British Standards (which do not cover design) refer to gate valves –

BS1218 : 1946 Sluice valves for waterworks purposes. 2–12 in (100–600 mm), 600 ft and 800 ft (183 and 244 m) head test pressures.

BS1414 : 1960 Flanged and butt welded end steel outside-screw and yoke wedge gate valves for the petroleum industry. Sizes 1–24 in (25–600 mm) Classes 150 to 2500.

BS1735 : 1960 Flanged cast iron gate valves. Classes 125 and 250 for the petroleum industry.

BS2591 : Part 1 : 1955 Glossary.

BS3464 : 1964 Cast iron gate valves for general purposes. Sizes ½–72 in (12.5–1 800 mm).

See also chapter on *Slide Valves.*

Penstocks

A PENSTOCK is a single faced valve consisting of an open frame and a door. This form of valve is normally located in tanks or channels as a means of controlling flow into a pipe. Many types are available to suit particular requirements and operating conditions. Basic alternatives are:-

(i) Penstocks for operating against pressure — *ie* pressure forcing the door onto the frame.

(ii) Penstocks for operating against off-seating pressure — *ie* pressure forcing the door away from the frame.

(iii) Penstocks designed to accommodate both seating and off-seating pressures.

Seating pressure can be accommodated by the use of side wedges only. Off-seating pressure requires the use of bottom wedges or top and bottom wedges — see Fig 1. Penstocks subject to sealing pressure normally seal tighter than penstocks for off-sealing pressure duties. Both types can be made virtually drop-tight with correct installation, distortion of the gate frame at the time of installation being the determining factor as far as leakage is concerned.

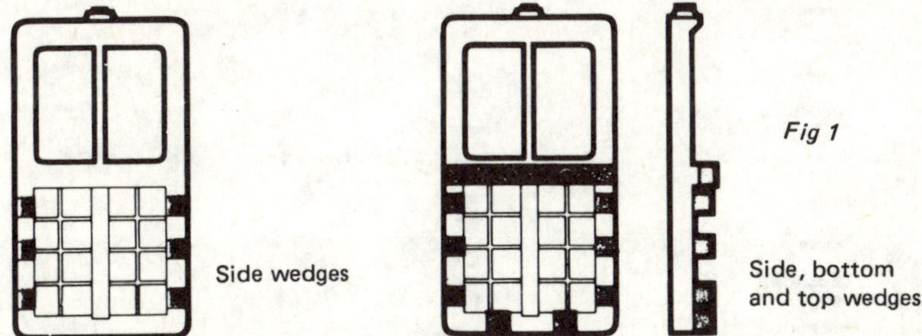

Fig 1

Penstock frames may be circular or rectangular. In the latter case a preferred proportion of width to depth is 2:3 for vertical form, and 4:3 for horizontal form. Frames and gates are commonly made of cast iron, although plastic materials (usually reinforced with steel) are also used for gates operating in aluminium, stainless steel or epoxy-coated frames for corrosive applications. Frames may be for channel or wall-mounted application. Sealing faces in suitable materials are embedded into both the frame and door surfaces.

Handwheel operation of the gate is normal, using a rising stem supported by a suitable headstock or bracket. The advantage of a rising stem is that the screw thread at the bottom of the stem is not usually immersed and is readily accessible for lubrication. A non-rising stem eliminates the need for a headstock and merely rotates through a nut in the penstock door. The threaded portion at the bottom is then usually immersed in the product being handled. Different systems for raising and lowering the gate may be employed on *modulating penstocks* used for flow control purposes.

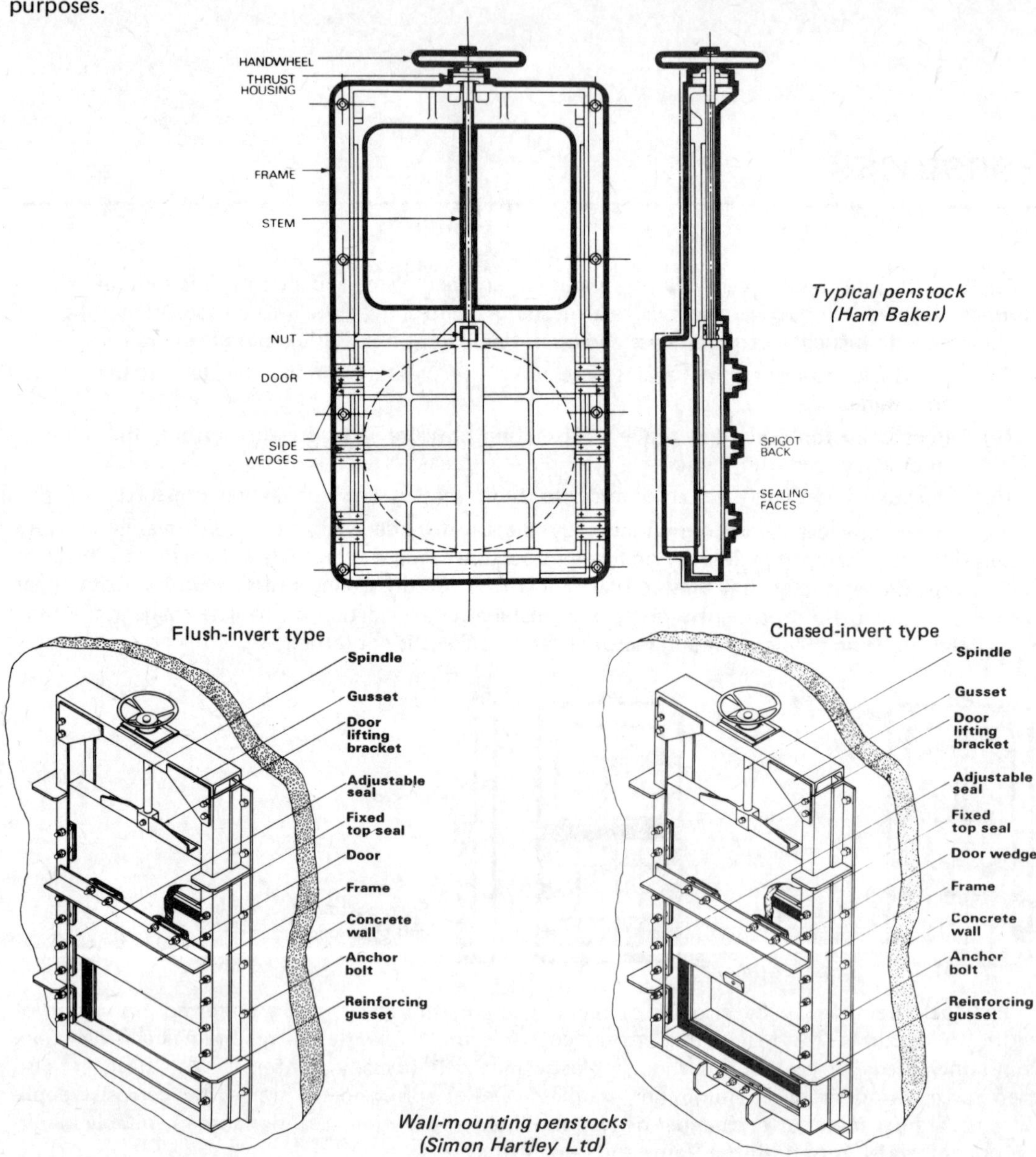

Typical penstock (Ham Baker)

Wall-mounting penstocks (Simon Hartley Ltd)

Blakeborough penstock for a power station intake.

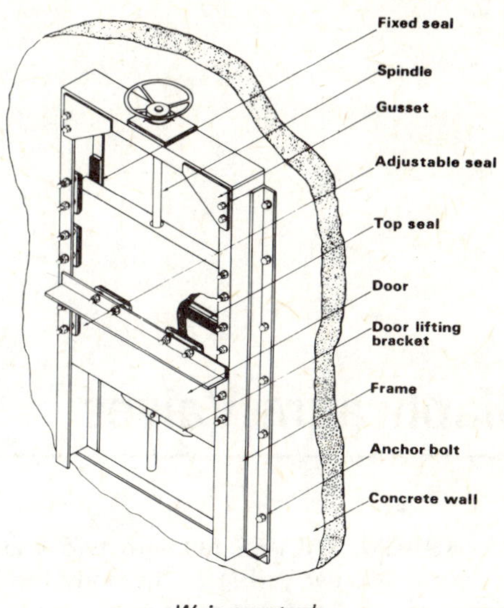

Weir penstock (Simon Hartley Ltd)

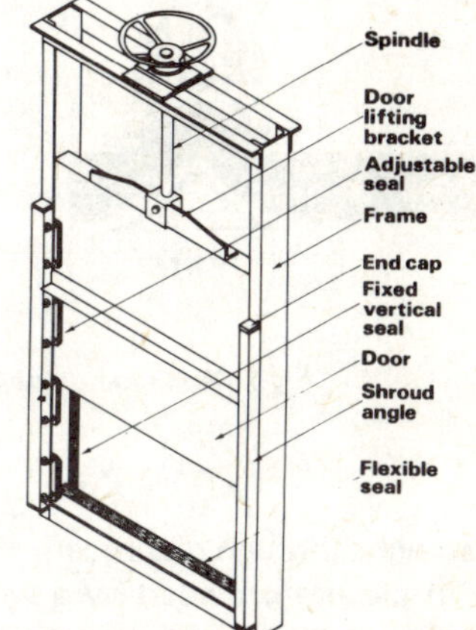

Channel penstock (Simon Hartley Ltd)

Discharge Through Penstocks

A penstock when opened represents a partial obstruction in an open channel over which liquid accelerates with a free liquid surface. Its performance is thus essentially similar to that of a weir where flow rate Q is proportional to width and velocity head. Working formulas are:

$$Q \text{ (Imp gallons/min)} = 0.13 \times \text{area (ft}^2) \times \sqrt{2g \times \text{head (feet)}}$$

$$Q \text{ (US gallons/min)} = 0.158 \times \text{area (ft}^2) \times \sqrt{2g \times \text{head (feet)}}$$

$$Q \text{ (m}^3\text{/sec)} = 0.7 \times \text{area (m}^2) \times \sqrt{2g \times \text{head (metres)}}$$

Diaphragm Valves

DIAPHRAGM VALVES fall into two main types — *weir* valves and *straight-through* valves. In the former geometry (Fig 1) the body has a dividing weir, above which is mounted an elastomeric diaphragm. In the closed position the diaphragm seats on the weir. In the open position, it is fitted to provide a streamlined flow through the valve body. Since the amount of lifting is variable, the valve can act both as a flow controller or a 'stop' valve.

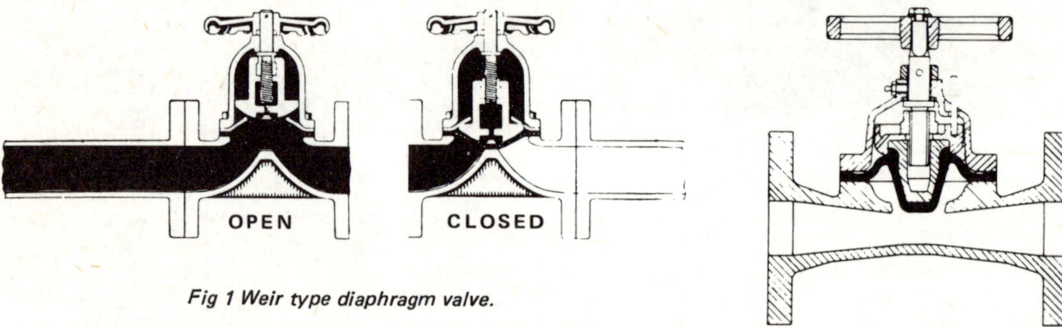

Fig 1 Weir type diaphragm valve.

Fig 2

The straight-through diaphragm valve (Fig 2) may have a parallel, top-tapered or venturi pattern body with closure provided by a wedge shaped projection of the diaphragm. Because of the full bore opening, it offers minimum resistance to flow in the open position, can pass suspended solids, and is capable of being rodded through for clearing any blockage.

Since the diaphragm isolates the moving parts, diaphragm valves are particularly suitable for handling aggressive fluids, as well as for 'clean fluid' applications, the type of elastomer being chosen accordingly. The body itself can also be lined for corrosive duties.

Particular advantages of the diaphragm valve are the glandless construction and absence of seating problems. Its main limitation is that the maximum service temperature and pressure are limited by the temperature/pressure rating of the elastomeric material.

Typical body materials are cast iron, malleable iron, bronze, gunmetal and stainless steel. Lined diaphragm valves normally have cast iron bodies lined with rubber, neoprene, polypropylene,

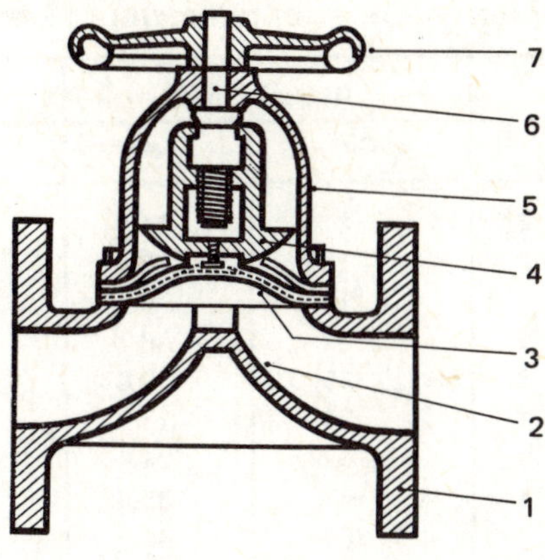

1. Body
2. Weir
3. Diaphragm
4. Diaphragm movement
5. Bonnet
6. Spindle
7. Handwheel

Nomenclature: Weir type diaphragm valve.

TABLE I – DIAPHRAGM MATERIALS

Material	Size		Temperature		Main Uses
	in	mm	°F	°C	
Butyl rubber	0.6–14	15–350	–22 to 134	–30 to 90	Acids and alkalis
Nitrile rubber	0.6–14	15–350	+14 to 134	–10 to 90	Oils, fats and fuels
Neoprene	0.6–14	15–350	–4 to 134	–20 to 90	Oils, greases, air and radio-active fluids
Natural/synthetic rubber	0.6–14	15–350	–40 to 134	–40 to 90	Abrasives, brewing and dilute mineral acids
White natural rubber	0.6–5	15–125	–31 to 134	–35 to 90	Foods and pharmaceuticals White diaphragm
White butyl	0.6–6	15–150	–22 to 212	–30 to 100	Natural colour, foodstuffs, plasticizers and pharmaceuti-cals
Viton®	0.6–14	15–350	+41 to 284	+5 to 140	Hydrocarbon acids, sulphuric and chlorine applications
Hypalon®	0.6–14	15–350	+32 to 134	0 to 90	Acid and ozone resistant
Butyl rubber	0.6–14	15–350	–4 to 248	–20 to 120	Hot water and intermittent steam services, sugar refining

®Trade Mark Du Pont

PTFE or glass. Typical diaphragm materials and their main uses are summarized in Table I. Reinforced diaphragm materials may be used for more arduous duties and are virtually standard for vacuum services.

Typical flow coefficients for weir-type and straight-through diaphragm valves are given in Tables IIA and IIB.

TABLE IIA – FLOW CO-EFFICIENTS FOR 'SAUNDERS' WEIR TYPE VALVES

Valve Size DN	Cast Iron		Rubber Lined		Halav/Glass Lined	
	Cv	Kv	Cv	Kv	Cv	Kv
15	5.8	1.6	*1.3	*4.4	6	1.7
20	11.5	3.2	9.4	2.6	12	3.4
25	17.4	5	14.5	4	18	5.6
32	26.5	7.5	22	6	28	8
40	43	12.2	35	9.8	45.4	12.8
50	84	21.4	70	17	88	22.5
65	126	30.5	102	24.4	132	32
80	180	44.4	147	35.5	186	46.6
100	320	77.7	264	62	336	81.6
125	420	108	348	86.4	444	113
150	600	147	504	117.6	630	154
200	1260	305	996	244	1320	320
225	1630	388	1320	310	1680	407
250	1990	500	1620	400	2076	525
300	2580	625	2088	500	2700	656
350	3840	833	3060	666	4020	875.7

*HSB bodies

TABLE IIB – FLOW CO-EFFICIENTS FOR 'SAUNDERS' KB TYPE STRAIGHT-THROUGH VALVES

Valve Size DN	Cast Iron		Rubber Lined		*ECTFE/Glass Lined	
	Cv	Kv	Cv	Kv	Cv	Kv
15	8.6	1.8	–	–	9	1.9
20	–	–	–	–	–	–
25	37.8	7.9	30.6	6.5	39	8.2
32	55.8	12.5	45.6	10.1	58.2	12.9
40	75	20	66	18	79	21
50	128	34	107	28	138	36
65	238	68	195	55	254	72
80	330	90	264	73	342	92
100	588	150	480	123	618	158
125	924	235	720	184	960	274
150	1680	385	1260	293	1800	545
200	2580	789	2196	665	2724	825
250	4020	1050	3420	900	4296	1100
300	6060	1510	4884	1221	6204	1550
350	10300	2800	9950	2360	–	–

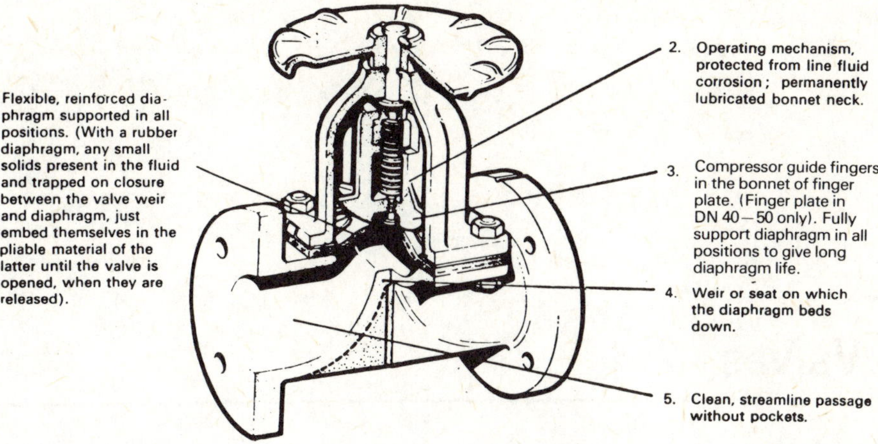

Saunders diaphragm valve.

Operation

Of the standard forms of manual operation available, the handwheel is preferred for most purposes. The design of handwheel and operating mechanism (spindle, compressor, bush or nut) varies according to the size and type of valve. Various shaped handwheels are also available. This facility, for instance, greatly assists the operator in identification in dark conditions or where the handwheel becomes slippery.

Various bonnet alternatives are applicable to handwheel operated valves. Extended spindle valves are readily available but existing valves can be simply converted by use of adaptor and extension. Other variants of bonnet assembly can be substituted for the handwheel operated design.

Normal bonnet assembly material is cast iron but alternative materials are offered by most manufacturers, *eg*

- (i) All iron and steel construction (for handling acetylene, ammonia and other similar fluids).
- (ii) All stainless steel (where there is severe atmospheric corrosion condition).
- (iii) Gunmetal (for medical oxygen, turbine oils and certain water services).
- (iv) Cast steel with rising spindle fitting (mainly used in oil refineries).
- (v) Silicon aluminium (for lightness and low temperatures).
- (vi) A variety of plastic materials (for general corrosion resistance).
- (vii) Epoxy coated cast iron (for corrosion resistance, attractive appearance).

For quick closure and/or opening lever-action valves can be used. A quarter-turn movement is preferred when the liner can also act as a valve position indicator — *eg* parallel with the pipeline in the fully open position and at right angles to the pipeline in the fully closed position.

Diaphragm valves are particularly suited for rapid closure/opening because the cushioning action of the flexible diaphragm minimizes the shock throughout the pipeline, compared with equally rapid but less resilient types of valve movement.

Pinch Valves

THE 'WORKING' element of a pinch valve is an elastomeric tube or sleeve which can be squeezed at its mid section by some mechanical system until ultimately the tube walls are pinched together producing full closure of the flow path.

In its simplest form it can consist merely of a length of elastomeric tube fitted with a pinch bar mechanism incorporating a closure stop to prevent over-pinching of the tube – Fig 1. More usually the moulded rubber tube is housed in a metal body which also incorporates the pinching mechanism (Fig 2). This can be a simple screw-operated mechanism where the pinch is applied only to one side of the tube; or a differential screw controlling two pinching mechanisms working in vertical opposition. The latter produces lower stress working of the tube.

Fig 1 Open-type pinch valve. (Robbins & Myers)

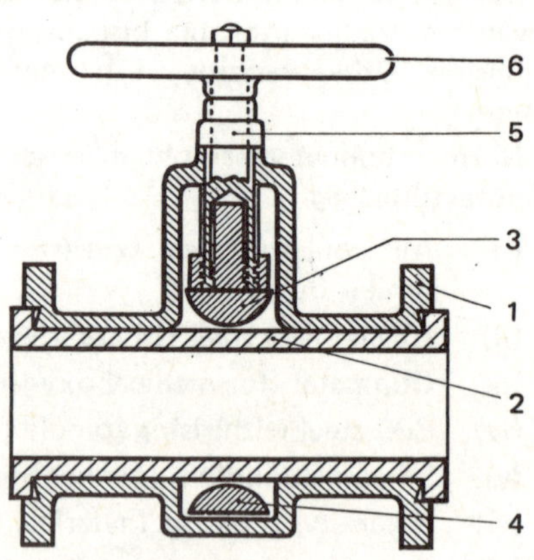

1. Body
2. Flexible tube
3. Upper pinch bar
4. Lower pinch bar
5. Spindle
6. Handwheel

Fig 2 Typical pinch valve

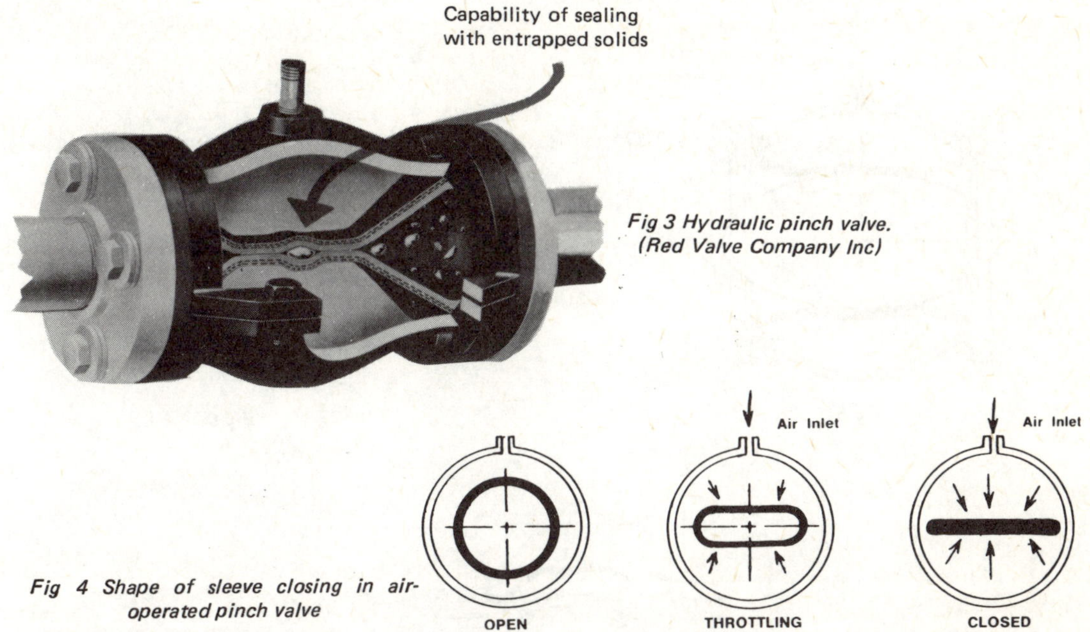

Fig 3 Hydraulic pinch valve. (Red Valve Company Inc)

Fig 4 Shape of sleeve closing in air-operated pinch valve

Other types of pinch valves dispose with mechanical mechanisms entirely, the tube being squeezed shut by air or hydraulic pressure injected directly into the body of the valve – Fig 3. With a regulated fluid pressure the valve may be used for throttling as well as shut-off (full closure) – Fig 4. The particular advantage of the fluid operated pinch valve is that it will still close tight over entrapped solids (making it particularly suitable for handling products with solids in suspension). Also since the tube is flexed under uniformly distributed pressure its life should be much longer than that of a similar tube working with a mechanical pinching system.

In common with the disphragm valve, the operating mechanism is not in contact with the working fluid at any time. Nor is the body. In this respect pinch valves have the advantage over diaphragm valves, unless the latter are rubber-lined or otherwise surface-protected. This exclusion of the working fluid from all parts excepting the sleeve itself makes it ideal for the handling of aggressive fluids and those which readily attack metal; and its straight-through characteristics mean it is suitable for the handling of slurries, pastes and semi-fluids generally, even those containing sizeable solid lumps in suspension.

Pinch valves with mechanical pinching mechanisms are normally operated by a handwheel and screw mechanism, but may equally well be driven by a powered actuator in larger sizes. Bodies are normally split horizontally to facilitate changing the tube when necessary, without removing the complete valve from the pipeline.

Another form of pinch valve is shown in Fig 5. In addition to a resilient sleeve this also incorporates a streamlined core onto which the sleeve closes to seal. This valve can be operated in a variety of modes. In mode 1 the valve remains closed when no pressure is present, due to the resilience of the sleeve – Fig 6. In the presence of pressure in the pipeline, the valve opens and remains open – Fig 7. It closes again in the absence of flow pressure. Line pressure of about 14 lb/in^2 (1 bar) is sufficient to hold the valve in the fully open position.

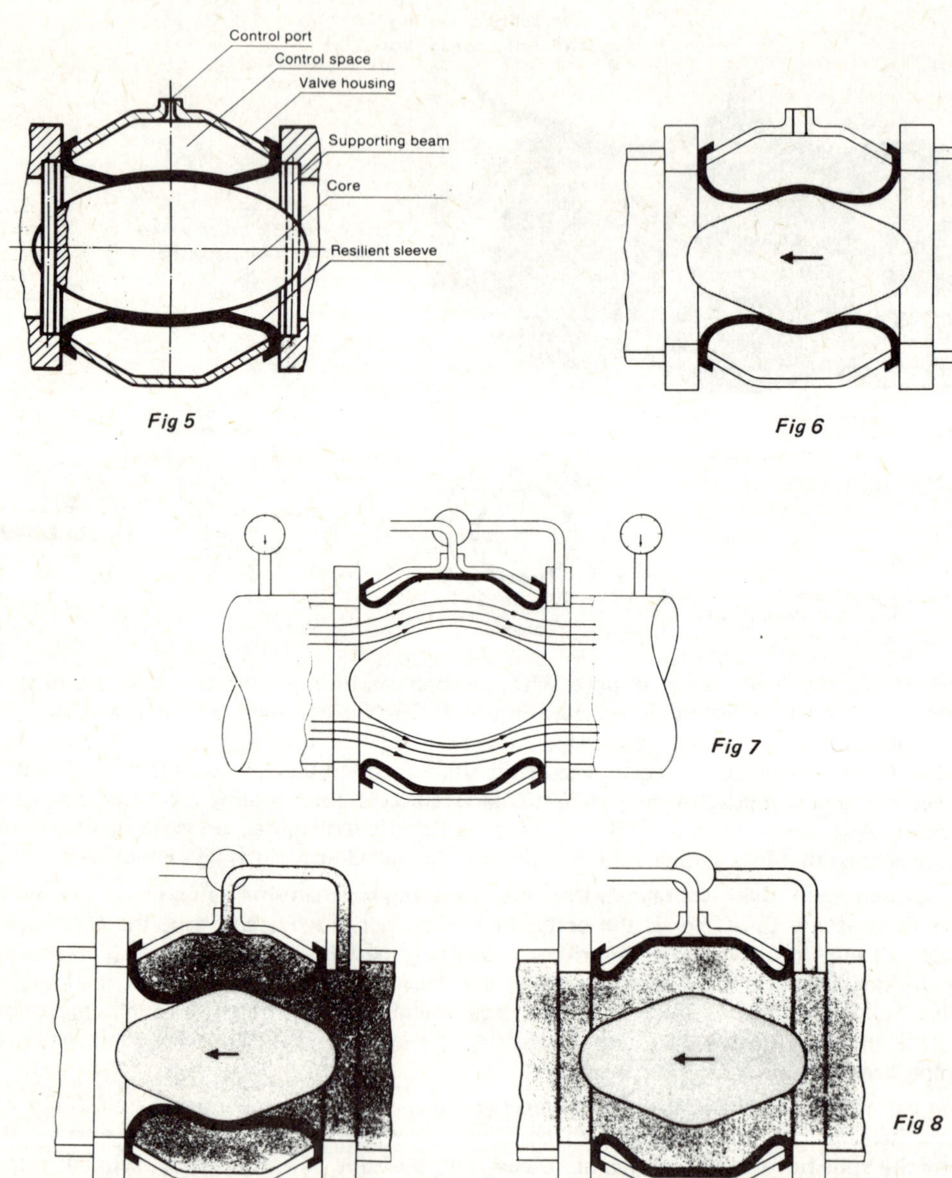

Fig 5

Fig 6

Fig 7

Fig 8

In mode 2 inlet pressure is tapped and fed to the control space between the sleeve and body, closing the valve. The valve opens when the operating pressure is relieved from the control space — Fig 8. At extremely low line pressures (circa 7 lb/in^2 (0.5 bar)) the valve remains closed and drop-tight, even when operating pressure is relieved from the control space.

In mode 3 the operating principle is the same, except that the control space is pressurized from an independent source — *eg* compressed air or a hydraulic supply.

Enclosed pinch valve

Air/hydraulic operated enclosed pinch valve.

Open pinch valve

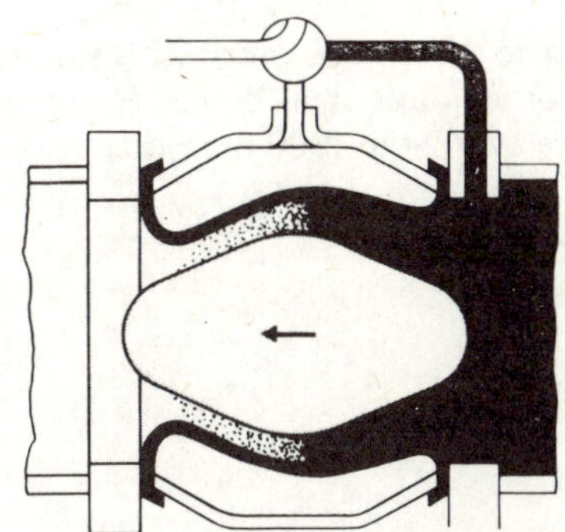

Fig 9

In mode 4 the flow is tapped to feed the control space which is only partially filled and then isolated, holding the sleeve in a partially closed position – Fig 9. In this mode the valve operates as a throttling valve or pressure regulation valve.

Various low hardness, high tensile elastomeric compounds are used for the tubes, choice being made on chemical resistance and/or abrasion resistance required, and service temperature. Typical materials used are:-

GRS – excellent abrasion resistance

BunaN – good resistance to solvents and hydrocarbons

Neoprene – good chemical and hydrocarbon resistance; in white compound also suitable for fast and dry application.

Butyl – good combination of chemical and temperature resistance; in white compound has good resistance to animal and vegetable oils.

Hypalon* – excellent chemical and temperature resistance.

PTFE – outstanding chemical and high temperature resistance.

Silicone – good combination of high and low temperature resistance.

FDA – excellent abrasion resistance; suitable for many food and dairy applications.

EPDM – excellent heat and chemical resistance.

*Du Pont trade name

Fluid Performance

A pinch valve presents full bore flow in the open position, with a straight, uninterrupted flow passage. Pressure drop or head loss under such conditions is thus minimal, related only to flow velocity and tube length, for a given fluid. There is no slamming when the valve closes against back pressure and the elastic nature of the tube tends to eliminate hammer (although this feature is absent in a hydraulically operated pinch valve).

Vacuum Services

Pinch valves tend to have limited suitability for vacuum services because of the tendency for the tube to collapse inwards. This is particularly true in the case of pinch valves with simple exposed tubes. Where the tube is enclosed in a body it is possible to adapt the valve for vacuum duties by applying a vacuum within the casing to balance the internal (vacuum) pressure.

Sizes & Ratings

Sizes commonly available range up to 12 in (300 mm), with standard bodies suitable for pressures up to 100 lb/in^2 (7 bar); or steel or stainless steel bodies for pressures up to 400 lb/in^2 (28 bar). Larger sizes are also available, those over 24 in (600 mm) diameter usually being individually made with fabricated steel bodies.

Screwdown Valves

THE GENERAL classification *screwdown valve* is taken (in the UK) to refer to all types of valves sealing by a disc or plug, *etc* and in which the sealing element is lifted from and lowered onto the valve seat by rotation of a threaded stem the axis of which is perpendicular to the valve seat. Mainly this embraces various types of stop valves – *eg* globe valves, oblique (or Y-) valves, gate valves, lift-type plug valves and angle valves. It also includes certain types of throttling valves, *ie* needle valves in particular.

Screwdown valves are also categorized as:-

(i) *inside screw,* where the threaded portion of the stem is fully enclosed within the bonnet.

(ii) *outside screw,* where the threaded portion of the stem is exterior to the bonnet and (usually) carried in a yoke. (See also Fig 1).

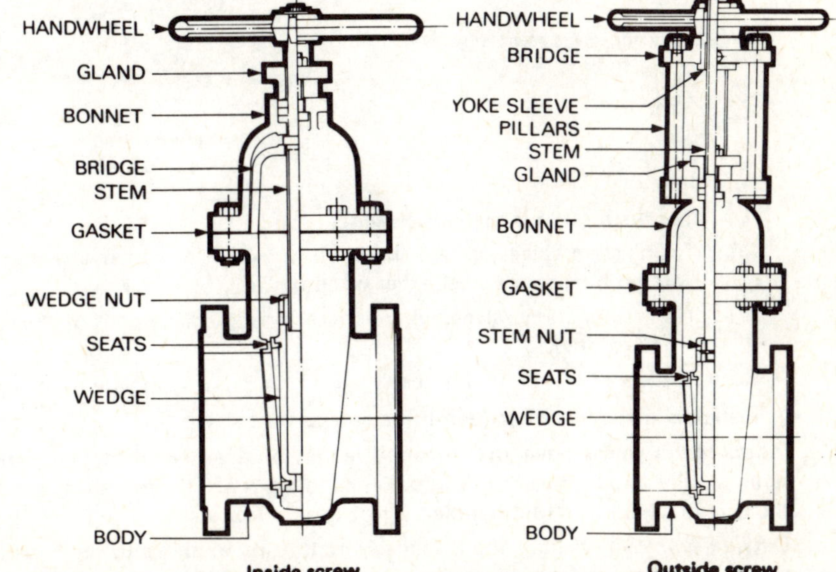

Fig 1 Outside and inside screw gate valves. (Ham Baker)

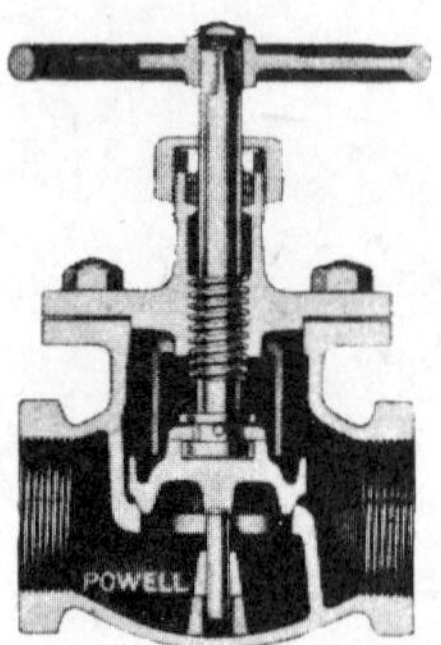

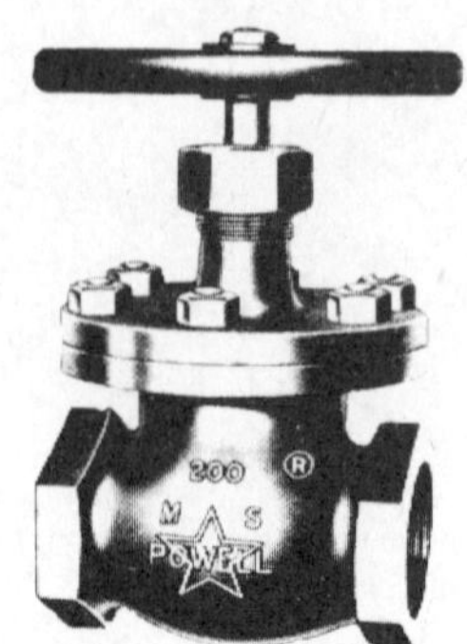

Inside screw rising stem valve.

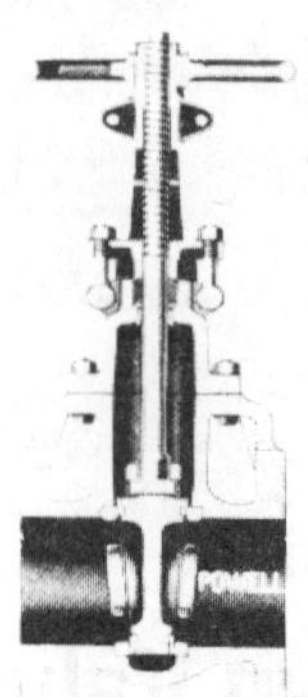

Outside screw rising stem valve.

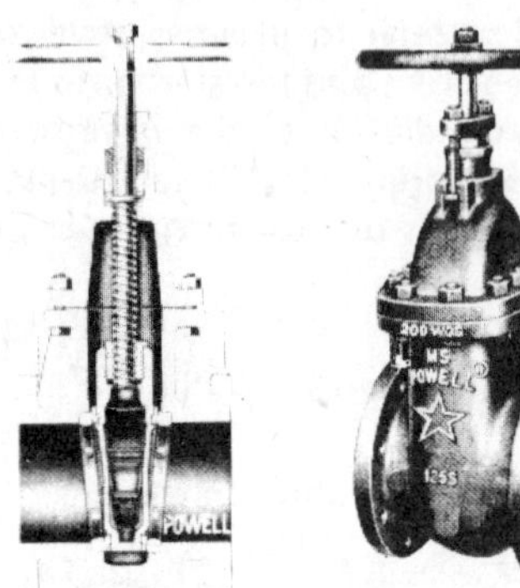

Inside screw non-rising stem valve.

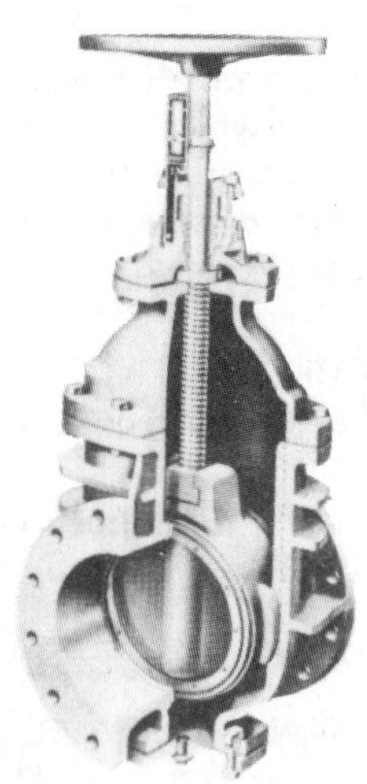

Inside screw gate valve. (Blakeborough Ltd)

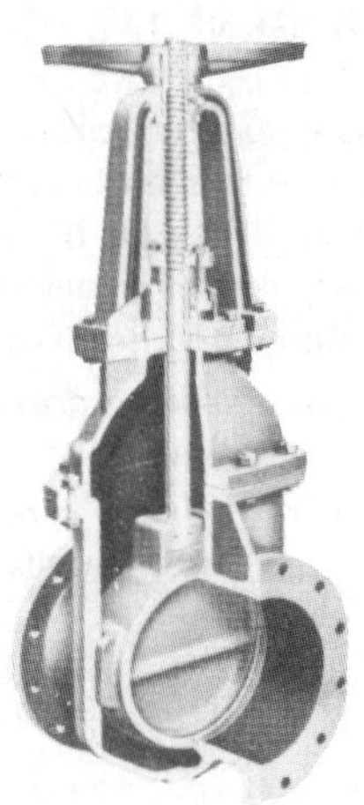

Outside screw gate valve. (Blakeborough Ltd)

A further distinction is made between:

(a) Rising-stem valves, where the stem moves in or out of the bonnet as the stem is rotated by a handwheel, lever or actuator.

(b) Non-rising stem valves, where there is no displacement of the stem along its axis when rotated.

Both inside screw and outside screw valves can be of rising-stem or non-rising stem type.

Principal differences between these categories are:

Inside screw valves have the threaded length of the stem protected from dirt, *etc.* However, the stem is fully exposed to the fluid being handled. Also it is more difficult to provide lubrication for the thread where the fluid handled is not itself a lubricant.

Outside screw valves have the threaded length fully exposed to the surroundings, hence can readily collect dirt and/or be subject to corrosion. Lubrication is easier because the threaded length is

Circle seal shut-off valves with extended stems.
(Tamo Ltd)

'Grasso' inside screw valve.

Screw-down stop valves.
(Blakeborough Ltd)

fully accessible, but again lubricant will promote collection of dirt on the threads. The threaded length of the stem is isolated from the fluid being handled by the stem glands, hence this type is more suitable for handling corrosive media, slurries, *etc.*

Rising stem valves provide a visual indication of the position of the valve disc or gate and thus the degree of valve opening. On the other hand adequate clearance is needed above the valve to accommodate the rising stem movement.

Non-rising stem valves have the advantage that they can be installed in positions where headroom is limited. Stem wear is also minimized, although there may be increased wear on the threaded length which lifts and lowers the valve disc or gate.

For corrosive duties, stem protection may be provided by a bellows seal applied either to an inside screw (usually) or an outside screw. See also chapter on *Seals and Packings.*

Slide Valves

SLIDE VALVES are one of the many variations of the gate type valve and were originally designed for use in the pulp and paper industry to overcome the problems in handling wet and dry fluidized solids in pipelines.

The slide valve, or knife or plate valve as it is sometimes called, in its original form was a simple rattle fit plate in a fabricated body. Opening or closing was achieved by pushing or pulling the plate to give a crude shut-off. Although design improvements have been made, this simple form of valve is still commonly used as a diverter or hopper outlet valve on dry powders or solids where a pressure seal is not required. In such applications the body of the valve will probably be of square or rectangular bore to suit the hopper outlet or duct on which it is fitted.

The logical advancement from this design was the fitting of resilient seals and an operating screw and handwheel to open and close the slide. With the improvement in sealing that this gives, the slide valve is suitable for a much wider range of services and can handle solids suspended in liquids or gases. Although they are simple valves there are a variety of designs on the market all sharing, to a greater or lesser degree, the following advantages:

(a) Short overall length.
(b) No wedging action.
(c) Thin closing member to cut through solids in the line.
(d) Substantially full round bore.
(e) Lightweight.
(f) Easy to power operate.
(g) Stem/slide connection not in contact with line fluid.
(h) Slide exposed for visual inspection when valve is open.
(j) Good regulating characteristics when fitted with specially-shaped slide and/or body bore.

The resilient seals, which can be fitted in the body or on the slide depending on individual design, are available in various materials to suit the fluid being handled. For most applications they will be in rubber, *eg* nitrile or Viton, but for higher temperatures or in the food and chemical industries it may be necessary to specify PTFE or other special elastomers.

3848

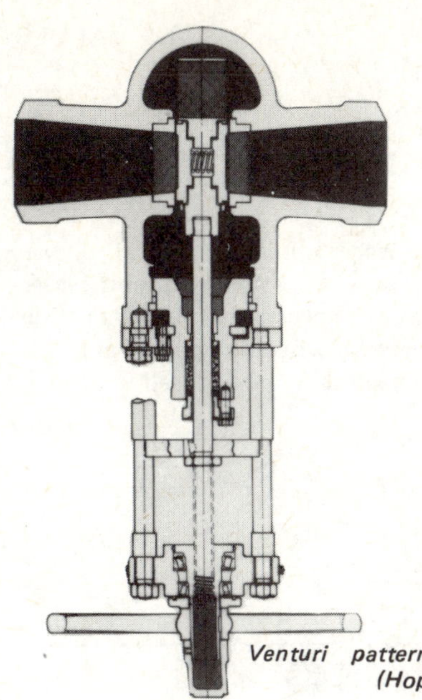

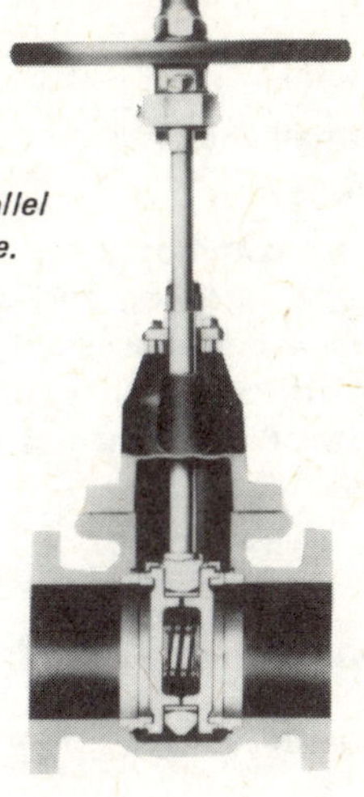

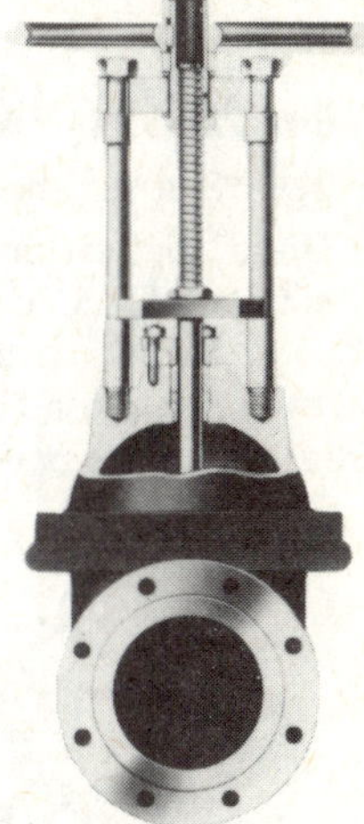

Hopkinson parallel slide gate valve.

Venturi pattern parallel slide valve. (Hopkinson Ltd)

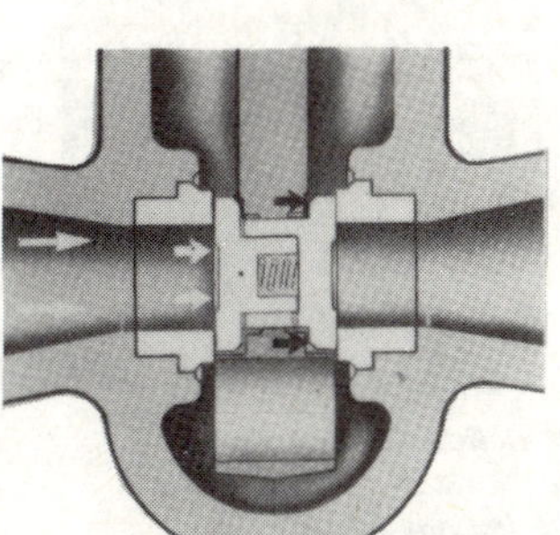

Valve closed
Fluid pressure (indicated by arrows) holds disc on outlet side in contact with seat.

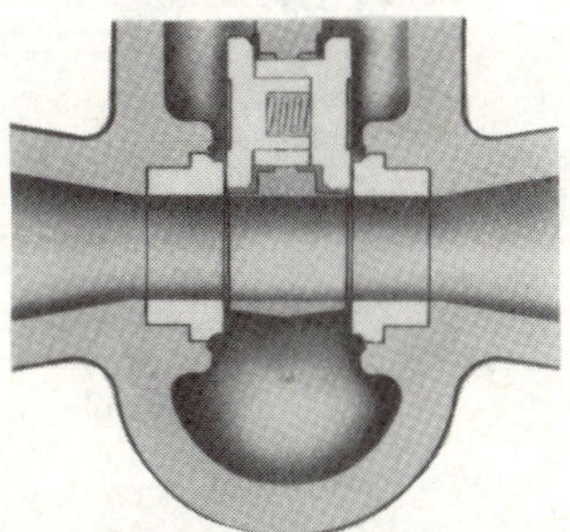

Valve open
Gives unobstructed flow. 'Eye-piece' bridges gap to complete venturi form passage and protect seat faces.

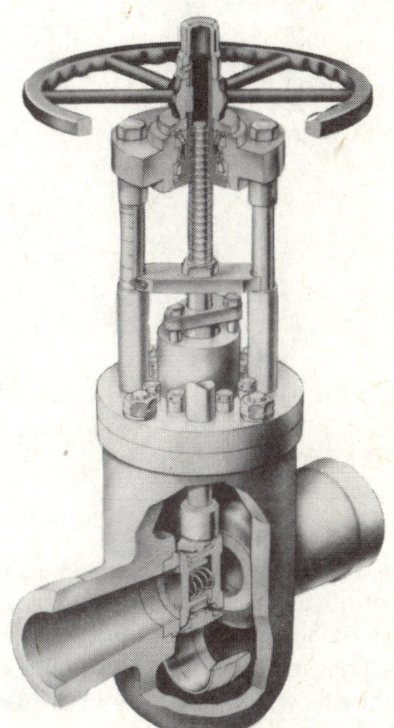

Venturi parallel slide gate valve. (Hopkinson Ltd).

The fact that there are no national or international standards for slide valve design has tended to increase the variety of individual designs on the market and, in some ways, this is advantageous as it gives the valve user more scope for finding a valve specifically designed to overcome his unique problems. Generally speaking, slide valves are used on services such as:

(i) Viscous media.

(ii) Dry powder.

(iii) Dry solids of small and large particle size.

(iv) Slurries and sludges.

These applications are commonly found in the water-works, sewage, mining, chemical, power generation, food, cement, brewing and other process industries as well as the pulp and paper industry for which the valve was originally designed. But this does not necessarily mean that slide valves are only used on difficult applications. They are equally successful when put on less arduous duties where the benefits of leak-tightness, space saving, price, *etc*, need to be considered.

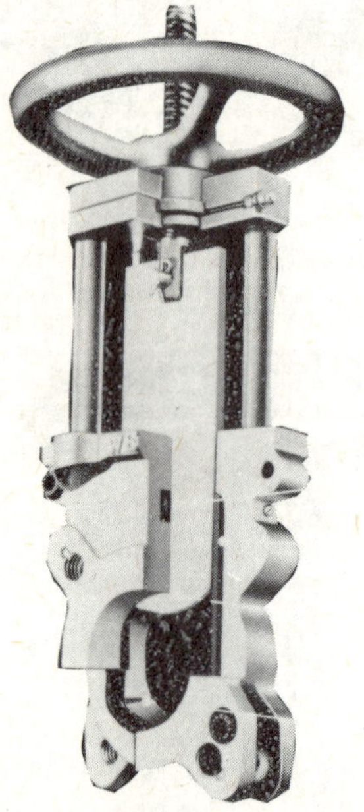

Fig 1 Patented Wey slide valve.

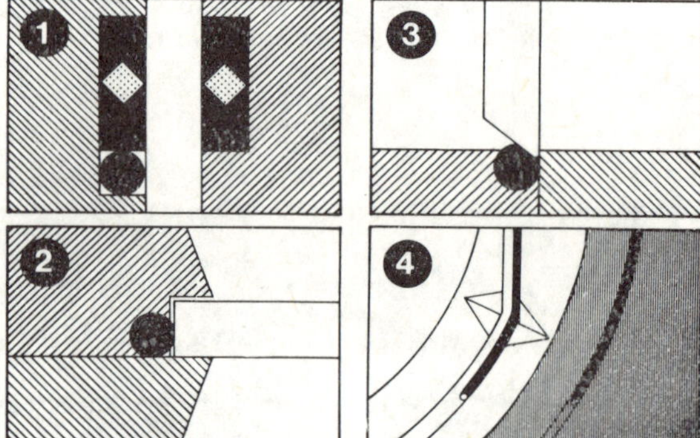

Fig 2 Features of the slide valve shown in Fig 1. 1. Built-in horizontal seal. 2. Internal seal located in valve body. 3. Valve slide with chamfered end. 4. Internal contours designed to ensure that deposits are flushed into the flow.

Fig 1 shows a typical design currently manufactured in the UK. It will be noted that the bore is clear of obstructions and that there are no pockets in which solids can collect, in addition to which the seal is housed in the body of the valve and shrouded by the body to avoid damage from solids in the line. In this design the seal seats on the edges of the slide, which means that the valve will seal equally well with flow in either direction and the slide has a chamfered leading edge which enables it to slice through any obstruction and seat effectively against the body seal (Fig 2). These features ensure bubble-tight shut-off against pressure or vacuum.

Unlike conventional gate valves, the slide passes through the body and a standard gland packing cannot therefore be used. In the design shown here, profiled transverse seals are employed. These comprise two lengths of V section rubber with the open sides of the V facing each other. During assembly, the diamond-shaped opening is filled with an oiled-fibre packing which is forced in and pushes the lips of the transverse seals into contact with the body and the slide. In this way, leakage to atmosphere is prevented. Also, by removing the screws on the sides of the body, more packing can be inserted at a later date to maintain a tight seal even if the valve is on stream and under pressure or vacuum. An optional extra on this design of valve is a set of scraper blades set below the transverse seals. During the opening phase, this enables the slide faces to be scraped clean over their whole width before they are drawn through the transverse seals and damage to the seals is thereby prevented. This is particularly important when handling sugar, sticky media such as honey, chocolate compound, glues, tenacious powders *etc*.

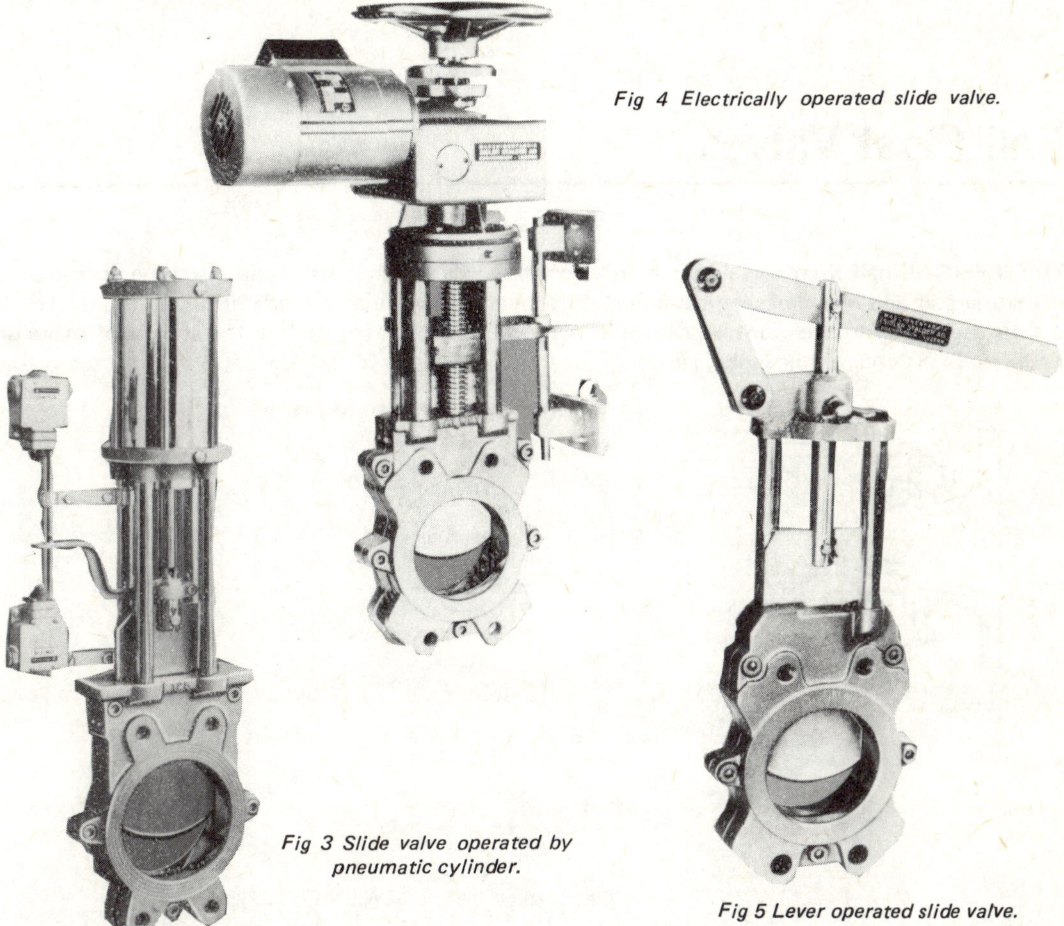

Fig 4 Electrically operated slide valve.

Fig 3 Slide valve operated by pneumatic cylinder.

Fig 5 Lever operated slide valve.

When power operation is required, slide valves are particularly easy to operate pneumatically as the pneumatic cylinder can be mounted directly on the valve pillars and the end of the piston rod connected direct to the valve slide (see Fig 3). Electric actuation is also widely used on slide valves (see Fig 4), and, to a lesser extent, hydraulic actuation. In addition, a very useful alternative to handwheel operation is the lever-operated slide valve (Fig 5) which gives a very quick open/close operation.

See also chapter on *Gate Valves.*

Ball Float Valves

THE TYPICAL ball float valve consists of a control valve of the piston and disc type operated by a floating ball and lever mechanism adjusted to open the valve at a predetermined liquid level. It is thus essentially a level control valve and is used mainly for controlling the supply of make-up water (*eg* in cisterns, water tanks, *etc*).

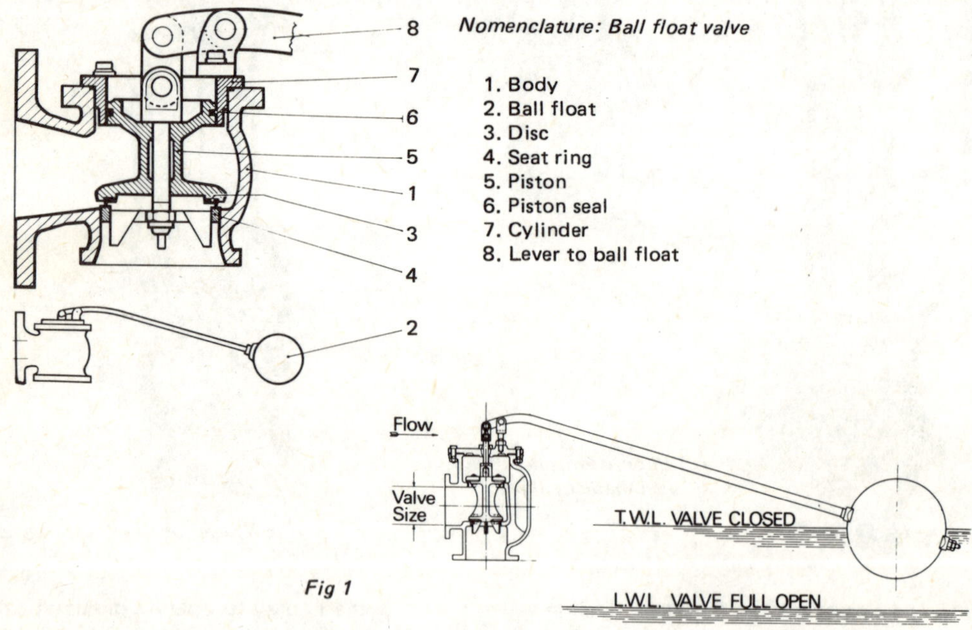

Fig 1

The most effective type is the equilibrium ball float valve — Fig 1 — so called because the upward and downward pressure forces are nearly balanced out (leaving just enough unbalance to eliminate hunting). The basic type finds widespread application. The body is usually of angle pattern with the inlet flanged and bolted to the inlet pipe flange with the tank wall sandwiched between.

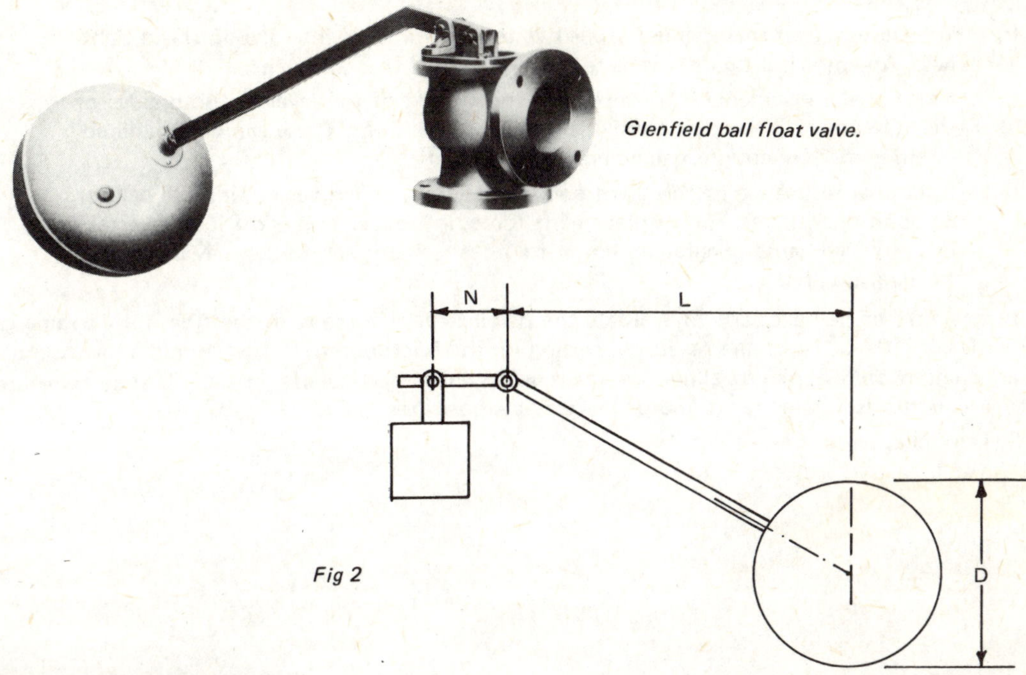

Glenfield ball float valve.

Fig 2

Design Geometry Calculations

Design geometry calculations for ball float valves can be tedious. As far as fluid forces are concerned there is an upward force at the valve position due to mains pressure tending to force the float downwards, which will normally be resisted by the displacement force generated by the float under equilibrium conditions – Fig 2. These equilibrium conditions correspond to the valve being held closed by a surplus of 'displacement' force. (In the event of the water level falling, of course, the valve will open to allow inflow of water until equilibrium conditions are restored).

Specifically the *valve force* (V_F) effective at the ball is given by:

$$V_F = \frac{0.7854\ P\ N}{L} \times d_2{}^2 - d_1{}^2$$

where P = supply water pressure
N = fulcrum, distance from valve
L = fulcrum, distance from ball float
d_1 = outside diameter of valve sealing face
d_2 = diameter of valve piston

The displacement force (D_F) is governed by the volume of the ball float (or diameter in the usual case of a circular float) and its depth of immersion. It is usual in the case of spherical ball floats to design for 50% immersion, when:

$$D_F = 0.01\ D^3 \text{ (to a close approximation)}$$

where D is the float diameter

Thus basic relationship is modified by:-

(i) The necessity of maintaining a positive closing force on the valve for fluid tightness. An empirical figure here is to make D_F equal to 1.25 x V_F.

(ii) Additional movements introduced by the weight of the lever on both sides of the fulcrum or pivot point and the weight of the float. These can be calculated, or eliminated by counterbalancing.

(iii) Frictional resistance of the pivot and rising part(s) of the valve. This will call for an additional increase in displacement force, *ie* float size. It is difficult to establish required values except on empirical lines as frictional, forces may be subject to change with age.

It may further be necessary to evaluate the displacement force with the float fully submerged (when $D_F = 0.02\ D^3$) and the resulting loading on the fulcrum pin. Design should allow adequate sheer strength on the pin to allow for this contingency, particularly in ball float systems used with level controls in large tanks and/or inaccessible positions.

See also chapter on *Level Controls.*

Flap Valves (Swing Check)

SWING CHECK valves is the preferred description for non-return valves consisting of a hinged disc, although they are also commonly called flap valves because of their geometric action. Their mode of working is obvious. With flow in one direction the disc hinges upwards to permit flow through the valve. With reverse flow the disc is held closed. Equally, spring pressure or mass effect normally holds the disc closed in the absence of flow. In some cases closure is also assisted by the use of a weighted lever.

Small size swing check valves for low pressure services may use an elastomeric disc with a square end clamped in position, eliminating any need for a separate hinge. A back-up plate is added to assist closure and also provide rigidity to the unsupported area of elastomer when the valve is closed. Without this the disc would tend to collapse and extrude through the port. Larger versions of flexible (elastomeric) flap valves are made in sizes up to 60 in (1 500 mm).

Larger size swing check valves are more usually made with discs or flaps of metal or composite materials, hinged at the top and sealing on a metal seal. The sealing surface is inclined at a small

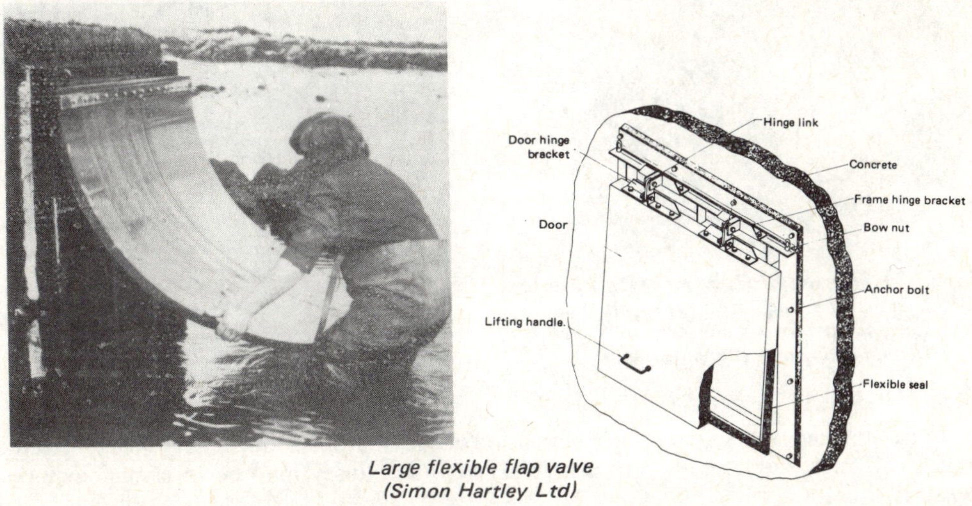

Large flexible flap valve
(Simon Hartley Ltd)

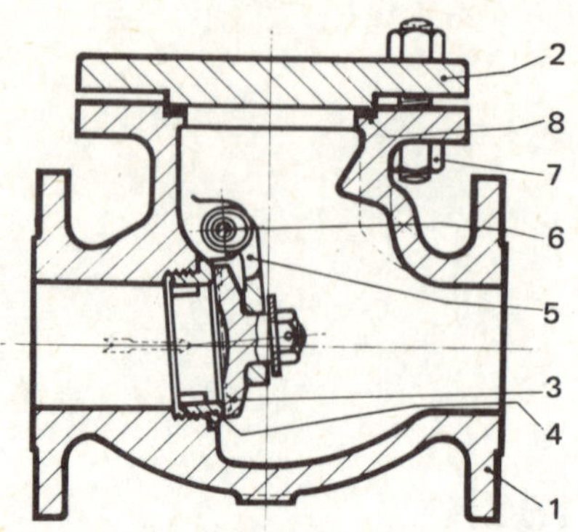

1. Body
2. Cover
3. Disc (flap)
4. Seat ring
5. Hinge
6. Hinge pin
7. Cover bolts
8. Cover seal

Fig 1 Swing check valve

angle to the vertical to assist opening, provide more positive sealing under back pressure and reduce shock when closing under high pressure. A typical valve design is shown in Fig 1. In larger valves the disc or flap may be double hung. Flap shapes are normally (but not necessarily) circular.

Swing check valves present relatively high resistance to flow in the open position as well as creating turbulence, since the flap 'floats' in the fluid stream. They may also tend to chatter in systems having frequent flow reversals. Swing check valves are normally used in horizontal pipelines. They can also be used in sectional pipelines with upward flow. They are not suitable for use in systems with pulsating flows. However, weight added to the flap or disc (or spring loading) can control the opening pressure as well as assisting closure under back pressure.

Commonly the body shape incorporates a 'dead' volume in which fluid can be trapped downstream of the flap once the valve is closed. If necessary a drain can be incorporated at this point, *eg* where it is desirable to drain a system completely on shut-down.

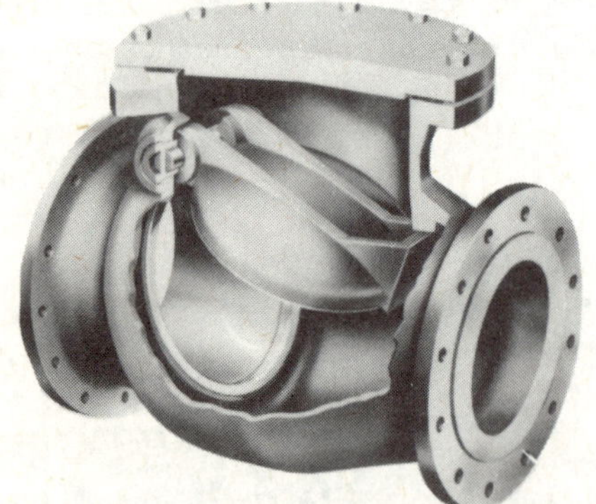

Blakeborough standard series swing check valve to BS 5150. One of the range of valves which includes special versions to combat the effects of reflux 'slamming' in water pumping mains.

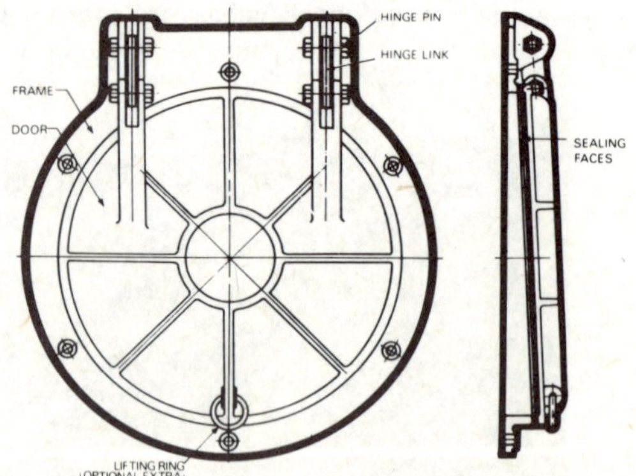

Typical large flap valve (Ham Baker)

Materials

Body materials used for swing check valves include cast iron, bronze, cast steel, forged steel, stainless steel, high duty alloys and also plastics. Valve discs may be in similar materials, or composites.

Rotary Valves

THE ROTARY VALVE is an obvious configuration, although not widely used. It can be classified as another type of spool valve where the ports run axially instead of circumferentially – Fig 1. Sealing in the closed position, valves are a close fit over a restricted contact area. Also friction can be high. In these respects it is inferior to a conventional spool valve, or even a plug valve.

A somewhat better configuration is to employ a 'solid' spool with drilled-through ports – Fig 2; or the rotary plate valve (see also chapter on *Spool Valves*).

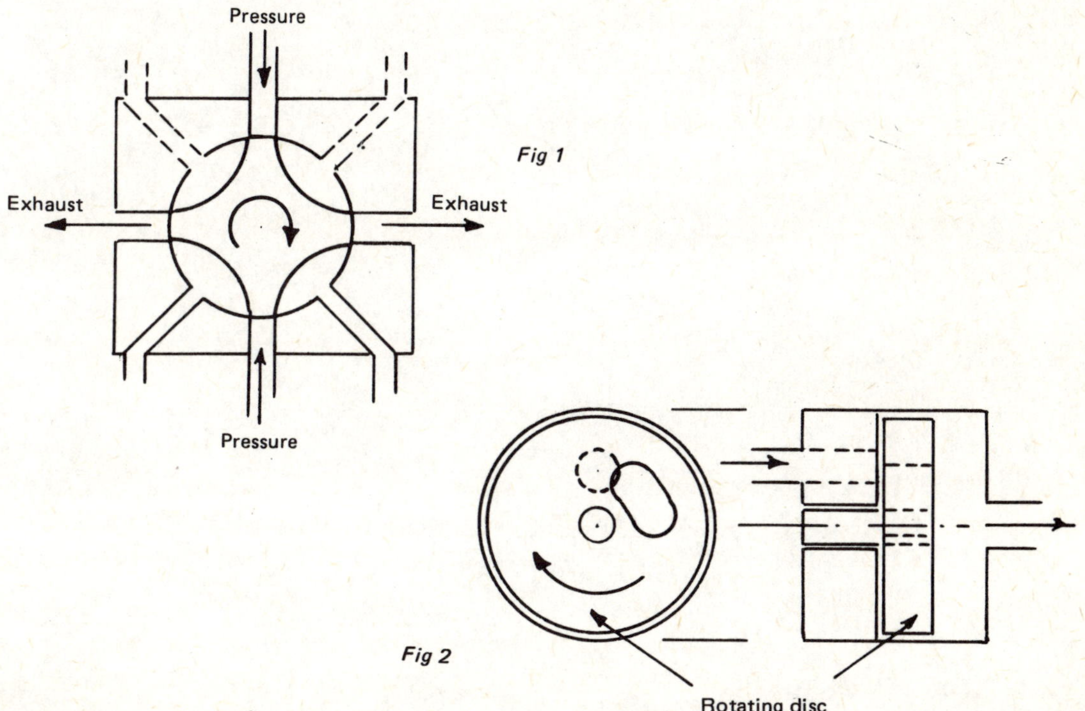

Fig 1

Fig 2

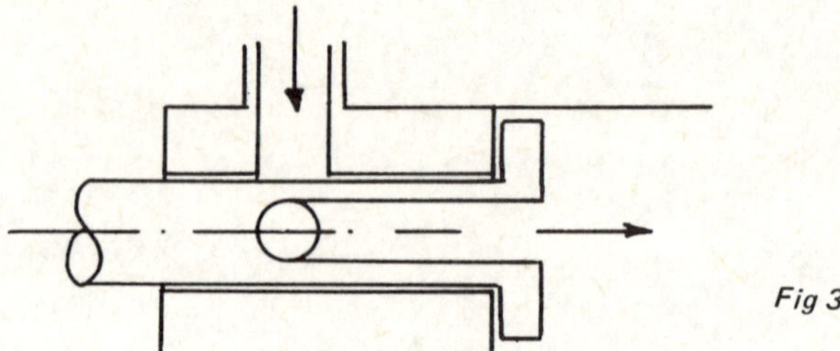

Fig 3

Geometric design is simplified to a degree with these latter types of valves, and both can readily be produced with over-lap or under-lap, as required, *eg* by enlarging the circumferential length of the rotating or stationary port, respectively. This type of valve is known as a disc valve or rotary disc valve.

A third basic configuration is shown in Fig 3 where a spindle or shaft rotates in a stationary housing. The shaft is drilled axially, and also vertically at the blind end to form a port whose position matches the inlet port in the stationary member. Again the port opening(s) can be modified in length to provide over-lap or under-lap if required.

A feature of both rotary valve types shown in Figs 2 and 3 is that they are capable of providing exact timing of port opening with continuous rotation of the rotary element.

Spool Valves

THE SIMPLEST form of spool valve to control both the direction of flow and flow rate is the three-way valve, but this will control the flow to only one side of the load and a differential area spool must be used. A more common design, shown in Fig 1, is the four-way valve whose symmetry results in improved linearity. However, there are three critical axial dimensions of the spool, and also of the sleeve. To reduce some of the manufacturing difficulties designers have used the following alternative designs.

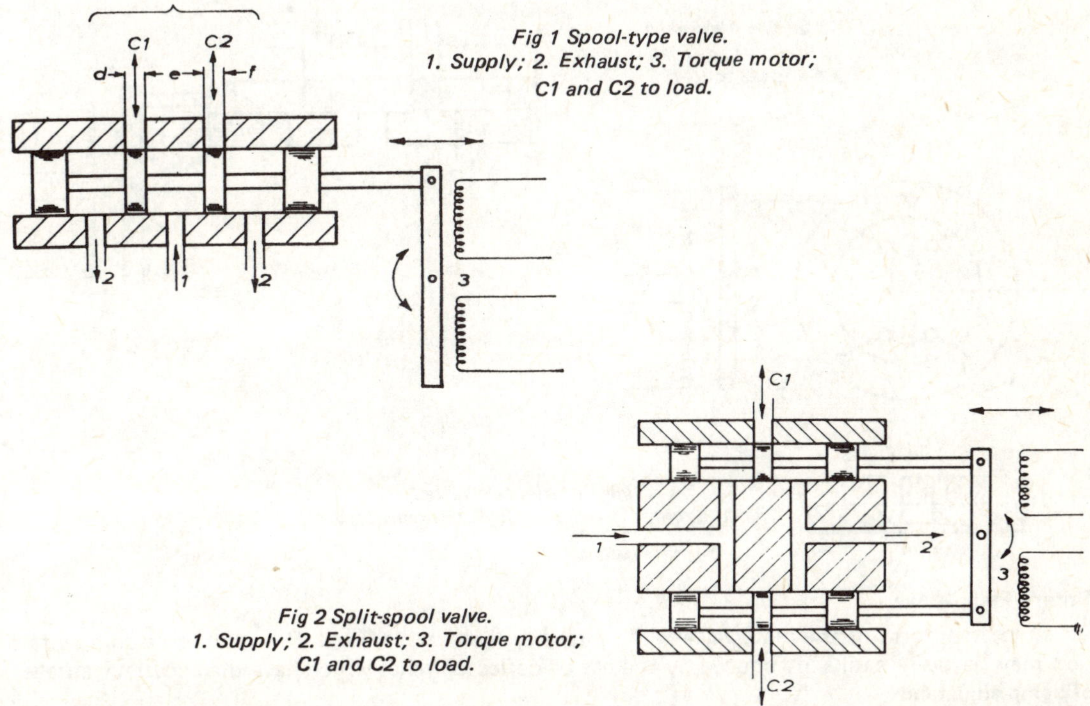

Fig 1 Spool-type valve.
1. Supply; 2. Exhaust; 3. Torque motor; C1 and C2 to load.

Fig 2 Split-spool valve.
1. Supply; 2. Exhaust; 3. Torque motor; C1 and C2 to load.

Split-Spool Valves

To reduce the number of critical axial dimensions of the four-way valve, two separate three-way valves may be linked together, as shown in Fig 2, and a means of zero adjustment can be provided in the links. Although manufacturing problems associated with porting are reduced, further difficulties are introduced in that two parallel bores have to be made, and there may be additional backlash in the linkage. Furthermore its weight is usually more than that of the simple four-way valve.

A design in which manufacturing difficulties associated with axial tolerances are eliminated, is the Elliott adjustable lap valve. This is a split, three-way valve in which the two spools are mounted back-to-back and are actuated by a central ram against spring restraint. The correct valve lap is obtained after assembly by axial adjustment of the sleeves.

Sliding Plate Valve

This design, shown in Fig 3, may be likened to an unwrapped spool (that is, two-dimensional), or even to the original D-type steam valve. It overcomes the difficulties associated with the manufacture of the bores in spool valves and hole-and-plug porting techniques (1) may be used. However, some manufacturers consider that the difficulties associated with the production of flat and parallel plates are greater than those of making spools and sleeves. Various methods are used to reduce friction forces; in some valves the sliding member is suspended on spring plates to prevent metallic contact, and others achieve hydrostatic pressure balance.

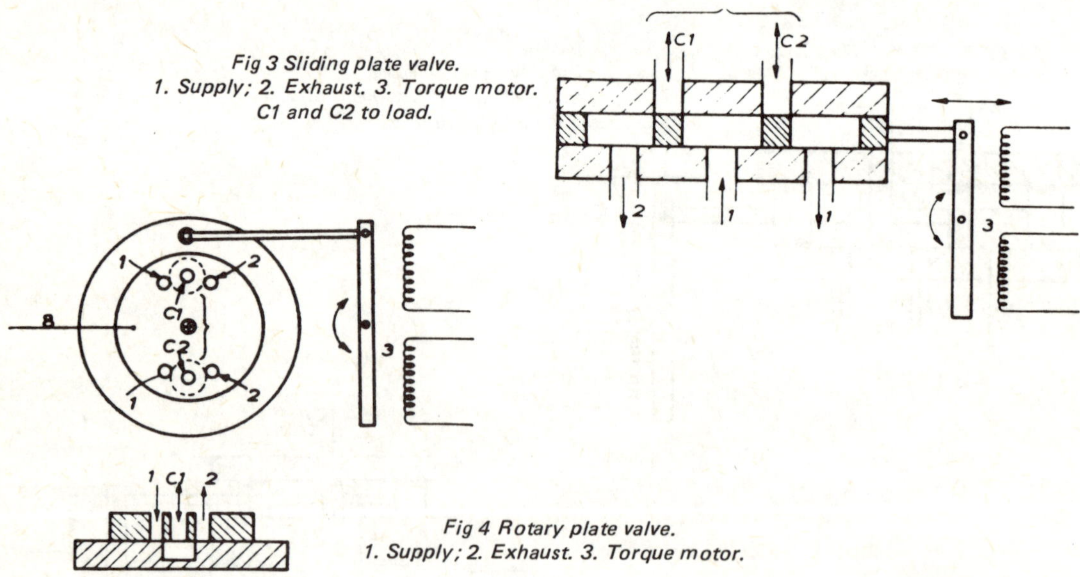

Fig 3 Sliding plate valve.
1. Supply; 2. Exhaust. 3. Torque motor.
C1 and C2 to load.

Fig 4 Rotary plate valve.
1. Supply; 2. Exhaust. 3. Torque motor.

Rotary Plate Valve

An alternative form of the plate valve is the rotary type shown in Fig 4. Reaction force compensation may be fairly easily introduced by the use of deflector vanes which have the added advantage of being adjustable.

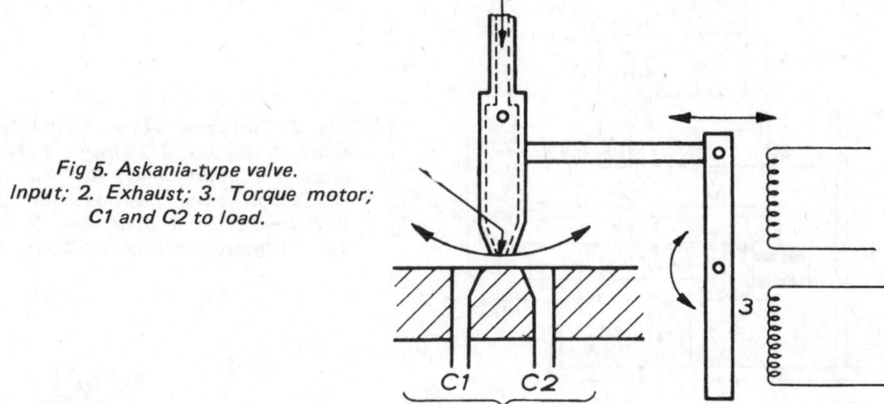

Fig 5. Askania-type valve.
1. Input; 2. Exhaust; 3. Torque motor; C1 and C2 to load.

Askania-type Valve

The main advantages claimed for this type of valve, shown in Fig 5, are that it is less susceptible to contamination clogging, and the ease of manufacture. Although it has been fairly extensively used for control purposes in low-pressure applications, its design for medium and high pressure systems presents a far more difficult problem since it is based on empirical methods. A comment has been made that large reaction forces leading to serious instability can occur, although the reasons for this are not known. This valve, whose action depends upon the conversion of kinetic energy of the jet into static pressure at the spool or ram, is referred to in the United States as the 'jet-pipe' design. However, it is probably preferable to avoid this term to prevent confusion with other nozzle or jet designs used in two-stage valves.

Two Stage Valves

Two stage valves were designed to overcome the practical limitations of power and response of single stage valves. Any of the previously mentioned single-stage valves may be used as the first stage of a two stage valve with a spool or a plate as the second stage, of which the following are examples.

Double Spool

In this valve, shown in Fig 6, the output from the first stage is used to control the motion of the second stage spool and hence produces delivery to the load. Feedback from the output stage is necessary to avoid the integrating effect of the two spools and is usually accomplished either electrically or mechanically.

Fig 6 Two-stage valve; double spool.
1. Supply; 2. Exhaust; 3. Torque motor; 4. Input force; 5. First stage; 6. Second stage position feedback.

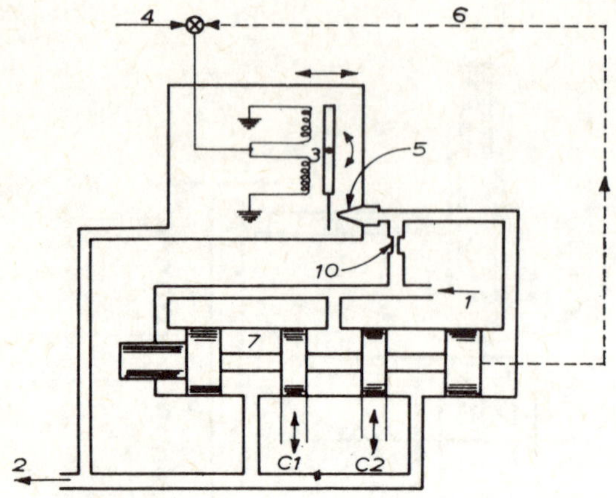

Fig 7 Two-stage valve; nozzle-flapper spool. 1. Supply; 2. Exhaust; 3. Torque motor; 4. Input force; 5. First stage; 6. Second stage position feed back; 7. Second stage; 8. Slide valve; 9. Load; 10. Restrictor; C1 and C2 to load.

Nozzle-Flapper Spool

The basic configuration of the single nozzle-flapper valve is shown in Fig 7, where it can be seen that a movement of the flapper towards the nozzle causes an increase in the pressure in the chamber between the nozzle and the restrictor which causes the second stage spool to move and thus deliver oil to the load. Movement of the flapper in the other direction will reduce the chamber pressure causing the spool to move in the opposite direction and produce reverse motion of the load. Thus the nozzle and flapper act as a variable impedance and a variable percentage of the supply pressure acts on the right-hand end of the second stage spool.

Double Nozzle-Flapper Spool

Representing a further development of the design previously described, this valve is shown in Fig 8. The flapper and nozzle control the pressure at both ends of the second stage spool and thus the operation of the first stage may be likened to that of the four-way valve. Although a fairly high quiescent flow is inevitable with this design the power loss is not significant for most applications.

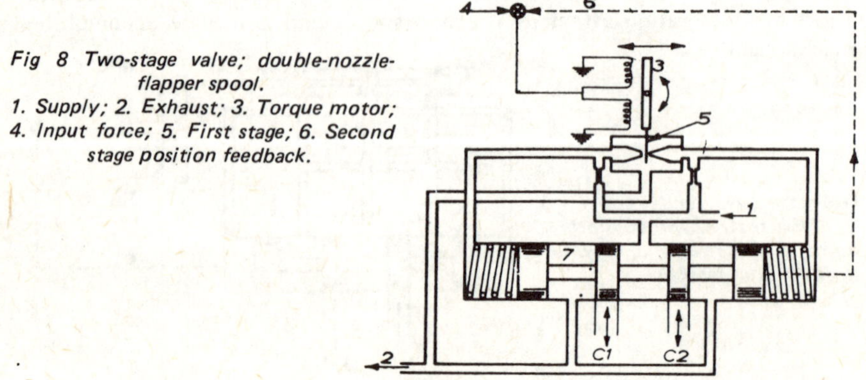

*Fig 8 Two-stage valve; double-nozzle-flapper spool.
1. Supply; 2. Exhaust; 3. Torque motor; 4. Input force; 5. First stage; 6. Second stage position feedback.*

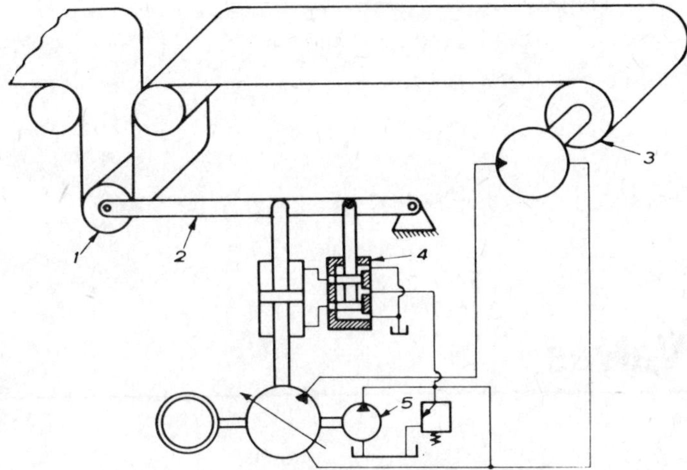

Fig 9 Controlling web tension with hydraulic servo-operated variable delivery pump and hydraulic motor drive 1. Dancing roll; 2. Lever; 3. Driver roll or 'in'-winder; 4. Servo-valve; 5. Boost pump.

The introduction of the nozzle-flapper device as the first stage, with its low inertia and short stroke was a major contribution in the field of two-stage valve design. The design shown in Fig 9 is basically that of a Moog valve and was one of the earliest of a family of valves having a nozzle-flapper as the first stage.

Solenoid Valves

A SOLENOID VALVE is basically a valve operated by a built-in actuator in the form of an electrical coil (or solenoid) and a plunger. The valve is thus opened or closed by an electrical signal, being returned to its original position (usually by a spring) when the signal is removed. Solenoid valves are produced in two modes – normally-open or normally-closed (referring to the state when the solenoid is not energized).

The solenoid itself may be operated by d.c. or a.c. A d.c. supply may be provided by a battery, d.c. generator or through a rectifier. An a.c. supply is normally taken from a.c. mains voltage, through a transformer if necessary.

There is an appreciable difference in the working characteristics of a solenoid supplied with d.c. and a.c. D.C. coils are slow in response time and can handle only low pressures. A.C. coils are quicker in response time and can handle higher pressures initially – see Fig 1. They can thus be cycled at faster rates, if required. Electrical losses are higher, however, and proportional to a.c. frequency. (The power losses in a solenoid operated by a.c. of 60 Hz frequency, for example, is higher than that of the same coil on a 50 Hz supply).

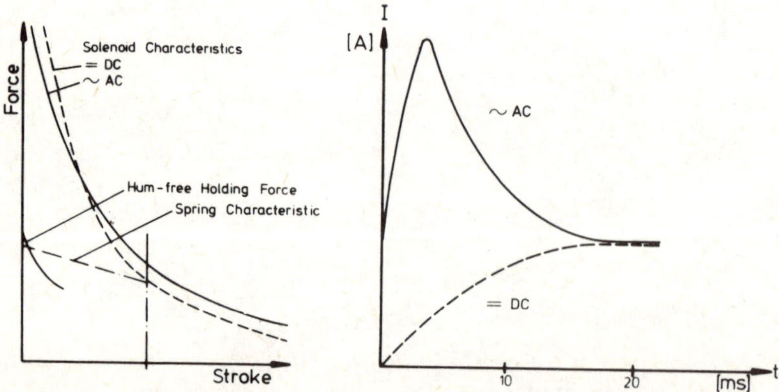

Fig 1 Force/stroke characteristics for a typical solenoid valve (left); and time-based comparison of solenoid engerization by a.c. and d.c. (Burkert Contromatic Ltd).

Return Spring Effect

With a 2-way *normally closed* valve both the spring force and the fluid inlet pressure act to close the valve. As a consequence the return spring can be made relatively weak, and in some designs eliminated entirely. The latter would require mounting the valve so that the solenoid was vertical, return action being by gravity plus fluid pressure.

With a 2-way *normally open* valve the spring holds the valve open, assisted by fluid pressure. The solenoid force must be sufficient to overcome both spring pressure and inlet pressure to close the valve.

3-way valves require an upper and a lower spring. The lower spring presses the valve against its seal opened by inlet pressure. The upper spring acts in a direction to force the valve open. The following are the combination of spring strengths required:

	lower spring	**upper spring**
3-way normally closed	strong	weak
3-way normally open	weak	strong
mixer valve	medium	medium
divider valve	strong	weak

Servo-assisted Valves

In general, direct solenoid operation is restricted to smaller sizes of valves, *ie* up to about ½ in (12 mm) bore size as a maximum. Larger solenoids needed for operating larger valves consume high levels of electricity and generate considerable heat. The *servo-assisted* valve offers a much more attractive proposition in such cases where a small solenoid is retained to operate a *pilot valve,* which in turn admits inlet pressure to an appropriate part of the valve to open the main valve. The pilot valve may be accommodated internally (internal pilot operated) or externally (external pilot operated).

For a servo-assisted valve to work there must be some differential pressure existing across the valve for it to operate properly. It will then operate against higher pressures with low electrical input, with the pressure rating of the servo-assisted valve the same as the pressure rating of the pilot valve, since both are subjected to the same line pressure.

Servo-assisted valves have slower response time than direct-acting solenoid valves, although in many designs this is adjustable.

There is a further general class of solenoid-operated valve known as a *semibalanced valve.* This is a double sealed valve with two plugs mounted on a common stem. The lower plug is slightly smaller than the upper plug. Line pressure is introduced below the lower plug and above the upper plug, creating a differential pressure to hold the valve on its seals, assisted by a spring if necessary. The solenoid force required to open the valve is then only that due to the differential pressure (and the spring force, if present).

Solenoid Enclosures

Various types of enclosures may be used for solenoid coils, ranging from general purpose enclosures to protect from indirect splashing and dust (*eg* NEMA type 1) through dust and watertight enclosures to full explosion-proof enclosures. Requirements in this respect are specific to the application and selected accordingly.

TABLE I – SUMMARY OF HYDRAULIC VALVE TYPES

Type or Name	Sub-type(s)	Remarks
Fixed restrictor	two-way	fixed orifice or capillary tube
(Fixed throttling valve)	one-way	spring-loaded poppet with orifice through poppet sharp-edged orifice and fixed end needle
Adjustable restrictor (adjustable throttling valve)	orifice	sharp edged orifice with adjustable needle
	screw-down	needle or plunger screwed down into orifice or port
	screw	two-way, with leakage past screw threads; or one-way with non-return valve and bypass.
Pressure-compensated flow restrictor	two restrictors in series	two-way; or one-way with non-return valve and bypass
	also with 'meter-in' characteristics	can be used in parallel or multiple combinations
	bypass type	gives maximum overall efficiency. Cannot be used in parallel
Flow divider	dual-orifice	to divide supply pressure equally between two outputs. Can be cascaded.
Priority	orifice	to maintain pressure in a priority circuit on multiple services fed from a single pump
Transfer	suction-operated	to short-circuit a cylinder on part of a hydraulically powered stroke
	pressure (or mechanically) operated	to short circuit a single-acting cylinder on a return stroke
Non-return (check)	ball poppet	limited performance with metal-to-metal seal or resilient face seal
	chamfered plunger damped	with metal-to-metal seal or resilient face seal usually plunger type (used with resilient seals)
Sequence	poppet	can also act as a non-return valve for return flow
	piston	tend to be less reliable due to spring friction
	drain-to-reservoir	commonly applied to piston type valves to prevent leakage from the inlet side causing premature operation
Locking	single	used for 'holding' actuator positions:
	double	normally employed with open centre systems
Shuttle	piston	three-port valves to provide duplicated line connections from alternative supplies
Spill-off	various	relief valve with speed control
Braking (deceleration)	tapered plunger	automatic deceleration of cylinder movements (similar to that otherwise obtained by cylinder cushions)
Shut-off (fuse)	(i) plunger (ii) quantity measuring (iii) fluid sampling	to shut off and isolate line in the event of excessive leakage developing

Note: flow-controlling valves may also be used for pressure control, or be combined with such valves.

Solenoid-operated Hydraulic Valves

Preference for the design of a solenoid-operated hydraulic valve is to use the solenoid for a 'push' operation, utilizing spring action for 'pull' motions. The solenoid must be powerful enough to override inertia and friction and also the spring and hydraulic forces. The latter may be extremely variable and not completely predictable, calling for a generous margin in the power of the solenoid and springs.

Solenoids may be of the 'dry' or 'wet' type. In general, 'wet' solenoids can be smaller for the same duty because of their lower static and dynamic friction. They also have the advantage that all moving parts are enclosed and lubricated, and seals between the solenoid and valve body are eliminated. They are also described as glandless valves.

The size of directly operated solenoid valves is generally restricted to flow rates up to about 10 gal/min (45 l/min) — *ie* 1/8 in (3 mm) and ¼ in (6 mm) nominal valve sizes. Many of these valves can be switched directly from static systems, the outputs usually being 24 V d.c. and 20–65 watts, depending on the system.

For higher flow rates demanding larger port sizes and larger spool diameters, directly operated solenoid valves become excessively bulky and costly. This can be overcome by pilot operation where a small solenoid valve is used as a piloting first stage in a large body valve whose spool is moved by differential pressure applied to its terminal surfaces.

The pilot valve controls pressure and exhaust flows in the piloting chamber and shifts the spool — a critical factor in design being the waterborne effects commonly associated with large, high pressure flows. Reaction time of the piloted valve is the sum of the pilot reaction time and the shifting time. It is usual to aim for a slow, controlled movement of the main spool, which can be achieved by incorporating restrictors between the pilot valve and piloting chamber. Where pilot pressure is derived from the pressure port of the valve itself it may be necessary to control the pilot pressure reducing valve in order to ensure constant operating times.

The sizing of piloted valves themselves requires careful consideration as regards pressure drop, especially where large flow rates are involved.

A typical layout of a miniature valve mounted pick-a-back on a pilot-operated valve is shown in Fig 2. The same spool arrangements are available as for pilot valves and the solenoid and pilot valve spools are matched. One point which must be watched is that pilot pressure is available if an open centre valve is used to unload the pump (this can be ensured by inserting a back pressure valve — a spring-loaded check valve is often used — as shown in Fig 3).

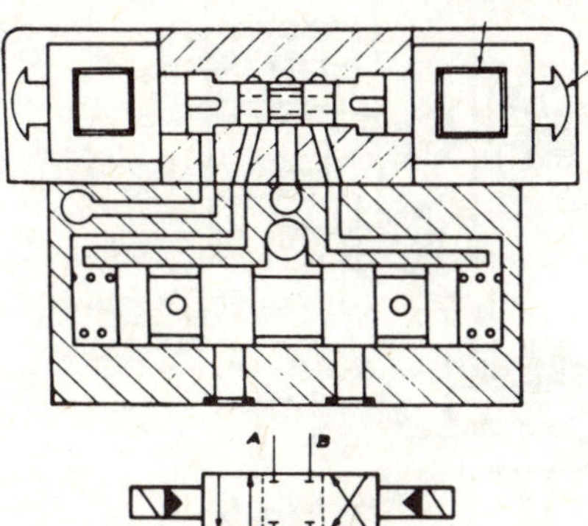

Fig 2 Sub-plate or gasket mounted solenoid-operated pilot-actuated valve. The upper valve is separated and can be used alone.

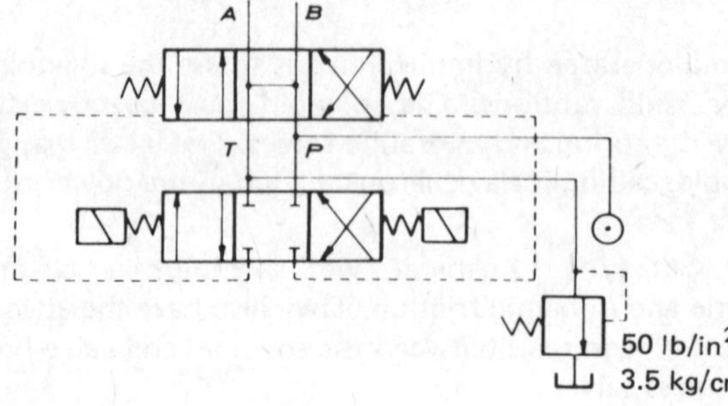

Fig 3 Pilot-operated valve circuit showing use of spring loaded check valve to maintain pilot pressure.

D.C. or A.C. Solenoids

D.C. solenoids are generally preferred to a.c. since d.c. operation is not subject to peak initial currents which can cause overheating and coil damage with frequent cycling or accidental spool seizure. A.C. solenoids are preferred, however, where fast response is required, or where relay-type electric controls are used. Response time with a.c. solenoid operated valves is of the order of 8–15 milliseconds, compared with the 30–40 milliseconds typical for d.c. solenoid operation.

Glandless Solenoid Valves

By arranging the solenoid armature to work in a sealed tube with the solenoid coil enveloping it, the sealing glands can be dispensed with, so simplifying the construction and eliminating one possible point of leakage.

This principle has been applied extensively to the smaller valves. A typical type is shown in Fig 4.

This valve is tee shaped with two ports opposite each other, whilst the third is at right angles to them. The plunger, usually of a corrosion resistant ferrous material, is spring biased so that when unenergized it closes the lower orifice, whilst leaving the other open. When energized, the plunger

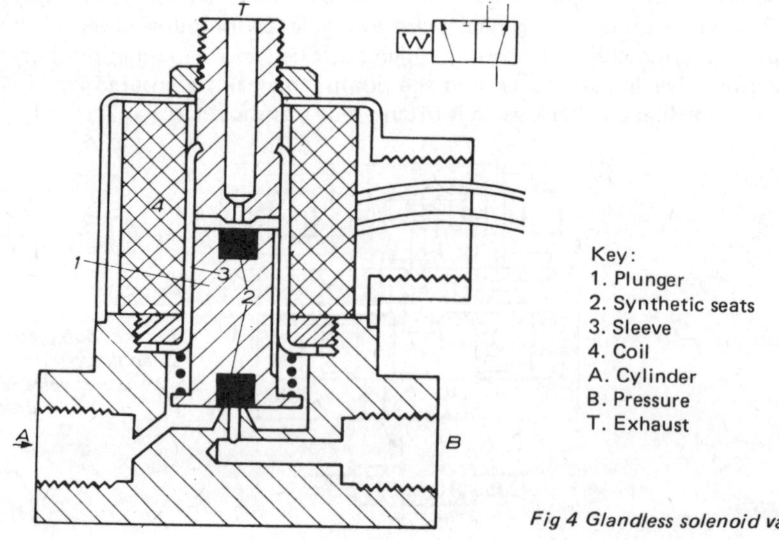

Fig 4 Glandless solenoid valves.

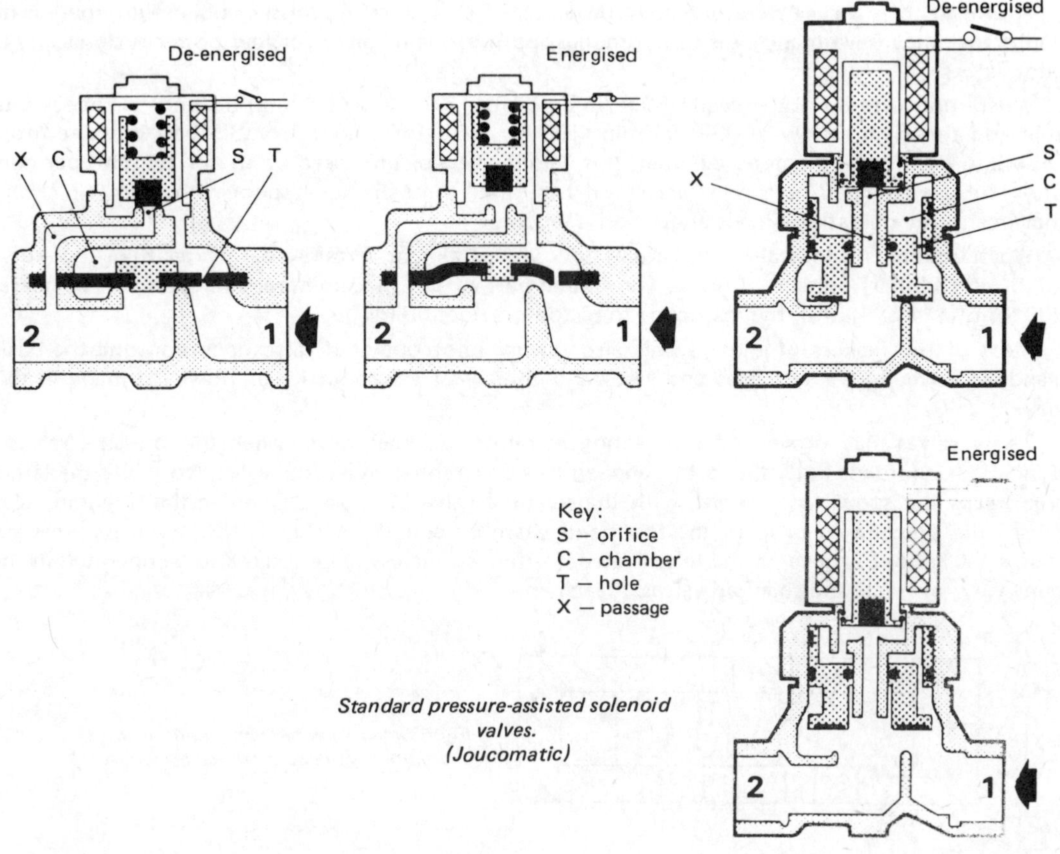

Standard pressure-assisted solenoid valves. (Joucomatic)

is pulled up so that the lower orifice is opened and the upper closed. If desired, the spring can be arranged to bias the plunger in the other direction.

The plunger is provided with plastic valve discs, usually of synthetic rubber or nylon. Because the plunger is unbalanced, the force due to the pressure must be limited and the size of orifice, and therefore the flow and pressure drop, is usually related to the pressure.

The maximum pressure is also related to the type of valve and may be as high as 3 000 lb/in^2 (210 bar) with a 1/32 in (0.85 mm) orifice. Flow depends on the allowable pressure drop and this in turn on the orifice size and fluid.

Sealing is normally 'bubble-tight' but this is to some extent dependent on the cleanliness of the fluid. Lubrication is not essential but if used with air the valve life is increased by air-line lubrication.

Glandless valves can be installed in any position and will withstand appreciable shock loads. Response time is extremely short, 5 milliseconds on a.c. and 10–15 milliseconds on d.c. and it is said that speeds of up to several hundred cycles per minute are possible.

For hazardous atmospheres most makers supply explosion-proof materials which are slightly heavier and bulkier than the standard type.

Although these valves were originally developed for aircraft and missile application, there is no doubt that they have many uses in hydraulics, both as main valves for low power systems and as pilot valves.

Where pressures do not exceed 250 lb/in^2 17.5 bar a 1/16 in (1.6 mm) diameter orifice is suitable and this gives a flow of 0.50 gal/min (100 in^3/min) for a 50 lb/in^2 (3.5 bar) pressure drop. On a 2 in (50.8 mm) diameter cylinder, this would give a piston speed of 30 in (760 mm) per minute, whilst when used as a pilot valve for 1 in (25.4 mm) diameter spool with ½ in (12.7 mm) movement, the operating time is about half a second.

When acting as pilot valve the actual flow would almost certainly be greater than that for a 50 lb/in^2 (3.5 bar) pressure drop, as for a large part of the time the pressure drop will be nearer 200 lb/in^2 (14 bar), until the resistance to piston or spool builds up.

Most of the makers of these valves also supply pilot operated valves incorporating the basic glandless valve; four-way valves and two- and three-way valves for larger flows are made in this way.

These valves may prove useful in acting as pilots to larger valves when the 'pressure release' principle is adopted. Fig 5 shows this applied to a differential area spool valve. Normally the larger area keeps the spool to the right, with the solenoid valve SC closed. Opening the solenoid valve causes the pressure to drop in the left hand chamber and the spring, with the excess pressure, causes the spool to move to the left. A two-position springless valve could also be operated in the same way, with jets and solenoid valves at each end.

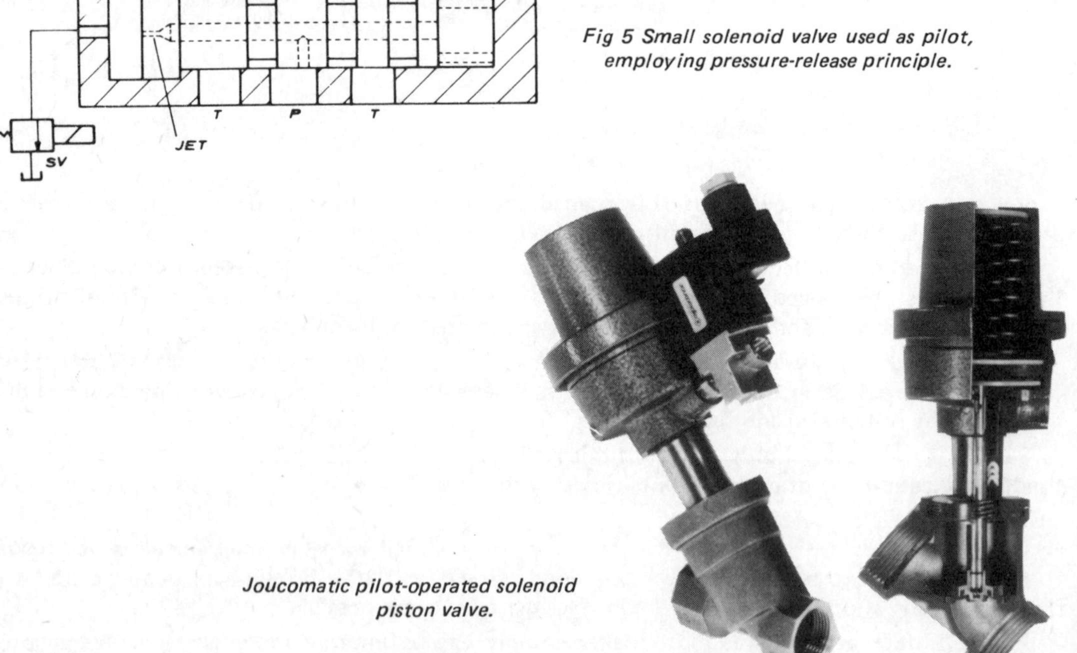

Fig 5 Small solenoid valve used as pilot, employing pressure-release principle.

Joucomatic pilot-operated solenoid piston valve.

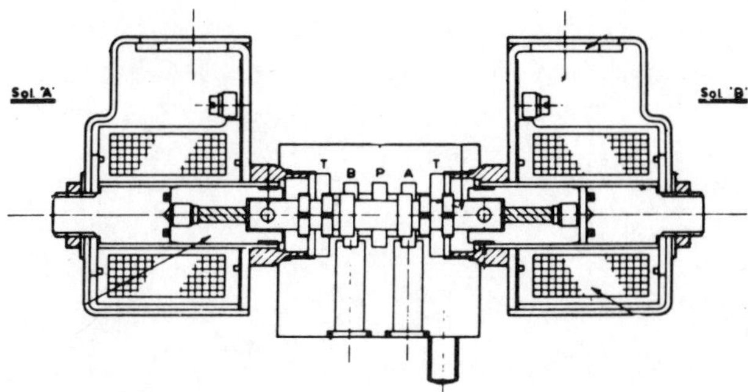

Fig 6 Glandless solenoid valve.
(Flui-Trol)

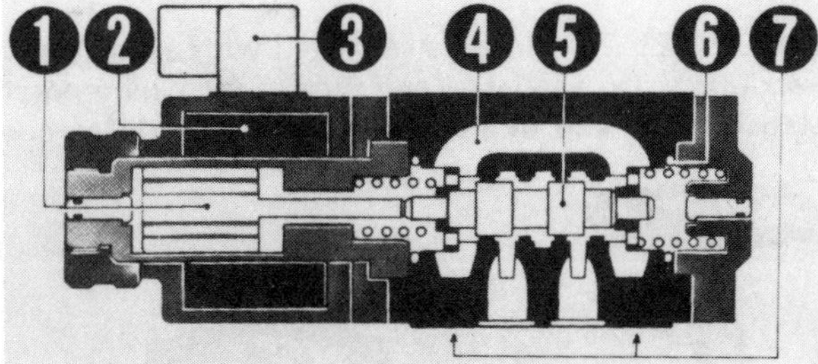

Key:
1. Wet solenoid
2. Encapsulated units
3. Plug and socket connectors.
4. Ratio body
5. Spine
6. 'O' ring seals
7. Mounting surface.

Fig 7 Wet solenoid valve.
(Atos)

Glandless Solenoid Valves — Spool Type

The construction of a four-way spring centred closed-centre double solenoid valve is shown in Fig 6. It has push solenoids with a spool movement of 0.040 in (1 mm) either side of the centre. It is suitable for pressures up to 2 000 lb/in^2 (140 bar) and has a flow of 2 gal/min (9 l/min) for a pressure drop of 30 lb/in^2 (2 bar) on light hydraulic oil at 80–100°F (27–38°C). It is gasket-mounted and when used as a pilot valve is bolted on top of the main valve.

Another example of a glandless valve with a 'wet' armature is shown in Fig 7. All seals are static O-rings.

Needle Valves

SMALL SIZES of globe valves fitted with a finely tapered plug are known as *needle valves.* This description also applies to any type of valve incorporating a tapered needle having axial movement relative to the axis of a concentric orifice and thus controlling the effective opening of the orifice.

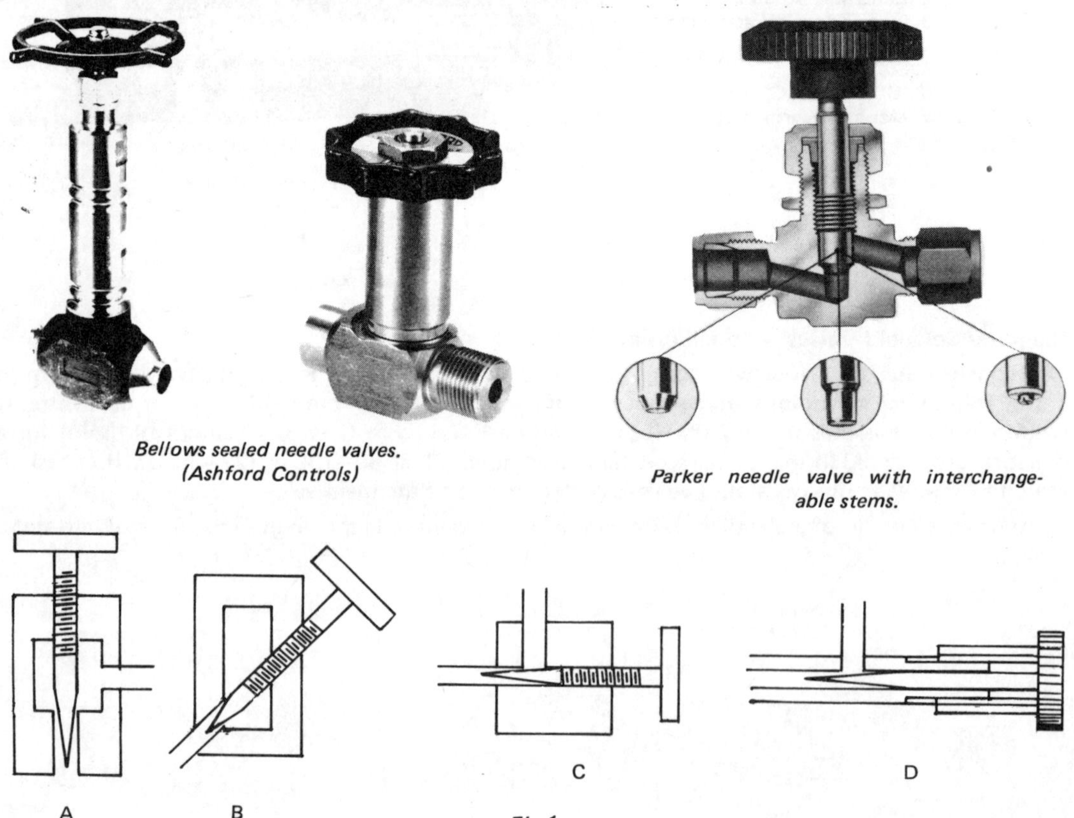

Bellows sealed needle valves. (Ashford Controls)

Parker needle valve with interchangeable stems.

Fig 1

Tee and oblique needle valves for process industries.
(Ashford Controls)

Three basic configurations are shown in Fig 1. A is a simple screwdown valve. B is an oblique version, offering a more direct flow path. C is another form where the controlled outlet flow is at right angles to the main flow (and may be distributed through one or more passages). In these basic versions a threaded needle is shown, the thread itself acting as a seal to eliminate leakage past the needle. This is normally quite satisfactory in very small sizes of needle, although a more-leak-tight arrangement is to mount the needle in an externally threaded endpiece — Fig 1D. This end piece can also act as a grip for adjustment of the needle. Sealing in this case can be further improved if necessary by incorporating an internal seal, such as an O-ring.

The other common form of needle valve is the float-controlled, carburettor type (valve).

See also chapter on *Globe Valves.*

Miscellaneous Valves

Membrane Check Valves

An original (patented) design of check valve is shown in Fig 1. This features a lipped elastomeric membrane as the working element, offering virtually unrestricted flow in the open position with a capability of passing suspended solids up to the full bore diameter. The membrane itself is held open to a full circular form by the flow, the circumference of the membrane in this condition being πD. Loss of head is thus minimal (*eg* directly comparable to that of a swing check valve).

With reverse flow the membrane assumes a closed position with the lips in mating contact (See Fig 2), *ie* the natural 'unloaded' form of the membrane. Closure is further assisted and maintained by the reverse flow impinging on the sides of the now wedge-shaped membrane. The length of seal in this case is ½ πD or substantially half that of a conventional check valve (*ie* the potential leakage path is reduced by half).

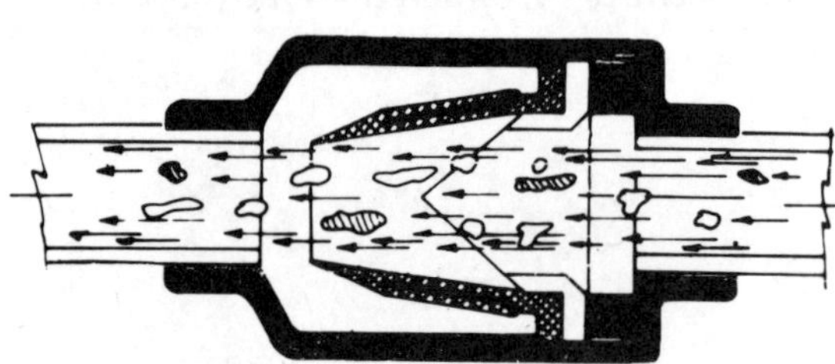

Fig 1 Ondastop membrane check valve. (Apparecchi E Macchine Idrauliche Speciali).

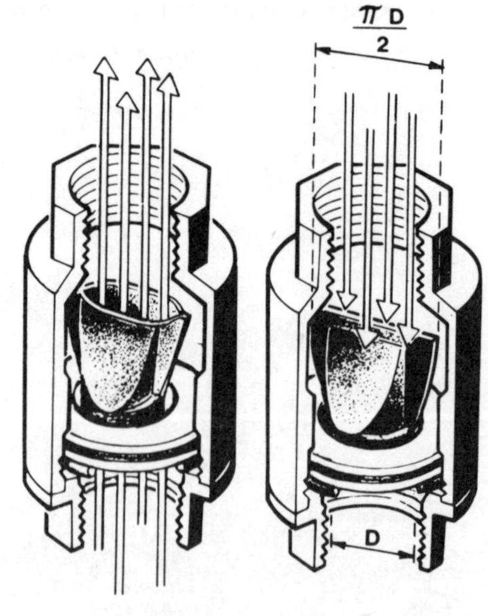

Fig 2

The membrane itself is not subject to elastic deformation, merely flexure, and thus has a long life, particularly if the fluid does not contain abrasive solids in suspension. When servicing is required, replacement of the membrane is a simple operation. No other servicing is necessary. Membrane elastomer is selected according to the product to be handled.

Due to the elastic nature of the closure this type of check valve cannot cause water hammer and is also noiseless in that it has no hinge or spring which can be excited into vibration in either the open or closed position.

Currently this valve is made in threaded form for water pipes up to 2 in (50 mm) diameter and flanged for water pipes from 2 in to 16 in (50 to 400 mm) diameter. Valves sizes up to 5 in (125 mm) have a single passage. Larger valves have several passages, each with its individual membrane (Fig 3). This solution of dividing the total flow into partial flows with smaller flow rates which never open or close strictly together eliminates water hammer in these larger valve sizes with higher flow rates.

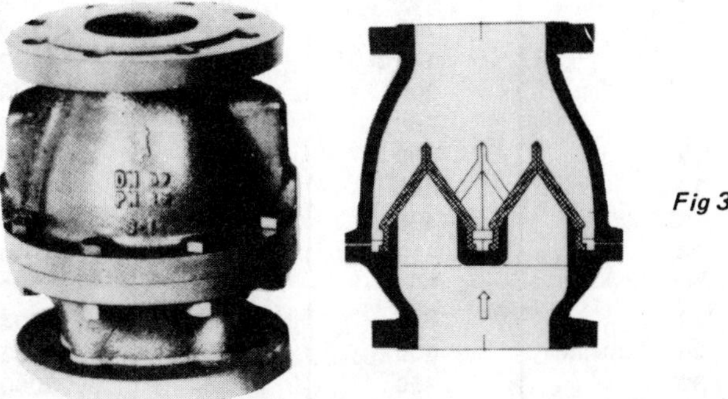

Fig 3

Eccentric Valves

The description *eccentric valve* is applied to *plug valves* having an eccentric motion against a resilient facing, *ie* as the eccentric plug rotates 90 degrees from open to closed it moves into a raised eccentric seat. The complete action can be followed from Fig 4.

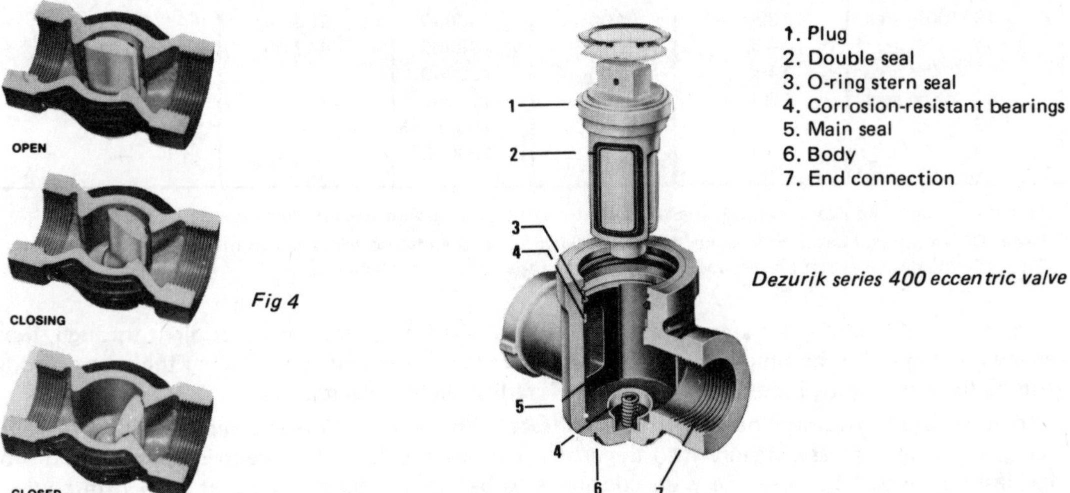

Fig 4

Dezurik series 400 eccentric valve

C_V VALUES FOR DE ZURIK ECCENTRIC VALVES

Flow in gal/min of water at 1 lb/in^2 pressure drop

Valve Size (in)	(mm)	Port Area %	C_V	Head Loss* (Feet of Pipe)	(Cm of Pipe)
½	15	210	10	1.55	47
¾	20	114	20	1.70	52
1	25	71	33	2.20	67
1¼	32	115	74	1.84	56
1½	40	84	74	3.92	119
2	50	94	148	3.68	112
2½	65	100	236	3.69	112
3	80	81	330	5.70	174
4	100	88	560	8.00	244
5 & 6	125 & 150	87	1180	14.80	451
8	200	89	2030	20.60	628
10	250	81	3130	28.00	853
12	300	83	4140	40.00	122
14	350	84	5500	38.00	1160
16	400	83	7300	45.00	1370
18	450	82	9600	48.50	1480
20	500	91	13000	47.20	1440
24	600	70	17500	79.00	2410
24 100% area	600	100	28000	50.00	1520
30	750	70	28000	101.00	3080
30 100% area	750	100	40000	70.00	2130
36	900	70	40000	131.00	3990
36 100% area	900	100	58000	64.00	1950
42	1100	70	58000	135.00	4120
42 100% area	1100	100	100000	143.00	4360
48	1200	—	—	—	—
48 100% area	1200	100	100000	255.00	7770
54	1400	70	100000	442.00	13470
54 100% area	1400	100	150000	—	—
60 100% area	1500	100	150000	—	—
66	—	82	150000	—	—
72	1800	70	150000	—	—

*Pressure drop in equivalent length (feet and centimetres) of standard weight steel pipe.

Note: C_V Values will be slightly higher for valves with screwed ends and for metal-to-metal seated valves. Sizing data is based on discharge into conduit rather than atmosphere.

In the open position, the segmented plug is out of the flow path. Flow is straight through, flow capacity is high. As the plug closes, it moves toward the seat without scraping the seat or body walls so there is no plug binding or wear. Flow is still straight through.

In the closed position, the plug makes contact with the seat. When furnished with a resilient facing, the plug is pressed firmly into the seat for deadtight shutoff. Eccentric plug and seat provide lasting shutoff because the plug continues to be pressed against the seat until firm contact is made.

Throttling characteristics of a valve of this type are generally excellent (see Fig 5), and shut-off in the closed position positive with air and gases as well as liquids.

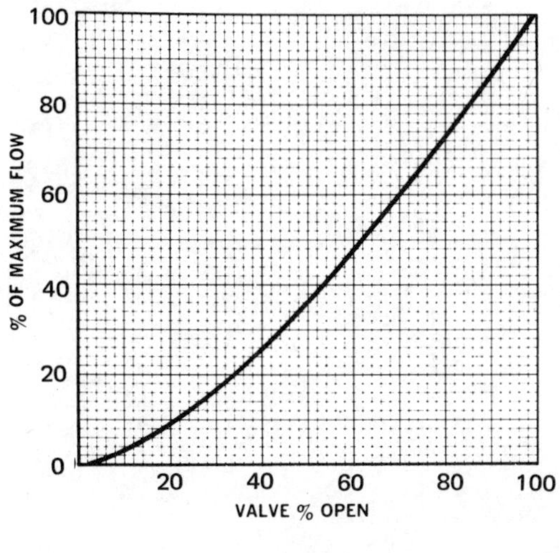

Fig 5

Lenticular Valve

The *lenticular valve* can be described as similar in concept to a ball valve except that the ball is replaced by a lens-shaped cup or lenticule – Fig 6. The cup is trunnion mounted and sweeps across a captive seal to effect shut-off. There is positive interference between cup and seal so that the valve seals without fluid pressure. At the same time the trunnion permits the cup to 'float' against the seal under positive fluid pressure, providing even more positive sealing.

This is a low cost shut-off valve for applications in the light duty, low pressure field, in sizes ranging from ¼ in to 1¼ in BSP (6 mm to 30 mm). Construction is normally brass body with a stainless steel cup and synthetic rubber seal. Maximum working pressure rating is 230 lb/in^2 (16 bar).

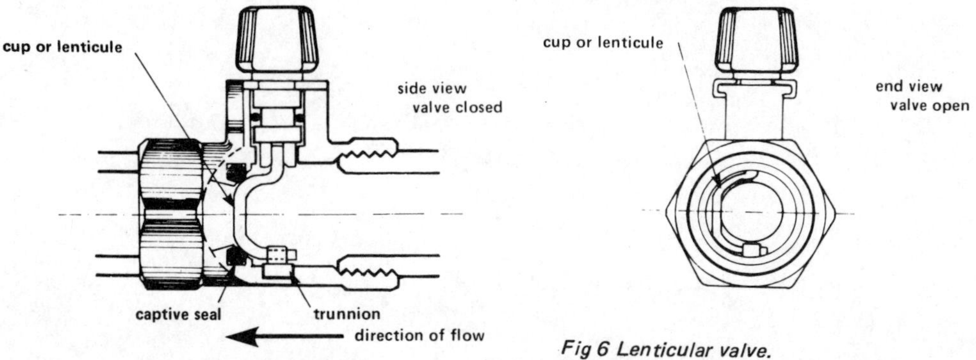

Fig 6 Lenticular valve.

SECTION 3

Performance

Flow of Liquids Through Pipes

LIQUIDS PUMPED or discharged through pipes under conventional pressures behave as incompressible fluids. Flow is assumed to fill the pipe section, basic flow rate (Q) and flow velocity (V) are directly related, viz

Q = V x pipe bore area (in consistent units)

Flow Velocity

Flow velocity is defined as the mean or average velocity at a given cross section. Due to frictional effects (and the fact that all real fluids possess viscosity), a velocity gradient will exist across a pipe section, ranging from zero at the point of contact with the pipe wall to a maximum at the centreline – Fig 1. The actual velocity *profile* may be smooth or irregular, depending on whether the flow is *laminar* or *turbulent,* respectively (see later).

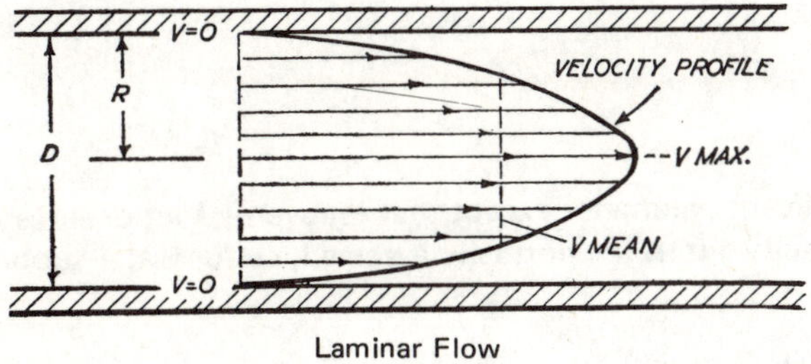

Laminar Flow

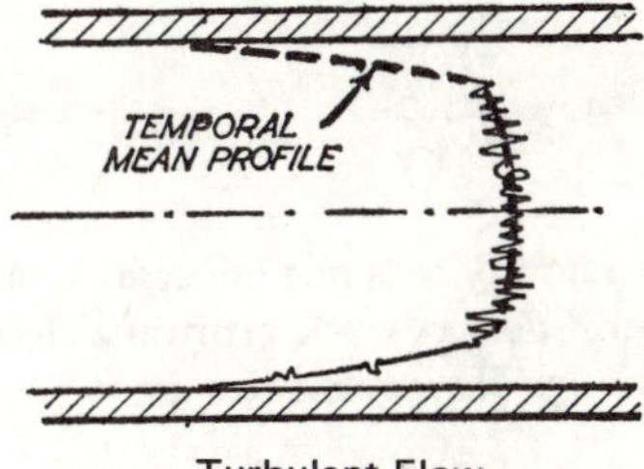

Turbulent Flow

Fig 1

The basic formula relating Q, V and pipeline diameter (d) is

$$Q = \pi/4.V.d^2 \quad \text{in consistent units}$$

In engineering units this becomes:

$$Q = k\,V.d^2$$

where k is a constant, depending on the units employed (see Table I).

TABLE I – FLOW VELOCITY BY CALCULATION

Flow Rate (Q) Unit	Flow Velocity ft/sec	Flow Velocity m/sec	Pipe Bore (d) Unit
Cubic inches/sec	$Q/10d^2$	$0.03\ Q/d^2$	inches
		$19.3\ Q/d^2$	millimetres
Cubic inches/minute	$0.0177\ Q/d^2$	$0.0053\ Q/d^2$	inches
		$3.42\ Q/d^2$	millimetres
Gallons per minute	$Q/2d^2$	$0.15\ Q/d^2$	inches
		$97\ Q/d^2$	millimetres
Litres per minute		$0.002\ Q/d^2$	inches
		$1.275\ Q/d^2$	millimetres
US gallons per minute	$4\ Q/d^2$	$0.133\ Q/d^2$	inches
		$86000\ Q/d^2$	millimetres
Tons of water per day	$44\ Q/d^2$	$13.3\ Q/d^2$	inches
Cubic metres of water per day		$8600\ Q/d^2$	millimetres
US barrels per day	$0.012\ Q/d^2$	$0.0036\ Q/d^2$	inches
		$2.32\ Q/d^2$	millimetres

Transposed forms of this equation are also useful for direct solutions for V and d, viz:

$$V = \frac{Q}{k.d^2}$$

$$d = \sqrt{\frac{Q}{kV}}$$

Flow velocity V is not necessarily a significant parameter except that it governs frictional losses. For general design work arbitrary flow velocity limits are normally assumed, *eg* for water supplies normal design flow velocities are

general services	4–10 ft/sec (1.2–3 m/sec)
water supplies	up to 7 ft/sec(up to 2 m/sec)
boiler feed	8–15 ft/sec (2.5–4.5 m/sec)

Pipe Sizing

Rather greater attention to limiting flow velocities is normally required on the suction side of pumps. Also frictional losses are proportional to fluid viscosity as well as flow velocity. More specific recommendations for flow velocities are given in Tables IIA and IIB. Again these are largely arbitrary figures based on providing suitable hydraulic conditions in suction pipes or generally acceptable levels of friction loss in delivery pipes (see also Table III).

TABLE IIA – RECOMMENDED SUCTION FLOW VELOCITIES

Pipe Bore		Water		Light Oils		Boiling Liquids		Viscous Liquids	
inches	mm	ft/sec	m/sec	ft/sec	m/sec	ft/sec	m/sec	ft/sec	m/sec
1	25	1.5	0.50	1.5	0.50	1.0	0.300	1.0	0.300
2	50	1.6	0.50	1.5	0.50	1.0	0.300	1.1	0.330
3	75	1.7	0.50	1.6	0.50	1.0	0.300	1.2	0.375
4	100	1.8	0.55	1.8	0.55	1.0	0.300	1.3	0.400
6	150	2.0	0.60	2.0	0.60	1.1	0.350	1.4	0.425
8	200	2.5	0.75	2.3	0.70	1.2	0.375	1.5	0.450
10	250	3.0	0.90	3.0	0.90	1.5	0.450	1.7	0.500
12	300	4.5	1.40	3.0	0.90	1.5	0.450	1.7	0.500
over 12*		5.0	1.50						

* General formula : pipe diameter (inches) $= \sqrt{\dfrac{\text{gal/min}}{10}}$

TABLE IIB – RECOMMENDED DELIVERY FLOW VELOCITIES

Pipe Bore		Water		Light Oils		Boiling Liquids		Viscous Liquids	
inches	mm	ft/sec	m/sec	ft/sec	m/sec	ft/sec	m/sec	ft/sec	m/sec
1	25	3.5	1.00	3.5	1.00	3.5	1.00	3.5	1.00
2	50	3.6	1.10	3.6	1.10	3.6	1.10	3.6	1.10
3	75	3.8	1.15	3.8	1.15	3.8	1.15	3.7	1.10
4	100	4.0	1.25	4.0	1.25	4.0	1.25	3.8	1.15
6	150	4.7	1.50	4.7	1.50	4.7	1.50	3.9	1.20
8	200	5.5	1.75	5.5	1.75	5.5	1.75	4.0	1.20
10	250	6.5	2.00	6.5	2.00	6.5	2.00	4.5	1.30
12	300	8.5	2.65	6.5	2.00	6.5	2.00	4.5	1.40
over 12*		10.0	3.00						

*General formula : pipe diameter (inches) $= \sqrt{\dfrac{\text{gal/min}}{20}}$

Accepting arbitrary values for flow velocities, the corresponding pipe size (d) required for a specified delivery (Q) follows from simple formula calculation. In the case of water, a general formula often used is:

$$\text{pipe diameter (inches)} = \sqrt{\frac{\text{gallons/min}}{10}}$$

In high pressure systems – *eg* hydraulic circuits using small bore pipes – pipe sizing is more critical and normally determined directly from a specified or nominal figure for *pressure drop.* This involves working an appropriate pressure drop formula as a solution for pipe bore.

TABLE III – PIPE BORE SIZE FOR GIVEN FLOW VELOCITY

FLOW VELOCITY ft/sec	m/sec	PIPE BORE (inches) FOR FLOW RATE IN gal/min	PIPE BORE (mm) FOR FLOW RATE IN l/min
1	0.3 3	$0.7\sqrt{Q}$	$8.4\sqrt{Q}$
	0.5		$6.5\sqrt{Q}$
2	0.6	$0.5\sqrt{Q}$	$6.0\sqrt{Q}$
3	0.9	$0.4\sqrt{Q}$	$4.9\sqrt{Q}$
	1.0		$4.6\sqrt{Q}$
	1.1		$4.4\sqrt{Q}$
4	1.2	$0.35\sqrt{Q}$	$4.2\sqrt{Q}$
5	1.5	$0.3\sqrt{Q}$	$3.8\sqrt{Q}$
6	1.8	$0.28\sqrt{Q}$	$3.4\sqrt{Q}$
	2.0		$3.3\sqrt{Q}$
7	2.1	$0.265\sqrt{Q}$	$3.2\sqrt{Q}$
8	2.4	$0.25\sqrt{Q}$	$3.0\sqrt{Q}$
	2.5		$2.9\sqrt{Q}$
9	2.75	$0.23\sqrt{Q}$	$2.8\sqrt{Q}$
10	3.0	$0.22\sqrt{Q}$	$2.6\sqrt{Q}$

The same technique may also be applied to fluid transport systems, especially if high pressures are involved or working conditions are critical. In this case, since relatively larger pipe sizes are normally involved, recommendations are commonly based on flow rates only, viz:

Suction lines – pressure drop 0.05 to 1 lb/in^2 per 100 feet. (0.0115 to 0.23 bar per 100 metres) depending on the available NPSH.

Delivery lines – 0.5 to 6 lb/in^2 per 100 feet (0.115 to 1.38 bar per 100 metres) depending on the flow rate, viz:

(a) 2 to 6 lb/in^2 per 100 feet for flow rates up to 100 gal/min
(0.46 to 1.4 bar per 100 metres for flow rates up to 450 l/min)

(b) 1.5 to 5 lb/in^2 per 100 feet for flow rates from 100 to 200 gal/min
(0.33 to 1.15 bar per 100 metres for flow rates from 450 to 900 l/min)

(c) 1 to 4 lb/in^2 per 100 feet for flow rates from 200 to 500 gal/min
(0.23 to 0.92 bar per 100 metres for flow rates from 900 to 2 250 l/min)

(d) 0.5 to 2 lb/in^2 per 100 feet for flow rates above 500 gal/min
(0.11 to 0.46 bar per 100 metres for flow rates above 2250 l/min).

TABLE IV – RECOMMENDED FLOW VELOCITIES BASED ON FLUIDS SG

Pipe Diameter		Power Driven Pumps						Turbine Driven Pumps					
		SG = 1.0		SG = 0.75		SG = 0.5		SG = 1.0		SG = 0.75		SG = 0.5	
inch	mm	ft/sec	m/sec	ft/sec	m/sec	ft/sec	m/sec	ft/sec	m/sec	ft/sec	m/sec	ft/sec	m/sec
2	50	6.00	1.80	7.00	2.10	7.5	2.30	5.00	1.50	5.50	1.70	6.00	1.80
3	75	7.00	2.10	8.00	2.40	8.5	2.60	5.50	1.70	6.00	1.80	6.50	2.00
4	100	8.00	2.40	9.00	2.75	10.0	3.00	6.00	1.80	6.50	2.00	7.00	2.15
6	150	9.00	2.75	10.00	3.00	12.0	3.65	6.50	2.00	7.00	2.15	8.00	2.40
8	200	10.00	3.00	11.25	3.40	13.0	4.00	6.75	2.10	7.50	2.30	8.50	2.60
10	250	11.00	3.25	12.00	3.65	14.0	4.20	7.00	2.15	7.75	2.35	9.00	2.75
12	300	11.50	3.50	12.50	3.80	14.5	4.40	7.00	2.15	8.00	2.40	9.25	2.80
14	350	11.75	3.60	13.00	4.00	15.0	4.50	7.00	2.15	8.00	2.40	9.50	2.90
16 and over	400	12.00	3.65	13.00	4.00	15.0	4.60	7.00	2.15	8.00	2.40	9.50	2.90

Pipe Sizing by SG of Fluid

The size of delivery lines on centrifugal pumps is sometimes based on economic flow velocity related both to the specific gravity of the fluid being handled and the type of driver. This is realistic in the sense that the power input required, and thus the cost of pumping, is directly proportional to fluid specific gravity, and economic flow velocity varies inversely to pump speed.

Recommended flow velocities are given in Table IV.

Critical Flow Velocity

Flow velocity is 'critical' in the sense that it is a major factor in determining the frictional losses of the flow. This is not necessarily significant for fluid transport applications involving flow velocities within the recommended ranges.

Flow velocity can, however, be a critical factor in practical applications involving the transport of solids in suspension in a fluid. The flow velocity will largely govern whether the solids are transported in suspension (homogeneous flow), or whether the solids tend to settle out forming sliding layers over a settled bed (heterogeneous flow).

To produce homogeneous flow it is necessary that the flow velocity should be greater than the fall velocity of the solids in the fluid. This sets critical or minimum flow velocity requirements for the handling of fluids containing solids in suspension, which can only be determined satisfactorily on empirical lines. Some specific recommendations are given in Table V.

See also chapter *Flow of Mixtures Through Pipes.*

Laminar and Turbulent Flow

Flow through pipes can be either *laminar* or *turbulent,* the flow condition being significant in affecting both the velocity gradient and the frictional losses. With laminar flow, frictional losses

TABLE V – MINIMUM FLOW VELOCITIES FOR SLURRIES

Type	Size of Solids (Mesh No)	Flow Velocity	
		ft/sec	m/sec
Fines	Over 200	3 – 5	1.00 – 1.50
Sands	200 – 20	5 – 7	1.50 – 2.00
Coarse	20 – 4	7 – 11	2.00 – 3.25
Sludge		11 – 14	3.25 – 4.25

are due to viscous drag and are independent of the condition of the pipe bore. With turbulent flow, viscous shear forces predominate and the condition of the boundary surface can materially affect the total friction.

The actual flow condition can be established by reference to a non-dimensional parameter, the Reynolds number (R_e) determined as:-

$$R_e = \frac{dV}{\nu}$$

where d = pipe bore
V = velocity of flow
ν = kinematic viscosity of the fluid
(in consistent units)

Note: the Reynolds number itself is dimensionless

In engineering units:-

$$R_e = \frac{7740dV}{\nu} \quad \text{or} \quad R_e = \frac{930dV}{\nu}$$

where d is in inches; V is in ft/sec; ν is in centistokes

where d is in cm; V is in m/sec; ν is in centistokes

In the case of clean cold water

$$R_e = 7740dV \text{ when d is in inches, V in ft/sec}$$
$$= 930dV \text{ when d is in cm, V in m/sec}$$

(See also Table VI).

TABLE VI – REYNOLD'S NUMBER FOR CLEAN COLD WATER

	PIPE BORE									
Inches (mm)	**1 (25)**	**1½ (40)**	**2 (50)**	**3 (75)**	**4 (100)**	**6 (150)**	**8 (200)**	**10 (250)**	**12 (300)**	**18 (450)**
Per 1 gal/min*	3800	2500	1900	1270	950	630	475	380	320	210
Per 1 l/min*	835	550	420	280	210	140	105	85	70	46

*Multiply by actual numerical value in either unit to give Reynold's Number

The same formulas apply for the calculation of Reynolds numbers for flow in non-circular pipes, substituting the equivalent hydraulic diameter for the circular diameter, viz:

$$\text{Equivalent hydraulic diameter} = 4 \times \left(\frac{\text{cross sectional flow area}}{\text{wetted parameter}}\right)$$

The quantity contained in the brackets is the *hydraulic radius* of a non-circular pipe.

Flow is *laminar* at Reynolds numbers up to 2 000; and *turbulent* at Reynolds numbers above approximately 4 000. In the transitional range (R_e = 2 000 – 4 000) flow can vary from laminar to turbulent, *ie* flow conditions are indeterminate.

In the case of laminar flow the velocity gradient will be linear. Maximum velocity at the centre of the bore will be of the order of 1.5 times the mean flow velocity.

With turbulent flow there is no clearly defined velocity profile. The temporal mean profile will be of the form shown in Fig 2, the actual profile varying with Reynolds number. At low Reynolds numbers the maximum flow velocity (at the centre of the bore) will be of the order of 2.0 times the mean velocity, reducing to about 1.25 times the mean velocity at higher Reynolds numbers.

The velocity profile is primarily of significance where a pilot tube or similar flow measuring device is inserted in the pipe as this will be subject to position error. There is no point in the cross section with turbulent flow where the local velocity is likely to be constant and fully predictable. This factor is not necessarily significant where measurements or calculations are based on flow rate, or mean flow velocity, the latter being determined directly from the flow rate and pipe bore.

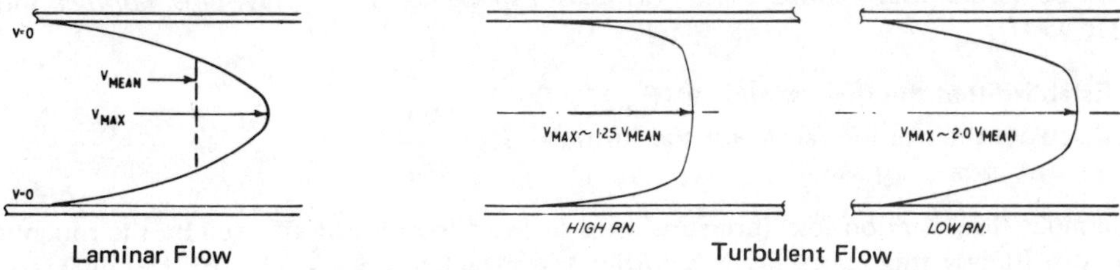

Fig 2 Velocity profiles for laminar and turbulent flow (temporal mean in case of turbulent flow).

Frictional Losses

Friction losses are calculated in terms of pressure drop or alternatively head loss. Figures for frictional losses are normally reduced to (friction) head equivalent per foot or per 100 feet of pipe and are then directly applicable to any length or aggregate length of straight run of pipe of the sizes concerned. Such data are available and presented both in graphical and tabular form for a wide range of pipe sizes, for water, oils and fluids. Agreement is not always good between such data originating from different sources. Many – particularly for water flow through pipes – are based on formulas more than half a century old and have a limited range of accuracy. Others are based on quite widely differing empirical coefficients, with similar limitations. If the reliability of the data available is suspect, or shows inconsistencies, more accurate solutions will be arrived at by working from basic principles.

In the case of laminar flow the Darcy-Weisbach formula can be applied in the form

$$\Delta P = f \frac{L\rho V^2}{2\,Dg}$$

where ΔP = pressure drop in lb/in^2
L = length of pipe in feet
V = flow velocity
D = bore of pipe in inches
ρ = mass density of fluid
f = a friction factor $= \dfrac{64}{\text{Reynolds number}}$

In the case of larger pipes (*eg* 1 in (25 mm) bore and above), and expressed in terms of flow rate (Q) rather than flow velocity (V), the following simplified formulas can be used:-

$$\Delta P = f \frac{Q^2 L}{K_1 D^5} \times \text{specific gravity of fluid}$$

where K_1 is a constant dependent on the units adopted for Q, L and D.

The corresponding formula for head loss (ΔH) is:

$$\Delta H = f \frac{Q^2 L}{K_2 D^5}$$

It must be noted that before these formulas can be used the *Reynolds number* must be determined to:-

(i) Establish that the flow *is* laminar ($R_e \leqslant 2\,000$)

(ii) Calculate the friction factor for the formula $\left(f = \dfrac{64}{R_e}\right)$

With laminar flow friction loss (pressure drop or head loss) is not affected by the roughness of the pipe bore (unless this appreciably modifies the effective bore size). The formulas are thus directly applicable to laminar flow in all types, construction and ages of circular pipes and non-circular pipes (the latter with R_e computed on the basis of their hydraulic radius).

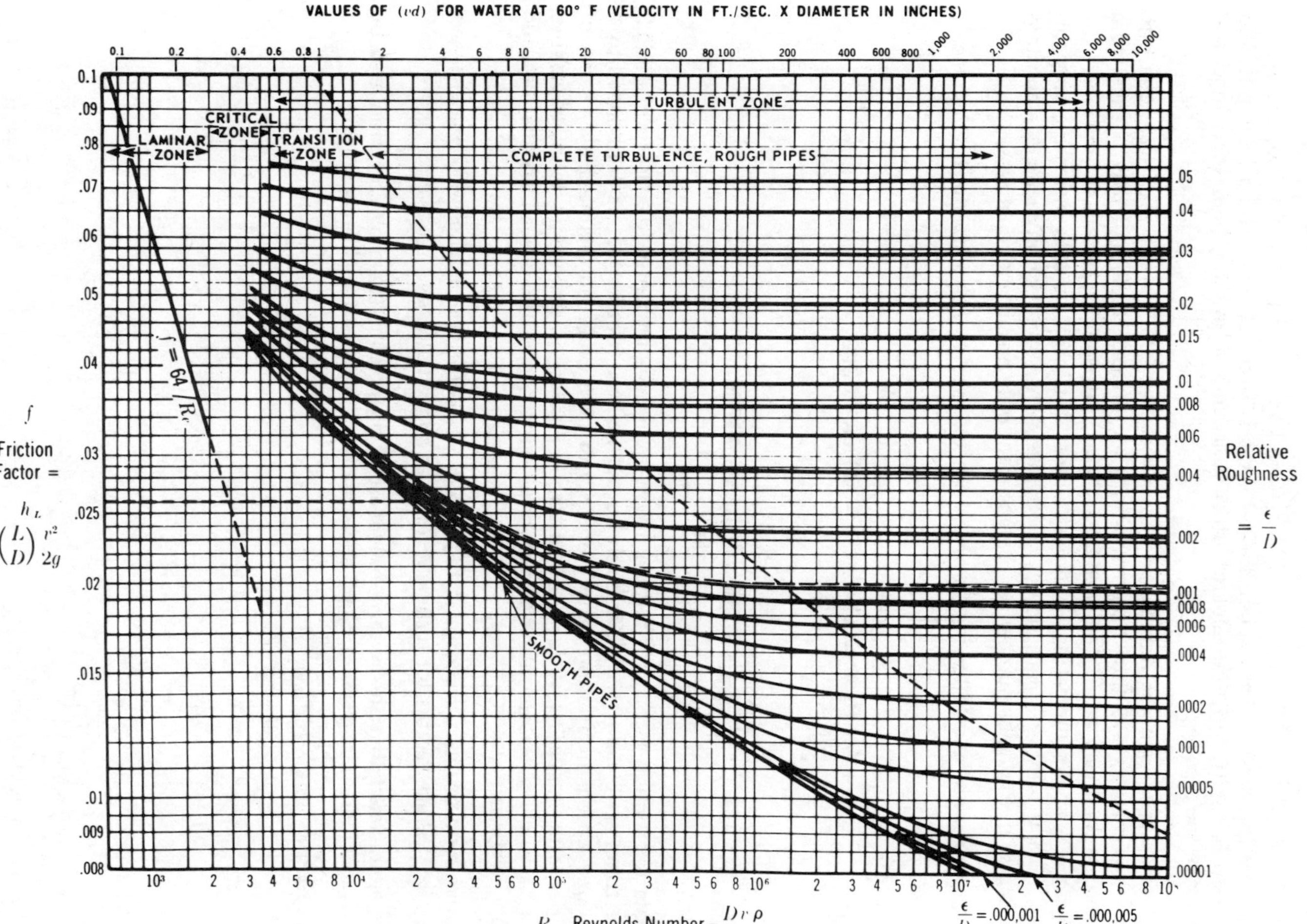

Fig 3 Friction factors for pipes.

Example : friction factor for pipe with relative roughness 0.001 at flow Reynolds Number of 30 000 = 0.026.

Turbulent Flow

In a majority of practical cases of pumped fluids flow is turbulent (*ie* $R_e \geqslant 4\,000$) and simple calculation of flow losses no longer applies. Basically the D'Arcy formula can be used, but with a different friction factor (f_{turb}), the value of which is dependent both on Reynolds number and surface roughness of the pipe bore.

Specifically when the Reynolds number of flow exceeds 2 000 there is a critical zone into which laminar flow may extend (but flow conditions are unpredictable and may change from laminar to turbulent, and vice versa); followed by a region of developing turbulent flow where the friction factor decreases with increasing Reynolds number but increases with increasing bore surface roughness. Finally fully turbulent flow is established, when the friction factor is a constant regardless of increase in Reynolds number and is dependent only on surface roughness.

For *smooth bore pipes* the following formulas can be used to calculate the friction factor directly

$$f_{turb} = \frac{0.3164}{R_e^{0.25}} \quad \text{for } R_e \text{ values between 4000 and 10000}$$

$$f_{turb} = 0.0032 + \frac{0.221}{R_e^{0.237}} \quad \text{for } R_e \text{ values over 10000}$$

Working data for friction factors for turbulent flow are usually presented in chart form – Fig 3 – where these zones are clearly seen. Effectively turbulent flow friction factors are bonded by a lower curve, representing the friction factor for smooth bore pipes, surmounted by a series of curves for pipes with increasing surface roughness. Roughness is defined in terms of *relative roughness* or e/D, where e is *absolute roughness,* or effective height of pipe wall irregularities, and D the pipe bore, in the same units.

An alternative form of chart is shown in Fig 4, together with typical values of e for different pipe materials, from which factors for turbulent flow can be read directly.

It must be appreciated that the relative roughness of pipes is increased by corrosion, also by age, due to encrustations. Data given in Figs 3 and 4 apply to new pipes in clean condition.

Basic formulas then applicable are:-

$$\frac{\Delta P}{L} = f_{turb} \cdot \frac{\rho V^2}{2D}$$

$$\frac{\Delta H}{L} = f_{turb} \cdot \frac{\rho V^2}{2D\omega}$$

where ρ is the fluid density

ω is the fluid specific weight.

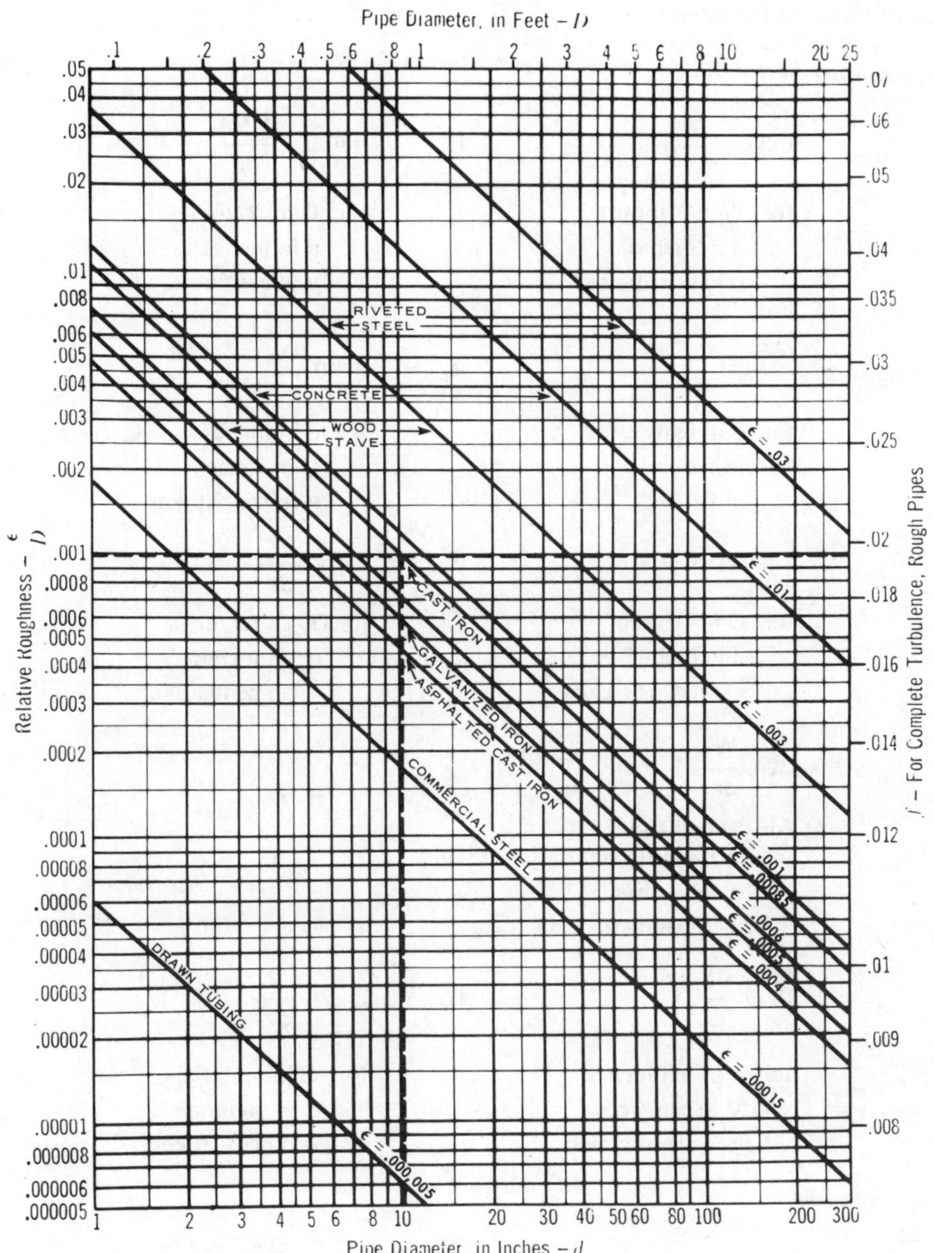

Fig 4 Friction factors and relative roughness for commercial pipes with fully turbulent flow (based on ASME data originated by L.F.Moody).
Example, for 10 inches diameter cast iron pipe, relative roughness (E/D) = 0.00085.
Friction factor = 0.0196

Practical formulas for calculations:

Reynolds number (R_e)

$$R_e = 6.31 \frac{W}{d\mu}$$

for W in lb/hour
d in inches
μ in centipoise

$$R_e = 0.482 \frac{Q \cdot SG}{d\mu}$$

for Q in ft^3/hr
d in inches
μ in centipoise

$$R_e = \frac{dV}{\nu}$$

for d in feet
V in ft/min
ν in ft^2/sec

$$R_e = 7740 \frac{dV}{\nu}$$

for d in inches
V in ft/sec
ν in centistokes

$$R_e = 1\,419\,000 \frac{Q}{d\nu}$$

for Q in ft^3/sec
d in inches
ν in centistokes

$$R_e = 3\,160 \frac{Q}{d\nu}$$

for Q in ft^3/hr
d in inches
ν in centistokes

$$R_e = 394 \frac{W\bar{V}}{d\nu}$$

for W in lb/hour
$\bar{V}$ (specific volume) in ft^3/lb
d in inches
ν in centistokes

$$R_e = 92.9 \frac{dV}{\nu}$$

for d in millinches
V in mm/sec
ν in centistokes

$$R_e = 2\,274 \frac{Q}{d\nu}$$

for Q in mm^3/hr
d in millimetres
ν in centistokes

Flow Velocity (V) :

$$V = 183.3 \frac{Q}{d^2}$$

for Q in ft^3/sec
d in inches

$$V = 0.408 \frac{Q}{d^2}$$

for Q in US gal/min
d in inches

$$V = 0.00389 \frac{Q \cdot SG}{\rho d^2}$$

for Q in ft^3/hr
ρ in lb/ft^3
d in inches
SG = specific gravity

$$V = 0.340 \frac{Q}{d^2}$$

for Q in Imperial gal/min
d in inches

$$V = 3\,350 \frac{Q}{d^2}$$

for Q in m^3/hr
d in millinches

Head Loss (Laminar Flow)

$$H_L(\text{feet}) = 0.0962 \frac{LV\mu}{d^2 \rho}$$

for L in feet
V in ft/sec
μ in centipoise
d in inches
ρ in lb/ft^3

$$H_L(\text{feet}) = 0.0393 \frac{LQ\mu}{d^4 \rho}$$

for L in feet
Q in US gal/min
μ in centipoise
d in inches
ρ in lb/ft^3

$$H_L(\text{feet}) = 0.0049 \frac{LW\mu}{d^4 \rho^2}$$

for L in feet
W in lb/hr
μ in centipoise
d in inches
ρ in lb/ft^3

$$H_L(\text{feet}) = 0.0328 \frac{LQ\mu}{d^4 \rho}$$

for L in feet
Q in Imperial gal/min
μ in centipoise
d in inches
ρ in lb/ft^3

$$H_L(\text{metres}) = 107 \frac{LV\mu}{d^2 \rho}$$

for L in inches
V in mm/sec
μ in centipoise
d in millinches
ρ in tonnes/m^3

$$H_L(\text{metres}) = 2\,670 \frac{LQ\mu}{d^4 \rho}$$

for L in inches
Q in mm^3/bar
d in millimetres
ρ in tonnes/m^3

Pressure Drop (Laminar Flow)

$$\Delta P\ (\text{lb/in}^2) = 0.000668 \frac{LV\mu}{d^2}$$

for L in feet
V in ft/sec
μ in centipoise
d in inches

$$\Delta P\ (\text{lb/in}^2) = 0.1225 \frac{LQ\mu}{d^4}$$

for L in feet
Q in ft^3/sec
μ in centipoise
d in inches

$$\Delta P\ (\text{lb/in}^2) = 0.000034 \frac{LW\mu}{d^4 \rho}$$

for L in feet
W in lb/hr
μ in centipoise
d in inches
ρ in lb/ft^3

$$\Delta P\ (\text{lb/in}^2) = 0.000273 \frac{LQ\mu}{d^4}$$

for L in feet
Q in US gal/min
μ in centipoise
d in inches

$$\Delta P\ (\text{lb/in}^2) = 0.0002275 \frac{LQ\mu}{d^4}$$

for L in feet
Q in Imperial gal/min
d in inches

$$\Delta P\ (\text{bar}) = 0.030 \frac{LV\mu}{d^2}$$

for L in inches
V in inches/sec
d in millinches

$$\Delta P\ (\text{bar}) = 9.05 \frac{LQ\mu}{d^4}$$

for L in inches
Q in litres/min
d in millinches

Head Loss (Turbulent Flow)

$$H_L\ (\text{feet}) = 0.1863\ f \frac{LV^2}{d}$$

where f = friction factor
L is in feet
V is in ft/sec
d is in inches

$$H_L\ (\text{feet}) = 6\,260\ f \frac{LQ^2}{d^5}$$

where f = friction factor
L is in feet
Q is in ft^3/sec
d is in inches

$$H_L \text{(feet)} = 0.0311\ f\ \frac{LQ^2}{d^5}$$

where f = friction factor
L is in feet
Q is in US gal/min
d is in inches

$$H_L \text{(feet)} = 0.026\ f\ \frac{LQ^2}{d^5}$$

where f = friction factor
L is in feet
Q is in Imperial gal/min
d is in inches

$$H_L \text{(feet)} = 0.000483\ f\ \frac{LW^2 \bar{V}^2}{d^5}$$

where f = friction factor
L is in feet
W is in lb/hr
$\bar{V}$ (specific volume) is in ft^3/lb
d is in inches

$$H_L \text{(feet)} = 0.01524\ f\ \frac{LB^2}{d^5}$$

where f = friction factor
L is in feet
B is in barrels (42 US gal) per hour
d is in inches

$$H_L \text{(metres)} = 0.041\ f\ \frac{LV^2}{d}$$

where f = friction factor
L is in metres
V is in metres/sec
d is in millimetres

$$H_L \text{(metres)} = 641\ 270\ f\ \frac{LQ^2}{d^5}$$

where f = friction factor
L is in metres
Q is in litres/min
d is in millimetres

Note: These formulas may be used for both laminar flow and turbulent flow, with appropriate friction factors.

Pressure Drop (Turbulent Flow)

$$\Delta P \text{ (lb/in}^2\text{)} = 0.001\ 294\ f\ \frac{L\rho V^2}{d}$$

where f = friction factor
L is in feet
ρ is in lb/ft^3
V is in ft/sec
d is in inches

$$\Delta P \text{ (lb/in}^2\text{)} = 4\ 315\ f\ \frac{L\rho Q^2}{d^5}$$

where f = friction factor
L is in feet
ρ is in lb/ft^3
Q is in ft^3/sec
d is in inches

$$\Delta P \text{ (lb/in}^2\text{)} = 0.000216\ f\ \frac{L\rho Q^2}{d^5}$$

where f = friction factor
L is in feet
ρ is in lb/ft^3
Q is in US gal/min
d is in inches

$$\Delta P \text{ (lb/in}^2\text{)} = 0.00018\ f\ \frac{L\rho Q^2}{d^5}$$

where f = friction factor
L is in feet
ρ is in lb/ft^3
Q is in Imperial gal/min
d is in inches

$$\Delta P\ (\text{lb/in}^2) = 0.00000336\ f\ \frac{LW^2\bar{V}}{d^5}$$

where f = friction factor
L is in feet
W is in lb/hr
$\bar{V}$ (specific volume) is in ft^3/lb
d is in inches

$$\Delta P\ (\text{lb/in}^2) = 0.0001058\ f\ \frac{L\rho B^2}{d^5}$$

where f = friction factor
L is in feet
ρ is in lb/hr
B is in barrels (42 US gal) per hour
d is in inches

$$\Delta P\ (\text{bar}) = 0.000001125\ f\ \frac{L\rho V^2}{d}$$

where f = friction factor
L is in metres
V is in metres/sec
d is in millimetres
ρ is in tonnes/m^3

$$\Delta P\ (\text{bar}) = 0.1613\ f\ \frac{L\rho Q^2}{d^5}$$

where f = friction factor
L is in metres
ρ is in tonnes/m^3
Q is in litres/min
d is in millinches

Note: These formulas may be used for both laminar flow and turbulent flow, with appropriate friction factors.

Other Formulas

Alternative charts or friction factor data may show substantially different values of friction factor for similar values of relative roughness. This is because the general formula is of the form –

$$\frac{\Delta H}{L} = f \cdot \frac{Q^2}{kD^5}$$

which includes the Reynolds number as a factor. The value of the friction factor (f) derived is thus adjusted accordingly.

Colebrook-White Equation

The Colebrook-White equation for transitional flow is extensively used for determining the hydraulic performance of sewage, drainage and effluent systems.

The general form of the equation is:

$$\sqrt{\frac{1}{\lambda}} = -2\log_{10}\left(\frac{k_s}{3.7D} + \frac{2.51}{Re\sqrt{\lambda}}\right)$$

where λ = friction coefficient $\dfrac{2g\ Di}{V^2}$

k_s = Linear measure of effective roughness (m)
D = Pipe internal diameter (m)
Re = Reynolds number,

$$\frac{VD}{\nu}$$

The equation expressed in engineering terms is:

$$V = -2\sqrt{(2g\ Di)}\ \log_{10}\ \frac{k_s}{3.7D} + \left(\frac{2.51\,\nu}{D\sqrt{(2g\ Di)}}\right)$$

where

V = Velocity (m/s)

g = Gravitational acceleration (9.81 m/sec^2)

i = Hydraulic gradient

ν = Kinematic viscosity of fluid (m^2/sec)

The Hydraulic Research Paper No 4 recommends values for the linear measure of effective roughness for the commonly used pipeline materials in various conditions. The values appropriate to ductile iron pipelines are given in Table VII.

TABLE VII – VALUES OF k_s FOR DUCTILE IRON PIPELINES

Service	Type of Pipe	Approximate Value of k_s (mm)
Raw and potable water pipelines	Coated spun ductile iron	0.06
	Cement mortar lined and bitumen lined spun ductile iron	0.03
Slimed sewers (pipe full roughness on sewers slimed to about half depth and running at velocities around 0.75 m/s	Coated spun ductile iron	6.00
	Cement mortar lined and bitumen lined spun ductile iron	3.00
Sewer rising mains	All types	3.0 at normal operating velocity of 1.1 m/s 1.5 at normal operating velocity of 1.3 m/s 0.6 at normal operating velocity of 1.5 m/s

The values of k_s shown above for raw and potable water pipelines are for pipes in a new condition. A higher value than 0.06 mm is used for coated spun ductile iron where waters are of the type which cause tuberculation.

Generally there is no significant deterioration with time of the linear measure of effective roughness k_s where cement mortar lined or bitumen lined pipes are conveying treated potable waters. However, conveying certain raw waters can lead to a build up of slime in the bores of all pipes and this will cause an increase in the value of k_s. The formation of these slime deposits is not deleterious to the linings and periodic cleaning of this type of main will restore the hydraulic performance virtually to that of the pipeline in its new condition.

The empirical Hazen-Williams formula has the advantage of simplicity, and, for determining the flow of raw or potable water at normal temperatures, it can be relied upon to give results of sufficient accuracy for all practical purposes. It is widely used for calculating flow in raw and potable water pipelines.

The formula can be conveniently expressed as:

$$V = 0.457 \times 10^{-2}\ C\ D^{0.63}\ i^{0.54}$$

or

$$Q = 0.359 \times 10^{-5}\ C\ D^{2.63}\ i^{0.54}$$

where

V	=	Velocity	— m/s
Q	=	Quantity	— l/s
D	=	Pipe internal diameter	— mm
i	=	Hydraulic gradient	— dimensionless
C	=	Hazen-Williams friction coefficient	— dimensionless

TABLE VIII — VALUES OF 'C' FOR DUCTILE IRON PIPELINES

Service	Type of Pipe	Approximate Value of 'C' for Pipe of Nominal Size DN				
		80	150	300	600	1200
Raw and potable water pipelines	Coated spun ductile iron	137	132	145	148	148
	Cement mortar lined and bitumen lined spun ductile iron	147	149	150	152	153

The values shown above are for pipes in a new condition. Lower values than those shown for coated spun ductile iron are used where waters are of the type which cause tuberculation.

By selecting the appropriate value for the coefficient 'C' the Hazen-Williams formula can be used for all types of pipe materials. Table VIII gives values of 'C' for ductile iron pipelines, the values shown are based on case studies and consequently account is automatically taken of losses due to irregularities at the joints.

Generally there is no significant deterioration with time of the friction coefficient 'C' where cement mortar lined or bitumen lined ductile spun iron pipes are conveying treated potable waters. However, conveying certain raw waters can lead to a build up of slime in the bores of all pipes and this will cause a reduction in the value of 'C'. The formation of these slime deposits is not deleterious to the linings and periodic cleaning of this type of main will restore the hydraulic performance virtually to that of the pipeline in its new condition.

The values of 'C' as given in Table VIII are approximately correct at a velocity of 1 m/sec. At other velocities the approximate corrections are given in Table IX.

TABLE IX – CORRECTIONS TO NOMINAL 'C' VALUES

Values of 'C' at 1 m/sec (3 ft/sec approx)	Velocities below 1 m/sec for each halving, re-halving etc, of velocity relative to 1 m/sec	Velocities above 1 m/sec for each doubling, re-doubling etc, of velocity, relative to 1 m/sec
'C' below 100	Add 5% to 'C'	Subtract 5% from 'C'
'C' below 100–130	Add 3% to 'C'	Subtract 3% from 'C'
'C' from 130–140	Add 1% to 'C'	Subtract 1% from 'C'
'C' above 140	Subtract 1% from 'C'	Add 1% to 'C'

Effect of Inclined Flow

Pressure drop (or equivalent head loss) due to flow is the same in straight pipes, whether the pipe run is horizontal, vertical or inclined. With vertical or inclined flow, however, pressure drop is modified by the difference in actual head involved. Here the Bernoulli theorem applies in defining the total energy at any particular point above any arbitrary horizontal datum plane. Total energy is equal to the sum of the elevation head, the pressure head and the velocity head, *ie*

$$\text{Total head } H = Z + \frac{P}{\rho} + \frac{V^2}{2g}$$

The total head (H) will be a constant for any point. Thus if the friction loss between points Z1 and Z2 on an inclined run (Fig 5) is expressed as a head loss Δh.

$$H \text{ at point Z1} = H \text{ at point Z2}$$

Thus

$$Z1 + \frac{P1}{\rho 1} + \frac{V1^2}{2g} = Z2 + \frac{P2}{\rho 2} + \frac{V2^2}{2g} + \Delta h$$

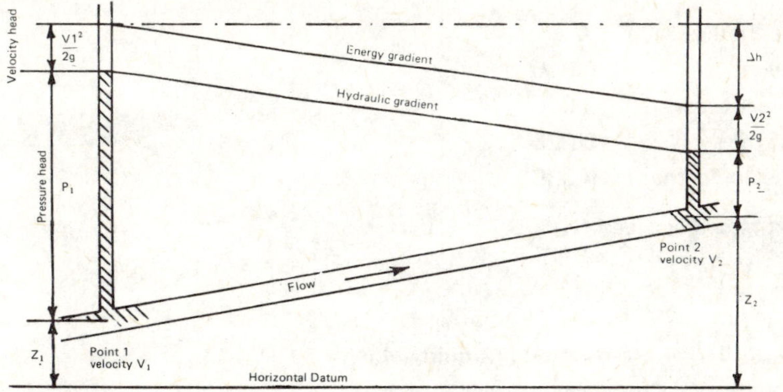

Fig 5

Confusion can be caused by reference to *hydraulic gradient* rather than head or pressure loss. Fundamentally steady flow conditions through a system can be analyzed in terms of a series of

Bernoulli equations appropriate to specific lengths of the system, from which may be derived energy curves and pressure curves, as shown simple in Fig 5. The energy curve, normally referred to as the energy gradient, shows the total energy at any point in the system. The pressure curve, normally referred to as the hydraulic gradient, shows the (pressure) head at any point in the system. The energy gradient will always drop in the direction of flow in the discharge side of the energy into potential energy, *eg* at a sudden expansion. Over a section not subject to changes, *eg* a straight length of pipe, the hydraulic gradient effectively represents the friction head loss and may be referred to as such. The term is rarely used in practical engineering calculations however.

Water Hammer

'Water hammer' is the name given to the distinctive 'knocking' noise which can develop in a closed pipe system when the flow velocity is suddenly changed — *eg* by the sudden opening or closure of a tap, valve or other flow control device. It can also be produced by other factors causing abrupt changes in flow velocity — *eg* sudden starting or stopping of a pump, or abrupt changes in speed of a pump feeding the system. The cause of the 'hammer' is a sudden pressure rise in the liquid, due to the rapid acceleration or deceleration imparted to it, which travels as a pressure wave along the length of pipe, and is reflected backwards and forwards. In addition to generating knocking noises or 'hammer', if the pressure rise is excessive it can cause damage to the piping or system components.

The pressure, velocity and time for the pressure wave to travel from one end of the pipe to the other can be determined from first principles, viz —

$$\text{Velocity of pressure wave (v)} = \frac{4660}{(1 + KD/t)}$$

where D = pipe diameter
t = pipe wall thickness — in same units
K = $\dfrac{\text{elastic modulus for liquid}}{\text{elastic modulus for pipe material}}$

Values of K for water and common pipe materials are

Pipe material	K
Steel	0.010
Wrought iron	0.0107
Cast iron	0.025
Asbestos/cement	0.088

Pressure rise, expressed as Head (H)

$$H = \frac{V\Delta V}{g}$$

where ΔV = reduction in liquid velocity

Time for pressure wave to travel length L

$$t = \frac{2L}{\Delta V} \text{ seconds}$$

Note that L is the length from the appliance concerned (producing the velocity change) to the ends of the pipe.

There are various methods of reducing the intensity of the pressure wave and thus the shock or degree of hammer. Since the magnitude of the pressure wave is directly proportional to reduction in flow velocity, it follows that –

(i) Lowering the flow velocity will be effective, *ie* reducing the flow rate for a given size of pipe or increasing the pipe size for a given flow rate, since ΔV must then be less.

(ii) Decreasing the rate of closure will have a similar effect if the flow stopping time is increased to several times the value of t (times of travel of pressure wave corresponding to instant closure). Surge suppressor valves are usually designed on these lines.

Other methods which can provide a cure for water hammer, rather than designing for the system parameters to avoid hammer, are

(i) air-injection, and

(ii) introducing a flexible element into the system.

Air aspiration is mainly applicable to larger pipelines, the entrained air so supplied acting as a cushion to absorb pressure surges. Air-relief valves can also be installed to relieve air and water during a surge.

Chatter

'Chatter' is rather like 'water hammer' in characteristics but is the result of elasticity in the fluid system. Such elasticity may be introduced by aeration of air entrainment. Under certain conditions axial oscillation of the fluid column may then develop in the system. In the case of high pressure systems, chatter may also arise from a mechanical fault, *eg* seal chatter.

Losses in Bends, Fittings, etc.

PRESSURE LOSSES in a piping system due to changes in the shape of the flow path or changes in cross section as produced by bends, valves, fittings, *etc*, can be evaluated in three different ways:-

(i) As a *resistance coefficient* (K) for the component involved.

(ii) As an *equivalent length* (L/D).

(iii) As a *flow coefficient* (C_V).

Resistance Coefficients

Specifically since a bend, valve or fitting, *etc*, presents additional resistance to flow there is a *velocity head loss* at that point which can be expressed directly as

$$h_L = K \frac{V^2}{2g}$$

where

h_L is the velocity head loss

K is the resistance coefficient for the component involved

Table I gives a range of worked out values.

Pressure drop (ΔP) can also be calculated directly from the resistance coefficient:

$$\Delta P = \frac{KV^2 w}{2g}$$

where

w is the specific weight of the fluid

For clean cold water: ΔP (lb/in^2) = $0.00673\ KV^2$

when V is in ft/sec

ΔP (bar) = $0.000044\ KV^2$

where V is in m/sec

TABLE I – HEAD LOSS IN FEET FROM RESISTANCE COEFFICIENT

Resistance Coefficient	FLOW VELOCITY – FEET PER SECOND (METRES PER SECOND)									
	1 (0.3)	2 (0.6)	3 (0.9)	4 (1.2)	5 (1.5)	6 (1.8	7 (2.1)	8 (2.4)	9 (2.75)	10 (3)
0.05	0.00080	0.003	0.007	0.013	0.019	0.030	0.039	0.050	0.063	0.077
0.1	0.00200	0.006	0.014	0.025	0.039	0.060	0.076	0.100	0.126	0.155
0.2	0.00300	0.012	0.028	0.050	0.078	0.112	0.152	0.199	0.252	0.310
0.3	0.00500	0.019	0.042	0.075	0.117	0.168	0.228	0.299	0.377	0.470
0.4	0.00600	0.025	0.056	0.100	0.155	0.224	0.304	0.398	0.503	0.620
0.5	0.00800	0.031	0.070	0.125	0.194	0.280	0.381	0.497	0.629	0.780
0.6	0.00900	0.037	0.084	0.149	0.233	0.335	0.457	0.597	0.755	0.930
0.7	0.01100	0.044	0.098	0.174	0.271	0.391	0.533	0.696	0.880	1.090
0.8	0.01200	0.050	0.112	0.199	0.311	0.447	0.609	0.796	1.006	1.240
0.9	0.01400	0.056	0.125	0.214	0.349	0.503	0.685	0.895	1.132	1.400
1.0	0.01600	0.062	0.140	0.249	0.388	0.559	0.761	0.995	1.258	1.550
1.1	0.01710	0.068	0.154	0.274	0.427	0.615	0.837	1.090	1.380	1.710
1.2	0.01860	0.074	0.168	0.299	0.466	0.671	0.913	1.190	1.510	1.860
1.3	0.02020	0.081	0.182	0.324	0.505	0.727	0.989	1.290	1.630	2.020
1.4	0.02170	0.087	0.196	0.349	0.544	0.783	1.065	1.390	1.760	2.170
1.5	0.02330	0.093	0.210	0.374	0.583	0.839	1.141	1.490	1.890	2.330
1.6	0.02480	0.099	0.224	0.398	0.621	0.895	1.217	1.590	2.020	2.480
1.7	0.02640	0.106	0.238	0.423	0.660	0.951	1.293	1.690	2.140	2.640
1.8	0.02800	0.112	0.252	0.448	0.699	1.007	1.369	1.790	2.270	2.800
1.9	0.02950	0.118	0.266	0.473	0.738	1.063	1.445	1.890	2.390	2.950
2.0	0.03106	0.124	0.280	0.497	0.776	1.118	1.522	1.990	2.520	3.110
2.1	0.03261	0.130	0.294	0.522	0.815	1.174	1.598	2.090	2.650	3.260
2.2	0.03416	0.136	0.308	0.547	0.854	1.230	1.674	2.190	2.780	3.420
2.3	0.03571	0.143	0.322	0.572	0.893	1.286	1.750	2.290	2.900	3.570
2.4	0.03726	0.149	0.336	0.597	0.932	1.341	1.826	2.390	3.030	3.730
2.5	0.03881	0.155	0.350	0.621	0.971	1.397	1.902	2.490	3.150	3.880
2.6	0.04036	0.161	0.364	0.646	1.009	1.453	1.978	2.590	3.280	4.040
2.7	0.04191	0.168	0.378	0.671	1.048	1.509	2.054	2.690	3.400	4.190
2.8	0.04346	0.174	0.392	0.696	1.087	1.565	2.130	2.790	3.530	4.350
2.9	0.04501	0.180	0.406	0.721	1.126	1.621	2.206	2.890	3.650	4.500
3.0	0.04659	0.186	0.419	0.746	1.165	1.677	2.283	2.990	3.770	4.660
3.1	0.04814	0.192	0.433	0.771	1.204	1.733	2.359	3.090	3.900	4.810
3.2	0.04969	0.198	0.447	0.796	1.243	1.789	2.435	3.190	4.030	4.970
3.3	0.05124	0.205	0.461	0.821	1.282	1.845	2.511	3.290	4.150	5.120
3.4	0.05279	0.211	0.475	0.845	1.321	1.901	2.587	3.390	4.280	5.280
3.5	0.05434	0.217	0.489	0.870	1.360	1.957	2.663	3.490	4.400	5.430
3.6	0.05589	0.223	0.503	0.895	1.398	2.013	2.739	3.590	4.530	5.590
3.7	0.05744	0.230	0.517	0.920	1.437	2.069	2.815	3.680	4.650	5.740
3.8	0.05899	0.236	0.531	0.945	1.476	2.125	2.891	3.780	4.780	5.900
3.9	0.06054	0.242	0.545	0.970	1.514	2.181	2.967	3.880	4.900	6.050
4.0	0.06212	0.248	0.559	0.994	1.553	2.236	3.044	3.980	5.030	6.210
4.1	0.06367	0.254	0.573	1.019	1.592	2.292	3.120	4.080	5.160	6.370
4.2	0.06522	0.260	0.587	1.044	1.631	2.348	3.196	4.180	5.290	6.520
4.3	0.06677	0.267	0.601	1.069	1.670	2.404	3.272	4.280	5.410	6.680
4.4	0.06832	0.273	0.615	1.093	1.709	2.460	3.348	4.380	5.540	6.830
4.5	0.06987	0.279	0.629	1.118	1.748	2.516	3.424	4.480	5.660	6.990
4.6	0.07142	0.285	0.643	1.143	1.786	2.572	3.500	4.580	5.790	7.140
4.7	0.07297	0.292	0.657	1.168	1.825	2.628	3.576	4.680	5.910	7.300
4.8	0.07452	0.298	0.671	1.193	1.864	2.684	3.652	4.770	6.040	7.450
4.9	0.07607	0.304	0.685	1.218	1.903	2.740	3.728	4.870	6.160	7.600

Footnote: For head loss in metres, factor by 0.3

TABLE I – HEAD LOSS IN FEET FROM RESISTANCE COEFFICINT (contd)

Resistance Coefficient	FLOW VELOCITY – FEET PER SECOND (METRES PER SECOND)									
	1 (0.3)	2 (0.6)	3 (0.9)	4 (1.2)	5 (1.5)	6 (1.8)	7 (2.1)	8 (2.4)	9 (2.75)	10 (3)
5.0	0.07765	0.311	0.698	1.243	1.942	2.795	3.806	4.97	6.29	7.77
5.1	0.07920	0.317	0.712	1.268	1.981	2.851	3.882	5.07	6.41	7.92
5.2	0.08075	0.323	0.726	1.293	2.020	2.907	3.958	5.17	6.54	8.08
5.3	0.08230	0.330	0.740	1.318	2.058	2.963	4.034	5.27	6.66	8.23
5.4	0.08388	0.336	0.755	1.342	2.097	3.019	4.110	5.37	6.79	8.39
5.5	0.08543	0.342	0.769	1.367	2.136	3.075	4.186	5.47	6.91	8.54
5.6	0.08698	0.348	0.783	1.392	2.175	3.131	4.262	5.57	7.04	8.70
5.7	0.08853	0.355	0.797	1.417	2.213	3.187	4.338	5.67	7.17	8.85
5.8	0.09008	0.361	0.811	1.442	2.252	3.243	4.414	5.77	7.30	9.01
5.9	0.09163	0.367	0.825	1.467	2.290	3.299	4.490	5.87	7.42	9.16
6.0	0.09318	0.373	0.839	1.491	2.329	3.354	4.567	5.97	7.55	9.32
6.1	0.09473	0.379	0.853	1.516	2.368	3.410	4.643	6.07	7.68	9.47
6.2	0.09628	0.385	0.867	1.541	2.407	3.466	4.719	6.17	7.81	9.63
6.3	0.09783	0.392	0.881	1.566	2.446	3.522	4.795	6.27	7.93	9.78
6.4	0.09938	0.398	0.895	1.590	2.485	3.572	4.871	6.37	8.06	9.94
6.5	0.10093	0.404	0.909	1.615	2.524	3.634	4.947	6.47	8.18	10.09
6.6	0.10248	0.410	0.923	1.640	2.562	3.690	5.023	6.57	8.31	10.25
6.7	0.10403	0.417	0.937	1.665	2.601	3.746	5.099	6.67	8.43	10.40
6.8	0.10558	0.423	0.951	1.690	2.640	3.802	5.175	6.76	8.56	10.56
6.9	0.10713	0.429	0.965	1.715	2.679	3.858	5.251	6.86	8.68	10.71
7.0	0.10871	0.435	0.979	1.740	2.717	3.913	5.328	6.96	8.80	10.87
7.1	0.11026	0.441	0.993	1.765	2.756	3.969	5.404	7.06	8.93	11.03
7.2	0.11181	0.447	1.007	1.790	2.795	4.025	5.480	7.16	9.06	11.18
7.3	0.11336	0.454	1.021	1.814	2.834	4.081	5.556	7.26	9.18	11.34
7.4	0.11491	0.460	1.035	1.839	2.873	4.137	5.632	7.36	9.31	11.49
7.5	0.11646	0.466	1.049	1.864	2.912	4.193	5.708	7.46	9.43	11.65
7.6	0.11801	0.472	1.063	1.888	2.950	4.249	5.784	7.56	9.56	11.80
7.7	0.11956	0.479	1.077	1.913	2.989	4.305	5.860	7.66	9.68	11.96
7.8	0.12111	0.485	1.091	1.938	3.028	4.361	5.936	7.76	9.81	12.11
7.9	0.12266	0.491	1.105	1.963	3.066	4.417	6.012	7.86	9.93	12.27
8.0	0.12424	0.497	1.118	1.988	3.105	4.472	6.089	7.96	10.06	12.42
8.1	0.12579	0.503	1.132	2.013	3.144	4.528	6.165	8.06	10.19	12.58
8.2	0.12734	0.509	1.146	2.038	3.183	4.584	6.241	8.16	10.32	12.73
8.3	0.12889	0.516	1.160	2.063	3.221	4.640	6.317	8.26	10.44	12.89
8.4	0.13044	0.522	1.174	2.087	3.260	4.696	6.393	8.36	10.57	13.04
8.5	0.13199	0.528	1.188	2.112	3.299	4.752	6.469	8.46	10.69	13.20
8.6	0.13354	0.534	1.202	2.137	3.338	4.808	6.545	8.55	10.81	13.35
8.7	0.13509	0.541	1.216	2.162	3.377	4.864	6.621	8.65	10.94	13.51
8.8	0.13664	0.547	1.230	2.187	3.416	4.920	6.697	8.75	11.07	13.66
8.9	0.13819	0.553	1.244	2.212	3.455	4.976	6.773	8.85	11.19	13.82
9.0	0.13977	0.559	1.258	2.237	3.493	5.031	6.850	8.95	11.32	13.98
9.1	0.14132	0.565	1.272	2.262	3.532	5.087	6.926	9.05	11.45	14.13
9.2	0.14287	0.571	1.286	2.287	3.571	5.143	7.002	9.15	11.58	14.29
9.3	0.14442	0.578	1.300	2.312	3.609	5.199	7.078	9.25	11.70	14.44
9.4	0.14597	0.584	1.314	2.336	3.648	5.255	7.154	9.35	11.83	14.60
9.5	0.14752	0.590	1.328	2.361	3.687	5.311	7.230	9.45	11.95	14.75
9.6	0.14907	0.596	1.342	2.386	3.726	5.367	7.306	8.55	12.08	14.91
9.7	0.15062	0.603	1.356	2.411	3.765	5.423	7.382	9.65	12.20	15.06
9.8	0.15217	0.609	1.370	2.435	3.804	5.479	7.458	9.75	12.33	15.22
9.9	0.15372	0.615	1.384	2.460	3.843	5.534	7.534	9.85	12.45	15.37
10.0	0.15530	0.621	1.398	2.485	3.882	5.589	7.612	9.94	12.58	15.53

Footnote: For head loss in metres, factor by 0.3

cont...

TABLE I – HEAD LOSS IN FEET FROM RESISTANCE COEFFICIENT (contd)

Resistance Coefficient	FLOW VELOCITY – FEET PER SECOND (METRES PER SECOND)									
	1 (0.3)	2 (0.6)	3 (0.9)	4 (1.2)	5 (1.5)	6 (1.8)	7 (2.1)	8 (2.4)	9 (2.75)	10 (3)
11	0.17083	0.68	1.54	2.74	4.27	6.15	8.37	10.9	13.8	17.08
12	0.18636	0.74	1.68	2.99	4.66	6.71	9.13	11.9	15.1	18.64
13	0.20189	0.81	1.82	3.24	5.05	7.27	9.89	12.9	16.3	20.19
14	0.21742	0.87	1.96	3.49	5.44	7.83	10.65	13.9	17.6	21.74
15	0.23295	0.93	2.10	3.74	5.83	8.39	11.41	14.9	18.9	23.30
16	0.24848	0.99	2.24	3.98	6.21	8.95	12.17	15.9	20.2	24.85
17	0.26401	1.06	2.38	4.23	6.60	9.51	12.93	16.9	21.4	26.40
18	0.17954	1.12	2.52	4.48	6.99	10.07	13.69	17.9	22.7	27.95
19	0.29507	1.18	2.66	4.73	7.38	10.63	14.45	18.9	23.9	29.51
20	0.31060	1.24	2.80	4.97	7.76	11.18	15.22	19.9	25.2	31.06

Footnote: For head loss in metres, factor by 0.3

The value of R for a given component is independent of friction factor or Reynolds number and is constant for all conditions of flow, including laminar flow. Equally, in theory at least, it should be the same for all sizes of a given component with similar geometry. In practice this is not so. Resistance coefficients are necessarily determined empirically, and can vary widely with pipe size. Some typical values are summarized in Table II.

Equivalent Length L/D

Equating velocity head loss derived from the Darcy formula with that derived from the resistance coefficient formula

$$(f \cdot L/D)\ \frac{V^2}{2g} = K \cdot \frac{V^2}{2g}$$

It follows that;

$$K = f \cdot L/D$$

If K is a constant for all conditions of flow (but modified by geometric dissimilarities in different sizes, as above), the value of L/D for any given component must vary inversely with the change in friction factor for different flow conditions. This rules out the effect of geometric dissimilarities in the sizes of valves, fittings, *etc*, of the same basic geometry. In other words the equivalent length or L/D value for any particular valve, fittings, *etc,* is a constant *for the same flow conditions,* and is valid for all sizes of that particular component – see Tables III and IV, and Fig 1.

TABLE II – RESISTANCE COEFFICIENT OF FITTINGS

FITTING	PIPE DIAMETER – INCHES (MILLIMETRES)									
	3/8 (10)	½ (12.5)	¾ (20)	1 (25)	1¼ (35)	1½ (40)	2 (50)	3 (75)	4 (100)	5 (125)
Integral pipe bend (turbulent flow) bend 3 x D			Approximately 0.04 (all sizes)							
Integral pipe bend (turbulent flow) bend 4 x D			Approximately 0.025 (all sizes)							
Integral pipe bend (laminar flow) bend 3 x D			Approximately 0.1 (all sizes)							
Integral pipe bend (laminar flow) bend 5 x D			Approximately 0.06 (all sizes)							
Integral pipe bend (laminar flow) bend 10 x D			Approximately 0.04 (all sizes)							
Standard 90 deg elbow (screwed)	2.40	2.10	1.70	1.50	1.25	1.15	1.00	0.80	0.70	0.55
Standard 90 deg elbow (flanged)	–	–	–	0.45	–	0.40	0.38	0.34	0.32	0.30
Large radius 90 deg elbow (screwed)	–	1.00	0.90	0.75	0.60	0.50	0.40	0.30	0.25	0.20
Large radius 90 deg elbow (flanged)	–	–	–	0.40	–	0.34	0.30	0.25	0.21	0.18
Standard 45 deg elbow (screwed)	0.39	0.37	0.36	0.35	0.34	0.32	0.30	0.29	0.28	0.27
Standard 45 deg elbow (flanged)	0.37	0.35	0.34	0.33	0.32	0.31	0.29	0.28	0.27	0.26
Large radius 45 deg elbow (screwed)	–	–	–	0.24	0.23	0.22	0.21	0.20	0.19	0.18
Large radius 45 deg elbow (flanged)	–	–	–	0.22	0.21	0.21	0.20	0.18	0.18	0.17
Return bend 180 deg (screwed)	2.40	2.10	1.70	1.50	1.38	1.25	0.96	0.78	0.68	0.58
Return bend 180 deg (flanged)	–	–	–	0.43	0.40	0.37	0.34	0.32	0.30	0.28
Large radius return bend 180 deg (screwed)	–		–	0.80	0.70	0.60	0.50	0.40	0.35	0.30
Large radius return bend 180 deg (flanged)	–	–	–	0.42	0.39	0.36	0.30	0.25	0.22	0.20
Tee – line flow (screwed)			Approximately 0.9 (all sizes)							
Tee – line flow (flanged)	–	–	–	0.26	0.24	0.22	0.18	0.16	0.14	0.13
Tee – branch flow (screwed)	2.50	2.40	2.00	1.80	1.60	1.50	1.40	1.20	1.10	1.00
Tee – branch flow (flanged)	–	·	–	1.00	–	0.90	0.80	0.72	0.64	0.62
Screw-down valve (straight)	–	12.00	–	10.00	–	–	8.00	7.00	6.00	5.00
Screw-down valve (right angle flow)	–	6.00	–	5.00	–	–	4.00	3.50	3.00	2.50
Gate valve (typical) (screwed)	–	–	–	–	0.23	0.20	0.17	0.15	0.13	0.11
Gate valve (typical) (flanged)	–	–	–	–	–	–	0.30	0.23	0.15	0.13
Gate valve – ¼ closed			0.8 to 0.2 this range							
Gate valve – ½ closed			4.0 to 0.8 this range							
Gate valve – ¾ closed			16.0 to 2.0 this range							
Globe valve (screwed)	–	15.00	–	12.50	–	–	8.50	7.50	6.50	6.00
Globe valve (flanged)	–	–	–	12.50	–	–	8.50	7.50	6.50	6.00
Swing check valve (screwed)	–	5.00	–	3.00	–	–	2.00	2.00	2.00	2.00
Swing check valve (flanged)			Typically 2.0 (all sizes)							
Foot valve			Typically 0.8 (all sizes)							
Basket strainer			Typically 1.5 to 1.0 this range							

TABLE III – EQUIVALENT STRAIGHT LENGTHS OF FITTINGS IN FEET

FITTING	PIPE DIAMETER – INCHES (MILLIMETRES)									
	6 (150)	8 (200)	10 (250)	12 (300)	14 (350)	161 (400)	18 (450)	20 (500)	24 (600)	36 (900)
Welded 90 deg L-Bends R/D = 0.5	19	25	32	28	44	50	56	–	–	–
R = Bend Radius 1.0	8	11	14	17	20	23	26	–	–	–
D = Pipe diameter 1.5	6	8	10	12	14	16	18	–	–	–
2.0	4	6	8	10	12	14	16	–	–	–
3.0	4	6	7	9	11	13	15	–	–	–
Standard 90 deg Elbow (Screwed)	12	16	–	–	–	–	–	–	–	–
Standard 90 deg Elbow (Flanged)	12	16	20	24	28	32	36	40	48	72
Large Radius 90 deg Elbow (Screwed)	–	–	–	–	–	–	–	–	–	–
Large Radius 90 deg Elbow (Flanged)	9	12	15	18	21	24	27	30	36	55
Standard 45 deg Elbow (Screwed)	–	–	–	–	–	–	–	–	–	–
Standard 45 deg Elbow (Flanged)	3	4	5	6	7	8	9	10	12	18
Large Radius 45 deg Elbow (Screwed)	–	–	–	–	–	–	–	–	–	–
Large Radius 45 deg Elbow (Flanged)	2	2.5	3.3	4	4.5	5.5	6	6.5	8	12
Return Bend 180 deg	36	47	62	73	–	–	–		–	–
Large Radius Return Bend 180 deg	19	26	33	39	–	–	–	–	–	–
Tee – Line Flow	30	40	50	60	70	80	90	100	120	180
Tee – Branch Flow	150	200	250	300	350	400	440	480	570	850
Cast Elbow (90 deg) Standard	16	20	26	32	36	42	46	52	63	94
Cast Elbow (90 deg) Large Radius	11	14	17	20	23	27	30	34	40	60
Screw-down Valve (Straight)	156	208	260	310	363	415	467	519	622	934
Screw-down Valve (Right Angled Run)	75	100	125	150	175	200	225	250	300	450
Gate Valve (Typical) (Flanged)	3.5	4.5	5.7	6.7	8.0	9.0	10.2	12.0	14.0	20.0
Gate Valve – ¼ Closed	19	26	33	39	–	–	–	–	–	–
Gate Valve – ½ Closed	100	130	160	190	–	–	–	–	–	–
Gate Valve – ¾ Closed	400	540	700	800	–	–	–	–	–	–
Globe Valve (Flanged)	160	214	267	320	373	427	480	534	640	960
Swing Check Valve (Screwed)	–	–	–	–	–	–	–	–	–	–
Swing Check Valve (Flanged)	40	53	67	80	93	107	120	134	160	240
Foot Valve (Typical)	12	16	20	24	28	32	36	40	48	72
Basket Strainer – (Typical)	11	14	16	18	21	24	30	35	–	–
Sudden Enlargement d/D = ¼	16	20	26	31	36	42	–	–	–	–
d/D = ½	11	15	18	22	25	29	–	–	–	–
d/D = ¾	3	4	6	7	8	10	–	–	–	–
Sudden Contraction d/D = ½	6	8	10	12	14	16	18	20	24	36
Entrance (Typical)	9	12	15	18	21	24	27	30	–	–

(d = Smaller pipe diameter
D = Larger pipe diameter)

TABLE IV – EQUIVALENT STRAIGHT LENGTHS OF PIPE IN METRES

Fitting		Fitting Diameter - mm															
		1000	800	600	500	400	300	200	150	100	80	65	50	40	32	25	15
Gate valve	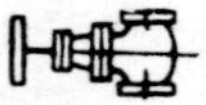	6	5	4	3	2.7	2.2	1.5	1	0.7	0.5	0.43	0.35	0.27	0.2	0.18	0.1
Non-return flap valve		110-200	90-170	70-130	60-110	50-90	35-70	25-50	20.35	13-25	10.20	8-15	6-12	5-9	4.7	3.6	2.4
Screw down valve	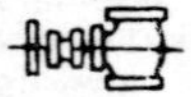	300	250	200	160	130	100	70	50	35	28	22	17	13	10	8	5
Screw down valve, Rt-angled		150	130	90	80	60	50	32	25	16	13	10	8	7	5	4	2.5
Bends and Elbows		18	15	12	10	8	6	4	3	2	1.7	1.4	1	0.8	0.6	0.5	0.3
		25	20	15	13	10	8	5	4	2.8	2	1.8	1.5	1	0.8	0.7	0.4
		12	10	7	6	5	4	2.5	2	1.5	1	0.9	0.7	0.5	0.4	0.3	0.2
		30	25	18	15	13	10	6.5	5	3.2	2.5	2	1.8	1.4	1	0.8	0.5
		75	60	50	40	33	25	17	13	8	6.5	5.5	4	3.2	2.6	2	1.4
Tees	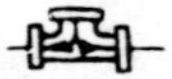	100	80	60	50	40	30	20	15	10	8	7	5	4	3.2	2.7	1.7
		70	58	45	35	30	24	15	12	8	6	5	4	3	2.5	2	1.2
		18	15	12	10	8	6	4	3	2	1.7	1.4	1	0.8	0.6	0.5	0.3

Taper connectors	$d/D = 3/4$	25	20	16	14	11	8	5.5	4	3	2.5	2	1.5	1.3	0.9	0.7	0.4
	$d/D = 1/2$	30	25	20	16	13	10	7	5	3.5	2.8	2.2	1.7	1.4	1	0.8	0.5
Abrupt 90° bend		60	50	40	35	28	20	14	10	7	5.5	4.4	3.5	3	2	1.8	1
Abrupt changes of section	$d/D = 3/4$	6	5	4	3.5	3	2	1.5	1	0.7	0.6	0.5	0.37	0.3	0.24	0.18	0.11
	$d/D = 1/4$	13	10	8	7	5	4	3	2	1.5	1.2	0.9	0.7	0.55	0.45	0.35	0.2
	$d/D = 1/2$	10	8	6	5	4	3.2	2.2	1.6	1.1	0.9	0.7	0.55	0.45	0.35	0.27	0.17
	$d/D = 1/4$	30	25	20	16	13	10	7	5	3.5	2.8	2.2	1.7	1.4	1	0.8	0.5
	$d/D = 1/2$	15	13	10	9	7	5.5	3.5	3	2	1.5	1.3	0.9	0.7	0.6	0.5	0.3
		15	13	10	8	7	5	3.5	2.5	1.8	1.5	1.2	0.9	0.7	0.5	0.4	0.25

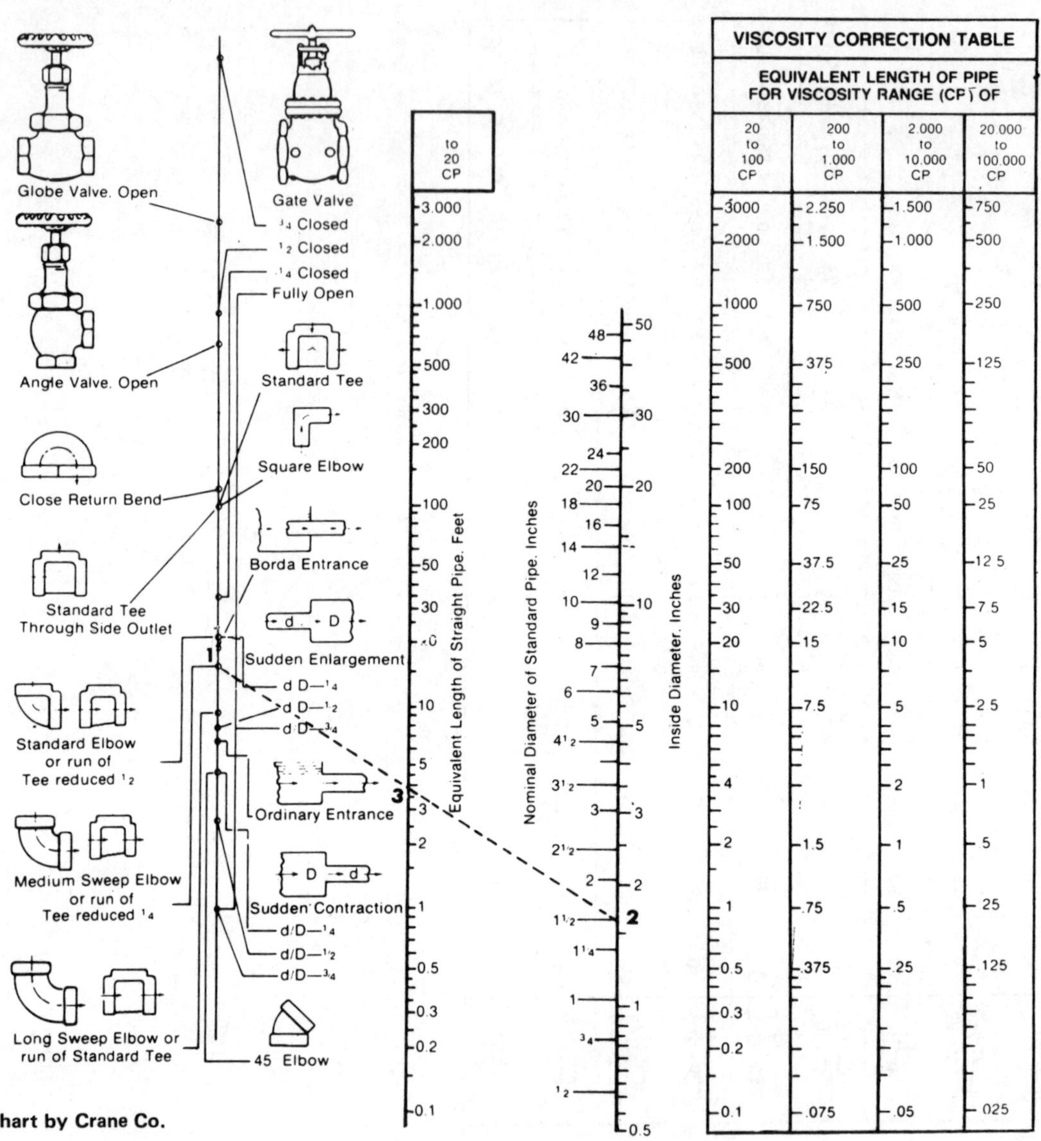

VISCOSITY CORRECTION TABLE			
EQUIVALENT LENGTH OF PIPE FOR VISCOSITY RANGE (CP) OF			
20 to 100 CP	200 to 1.000 CP	2.000 to 10.000 CP	20.000 to 100.000 CP
3000	2.250	1.500	750
2000	1.500	1.000	500
1000	750	500	250
500	375	250	125
200	150	100	50
100	75	50	25
50	37.5	25	12 5
30	22.5	15	7 5
20	15	10	5
10	7.5	5	2 5
4		2	1
2	1.5	1	5
1	.75	.5	25
0.5	.375	.25	.125
0.3			
0.2			
0.1	.075	.05	.025

Chart by Crane Co.

Fig 1 Friction loss in valves and fittings.

Flow Coefficient C_V

For valves (and particularly control valves) it is often more convenient to express the capacity and flow characteristics in terms of a *flow coefficient* (C_V), defined as the flow of water at 60°F in US gallons/minute at a pressure drop of 1 lb/in^2 across the valve.

Note: this definition of C_V based on the US gallon and pressure drop in lb/in^2 is used both in countries employing English units and those normally employing metric units. The metric 'equivalent' is called the *flow factor,* designated K_V and defined as the number of cubic metres per hour of water at 20°C which will flow with a pressure drop of 1 kg/cm^2 (1 bar). This gives somewhat different values for the same condition. Equivalents are:

$$K_V = 0.853\ C_v$$
$$C_v = 1.16\ K_V$$

In terms of the Darcy equation:

$$C_v = \frac{29.9d^2}{\sqrt{f \cdot L/D}} = \frac{29.9d^2}{\sqrt{K}}$$

The flow through a valve handling liquids with a reasonably similar viscosity to that of water can then be determined as

$$Q = C_v \sqrt{\Delta P \frac{62.4}{w}}$$

$$= 7.9C_v \sqrt{\frac{\Delta P}{w}} \quad \text{or} = C_v \sqrt{\frac{\Delta P}{sg}}$$

where

w is the specific weight of the fluid (lb/ft^3)

sg is the specific gravity of the fluid

ΔP is the pressure drop across the valve

The pressure drop across a valve can be determined from the same formulas rearranged.

$$\Delta P = \frac{w}{62.4}\left(\frac{Q}{C_v}\right)^2 = sg\left(\frac{Q}{C_v}\right)^2$$

Working Formulas

These formulas are expressed in terms of the *resistance coefficient* (K) where $K = f\,\frac{L}{D}$

Head Loss Through Valves and Fittings:

$$H_L\ (\text{feet}) = 522\,\frac{KQ^2}{d^4}$$

where

Q is in ft^3/sec

d is in inches

$$H_L \text{ (feet)} = 0.00127 \frac{KB^2}{d^4}$$

where
B is in barrels (42 US gallons) per hour
d is in inches

$$H_L \text{ (metres)} = 1877197 \frac{KQ^2}{d^4}$$

where
Q is in m^3/sec
d is in millimetres

$$H_L \text{ (metres)} = 6852 \frac{KQ^2}{d^4}$$

where
Q is in litres/min
d is in millimetres

$$H_L \text{ (feet)} = 0.00259 \frac{KQ^2}{d^4}$$

where
Q is in US gal/min
d is in inches

$$H_L \text{ (feet)} = 0.00216 \frac{KQ^2}{d^4}$$

where
Q is in Imperial gal/min
d is in inches

$$H_L \text{ (feet)} = 0.0000403 \frac{KW^2 \bar{V}^2}{d^4}$$

where
W is in lb/bar
$\bar{V}$ (specific volume) is in ft^3/lb

Pressure Drop Through Valves and Fittings (using resistance coefficient K)

$$\Delta P \text{ (lb/in}^2\text{)} = 0.0001078\, K\rho V^2$$

where
ρ is in lb/ft^3
V is in ft/sec

$$\Delta P \text{ (lb/in}^2\text{)} = 3.62 \frac{K\rho Q^2}{d^4}$$

where
ρ is in lb/ft^3
Q is in ft^3/sec
d is in inches

$$\Delta P \text{ (lb/in}^2\text{)} = 0.00000882 \frac{K\rho B^2}{d^4}$$

where
ρ is in lb/ft^3
B is in barrels (42 US gal) per hour
d is in inches

$$\Delta P \text{ (bar)} = 0.000000112\, K\rho V^2$$

where
ρ is in tonnes/m^3
V is in metres/sec

$$\Delta P \text{ (bar)} = 0.1729 \frac{K\rho Q^2}{d^4}$$

where
ρ is in tonnes/m^3
Q is in litres/min
d is in millimetres

$$\Delta P \text{ (lb/in}^2\text{)} = 0.00000003\, K\rho V^2$$

where
ρ is in lb/ft^3
V is in ft/min

$$\Delta P \text{ (lb/in}^2\text{)} = 0.000018 \frac{K\rho Q^2}{d^4}$$

where
ρ is in lb/ft^3
Q is in US gal/min
d is in inches

$$\Delta P \text{ (lb/in}^2\text{)} = 0.000015 \frac{K\rho Q^2}{d^4}$$

where
ρ is in lb/ft^3
Q is in Imperial gal/min
d is in inches

$$\Delta P\ (\text{lb/in}^2) = 0.00000028\ \frac{KW^2\bar{V}}{d^4}$$

where
W is in lb/bar
$\bar{V}$ (specific volume) is in ft³/lb

Pressure Drop Through Valves and Fittings (using flow coefficient C_V)

$$\Delta P\ (\text{lb/in}^2) = \frac{\rho Q^2}{62.4\ C_V^2}$$

where
Q is in lb/ft³
ρ is in lb/ft³
Q is in US gal/min

$$\Delta P\ (\text{bar}) = \frac{\rho Q^2}{223\ C_V}$$

where
ρ is in tonnes/m³
Q is in litres/min

$$\Delta P\ (\text{lb/in}^2) = \frac{\rho Q^2}{90\ C_V^2}$$

where
ρ is in lb/ft³
Q is in Imperial gal/min

Discharge Through Pipes and Fittings

$$Q\ (\text{ft}^3/\text{sec}) = 0.0438\ d^2\sqrt{\frac{H_L}{K}}$$

$$= 0.525\ d^2\sqrt{\frac{\Delta P}{K\rho}}$$

where
d is in inches
H_L is in feet
ΔP is in lb/in²
ρ is in lb/ft³

$$Q\ (\text{US gal/min}) = 19.65\ d^2\sqrt{\frac{H_L}{K}}$$

$$= 236\ d^2\sqrt{\frac{\Delta P}{K\rho}}$$

where
d is in inches
H_L is in feet
ΔP is in lb/in²
ρ is in lb/ft³

$$Q\ (\text{Imperial gal/min}) = 16.375\ d^2\sqrt{\frac{H_L}{K}}$$

$$= 197\ d^2\sqrt{\frac{\Delta P}{K\rho}}$$

where
d is in inches
H_L is in feet
ΔP is in lb/in²
ρ is in lb/ft³

$$Q\ (\text{lb/hr}) = 197.6\ \rho\ d^2\sqrt{\frac{H_L}{K}}$$

$$= 1891\ d^2\sqrt{\frac{\Delta P\rho}{K}}$$

where
d is in inches
ρ is in lb/ft³
H_L is in feet
ΔP is in lb/in²

$$Q\ (\text{litres/min}) = 0.1357\ d^2\sqrt{\frac{H_L}{K}}$$

$$= 6.06\ d^2\sqrt{\frac{\Delta P}{K\rho}}$$

where
d is in millimetres
H_L is in metres
ΔP is in bar
ρ is in tonnes/m³

None of these formulas is strictly valid for predicting valve performance with viscous fluids or compressible fluids (*ie* air or gases).

Losses in Bends

Losses in bends are difficult to evaluate other than on purely empirical lines. Attempts to rationalize resistance coefficients in terms of relative radius (ratio of bend radius to internal pipe diameter) are generally unsuccessful, but values are fairly well established for standard bends, including mitre bends. Alternative data are presented in terms of equivalent length or L/D. See Tables III and IV and Fig 1.

Contractions and Enlargements

Flow through gradual changes in pipe sections (contractions or enlargements) can be analyzed from first principles using the Bernoulli equation (Bernoulli constant), with the subsequent introduction of an empirical coefficient to take into account frictional losses introduced with a real fluid. In problems involving closed circuits the potential head can normally be ignored in view of the inherently higher value of velocity and frictional head so that the original equation can be reduced to

$$\frac{P}{\rho} + \frac{V^2}{2g} = \text{a constant}$$

This is often more convenient to use in the form

$$V^2 + \frac{2Pg}{\rho} = \text{a constant}$$

Applying this condensed equation to steady frictionless flow along a length of horizontal pipe which contracts in cross section, the following applies:-

$$V_1{}^2 + \frac{2P_1 g}{\rho} = V_2{}^2 + \frac{2P_2 g}{\rho}$$

ie $$P_1 - P_2 = \frac{\rho}{2g}(V_2{}^2 - V_1{}^2)$$

This form of equation also shows that a steadily contracting section followed by a widening section can be used as a principle for flow velocity measurement by measuring the difference in pressure at two extreme sections, as in the venturi – Fig 2. If x expresses the ratio A_2/A_1 it follows since the volume of flow is the same at both sections that $V_1 = x \cdot V_2$ whence rearranging the equation above and rewriting $(P_1 - P_2)$ as ΔP.

$$V_1 = \sqrt{\frac{2g\Delta P}{\rho} \cdot \frac{x^2}{(1 - x^2)}}$$

This is the basic formula for the theoretical design of a venturi velocity meter. In practice the formula is modified by the introduction of a calibration coefficient to take into account frictional losses, *etc* neglected in the Bernoulli equation.

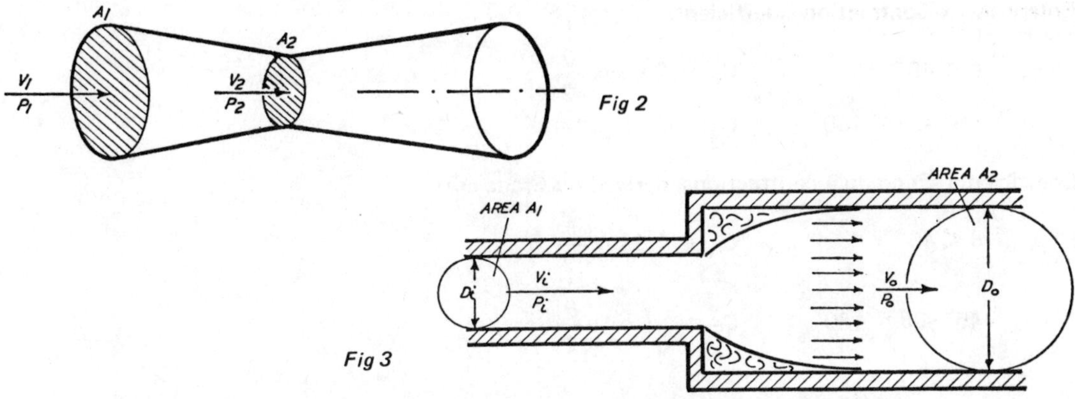

Fig 3

Sudden Enlargements (Fig 3)

Again flow conditions can be analyzed from first principles, although for engineering calculations it is only necessary to determine the velocity head loss from the corresponding resistance coefficient K_1. This can be determined as

$$K_1 = \left(1 - \frac{d_1{}^2}{d_2{}^2}\right)^2$$

where

d_1 = internal diameter of smaller pipe

d_2 = internal diameter of larger pipe

This formula is also quoted in the form

$$K_1 = (1 - \beta^2)^2$$

when

$$\beta = d_1/d_2$$

Sudden Contractions (Fig 4)

In this case the resistance coefficient is one half that for a sudden enlargement, viz:

$$K_2 = 0.5\left(1 - \frac{d_1{}^2}{d_2{}^2}\right)^2$$

or

$$K_2 = 0.5\ (1 - \beta^2)^2$$

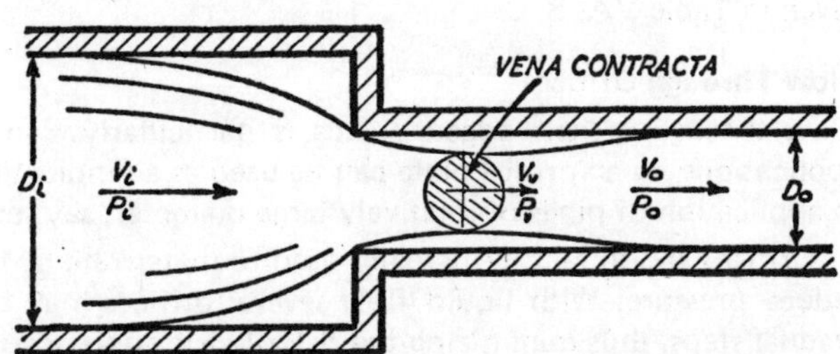

Fig 4

Enlargement/Contraction Coefficients

$$\theta \leqslant 45^\circ \qquad C_e = 2.6 \sin \frac{\theta}{2}$$

$$45^\circ < \theta \leqslant 180^\circ \qquad C_e = 1$$

Coefficients for gradual contractions, derived by Crane are:-

$$\theta \leqslant 45^\circ \qquad C_c = 1.6 \sin \frac{\theta}{2}$$

$$45^\circ < \theta \leqslant 180^\circ \qquad C_c = \sqrt{\sin \frac{\theta}{2}}$$

where

θ is the angle of divergence

These formulas can also be expressed in terms of the larger pipe and have been extended to define resistance coefficients (K_3) for both sudden and gradual enlargements and contractions, viz:

Enlargement

$$45^\circ < \theta \leqslant 180^\circ \qquad K_3 = \frac{(1-\beta^2)^2}{\beta^4}$$

$$0 \leqslant 45^\circ \qquad K_3 = \frac{2.6 \sin \frac{\theta}{2} (1-\beta^2)^2}{\beta^4}$$

Contraction

$$45^\circ < \theta \leqslant 180^\circ \qquad K_3 = \frac{0.5 \sin \sqrt{\frac{\theta}{2}} (1-\beta^2)}{\beta^4}$$

$$0 \leqslant 45^\circ \qquad K_3 = \frac{0.8 \sin \frac{\theta}{2} (1-\beta^2)}{\beta^4}$$

Resistance coefficients and equivalent strength lengths applicable to changes of cross section are given in Table VA, B, C.

Flow Through Orifices

Flow through a sharp-edged orifice is particularly significant as this has a number of practical applications, *eg* an orifice plate can be used as an indicating flow meter, though normally restricted in application of pipes of relatively large diameter, say, over 2 inches (5 cm).

Orifices are, in fact, principally used to meter rate of flow. They are also used to restrict flow or reduce pressure. With liquid flow several orifices may be used in sequence to reduce pressure in gradual steps, thus minimizing the risk of cavitation occurring.

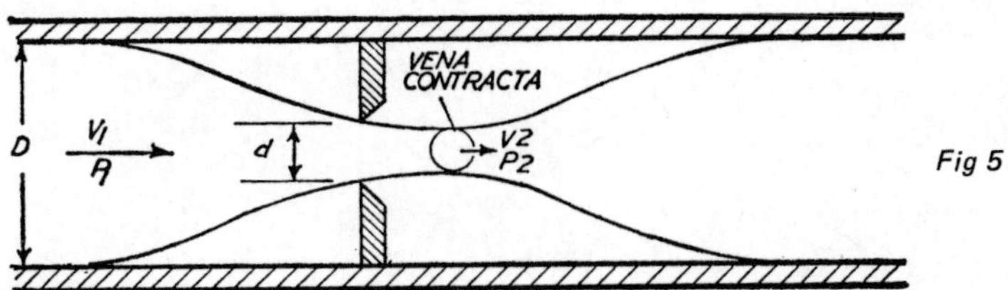

Fig 5

Simple analysis is restricted to streamlined flow, the flow pattern being of the form shown in Fig 5. The streamlines converge on approaching the orifice and continue to converge after passing through the orifice, reaching a minimum cross sectional area, known as the *vena contracta,* downstream of the orifice before diverging again. For small circular orifices the downstream position of the *vena contracta* is of the order of half the diameter of the orifice. At the *vena contracta* all the streamlines are perpendicular to the plane of the orifice.

TABLE VA – CHANGE OF CROSS SECTION : EQUIVALENT STRAIGHT LENGTHS

PIPE SIZE (d) in	1	2	3	4	5	6	8	10	12	14	16	18	20	24
mm	25	50	75	100	125	150	200	250	300	350	400	450	500	600
Expansion d/D = 0.25	3	5	8	11	14	16	20	26	31	36	42	48	55	65
0.5	2	4	6	7	9	12	15	18	22	25	29	34	46	52
0.75	5	1	2	2	3	3	4	6	7	8	10	12	15	20
Contraction d/D = 0.5	1	2	3	3	4	6	8	10	12	14	16	18	20	26

TABLE VB – CHANGE OF CROSS SECTION : RESISTANCE COEFFICIENT

Change of Section	RATIO OF SMALLER PIPE DIAMETER TO LARGER PIPE DIAMETER (d/D)								
	0.1	0.2	0.3	0.4	0.5	0.6	0.7	0.8	0.9
10 degree Taper				0.35	0.25	0.20			
20 degree Taper				0.15	0.12	0.10			
Sudden Expansion	2.0								0.15
Sudden Contraction	0.5	0.45	0.45	0.45	0.45	0.4	0.3	0.2	

TABLE VC – INLETS AND OUTLETS : RESISTANCE COEFFICIENT

Change of Section	Typical Value
Inlet: Abrupt	0.5
Gradual	Not less than 0.5
Projecting tube	1.0
Outlet: Abrupt	1.0
Gradual	Not less than 0.12

Balancing of Hydronic Systems

EXCESS CONSUMPTION of heating energy is the result of temperature variations in a building and the generally incorrect approach taken to solve the problem. If, for example, one room is too warm and another too cold, the "necessary adjustments" are often made in the room that is too cold – the result being that the average temperature of the building, which to start with was probably quite sufficient had it been balanced out, now becomes too high. See Figs 1 and 2.

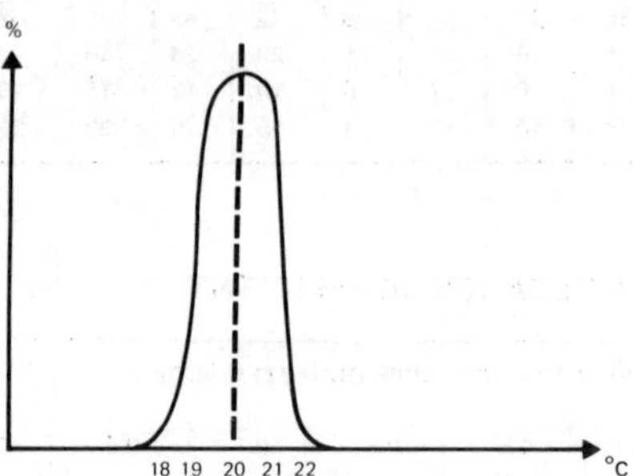

Fig 1 Example of correctly balanced building. Average temperature 20° C.

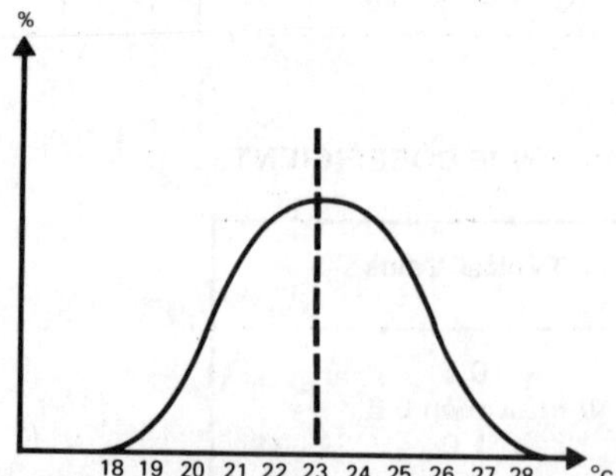

Fig 2 Example of wrongly balanced building. Average temperature 23° C.

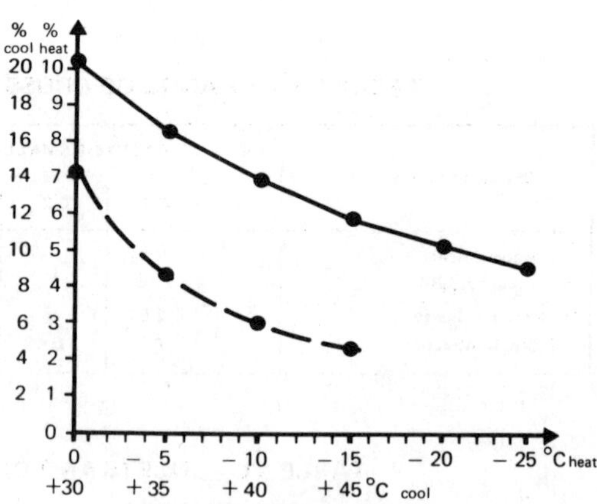

Fig 3 Energy cost saving in % for each degree higher indoor temperature at various maximum outdoor temperatures in a cooling system.
Energy cost saving in a heating system in % for each degree lower indoor temperature at various minimum outdoor temperatures.

Obviously, this higher average temperature implies a higher consumption of energy. How much higher depends on the specific heat requirement of the geographical area in degree-days. Generally speaking, however, one can say that each degree above +68°F (+20°C) indoors means heating costs which are about 5 or 6% higher than normal. See Fig 3.

If it is a cooling plant the conditions and the means are of course reversed. But the waste of energy is often increased.

In many cases, excess temperature is even ventilated away, or the windows are opened, which can mean an average temperature which is more than 7–9 deg F (4–5 deg C) too high.

Energy conservation measures have always been of considerable interest and the balancing of a hydronic system is particularly interesting since only simple measures are needed and they give quick and very evident results. Savings of 20–30% are not unusual.

Pump Energy Waste

It is often forgotten that pump energy also costs money. In many cases, the pumps are oversized.

In heating systems, this is not always so very significant because the temperature differences are often high and therefore a relatively small amount of water is being circulated. However, particularly in refrigeration systems with their lower temperature differences, pump energy waste can add a lot to the operating costs.

Another essential difference between heating and cooling systems is that energy losses in the systems are converted to heat; this works to the benefit of a heating system but necessitates an increase in capacity for a cooling system.

Over-dimensioning is generally a consequence of the following:

1. When the pump was chosen, the designer was uncertain of the pressure drop in the system (boiler, heating batteries, valves *etc*) because the tender covering the components had not been finally accepted.
2. Insufficient data concerning the pressure drop in the piping system.
3. General safety factors. However to facilitate balancing, it is wise to increase the size of pump slightly, but not to exaggerate.

Modifications to an over-dimensioned system (for example, by changing the pump or the impeller), do not always lead to lower overall costs, although such measures do often prove worthwhile.

For some time now, it has been taken for granted that a hydronic system in a new building must be balanced. The question being discussed is how to balance it and who is to balance it – the application engineer, the heating contractor, or a firm specializing in the balancing of hydronic systems.

Construction

To skip the piping calculations entirely and simply specifiy that the system shall be balanced without detailing the means or the way to do the work implies a significant amount of extra work for the person who is to do the balancing. Furthermore, balancing is not a universal solution which will make a poorly designed system function adequately.

When designing the system, care must be taken to arrange clearly demarcated sections.

Calculations

For newly built facilities in Sweden, the SBN 80 39:32 specifies that the method of balancing and the presetting values and water flow values for the balancing valves must be specified on the building permission documents. This necessitates complete piping calculations and the determination of pressure losses in the heating system.

The Kv-values for the valves can, in the case of radiator valves, generally be adjusted directly by means of the presetting unit which is marked off in Kv, although this only applies in a limited number of manufacturers' valves.

Balancing valves must be incorporated in all branches to avoid having to balance the radiator valves in one branch against the balancing valves in other branches, (Fig 4).

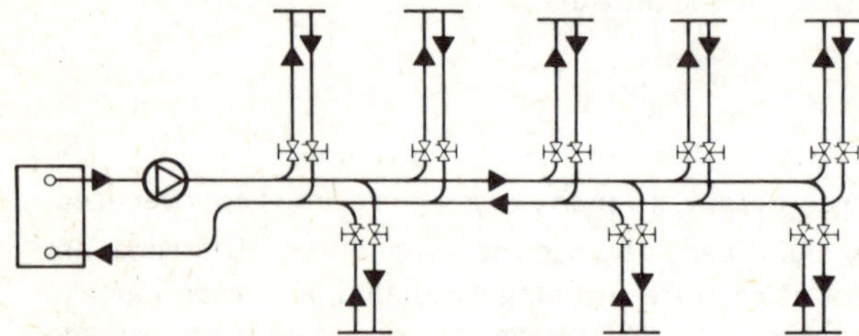

Fig 4 Balancing valves incorporated in all branches to avoid having to balance the radiator valves in one branch against the balancing valves in other branches.

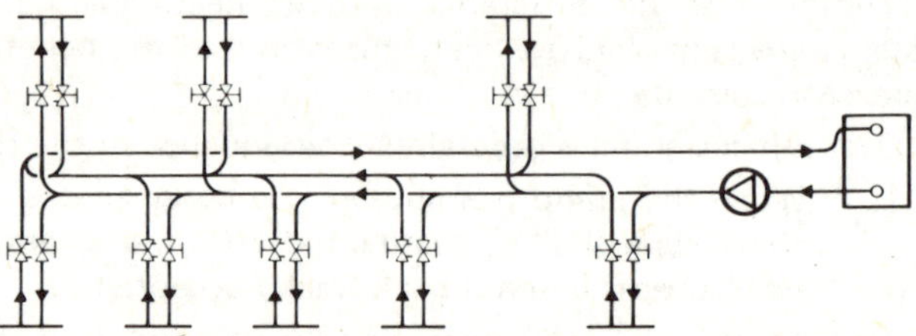

Fig 5 The Tischelmann reverse return system can simplify many balancing problems. The exclusion from such a system of balancing valves will, however, lead to imbalance, since different radiators and heaters or cooling units do not have the same output or pressure drop, but the pressure differences in the system between branches will be lower. This means that balancing will be easier. The Tischelmann reverse return system also has benefits to offer when balancing the sub circuits.

Reverse return mains, according to the Tischelmann system, can simplify many balancing problems. The exclusion from such a system of balancing valves will, however, lead to imbalance, since different radiators and heaters or cooling units do not have the same output or pressure drop, but the pressure differences in the system between branches will be lower. This means that balancing will be easier. The Tischelmann reverse return system also has benefits to offer when balancing the sub circuits, (Fig 5).

Balancing

Before commencing to balance a system, all valves must be opened fully. This applies particularly to thermostatic radiator valves and two-way control valves. This type of valve operates with varying flows and unless it is ensured that the valve is fully opened, it may just have closed automatically as balancing was commenced.

A thorough knowledge of the system is also important before the commencement of balancing. The information required includes the following:

A. Drawings with hydronic sketches.

B. Data concerning flows and pressure drops across heat generator, batteries, radiator heaters, balancing and control valves.

C. Pumps data – pump diagram.

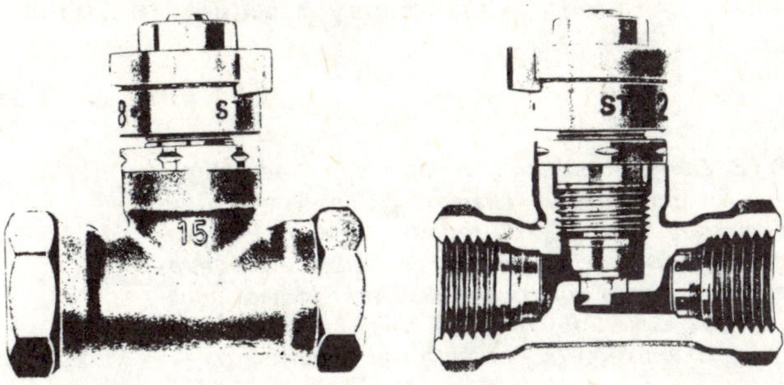

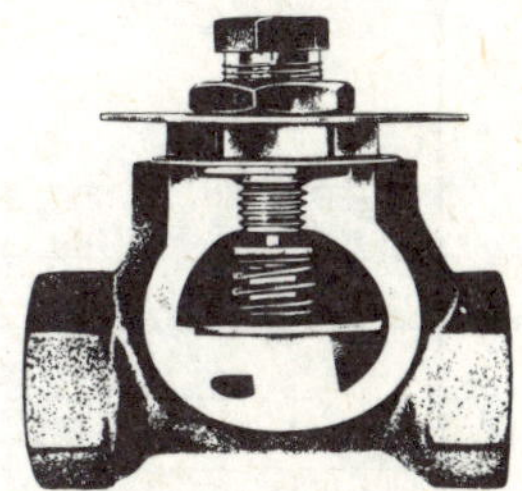

Balancing and shut-off valves (Tour & Andersson AB)

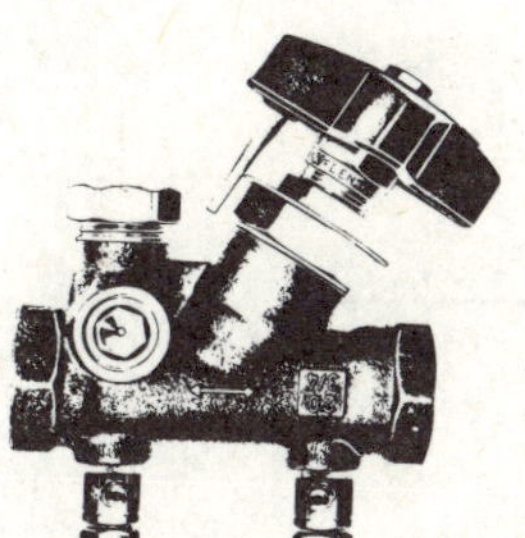

Example of balancing valves. (Tour & Andersson AB)

The Desk Method

A well-defined system of not too great a complexity generally requires only one single adjustment of all valves used, including radiator valves and balancing valves, in accordance with the values specified by the drawings. Control measurements should, however, be made on one or more extremity branches (furthest away from the pump) as well as on a few central branches. The flow deviations found in these control measurements should not exceed 10% of the volume or 20% of the pressure. If, after an adjustment as described above, there are still temperature deviations of more than 2 deg F (1 deg C) in individual rooms, and if these cannot be related to temporary fluctuations in the heating or cooling load, the explanation will be found in one of the following.

A. Incorrect calculation of the heating/cooling facility.

B. Incorrect design/installation of heating/cooling facility.

C. Bad building (insufficient sealing, draughts *etc*).

Temperature Measurement Method

This method, which is only applicable to heating systems, is based on the fact that each radiator/heater is dimensioned according to the same temperature drop with an equal outdoor temperature.

As a consequence, the system can be balanced by measuring the temperature drop at the pump and then adjusting the balancing valves so that the temperature drop is the same at the pump as it is over each branch.

To achieve acceptable accuracy with this method, the outdoor temperature must be almost constant throughout the entire balancing process and, in addition, below 34°F (1°C). It is often small temperature drops that are being measured and therefore the temperature differences become even smaller. For this reason the system is at times less than exact – see also Fig 6.

The temperature method can also save time if it is used as a preliminary stage prior to the proportionate balancing method described below.

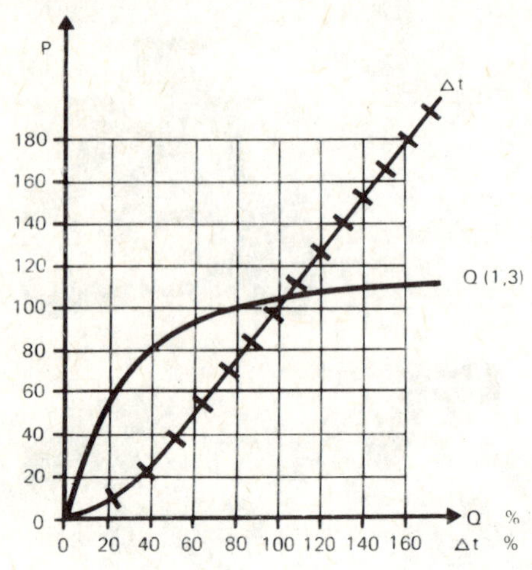

Fig 6 The heat emission variations in % as a function of temperature changes Δt in % and flow changes in % (at 80–60 radiator system). The diagram shows that a per cent deviation in temperature will give a significant deviation in heat emission. A deviation in flow will influence heat emission to a much lower extent.

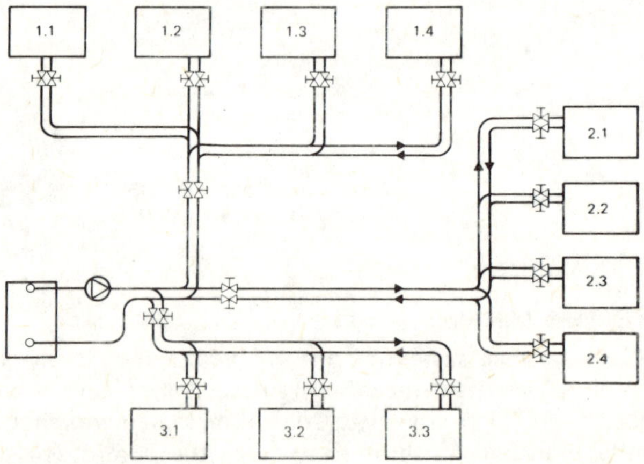

Fig 7

Proportionate Balancing Method

This method is one of the most frequently used and it is suitable for old facilities as well as for new ones.

The procedure is to measure the pressure drop and to move proportionally from branch line to branch line. This is done as follows.

1. Set all radiators and balancing valves according to the drawings specifications – or open them if no specifications are available, (Fig 7).
2. Start by measuring branch lines 1.1, 1.2, 1.3 and 1.4 of sub-system 1 and determine the proportionate flow rate, that is to say the relationship between measured flow rate and design flow rate. If flow in 1.1 is 1 500 lit/h and the design flow rate is 1 000 lit/h, the proportionate flow rate will be 1.5.

3. Presume that 1.1 has the lowest proportionate flow rate, 1.2 the next lowest and so on. In this case, leave the valve of branch 1.1 open and balance 1.2 to give the same proportionate flow rate as 1.1 (within the tolerance applying). These two branches are now balanced and you can continue with 1.3 balancing it against 1.2 until they both have the same proportionate flow rate. 1.1 need not be checked. It will change in direct proportion to 1.2 and remain in balance with 1.2. Then continue with 1.4 in the same manner.

   Since 1.4 is the last branch line of sub-system 1, this means sub-system 1 is now ready and if any flow changes occur in the total system, the branch lines of sub-system 1 will be altered by the same proportional amount to a new common proportionate flow rate.
4. Proceed in a similar manner with sub-systems 2 and 3.
5. Leave the balancing valve open in the sub-system which has the lowest proportionate flow rate and balance the other sub-systems as described above.
6. The final stage is to determine whether or not the pump is supplying too large a quantity of water. If it is giving too much water, it can be throttled by means of the balancing valves, by altering the pump speed, or by changing the impeller. In the case of large oversizes, it is often best to reduce the speed of the pump or change the impeller.

   Some other examples are given in Figs 8, 9 and 10.

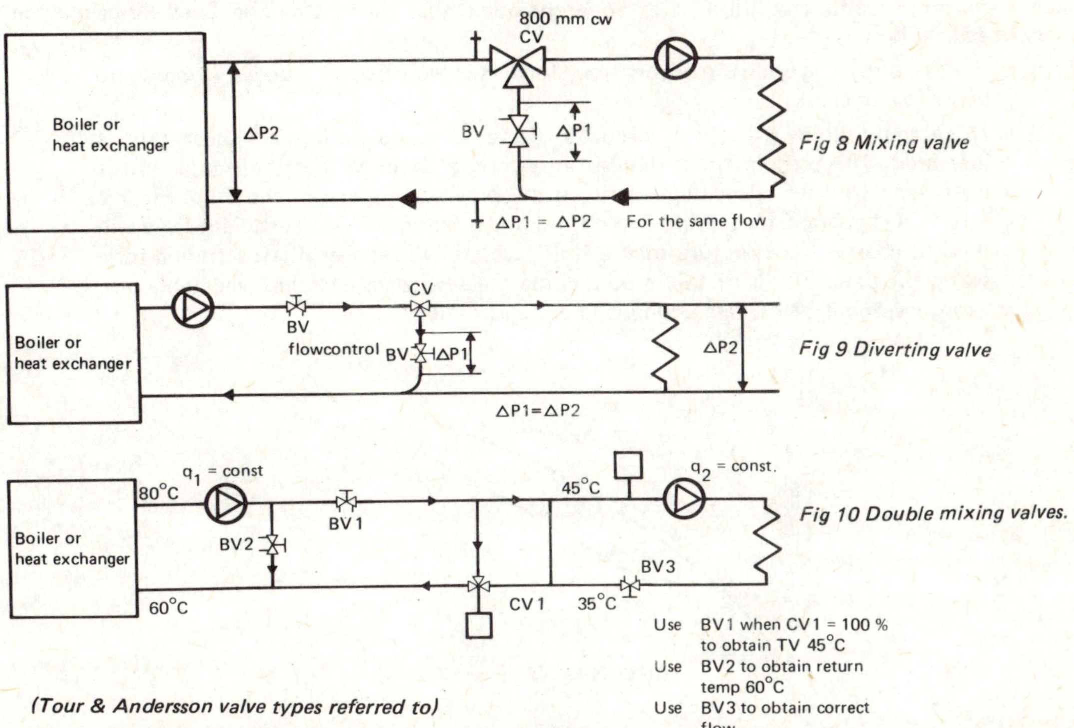

Fig 8 Mixing valve

Fig 9 Diverting valve

Fig 10 Double mixing valves.

(Tour & Andersson valve types referred to)

Flow of Mixtures Through Pipes

STANDARD FORMULAS for pipeline performance calculations (*eg* for pressure drop or head loss) incorporate fluid viscosity as a constant parameter, *ie* as a specific value dependent on the working temperature of the fluid involved. Such calculations are valid only for *Newtonian* fluids where viscosity remains constant with agitation or change in shear rate. Typical Newtonian fluids include water, aqueous solutions, mineral oils, hydrocarbons, syrups and resins (some).

Various other types of fluid are essentially *non-Newtonian* in characteristics, when the viscosity value under any specific conditions is an apparent one rather than a true one. Such fluids may be categorized as follows:

(i) Fluids containing solids in suspension, further categorized as *slurries, sludges* and *pulps* (paper stock).

(ii) *Thixotropic* fluids – where viscosity decreases as agitation or shear rate is increased. Thixotropic fluids exhibit a hysteresis effect in that their apparent or instantaneous viscosity is dependent on the previous history of the fluid. Fig 1 is a typical rheogram for a thixotropic fluid under laminar flow. Turbulent flow will tend to change the structure of the fluid, which will recover if left standing for a sufficient time. Fluids of this type include greases, soaps, starches, vegetable oils, varnishes, some resins, tars, asphalts, glues and some inks.

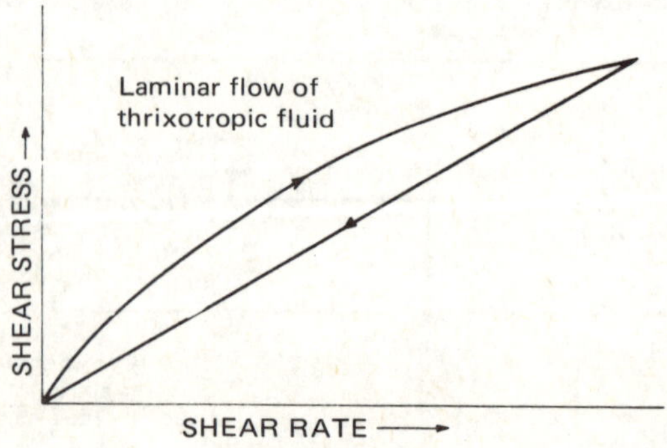

Fig 1

(iii) *Colloidal* fluids – which behave like thixotropic fluids but will not recover their original viscosity when agitation is stopped. Fluids of this type include colloidal solutions of soaps in water and oils, lotions, shampoos and gelatinous compounds.

(iv) *Dilatent* fluids – where viscosity *increases* as agitation or shear rate is increased. Fluids of this type include clays and some slurries.

(v) *Rheopectic* fluids – where viscosity increases with increasing agitation in shear rate up to a maximum value at any constant rate of agitation.

(vi) *Plastic and pseudo-plastic* fluids – where viscosity decreases with increasing shear rate, but initial viscosity may be so high as to prevent start of flow in a normal pumping system. These are also known as *Bingham* fluids.

Strictly speaking, only plastic fluids are true Bingham fluids and include such products as drilling muds, thick mineral slurries and sewage sludge. Pseudo-plastic fluids exhibit a different shear rate – shear stress relationship. Fluids in this category include paper stock, detergent slurries, some paints and lacquers, some mineral slurries, mayonnaise, and cellulose acetate in acetone. A further sub-category of such fluids is known as *yield pseudo-plastic,* typical products of this type being clay-water suspensions and polymer solutions.

Complex Mixture Flow

Complex mixture flow may be *homogeneous, pseudo-homogeneous, heterogeneous* or *complex,* according to the phase(s) involved and the size of the solids involved – see Fig 2. Homogeneous flow applies only in pure liquid flow. Simple mixtures involve two-phase flows; complex mixtures multi-phase flows. In pseudo-homogeneous flow the solids are present in finely divided, highly dispersed form with almost uniform dispension in the carrier phase. The whole mixture then tends to behave as a single-phase fluid.

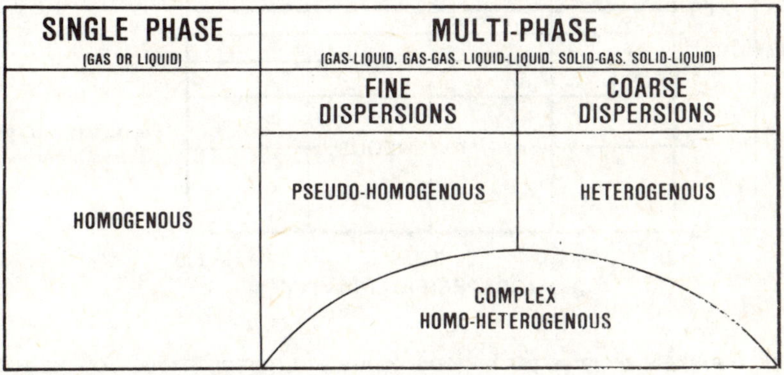

SINGLE PHASE (GAS OR LIQUID)	MULTI-PHASE (GAS-LIQUID. GAS-GAS. LIQUID-LIQUID. SOLID-GAS. SOLID-LIQUID)	
	FINE DISPERSIONS	COARSE DISPERSIONS
HOMOGENOUS	PSEUDO-HOMOGENOUS	HETEROGENOUS
	COMPLEX HOMO-HETEROGENOUS	

Fig 2 Regimes for homogeneous and heterogeneous flow.

With increasing size and/or quantity of solids, dispersion is coarser, yielding *hetergeneous* behaviour, *ie* with a pronounced solids concentration gradient along the vertical axis of the pipe. The actual velocity of flow then becomes a critical parameter.

With *complex* flow some of the solids content behave heterogeneously in pseudo-homogeneous flow, *ie* the flow can be described as *homo-hetergeneous.* There are thus two separate sources of friction and pressure drop.

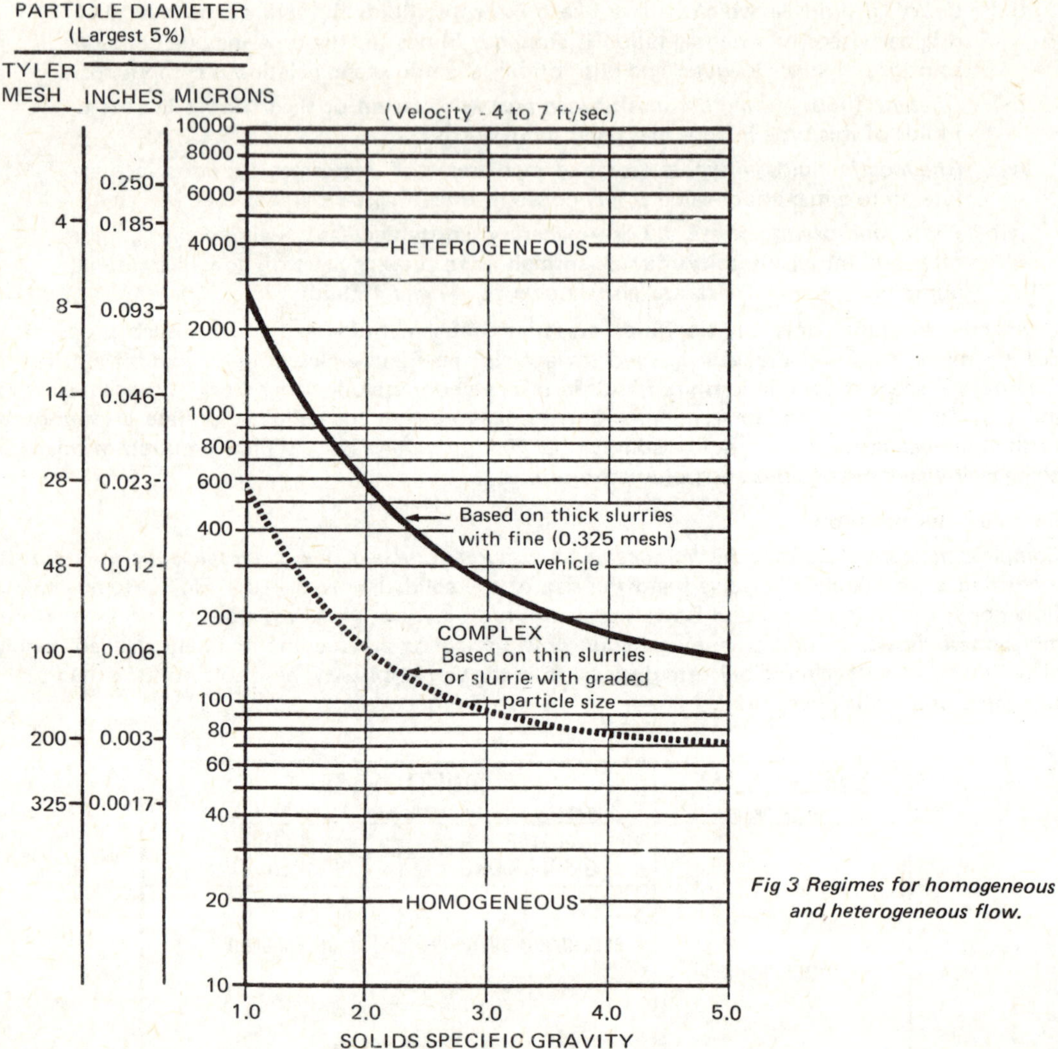

Fig 3 Regimes for homogeneous and heterogeneous flow.

Fig 3 shows the likely regimes for heterogeneous and homogeneous flow with typical slurries related to particle size and solids specific gravity for flow velocities in the range 4 – 8 ft/sec (1.2 – 2.6 metres/sec).

Homogeneous Flow

Common practice with slurries is to use the Fanning friction factor to estimate frictional losses. This is quarter the value of the D'Arcy-Weisbuck friction factor. However this straightforward approach does not take into account the fact that solids present have the effect of suppressing turbulence which can reduce the actual friction factor by up to 15 per cent, depending on the type of slurry. Empirical formulas can thus be more realistic.

Pseudo-Homogeneous Flow

Pseudo-homogeneous flow is considered to exist where there is no measurable solids concentration gradient along the vertical axis of the pipe. For any given mixture this is related to the flow velocity. Below the critical velocity flow will be heterogeneous; above the critical velocity flow will be pseudo-homogeneous.

Specifically the flow condition can be expressed in terms of the C/C_A ratio where C is the solids concentration measured at an arbitrary part near the top of the pipe (usually 8% of the pipe diameter); and C_A is the solids concentration at the centre of the pipe.

If these two values are equal (*ie* $C/C_A = 1$), flow will be homogeneous. Progressively lower values represent pseudo-homogeneous flow, degenerating into heterogeneous flow. The actual value of C/C_A is influenced by particle size and concentration, as well as flow velocity. In a mixture of solids, finer particles will have a high C/C_A and coarser particles a low C/C_A. As a general guideline a C/C_A of 0.8 or greater is necessary to maintain pseudo-homogeneous flow.

Heterogeneous Flow

With C/C_A values below 0.8 flow will be a mixture of pseudo-homogeneous and heterogeneous – *eg* fewer particles remaining in suspence with coarser particles tending to settle out. Flow will be fully heterogeneous at C/C_A values of 0.1 or less. With heterogeneous flow inertia; effects are far more significant that viscous effects. Also there may be several different flow patterns ranging from symmetric suspension through asymmetric suspension to sliding bed (solids sliding along the bottom of the tube), then stationary bed, finally leading to plugging (pipe blockage).

Friction losses for heterogeneous flow are commonly based on the Durand formula, although empirical formulas are also used (see later). The Durand formula for the friction factor (f_h) for heterogeneous flow is

$$f_h = f_l \left(1 + \frac{150\,Dg}{V^2}\right) \left(\frac{\rho S - \rho l}{\rho 1}\right) \left(\frac{1}{\sqrt{C_D}}\right)^{3/2} \cdot S_V$$

where

f_l = friction factor for liquid (dimensionless)
D = pipe diameter, feet
V = flow velocity, ft/sec
ρl = density of liquid (lb/ft^3)
ρS = density of solids (lb/ft^3)
C_D = drag coefficient (dimensionless)
S_V = volume fraction of solids (dimensionless)

Homo-heterogeneous Flow

With homo-heterogeneous flow some of the solids behave heterogeneously in a homogeneous vehicle. This is a condition commonly encountered in practice where the carrier fluid contains a mixture of particle sizes. To determine friction losses in this case it is necessary to split the solids content into fractions of different size and into homogeneous and heterogeneous portions – *eg* based on the respective C/C_A ratios.

Thus, taking each sign fraction in turn and determining its C/C_A ratio, multiply this by volume concentration for that fraction will give the proportion of that fraction having homogeneous flow. The remainder will be heterogeneous flow. Each fraction is split into homogeneous and heterogeneous flow in a similar manner. Friction losses are then calculated for each flow.

Transition Velocity

Normally all slurry pipeline systems operate with turbulent flow. Operating under laminar flow conditions will allow some settlement which in time can lead to unstable flow conditions, or even blockage. Flow velocities must therefore be above the transition velocity, determined by the critical Reynold's number.

Transition velocities for Bingham plastic fluids are conveniently related to a dimensionless *Hedstrom number* (N_{HI}), where

$$N_{HI} = \text{Reynold's number} \times \text{plastic number}$$

$$= \left(\frac{\rho VD}{\eta}\right) \left(\frac{\frac{T_o g}{\eta}}{V/D}\right)$$

where

ρ	=	slurry density, lb/ft^3
V	=	flow velocity ft/sec
D	=	pipe diameter, feet
T_o	=	yield spec lb/ft^2
g	=	acceleration of gravity
η	=	coefficient of rigidity (lb/ft sec)

The relationship between critical Reynold's number and Hedstrom number is shown in Fig 4 and is closely followed by most slurries. Mud and clay slurries can be the exception.

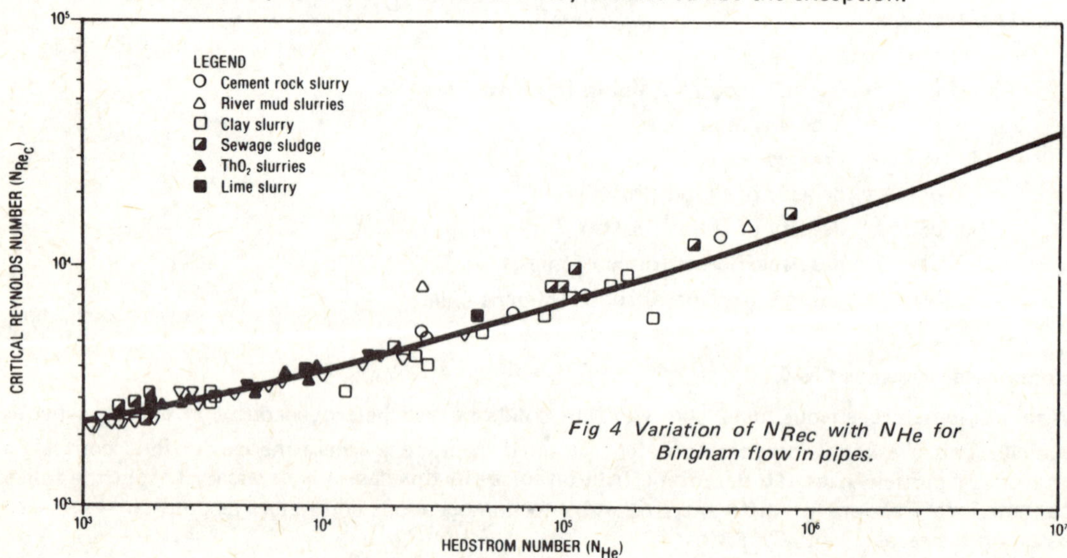

Fig 4 Variation of N_{Rec} with N_{He} for Bingham flow in pipes.

Critical Deposition Velocity

The critical deposition velocity relative to heterogeneous flow is given by Durand as

$$V_{crit} = K \left[\frac{2g\,D\,(\rho_s - \rho_l)}{\rho_l}\right]^{1/2}$$

where

K is an empirical constant

g = acceleration of gravity

D = pipe diameter, feet

ρ_s = density of solids, lb/ft^3

ρ_l = density of liquid, lb/ft^3

In all such cases [(i) through (vi)] performance calculations can be based on a *pseudo viscosity* or equivalent Newtonian viscosity.

Slurries

Slurries are liquids (usually water) containing abrasive solids in suspension, resulting in an increase in specific gravity over the carrier fluid. Slurries are categorized by the size of the solids as *fines, sands,* and *coarse.*

Slurries behave as non-Newtonian fluids with an apparent viscosity depending on the degree of suspension, which in turn is dependent on the flow rate. This may be generally related to a *fall velocity* or the minimum flow rate necessary to maintain the solids in suspension and prevent them from settling out. This, in turn, depends on the size of the solids, and also their concentration. Approximate flow velocities to retain solids in suspension in water for various classes of slurries are

Fines	(particle size 75 μm or less)	3 ft/sec	(0.9 m/sec)
Sands	(particle size 75 to 850 μm)	5 ft/sec	(1.5 m/sec)
Coarse	(particle size 850 to 5 000 μm)	7 ft/sec	(2.1 m/sec)

These empirical figures are based on a solids content of 30% to 35% by weight and solids of specific gravity 2.5 to 3.0. (See also Table I and Fig 5).

TABLE I – SOME TYPICAL SLURRIES

Slurry	Proportion of Solids by Weight
Alumina	up to 50%
Crushed chalk	up to 68%
Clay	up to 60%
Coke fines	up to 55%
Gravel	up to 25%
Lime	up to 65%
Magnetite	up to 60%
Sand	up to 60%
Soda ash	up to 60%

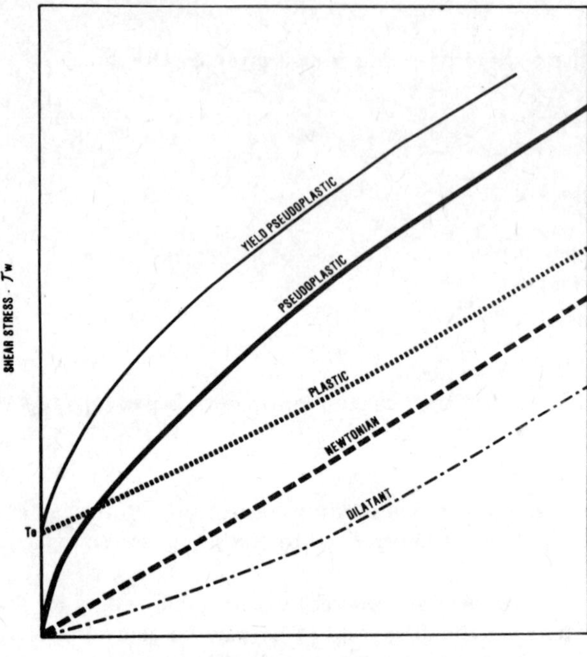

Fig 5 Fluid classification of slurries.

Solids which form an intimate mixture yield a homogeneous fluid, in which case the pump performance is largely determined by the specific gravity and viscosity of the homogeneous mixture, which behaves as a normal fluid. Solids in suspension, however, form non-homogeneous mixtures and flow is then heterogeneous, with particles tending to slide at the surface of a 'bed' of solids. This 'bed' is only carried fully suspended if the fluid velocity is greater than the settling velocity of the solids involved. At lower fluid velocities there will be a corresponding degree of settlement producing a sliding rather than a suspended 'bed'.

The quantity of water required to deliver a specific quantity of solids when pumping slurries can be determined from

$$\text{water quantity} = K \times T \left(\frac{W}{R} + \frac{1}{sg_D} \right)$$

where T = weight of dry solids per hour
W = percentage of water
R = percentage of dry solids
For water quantity in litres and T in metric tonnes K = 1.06
For water quantity in Imperial gallons and T in Imperial tons K = 3.75
For water quantity in US gallons and T in US tons K = 4.02
sgD = specific gravity of dry solids.

Specific gravity figures for suspensions of solids in water are given in Table II.

TABLE II – SPECIFIC GRAVITY OF SUSPENSIONS OF SOLIDS IN WATER

Percentage by Weight of Solids	Ratio Water to Solids	Specific Gravity of Dry Solids																	
		2.0	2.1	2.2	2.3	2.4	2.5	2.6	2.7	2.8	2.9	3.0	3.1	3.2	3.3	3.5	4.0	4.5	5.0
		Specific Gravity of Solution																	
10	9:1	1.05	1.05	1.06	1.06	1.06	1.06	1.06	1.07	1.07	1.07	1.07	1.07	1.07	1.07	1.08	1.08	1.09	1.09
15	5.66:1	1.08	1.08	1.09	1.09	1.09	1.10	1.10	1.10	1.11	1.11	1.11	1.11	1.11	1.12	1.12	1.13	1.13	1.14
20	4:1	1.11	1.11	1.12	1.12	1.13	1.14	1.14	1.14	1.15	1.15	1.15	1.16	1.16	1.16	1.17	1.18	1.19	1.19
15	3:1	1.14	1.15	1.15	1.16	1.17	1.18	1.18	1.19	1.19	1.19	1.20	1.20	1.21	1.21	1.22	1.23	1.24	1.25
30	2.33:1	1.17	1.18	1.19	1.20	1.21	1.22	1.23	1.23	1.24	1.24	1.25	1.25	1.26	1.26	1.27	1.29	1.31	1.31
35	1.87:1	1.21	1.22	1.23	1.25	1.26	1.27	1.28	1.28	1.29	1.30	1.30	1.31	1.32	1.32	1.33	1.35	1.37	1.39
40	1.5:1	1.25	1.26	1.28	1.29	1.30	1.32	1.33	1.34	1.35	1.36	1.36	1.37	1.38	1.39	1.40	1.43	1.45	1.47
45	1.22:1	1.29	1.30	1.32	1.34	1.36	1.37	1.38	1.40	1.41	1.42	1.43	1.44	1.45	1.46	1.47	1.51	1.54	1.56
50	1:1	1.33	1.35	1.37	1.39	1.41	1.43	1.44	1.46	1.47	1.49	1.50	1.51	1.52	1.53	1.55	1.60	1.63	1.67
55	0.91:1	1.37	1.38	1.41	1.43	1.44	1.49	1.51	1.53	1.55	1.56	1.58	1.59	1.61	1.62	1.65	1.70	1.75	1.79
60	0.67:1	1.43	1.46	1.48	1.51	1.54	1.56	1.58	1.61	1.63	1.65	1.67	1.68	1.70	1.72	1.75	1.82	1.87	1.92
65	0.54:1	1.48	1.51	1.55	1.58	1.61	1.64	1.67	1.69	1.72	1.74	1.76	1.79	1.81	1.83	1.87	1.95	2.03	2.08
70	0.43:1	1.54	1.57	1.62	1.65	1.69	1.72	1.75	1.79	1.82	1.85	1.88	1.90	1.93	1.95	2.00	2.10	2.20	2.27
75	0.33:1	1.60	1.65	1.69	1.73	1.78	1.82	1.86	1.90	1.93	1.97	2.00	2.03	2.06	2.09	2.15	2.29	2.40	2.50

Frictional Losses

There is no complete agreement on the method of calculating the frictional losses in pipes carrying fluids with solids in suspension. Generalized data can give extremely inconsistent results when applied to individual systems particularly if localized areas exist where the fluid velocity may be less than the minimum needed to keep the solids in suspension. It can thus prove difficult to estimate for centrifugal pumps the total head to be supplied by the pump and thus to determine the most efficient working point.

A general formula which can be used is:

$$\frac{\Delta P_S}{\Delta P_W} - 1 = KC_V \qquad \frac{D\,(sg - 1) \times V_S}{\sqrt{d\,(sg - 1) \times V^2}}$$

where the constant K is determined from empirical data.

where

ΔP_S = pressure drop when transporting solids

ΔP_W = pressure drop for water

C_V = concentration of solids by volume

D = pipe diameter

d = mean particle diameter

sg = specific gravity of solids

V_S = settling velocity of solids (in still water)

V = flow velocity

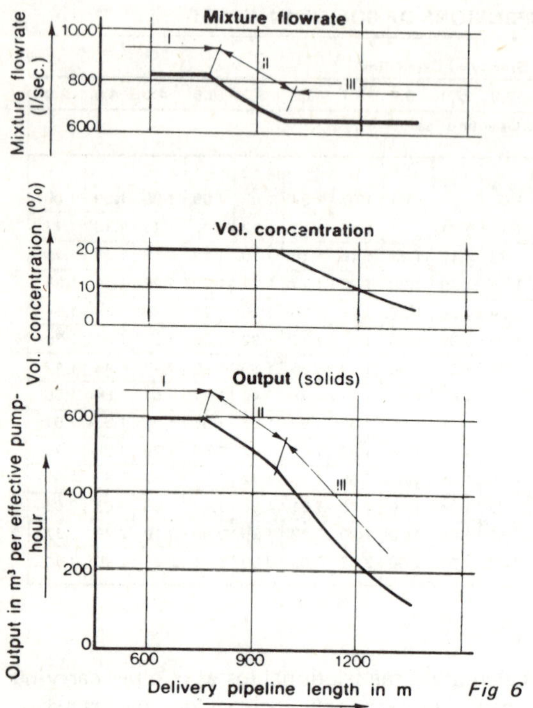

Fig 6

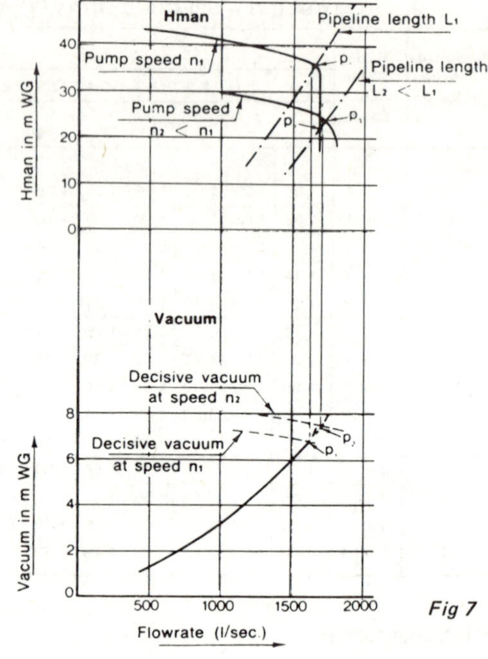

Fig 7

Output

Typically the output from a slurry pump will follow the characteristics shown in Fig 6, with the output curve consisting of three zones.

(i) delivery distance

(ii) delivery distance within the normal working range of the pump

(iii) loss of output over any further delivery distance.

The first zone is critical in that it determines the suction conditions. At a high flowrate, when the upper limit is reached, the vacuum becomes critical, *ie* the flowrate at the attainable mean specific gravity of the mixture is so high that the corresponding vacuum on the suction side of the pump installation is equal to the critical vacuum, which may not be exceeded and thus consistutes the limit of the attainable vacuum. Fig 7 shows the critical vacuum applying to a delivery pipeline length L1.

The critical vacuum is reached at working point P1. If the critical vacuum is exceeded – for example, as a result of shortening the delivery pipeline, with the result that a lower resistance curve applies, *eg* L2 in Fig 7 – cavitation ensues.

If it is desired to operate with a shorter delivery pipeline, the speed n1 of the pump must be reduced to the point where the intersection of the corresponding pump characteristic and that of the shorter pipeline falls within the normal working range of the pump. This implies that the point of intersection (working point P3) must coincide with a flowrate at which the vacuum isslightly below the critical level corresponding to the lower pump speed. This is shown in Fig 7. The two working points P1 and P3 have a virtually equal flowrate. If the delivery distance is less specific gravity of the mixture remains virtually unchanged, the output at these (too) short delivery distances will also remain virtually constant as shown by the horizontal section 1 of the output group (Fig 6).

Critical Velocity

If the delivery pipeline is lengthened, the specific gravity of the mixture to be pumped must be reduced as soon as the critical velocity (the second bend in the output curve) is reached. This is necessary to avoid a subcritical situation, leading to sedimentation. At what is in effect an excessive delivery distance, pumping a mixture of such specific gravity as to cause sedimentation in the pipeline is to risk total blockage. In practice, this danger can be averted by admitting more water. The specific gravity must be reduced just sufficiently to restore a supercritical situation.

In approximate terms, it can be stated that the critical flowrate of a soil/water mixture at the reduced sg differs little from the critical flowrate at the highest attainable mean mixture sg. This implies that, when delivering into (too) long pipelines, the specific gravity must be reduced when a certain flowrate, which is virtually constant, is reached. This is the flowrate corresponding to the critical velocity. Increasing the delivery distance will therefore result in a lower output of solids.

When pumping mixtures of water and fine sand (less than 75 μm), or clayey soil and silt, or combinations of these materials, the critical velocity in the delivery pipe will be very low. Moreover the resistance offered by the pipeline is less than would be the case during the transport of mixtures of water and coarser sand under comparable circumstances. As a result, the bend in the output curve, between line section II (the actual working range of the pump) and section III (the area relating to (too) long delivery distances), will coincide with an extremely high delivery distance value, or will be missing altogether. The output curve will then be as shown in Fig 8 and consist of only two sections.

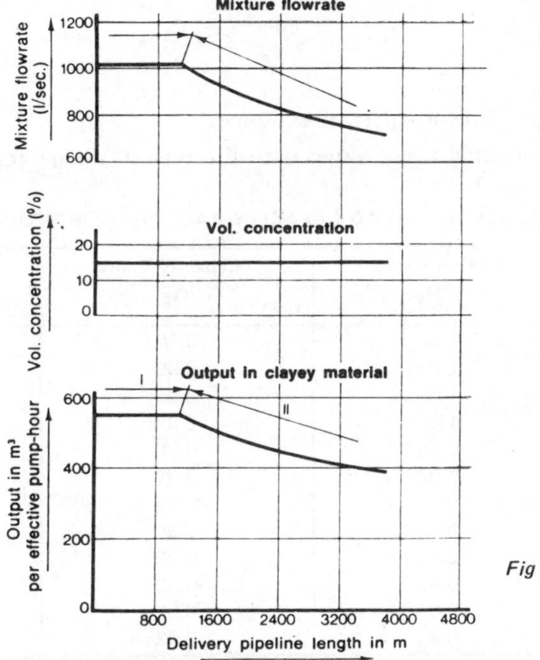

Fig 8

Sludge

Sludge is defined as a liquid (usually water) containing large solids with a particle size of 6 mm (¼ in) or greater, the solids being soft rather than abrasive in nature. These solids may be further described as 'stringy', 'clogging', *etc*, although a more useful classification would be 'soft' and 'hard' sludges, since if a sludge is defined by particle size only the solids may be hard and abrasive in some cases. Sludges may also contain a high proportion of smaller solids or sand which could have an abrasive effect, affecting pump material choice, clearances, *etc.* Thus sludges may have some of the characteristics of slurries. Sewage, on the other hand, mostly involves soft solids.

Frictional Factor

Frictional losses involved in the transport of sludges and sewage are difficult to evaluate other than on empirical lines. However, where the mixture is reasonably homogeneous a friction factor may be calculated on the basis of a pseudo-Reynolds number. This takes the form

$$R_e = \frac{AQ}{C^x d^y}$$

where Q is the flow rate
A is an empirical factor
C is the consistency of the mixture
d is the solids diameter

Values of the exponentials x and y are determined empirically for different fluids. Typical values for pulps are $x = 1.157$ and $y = 1.795$.

The friction factor, for insertion in standard function formulas, is then determined as:-

$$fs = \frac{K}{(R_e^{\,z})}$$

where K is a constant for a particular sludge.

The value of the exponential z also varies with the type of sludge, but is typically of the order of 1.63.

TABLE III – HEAD-CAPACITY FACTORS FOR STOCK

Stock Consistency %	Head Factor H_F	Capacity Factor Q_F	Head-Capacity Factor $H_F \cdot Q_F$
1.0	1.00	0.99	0.99
2.0	1.00	0.99	0.99
2.5	1.00	0.98	0.98
3.0	1.00	0.97	0.97
3.5	0.99	0.96	0.95
4.0	0.98	0.92	0.90
4.5	0.97	0.87	0.85
5.0	0.95	0.80	0.76
5.5	0.93	0.72	0.67
6.0	0.90	0.62	0.56
6.5	0.87	0.52	0.45
7.0	0.83	0.42	0.35

Paper Stock (Pulp)

Paper stock is basically in the form of sludge with a specific type of solids (paper pulp). With low consistencies (*ie* less than 1% pulp by weight), flow and friction losses can be calculated as for water. With higher consistencies there is an increasing derating of pump performance – see Table III.

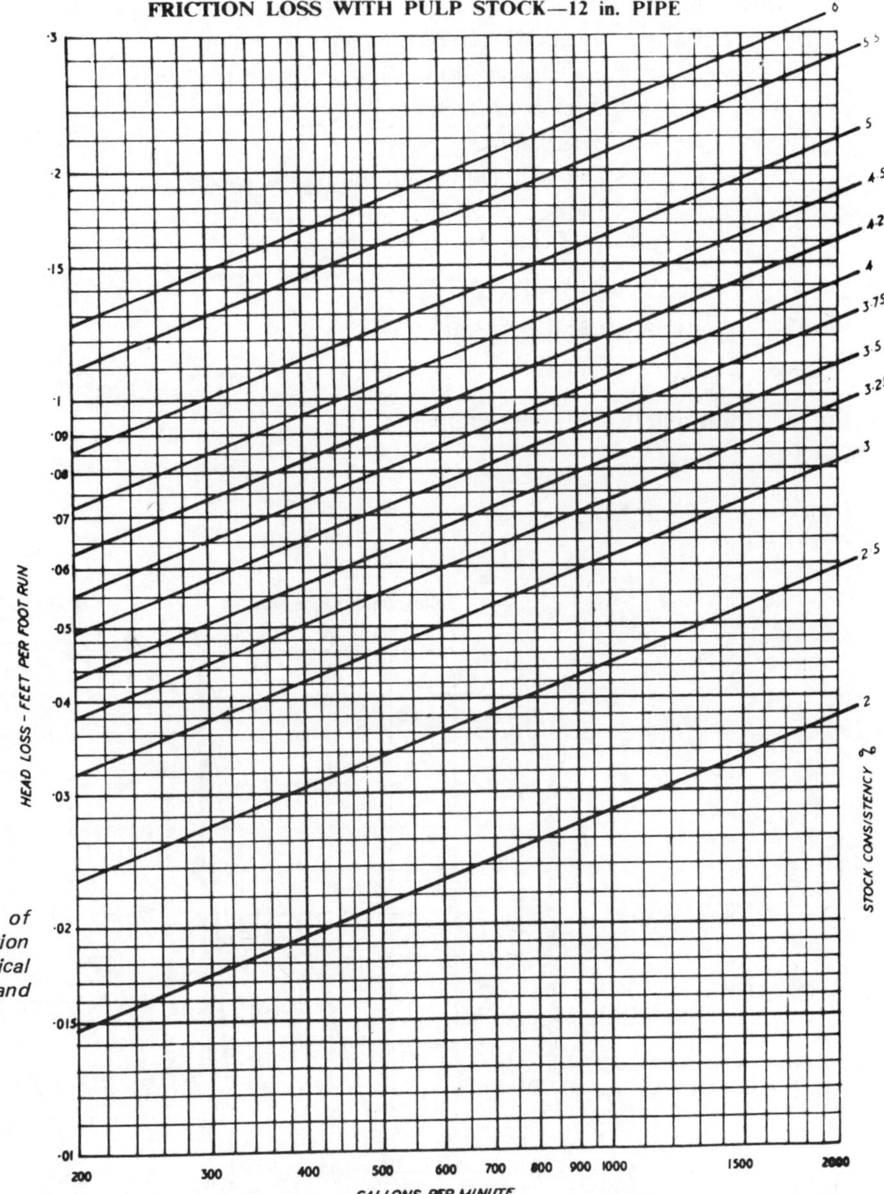

Fig 9 Determination of head-capacity correction factors. (a) chemical stock. (b) mechanical and reclaimed stock.

Frictional losses increase rapidly with stock consistencies above about 2% by weight. A friction factor can be determined based on a pseudo-Reynolds number, viz:

$$f = \frac{K}{{R_e}^x} = \frac{K^1 Q}{C^y D^z}$$

where K = an empirical constant, dependent on the type of pulp

R_e = psuedo-Reynolds number

K^1 = is a constant depending on the units employed

C = stock consistency %

D = pipe diameter

x, y and z are exponents with the typical values

x = 1.63

y = 1.16

z = 1.8

For Q in gallons/minute and D in feet

K^1 = 17.2

For Q in litres/minute and D in centimetres

K^1 = 8.18

(See also Figs 9 and 10).

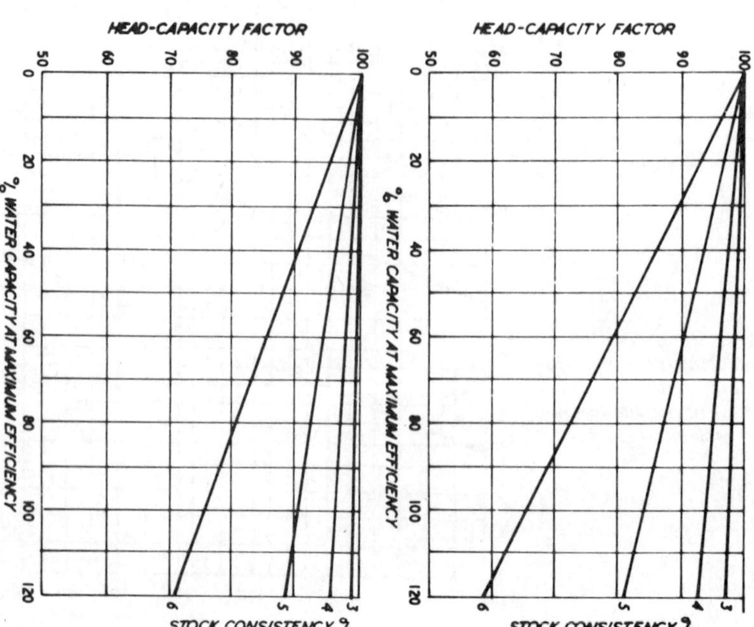

Fig 10 Friction loss with pulp stock – 12 inch pipe.

Compressible Flow in Pipes

AIR, STEAM and gases are all *compressible fluids* and the D'Arcy equation used for determining pressure and head losses with liquid flow is no longer applicable because the density of gases and vapours changes considerably with changes of pressure. However for simplified general engineering calculations not requiring great accuracy, liquid (Darcy) flow formulas may be used if the pressure drop involved is less than 10% of the inlet pressure. Use of such formulas is also sometimes extended to pressure drops up to 40% of the inlet pressure, provided in this case the specific volume is taken as the average of the upstream and downstream conditions.

The real flow of a pressurized gas through pipes differs appreciably in a number of important characteristics from the flow of liquids in pipes. Pressure, for example, drops at an increasing rate along the pipe, rather than with a constant pressure gradient. At the same time velocity tends to increase up to a maximum defined by $V = \sqrt{kgRT}$ for air, but subject to a limiting or maximum length, which must correspond to the end of the pipe. At this point the pressure gradient is infinite, *ie* the pipe is effectively closed. In this equation k is the ratio of specific heats at constant pressure to constant volume, R the individual gas constant, and T the absolute temperature (degrees Rankine). An alternative formula is $M = 1/\sqrt{k}$ (=0.845 for air), where M is the Mach number.

The general equation may be written in the same form as the Darcy equation for fluid flow, but with the addition of an extra term representing the pressure drop required to increase the flow momentum:

$$\frac{\Delta P}{L} = \frac{f}{D} \cdot \frac{\rho V^2}{2} + \beta\rho\overline{Q}\,\frac{dV}{dL}$$

where ΔP is the pressure drop

L is the pipe length

f is a constant friction factor (but dependent on surface roughness)

D is the pipe diameter

ρ is the gas density

$\overline{Q}$ is the specific volume of gas

dV/dL is the velocity gradient

β is a factor of the order of unity and normally taken as 1.0 (*ie* can be eliminated from the equation).

Actual flow conditions may range from adiabatic to isothermal. Adiabatic conditions are only likely to apply in short, well insulated pipes where no appreciable heat is transferred to or from the pipe. Isothermal flow or flow at constant temperature is commonly assumed as more consistent with normal practice, especially for long pipes. In fact, most practical pipelines will generate polytropic flow conditions, virtually impossible to analyze. The assumption of isothermal flow is thus a practical compromise.

Isothermal Flow

With isothermal flow, a formula developed from basic principles is

$$\frac{\Delta P}{L} = \frac{\frac{f}{D} \cdot \frac{\rho V^2}{2}}{kM^2 - 1}$$

The friction factor is dependent on the Reynolds number of flow and pipe roughness. It can be assumed independent of Mach number.

The Reynolds number will be constant for isothermal flow, but may vary with adiabatic or isentropic flow (and certainly with diabatic flow), in which case a mean value can be assumed.

The Reynolds number is a dimensionless quantity, given arithmetically by

$$R_e = \frac{\rho V D}{\mu}$$

$$= \frac{VD}{\nu}$$

where

ρ is the mass density
μ = viscosity
ν = kinematic viscosity

} in units consistent with velocity (V) and pipe diameter (D)

For air at standard temperature $\nu = 1.568 \times 10^{-4}\ ft^2/sec$

Thus $R_e = 531.5\ VD$
$= 530\ VD$ (with sufficient practical accuracy)

where

V is in ft/sec
and D is in inches

The friction factor (f) is common for all fluids (*ie* gases and liquids) and is normally determined from empirical charts. For laminar flow, the friction factor is dependent only on Reynolds number and is numerically equal to $64/R_e$, laminar flow being defined by the Reynolds number not exceeding 2 000. For turbulent flow and smooth bore pipes, the Reynolds number can be calculated from the following empirical formula –

$$f = \frac{0.3164}{R_e^{0.25}}$$

As the Mach number (M) approaches 0, the denominator in this equation comes closer and closer to unity, reducing the equation to the same as that for liquid flow. There is thus some justification for using the Darcy formula for compressible flow calculations (as mentioned initially) as $M \to 0$ (*ie* consistent with short lengths of pipes with resulting low pressure drops). This also implies that at very low Mach numbers, compressible flow can be treated as incompressible. Pressure gradients ($\Delta P/L$) will, in fact, be within 5% of incompressible flow values at Mach numbers up to about 0.18 for air.

A complete working formula for isothermal flow, expressed in terms of *mass flow rate* (Qm) is:

$$Qm^2 = \left[\frac{144gA^2}{\overline{V}_1\left(\frac{fL}{D} + 2\log\frac{P_1}{P_2}\right)}\right] \left[\frac{P_1^{\,2} - P_2^{\,2}}{P_1}\right]$$

where

A	=	cross sectional area of pipe, ft^2
g	=	acceleration of gravity = 32.2 ft/sec^2
$\overline{V}_1$	=	specific volume of gas, ft^3/lb
L	=	length of pipe, feet
f	=	friction factor
D	=	pipe bore, feet
P_1	=	absolute pressure (entry)
P_2	=	absolute pressure (exit)

For long gas pipelines (where velocity gradient can be ignored) this reduces to:

$$Qm^2 = \left(\frac{144gDA^2}{\overline{V}_1 f \cdot L}\right) \cdot \left(\frac{P_1^{\,2} - P_2^{\,2}}{P_1}\right)$$

Or expressed in terms *volume flow rate* (Q_V) in ft^3/sec

$$Q_V = 114.2\sqrt{\left(\frac{P_1^{\,2} - P_2^{\,2}}{fL\,T \cdot sg}\right) \cdot D^5}$$

where

T is the absolute temperature (degrees Rankine)

sg is the specific gravity of the gas

Other working formulas of similar derivation are:

Weymouth formula:

$$Q_V = 28.0D^{2.667}\sqrt{\left(\frac{P_1^{\,2} - P_2^{\,2}}{L\,sg}\right)\frac{520}{T}}$$

Panhandle formula:

$$Q_V = 36.8ED^{2.6182} \left(\frac{P_1{}^2 - P_2{}^2}{L}\right)^{0.5394}$$

where

E is an efficiency factor, normally taken as 0.92 for average conditions

The Panhandle formula is widely used for natural gas pipelines from 6 to 24 inch diameter.

Limiting Values

Maximum possible velocity (V^*) in a pipe is source velocity (M = 1), given directly by

$$V^* = 12\sqrt{kg\, PV_V}$$

where

V^* is in ft/sec

k is the ratio of specific heats of the gas

P is the absolute pressure, lb/in^2

V_V is the specific volume of gas, ft^3/lb

Maximum pressure (P^*) can be determined on the basis of continuity, viz:

$$V_r\, P_1 = VP = V^*\, P^*$$

Hence

$$\frac{P^*}{P_1} = \frac{V_1}{V^*} = \frac{M_1}{M^*} = M_1\sqrt{k}$$

where the suffix 1 refers to initial contribution

Hence (and since the velocity of sound is constant under isothermal conditions)

$$P^* = P_1 \cdot M_1\ k$$

Unity Length (L^*) can be determined from the general equation

$$\frac{fL^*}{D} = \left(\frac{1}{kM_1{}^2} - 1\right) - \log_e \frac{1}{kM_1{}^2}$$

or

$$L^* = \frac{D}{f}\left(\left(\frac{1}{kM_1{}^2} - 1\right) - \log_e \frac{1}{kM_1{}^2}\right)$$

The implication of this is that velocity of gas flow can go on increasing in a pipe up to a maximum Mach number of $1/\sqrt{k}$. This increase ceases at a limiting length of pipe (L^*), which must be at the end of the pipe. If the actual length of pipe is greater than L^*, initial conditions will have to be adjusted to reduce M_1 so that the actual pipe length is less than, or equal to, L^*.

At the same time the limiting pressure ratio (P^*/P_1) and limiting length (L^*) are dependent only on initial velocity (Mach number) and the k value of the gas.

Pressures and lengths between two sections of a pipeline can thus be expressed as follows, knowing the Mach numbers at each section:

$$\frac{P_2}{P_1} = \frac{M_1}{M_2}$$

$$\frac{fL}{D} = \left(\frac{fL^*}{D}\right)_{M_1} - \left(\frac{fL^*}{D}\right)_{M_2}$$

Adiabatic Flow

Similar treatment applies, although the corresponding formulas are more complicated and may require working as a series of approximations in order to reach real values in particular cases. In general, the pressure, temperature and velocity will always be slightly less than those for isothermal flow, but the limiting length will be similar. The differences are usually small enough to be negligible, except at higher Mach numbers, and thus, for simplification of calculations, isothermal formulas can be used for subsonic adiabatic flow.

Limiting Velocity:

This is given directly by:-

$$\frac{V^*}{V_1} = \frac{1}{M_1}\sqrt{\frac{2(1 + \frac{k-1}{2}M_1^2)}{k+1}}$$

Limiting Temperature:

$$\frac{T^*}{T_1} = \frac{2(1 + \frac{k-1}{k2}M_1^2)}{k+1}$$

Limiting Length:

$$L^* = \frac{D}{F}\left(\frac{1-M_1^2}{kM_1^2}\right) + \frac{k+1}{2k}\log_e\left[\frac{(k+1)M_1^2}{2(1 + \frac{k-1}{2}M_1^2)}\right]$$

Limiting Pressure:

$$\frac{P^*}{P_1} = M_1\left[\frac{2(1 + \frac{k-1}{2}M_1^2)}{k+1}\right]$$

Stagnation State

Flow is possible between two extremes. At one extreme, velocity is zero and temperature is a maximum, since all the kinetic energy is converted to enthalpy. The speed of sound is also a maximum (stagnation point or stagnation state). At the other extreme, the velocity is a maximum and the temperature falls to absolute zero, all the enthalpy being converted into kinetic energy. The speed of sound is then zero (zero temperature state). Between these extremes the practical flow may be subsonic, transonic, or supersonic – see Fig 1 – although the zero temperature state can never be reached (*ie* is a hypothetical condition).

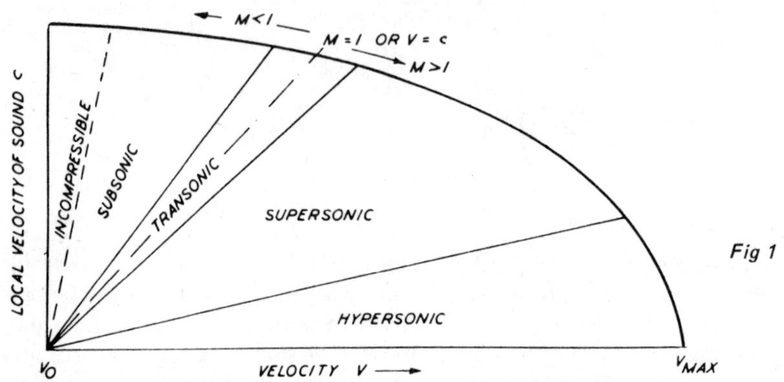

Fig 1

At the stagnation state

$$\frac{O^2}{2} + h_o = \frac{V^2}{2} + h$$

or

$$h_o = \frac{V}{2} + h$$

From the general gas relationship it follows that the stagnation temperature (T_o) is given by

$$T_o = T + \frac{V^2}{2sp}$$

where sp = specific heat at constant pressure

Alternatively,

$$\frac{T_o}{T} = 1 + \frac{k - 1}{2} \times M^2$$

where M = Mach number = $\frac{V}{c}$

The stagnation pressure can be derived as

$$\frac{P_o}{P} = \left(\frac{T_o}{T}\right)^{\frac{k}{k-1}}$$

$$= \left(1 + \frac{k-1}{2} M^2\right)^{\frac{k}{k-1}}$$

This may be expanded in the form

$$P_o = P + \frac{\rho V^2}{2} \left\{ 1 + \frac{M^2}{4} + \frac{2-k}{24} M^4 + \ldots\ldots\ldots\ldots \right\}$$

This can be compared with the equation for incompressible flow

$$P_o = P + \frac{\rho V^2}{2}$$

The difference between these two equations represents the effect of increased gas density due to compressibility, generally termed the *compressibility factor.* Values of the compressibility factor range from unity at very low Mach numbers (where there are no compressibility effects), up to 1.276 as the Mach number approaches 1 (velocity approaches the speed of sound in the gas). Besides increasing the dynamic pressure of compressible flow, compared with incompressible flow, the rising value of the compressibility factor can also affect the flow velocity through ducts with varying area. The relationship between area and velocity changes is, in fact, a function of the local Mach number, and can be rendered in the form

$$\frac{dA}{dV} = \frac{A}{V} (M^2 - 1)$$

With subsonic flow, a decrease in area produces an increase in flow velocity and vice versa (similar to incompressible flow) *ie,* area and Mach number changes are opposite. The flow velocity may be sonic only at a constant section. With supersonic flow, a decrease in flow area produces a decrease in flow velocity, and vice versa, *ie,* area and Mach number changes are the same.

Flow from Stagnation Conditions

Gas compressed and stored in a reservoir is essentially under stagnation conditions, where velocity is zero and the pressure and temperature are known (or can be determined). Where the reservoir is used as a supply, the velocity, temperature and pressure at any other section of flow are determined basically from the following relationships:-

Velocity at any arbitrary section:

$$V = \sqrt{2sp\, T_o \left(1 - \frac{P}{P_o}\right)^{\frac{k-1}{k}}}$$

Alternatively, for adiabatic flow, the velocity at any section can be determined from the temperature at that section

$$V = \sqrt{2sp\, T_o \left(1 - \frac{T}{T_o}\right)}$$

Flow rate

Flow rate can be determined as the mass flow, *ie,* mass flow = $VA\rho$ or directly as the product of V and A in numerically consistent units

$$\text{Dimensions are } \frac{L}{T} \times L^2 = \text{flow rate} = \frac{L^3}{T}$$

Pressure at any arbitrary section:

$$P = \frac{P_o}{\left(1 + \frac{k-1}{2} M^2\right)^{\frac{k}{k-1}}}$$

Temperature at any section:

$$T = \frac{T_o}{1 + \frac{k-1}{2} \cdot M^2}$$

At any (constant) section where the flow is sonic the flow conditions are described as critical, yielding a critical temperature (T^*) and a critical pressure (P^*), where

$$\frac{T^*}{T_o} = \frac{2}{k+1} \quad \text{(adiabatic or isentropic flow)}$$

$$\frac{P^*}{P_o} = \left(\frac{2}{k+1}\right)^{\frac{k}{k-1}} \quad \text{(isentropic flow only)}$$

Note : For air, where k = 1.4, the value of critical pressure is $\left(\frac{2}{2.4}\right)^{\frac{1.4}{0.4}} = 0.528$

That is, the critical pressure is 52.8% of P_o. Similarly the critical temperature can be calculated as 83.3% of T_o.

Critical area:

The relationship between the critical area (A^*) or throat area where $\mu = 1$ and the area of any other section (A) is given by

$$\frac{A}{A^*} = \frac{1}{M} \left(\frac{1 + \frac{k-1}{2} \cdot M^2}{\frac{k+1}{2}} \right)^{\frac{k+1}{2(k-1)}}$$

$$= \frac{1}{M} \left(\frac{1 + 0.2\,M^2}{1.2} \right)^3 \quad \text{for air}$$

Nozzle Flow

Flow at the throat of a nozzle, supplied by a reservoir or similar source under stagnation conditions, will be sonic if the critical pressure is greater than the receiver pressure – see Fig 2a. This means that the flow will be critical. The flow velocity follows from calculating the critical temperature, from which

$$\text{flow velocity} \quad c = 49\sqrt{T^*} \quad \text{ft/sec}$$

(The velocity of sound in air = $c = 49\sqrt{T}$ feet per second where T is the absolute temperature in degrees Rankine).

If the critical pressure is less than the receiver pressure, then the flow cannot be critical – Fig 2b. In this case the flow will be subsonic and the exit pressure will equal the receiver pressure. The temperature can be calculated from the general formula, or from

$$T_1 = T_o \left(\frac{P_1}{P_o} \right)^{\frac{k-1}{k}}$$

The velocity is likewise calculated from the general formula.

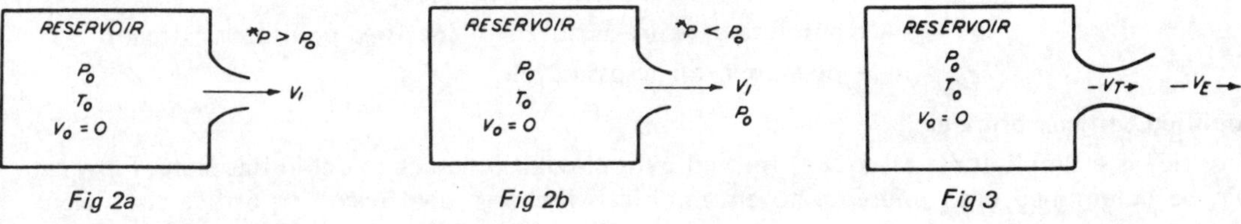

Fig 2a *Fig 2b* *Fig 3*

Similar analysis applies where the nozzle is of convergent-divergent form – Fig 3. In this case it is first necessary to establish whether the flow is critical or not (at the throat). The throat velocity can be determined accordingly; and from this the final exit velocity from the divergent section.

Flow through a nozzle can also be rendered directly in terms of flow rate and a discharge co-efficient, this being a convention for engineering calculations. The complete nozzle formula is:-

$$\text{mass flow} = A \cdot E\,c \cdot \delta \cdot d^2 \sqrt{h}\,\sqrt{P_2/T}$$

where A is a constant depending on the units employed

E = coefficient for the velocity of approach

$$= \frac{1}{\sqrt{1 - m^2}}$$

$$\text{where } m = \frac{\text{cross sectional area of nozzle}}{\text{cross sectional area upstream}} = \frac{A_2}{A_1}$$

c = nozzle coefficient

δ = expansibility factor allowing for the change in air density which occurs during acceleration through the nozzle

$$= \frac{1 - 0.07h}{13.6\,P_2} \quad \text{for values of m circa 0.16}$$

(and h in inches wg and P_2 in inches of mercury)

d = diameter of nozzle

h = pressure drop across nozzle

P_2 = absolute pressure on downstream side of nozzle

T = absolute temperature on downstream side of nozzle

If T is in °R, P_2 in inches of mercury, h is in inches wg and d is in inches, a value of A = 0.1148 gives the mass flow in units of lb/sec.

For a specific nozzle profile the formula can be simplified by the use of a nozzle constant appropriate to that particular geometry and nozzle size. Rendered as a solution for conventional flow rate (Q)

$$Q = K\,(T_1/P_1)\sqrt{h}\quad\sqrt{P_2/T_2}$$

where K = nozzle constant

T_1 = absolute temperature at specified inlet point

T_2 = absolute temperature at nozzle or specified point downstream

P_1 = absolute pressure at specified inlet point

P_2 = absolute pressure at nozzle or specified point downstream

h = pressure drop across nozzle

Simplified Orifice Formulas

An orifice is a simple form of nozzle, formed by a circular hole cut in a thin flat plate. Flow can again be determined with reference to an empirical discharge coefficient, or orifice coefficient. This will be much lower than for nozzles because of the less streamlined flow but, due to the simpler form of the nozzle, will be less subject to variation. Thus nozzle coefficients may vary between 0.90 (or less) and 0.995, depending on size and geometry, whereas an orifice coefficient can be expected to be of the order of 0.61, regardless of size, and differing only if the orifice has a well rounded, as opposed to a sharp, entry.

Very much simplified formulas can therefore be applied to assess the discharge of air through orifices, and the following are generally satisfactory for straightforward engineering calculations:-

(1) For upstream pressures above 14.7 lb/in^2g

Q (ft^3/min) for sharp-edged orifice $= \dfrac{218 \times A \times P_u}{\sqrt{460 + T}}$ (a)

or $= \dfrac{172 \times d^2 \times P_u}{\sqrt{460 + T}}$

Q (ft^3/min) for rounded entrance orifice $\triangleq \dfrac{417 \times A \times P_u}{\sqrt{460 + T}}$ (b)

or $\triangleq \dfrac{263\ 5 \times d^2 \times P_u}{\sqrt{460 + T}}$

(2) For upstream pressures below 14.7 lb/in^2g

Q (ft^3/min) for sharp-edged orifice $= \dfrac{210 \times A \times P_u}{\sqrt{460 + T}}$ (c)

or $= \dfrac{166 \times d^2 \times P_u}{\sqrt{460 + T}}$

Q (ft^3/min) for rounded entrance orifice $\triangleq \dfrac{324 \times A \times P_u}{\sqrt{460 + T}}$ (d)

or $\triangleq \dfrac{255 \times d^2 \times P_u}{\sqrt{460 + T}}$

where

A = orifice area, square inches
P_u = upstream pressure, lb/in^2g
d = orifice diameter, inches
T = upstream air temperature °F

Note that for an upstream air temperature of 60°F these formulas further simplify to:-

1 (a)	Q (ft^3/min)	$= 11.9\ A \cdot P_u$	2 (c)	Q (ft^3/min)	$= 11.5\ A \cdot P_u$
	or	$= 9.4\ d^2\ P_u$		or	$= 9.05\ d^2\ P_u$
(b)	Q (ft^3/min)	$\triangleq 18.3\ A \cdot P_u$	(d)	Q (ft^3/min)	$\triangleq 17.7\ A \cdot P_u$
	or	$\triangleq 14.4\ d^2\ P_u$		or	$\triangleq 13.92\ d^2\ P_u$

Steam Flow Calculations

WATER FREEZES at 32°F (0°C) and boils at 212°F (100°C) under normal atmospheric pressure. 32°F (0°C) is the reference phase for zero heat content. Since the actual boiling point is dependent on ambient pressure, this can be designated t_1. The heat required to raise the temperature of water from 32°F to t_1, marking the onset of vaporization, is known as the *sensible heat* (h). At atmospheric pressure the sensible heat of water is 150.17 Btu per pound.

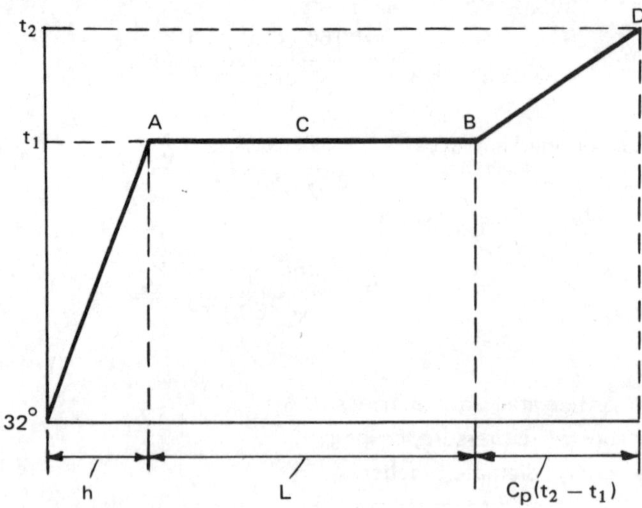

Fig 1

If heat continues to be applied to the water the process of vaporization (boiling) goes on until all the water has been transformed to steam – AB in Fig 1. During this period the temperature remains constant. The heat absorbed is called the *latent heat of vaporization* (L), so that at point B

$$\text{total heat absorbed (H)} = h + L$$

Again at atmospheric pressure L = 970.3 Btu per pound, so that

$$H = 150.17 + 970.3$$
$$= 1150.5 \text{ Btu per pound}$$

At any intermediate point (C) between A and B, the stage of vaporization can be expressed as the *dryness fraction* (q), or ratio of latent heat at that point to that necessary to produce a state of *dry saturation* (point B).

Thus at any point C;

$$H = h + qL$$

If after point B more heat is added to the dry saturated steam, the steam is said to be *superheated.* Provided the steam is subject to unrestricted expansion the pressure remains the same, but the temperature rises to t_2. The degree of superheat is then $Cp(t_2 - t_1)$; or total heat absorbed to superheat steam to temperature t_2 is:

$$H = h + L + Cp\,(t_2 - t_1)$$

The approximate value of Cp is 0.48. Specific values are obtained from steam tables.

Steam can thus exist in three forms – *dry saturated steam,* free of any water particles in suspension; *wet saturated steam,* containing water particles in suspension; and *superheated steam.* The higher the degree of superheat the steam possesses, the more closely its characteristics resemble those of a perfect gas.

Steam Flow Through Pipes

The original (Babcock) formula for determining the quantity of steam able to pass through a particular pipe with a specified pressure drop is:-

$$W = 87.5\sqrt{\frac{D(P_1 - P_2)d^5}{L(1 + 3.6/d)}}$$

where W = quantity of steam, lb/min
D = density of steam, lb/ft^3
P_1 = initial pressure of steam, lb/in^2
P_2 = final pressure of steam at end of pipe, lb/in^2
d = inside diameter of pipe, inches
L = length of pipe, feet

The formula can also be rewritten as a solution for pressure drop

$$(P_1 - P_2) = 0.0001306\ \frac{W^2 L(1 + 3.6/d)}{D \times d^5}$$

These formulas are applicable both to saturated steam and superheated steam.

A further formula derived from this gives the size of pipe required for a specific mean velocity of steam.

$$d = 1.75 \sqrt{\frac{W}{V \times D}}$$

where V = steam velocity, ft/sec

Typical values of V are 70 – 100 ft/sec for exhaust steam
100 – 150 ft/sec for saturated steam
130 – 200 ft/sec for superheated steam

Such formulas may still be used for simplified (approximate) calculations, but modifications of the D'arcy equation are now normally employed, viz:-

$$(P_1 - P_2) = \Delta P = W^2 \left(\frac{0.000336f}{d^5}\right) \cdot \bar{V}$$

$$= W^2 \left(\frac{33600f}{d^5}\right) \cdot \bar{V} \times 10^{-9}$$

where W is the flow rate, lb/hour
$\bar{V}$ is the specific volume, ft^3/lb
f is the friction factor appropriate to the pipe and flow velocity

Working is further simplified by calculating a discharge factor (C_1):

$$C_1 = W^2 \times 10^{-9}$$

and a size factor (C_2):

$$C_2 = \frac{336000f}{d^5}$$

(Tabular data are available for determining C_2).

Then:

$$\Delta P = C_1 C_2 \bar{V}$$

$$= \frac{C_1 C_2}{\rho}$$

where ρ is the mass density of the steam, lb/ft^3
and is dependent on temperature and pressure

$$W = \sqrt{C_1 \times 10^{-9}}$$

$$d = \sqrt[5]{\frac{35600f}{C_2}}$$

$$Q = \frac{W}{4.8}$$

where Q = flow rate, ft^3/min at STP

Sizing of Condensate-Return Lines

Basic Considerations

The diameter of the pipeline between the heat exchanger and steam trap is normally chosen to fit the nominal size of the trap.

When choosing the diameter of the condensate line downstream of the trap flashing has to be considered. Even at very low pressures the volume of flash steam is many times that of the liquid if the condensate is at saturation temperature upstream of the trap (*eg* during flashing from 17 to 14.4 lb/in^2 (1.2–1 bar) the volume increases approximately seventeen times.

In these cases it is possible to dimension the condensate line in accordance with the amount of flash steam formed. The flow velocity of the flash steam should not be too high otherwise water-hammer (by the formation of waves), flow noises and erosion may occur.

A flow velocity of 50 ft/sec (15 m/sec) at the end of the pipeline before the inlet into the collecting tank or flash vessel is a useful empirical value. The inside diameter of the pipeline required can be taken from Table I.

For long pipelines (over 300 ft or 100 metres) and large condensate flowrates the pressure drop should be calculated to avoid the back pressure becoming too high. The velocity of the flash steam may be used in the calculations – see Table I and Fig 2.

When the condensate is mainly in the liquid state (*eg* high degree of undercooling, extremely low pressure) the flow velocity of the condensate should, if possible, be rated at 1.64 ft/sec (0.5 m/sec) or higher. The pipeline diameter can be chosen from the chart – Fig 3.

If the condensate is pumped, the condensate in the pump discharge line can only be in the liquid phase. For choosing the pipeline diameter the mean velocity can be rated at 5 ft/sec (1.5 m/sec). Again Fig 3 may be used.

Calculation of Condensate Flowrates

Basic Formulae

If the amount of heat required in kcal/h is known (indicated on the name plate of the heat exchanger) or easy to determine, the condensate flowrate $\dot{M}$ can be calculated

$$\dot{M} = 1.2 \frac{kcal/h}{500} \ (kg/h)$$

The quotient 500 is the latent heat of steam (kcal/kg) for medium pressures. The factor 1.2 is added to compensate for the heat losses.

TABLE I – CONDENSATE-LINE SIZING (BASED ON FLASH STEAM)

State of condensate before flashing		Pressure at end of condensate line (bar a)																					
Pressure bar a	Related boiling temperature °C	0.2	0.5	0.8	1.0	1.2	1.5	2.0	2.5	3.0	3.5	4.0	4.5	5.0	6	7	8	9	10	12	15	18	20
1.0	99	35.7	16.0	7.4	—	—	—	—	—	—	—	—	—	—	—	—	—	—	—	—	—	—	—
1.2	104	37.9	18.0	10.0	6.1	—	—	—	—	—	—	—	—	—	—	—	—	—	—	—	—	—	—
1.5	111	40.1	20.6	12.9	9.5	6.8	—	—	—	—	—	—	—	—	—	—	—	—	—	—	—	—	—
2.0	120	44.2	23.5	15.8	12.6	10.3	7.6	—	—	—	—	—	—	—	—	—	—	—	—	—	—	—	—
2.5	127	46.8	25.5	17.7	14.5	12.3	9.2	5.3	—	—	—	—	—	—	—	—	—	—	—	—	—	—	—
3.0	133	48.8	27.1	19.2	16.0	13.9	10.7	7.3	4.5	—	—	—	—	—	—	—	—	—	—	—	—	—	—
3.5	138	50.4	28.4	20.4	17.1	15.0	11.9	8.5	6.0	3.8	—	—	—	—	—	—	—	—	—	—	—	—	—
4.0	143	52.0	29.6	21.5	18.2	18.0	12.9	9.7	7.3	5.3	3.5	—	—	—	—	—	—	—	—	—	—	—	—
4.5	147	53.3	30.5	22.3	19.0	16.9	13.7	10.5	8.1	6.3	4.7	3.0	—	—	—	—	—	—	—	—	—	—	—
5	151	54.3	31.5	23.1	19.8	17.7	14.4	11.2	8.9	7.1	5.6	4.2	2.8	—	—	—	—	—	—	—	—	—	—
6	155	55.7	32.3	23.9	20.5	18.4	15.2	11.9	9.6	7.9	6.5	5.1	4.0	2.7	—	—	—	—	—	—	—	—	—
7	158	56.5	33.0	24.5	21.1	18.9	15.7	12.4	10.1	8.4	7.0	5.7	4.6	3.5	2.1	—	—	—	—	—	—	—	—
8	170	59.9	35.5	26.7	23.1	20.9	17.6	14.2	11.9	10.2	8.9	7.7	6.7	5.8	4.8	4.0	—	—	—	—	—	—	—
9	175	61.3	36.4	27.5	23.9	21.7	18.3	14.9	12.6	10.9	9.5	8.4	7.4	6.6	5.5	4.8	2.4	—	—	—	—	—	—
10	179	62.3	37.2	28.2	24.6	22.3	18.9	15.5	13.1	11.4	10.0	8.9	7.9	7.1	6.0	5.3	3.3	2.1	—	—	—	—	—
12	187	64.4	38.7	29.5	25.7	23.5	19.9	16.5	14.1	12.3	11.0	9.8	8.9	8.0	7.0	6.2	4.5	3.6	2.8	—	—	—	—
15	197	66.9	40.5	31.0	27.2	24.8	21.5	17.7	15.2	13.4	12.0	10.8	9.9	9.1	8.0	7.2	5.6	4.8	4.2	2.9	—	—	—
18	206	69.0	42.0	32.3	28.4	26.0	22.3	18.7	16.2	14.3	12.9	11.7	10.8	9.9	8.8	8.0	6.5	5.7	5.1	3.9	2.5	—	—
20	211	70.2	42.9	33.0	29.0	26.6	22.9	19.2	16.7	14.8	13.4	12.2	11.2	10.4	9.2	8.4	7.0	6.2	5.6	4.4	3.1	1.7	—
25	223	72.9	44.8	34.7	30.6	28.1	24.2	20.4	17.9	15.9	14.5	13.2	12.2	11.4	10.2	9.3	7.9	7.1	6.5	5.4	4.2	3.1	2.5
30	233	75.1	46.3	36.0	31.8	29.2	25.3	21.4	18.8	16.8	15.3	14.0	13.0	12.1	10.9	10.0	8.6	7.8	7.2	6.1	4.9	4.0	3.4
35	241	76.8	47.5	37.0	32.7	30.1	26.1	22.1	19.5	17.5	15.9	14.6	13.6	12.7	11.4	10.5	9.2	8.4	7.8	6.7	5.5	4.5	4.0
40	249	78.5	48.7	38.0	33.6	31.0	26.9	22.9	20.1	18.1	16.5	15.2	14.1	13.2	12.0	11.0	9.7	8.6	8.2	7.1	6.0	5.0	4.5
45	256	80.0	49.7	38.8	34.4	31.7	27.5	23.5	20.7	18.6	17.0	15.7	14.6	13.7	12.4	11.4	10.1	9.3	8.6	7.5	6.3	5.4	4.9
50	263	81.4	50.7	39.6	35.2	32.5	28.2	24.1	21.2	19.1	17.5	16.2	15.1	14.2	12.8	11.8	10.5	9.6	9.0	7.9	6.7	5.7	5.2

To determine the actual diameter the above values must be multiplied with the following factors:

kg/h	100	200	300	400	500	600	700	800	900	1000	1500	2000	3000	5000	8000	10 000	15 000	20 000
Factor	1.0	1.4	1.7	2.0	2.2	2.4	2.6	2.8	3.0	3.2	3.9	4.5	5.5	7.1	8.9	10.0	12.2	14.1

Bases for determining the **inside** pipe diameter: 1. The flash steam amount only is being considered
2. The flow velocity of the flash steam is assumed to be 15 m/s.

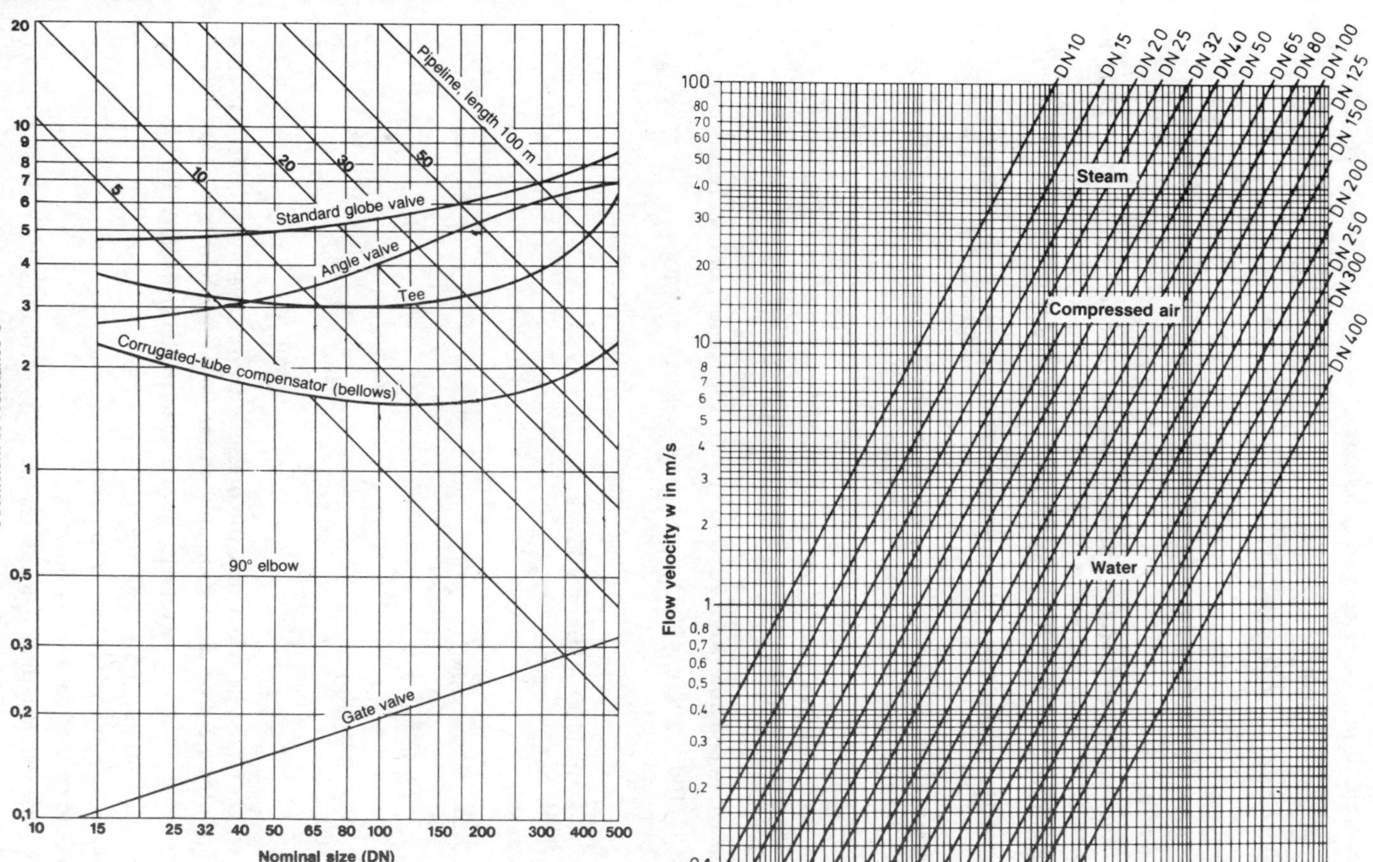

The coefficients of resistance C of all pipeline components of the same size are read in the above chart. The total pressure drop Δp in bar can be determined with the sum of all individual components (ΣC) and the operating data, see Fig 5.

Fig 2 Pressure drops in steam lines. (Gestra UK Ltd)

Fig 3 Volume flowrates in pipelines (Gestra UK Ltd)

In SI units the condensate flowrate is calculated as follows:

$$\dot{M} = 1.2 \quad \frac{W}{2000} \quad \cdot \quad \frac{3600}{1000}$$

Hence

$$\dot{M} \approx 1.2 \quad \frac{W}{560} \quad (kg/h)$$

W is the amount of heat required in Watts or Joule per second (J/s) and the quotient 2000 the latent heat of steam (kJ/kg) for medium pressures.

If the amount of heat Q required per hour is not known, it can be calculated from the weight $\dot{M}$ of the product to be heated in one hour, the specific heat ($c = \frac{kcal}{kg\ K}$ or in SI units $c = \frac{J}{kg\ K}$) and the difference between initial temperature t_1 and final temperature t_2 ($\Delta t = t_2 - t_1$):

$$Q = \dot{M} \cdot c \cdot \Delta t \ (kcal/h)$$

or in SI:

$$Q = \dot{M} \cdot \frac{c}{3600} \cdot \Delta t \ (W)$$

Example:

50 kg of water to be heated in one hour from 20°C to 100°C. The amount of heat required is:

$$Q = 50 \cdot 1 \cdot (100 - 20) = 4000 \ kcal/h$$

or in SI:

$$Q = 50 \cdot \frac{4187}{3600} \cdot (100 - 20) = 4652 \ W$$

The amount of condensate is:

$$\dot{M} = 1.2 \cdot \frac{4000}{500} = 9.6 \ kg/h$$

or in SI:

$$\dot{M} = 1.2 \cdot \frac{4652}{560} = 9.97 \ kg/h$$

Sizing of Steam Lines

When sizing steam lines, care must be taken that the pressure drop between the boiler and steam users is not too high. The pressure drop depends mainly on the flow velocity of the steam.

The following empirical values for the flow velocity have proven to be satisfactory.

Saturated steam lines 20 – 40 m/s (65 – 130 ft/sec)

Superheated steam lines 35 – 65 m/s (115 – 215 ft/sec)

The lower figures should be used for smaller flowrates.

For a given flow velocity the required pipe diameter can be chosen from the chart – Fig 4.

The pressure drop can be calculated from the charts Fig 2 and Fig 5.

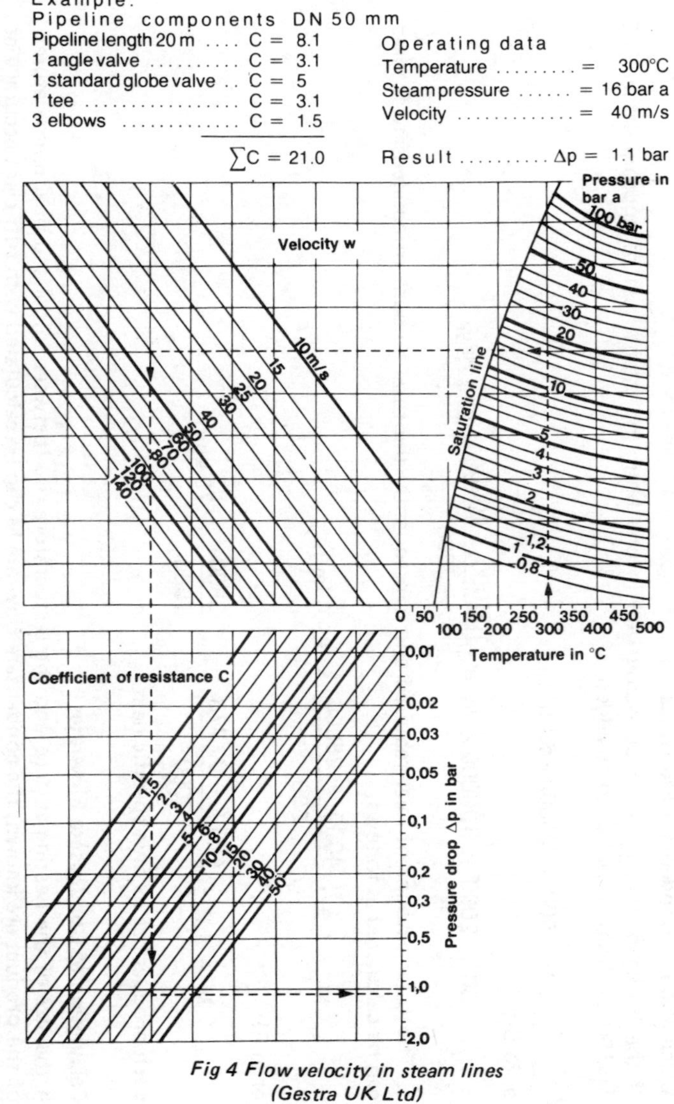

Fig 4 Flow velocity in steam lines (Gestra UK Ltd)

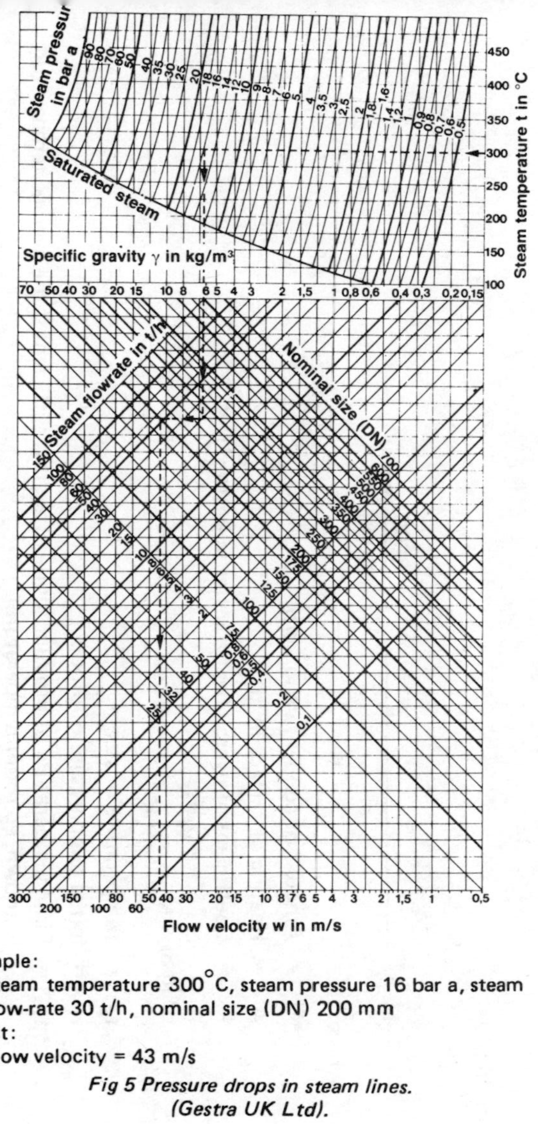

Example:
Steam temperature 300°C, steam pressure 16 bar a, steam flow-rate 30 t/h, nominal size (DN) 200 mm

Result:
Flow velocity = 43 m/s

Fig 5 Pressure drops in steam lines. (Gestra UK Ltd).

Calculation of Condensate Flowrates

If the 50 kg of water are to be vapourized in one hour the latent heat of approximately 500 kcal/kg or 2000 kJ/kg has to be added

$$50 \cdot 500 = 25000 \text{ kcal/h}$$

or in SI:

$$50 \cdot 2000 = 100000 \text{ kJ/h} = \frac{100000 \cdot 1000}{3600} = 27778 \text{ W}$$

The total amount of heat required and consequently the total amount of condensate formed can be calculated as follows:

$$\dot{M} = 1.2 \cdot \frac{4000 + 25000}{500} = 69.6 \text{ kg/h}$$

or in SI:

$$\dot{M} = 1.2 \cdot \frac{4652 + 27778}{2000} \cdot \frac{3600}{1000} = 70 \text{ kg/h}$$

Each product has its own specific heat.

Calculation of Condensate Flowrates

If the size of the heating surface and the temperature rise (between initial and final temperatures) of the product are known, the condensate flowrate $\dot{M}$ can be calculated with sufficient accuracy as follows:

$$\dot{M} = \frac{A \cdot k \left(t_s - \frac{t_1 + t_2}{2}\right)}{r} \text{ (kg/h)}$$

or in SI:

$$\dot{M} = \frac{A \cdot k \left(t_s - \frac{t_1 + t_2}{2}\right)}{r} \cdot \frac{3600}{1000} \text{ (kg/h)}$$

where:

$\dot{M}$ = Amount of condensate in kg/h

A = Heating of surface in m^2

k = Coefficient of overall heat transfer in kcal/m^2 h K

or in SI in $\frac{W}{m^2 K}$

t_s = Temperature of steam

t_1 = Initial temperature of product

t_2 = Final temperature of product (quite often it is sufficient if the average temperature is known, *eg* room temperature)

r = Latent heat in kcal/kg or kJ/kg (approximation for medium pressures 500 kcal/kg or in SI 2000 kJ/kg

A few empirical values for the coefficient of overall heat transfer k are given as follows:-

	$\frac{\text{kcal}}{\text{m}^2\text{h K}}$	$\frac{\text{W}}{\text{m}^2\text{K}}$
Insulated steam line	0.5 – 2	0.6 – 2.4
Non-insulated steam line	7 – 10	8 – 12
Unit heater with natural circulation	4 – 10	5 – 12
Unit heater with forced circulation	10 – 40	12 – 46
Jacketed boiling pan with agitator	400 – 1300	460 – 1500
As above with boiling liquid	600 – 1500	700 – 1750
Boiling pan with agitator and heating coil	600 – 2100	700 – 2400
As above with boiling liquid	1000 – 3000	1200 – 3500
Tubular heat exchanger	250 – 1000	300 – 1200
Evaporator	500 – 1500	580 – 1700
As above with forced circulation	800 – 2600	900 – 3000

Boiler Feed Calculations

BOILER FEED pumps have to deliver hot water at temperatures exceeding 100°C from closed feed tanks, with a steam cushion of a minimum saturated vapour pressure at a given temperature of the feed water. A typical layout is shown in Fig 1, when the geodetic positive suction lift is given by

$$H_{gs} = \Delta H_c + h_{zs}$$

where

ΔH_c = cavitation margin

h_{zs} = pressure losses in suction pipe

The positive suction lift (h) must be equal to or greater than H_{gs}.

In smaller size boiler feed pumps the calculated H_{gs} value will usually be increased by the difference of saturated vapour pressure at t_{max} and [t_{kmax}] converted to metres [of water column], where t_{kmax} is the temperature of water from the balancing device. When planning larger boiler

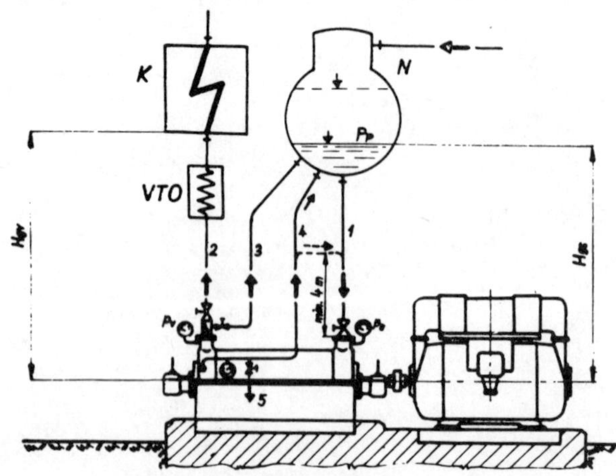

Fig 1 Layout diagram of a boiler feed pump (Intersigma)

- K — steam boiler
- N — feed water tank
- VTO — high pressure heater
- 1 — inlet piping
- 2 — discharge piping
- 3 — leak-off
- 4 — balancing piping from balancing device
- 5 — discharge from balancing device
- p_p — saturated steam pressure of feed water at respective temperature
- p_s — suction branch manometric pressure
- p_v — final resultant pressure of boiler feed pump in discharge branch
- H_{gv} — geodetic delivery head
- H_{gs} — geodetic suction head

feed pumps, check calculations of the intake piping are recommended, particularly with regards to the positive suction lift. This is mainly to be done in the operation of the so called stepping de-aeration power s.

To meet potential requirements in individual projects regarding small positive suction lifts, a so called 'booster pump' is installed before the boiler feed pump. The positive suction life of the booster pump is determined in the same way as for the boiler feed pump.

Minimum pressure in the suction branch (p_s) of the boiler feed pump and/or the booster pump is given by the relation:

$$p_s = p_p + \frac{(h - H_{zs})\,\gamma}{10}\ \text{(bar)}$$

where

p_p = vapour tension pressure at a given temperature (bar)

γ = specific weight of water at a given temperature (kg/dm^3)

$$p_s = p_p + \frac{(h - H_{zs})\,\gamma}{148}\ (\text{lb/in}^2)$$

where

h and H_{zs} are in feet

γ is in lb/ft^3

Pressure in the discharge branch (p_v) is given by:

$$p_v = p_k + \frac{h_{zv} \times \gamma}{10}\ \text{(bar)}$$

where

p_k = steam generator (boiler) pressure (bar)

h_{zv} = pressure losses in delivery piping from the boiler feed pump branch as far as the boiler (m wg)

$$p_v = p_k + \frac{h_{zv} \times \gamma}{148}\ (\text{lb/in}^2)$$

where

h_{zv} is in feet

and γ is in lb/ft^3

Boiler Feed Pump Head

The head to be generated by the boiler feed pump should be calculated from the required pressure in the discharge branch of the boiler feed pump using the following relation:

$$H = \frac{p_v - p_s}{\gamma} \times 10\ \text{(m wg)}$$

$$H = \frac{p_v - p_s}{\gamma} \times 1\,755\ \text{(in wg)}$$

where

pressures are in lb/in^2

and γ in lb/ft^3

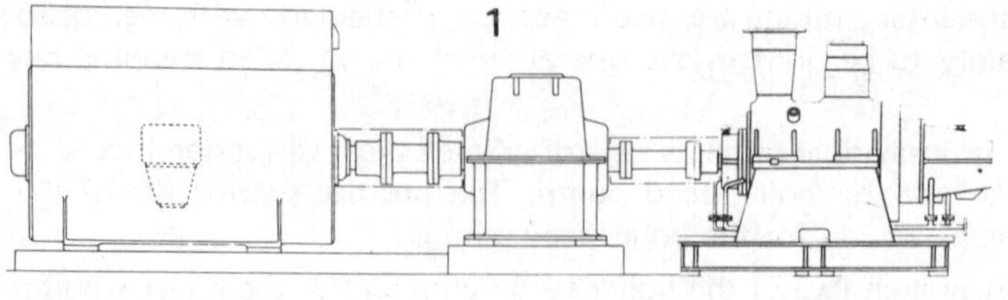

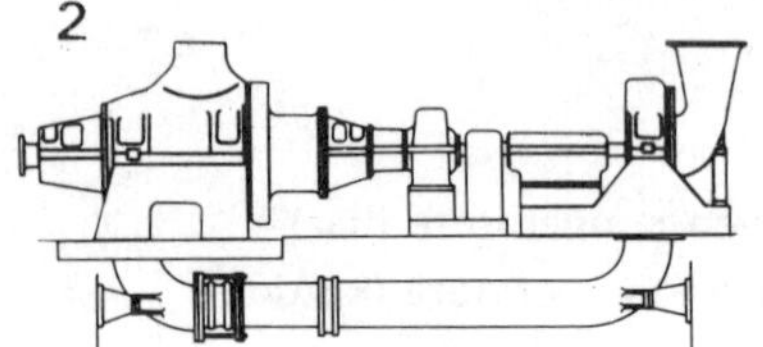

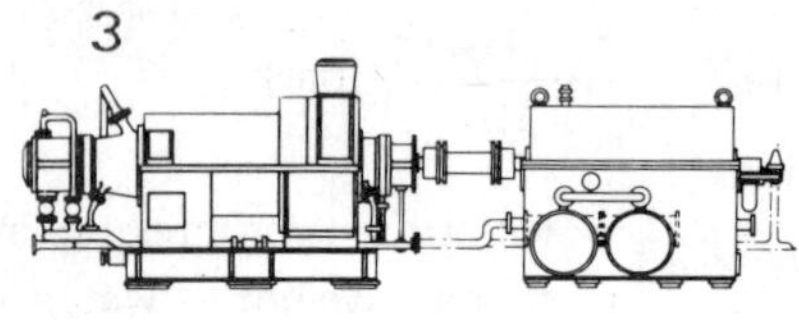

1. High speed boiler feed pump with electric motor and gearbox.
2. Tandem set of turbine-driven main boiler feed pump, gearbox and booster pump.
3. Compensated boiler feed pump with fluid coupling.
4. Tandem set of booster pump, motor, gearbox and main boiler feed pump.

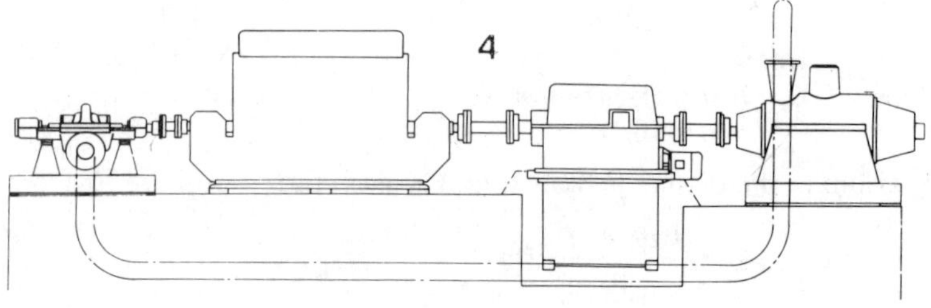

Arrangements of boiler feed pumps. (Weir).

However, the project engineer should take into account that the pump head in (m wg) will change its characteristics consistently regardless of the feed water temperature.

The specific weight of the water will change with its temperature. If a booster pump is to be installed before the boiler feed pump, the same relation should be applied for calculating the pump head as that used for the calculating the pump head when no booster pump is involved, so that:

$$H = \frac{p_v - p_s}{\gamma} \times 10 \text{ (m wg)}$$

$$H = \frac{p_v - p_s}{\gamma} \times 1755 \text{ (in wg)}$$

The only difference will be in establishing the magnitude of pressure in the inlet branch of boiler feed pump which can be calculated by the relation:

$$p_s = p_{vn} - \frac{h_{zp} \times \gamma}{10} \text{ (bar)}$$

where

p_{vn} = discharge branch pressure of booster pump (bar)

h_{zp} = head losses in the piping between the booster pump and boiler feed pump (m)

$$p_s = p_{vn} - \frac{h_{zp} \times \gamma}{148} \ (\text{lb/in}^2)$$

for h_{zp} in ft,
γ in lb/ft^3

Condensate collection equipment using Girdlestone pumps as installed at Shellstar Fertilizer complex in Cheshire. (Girdlestone Pumps Ltd).

Sigmund pulsometer type CA/CB boiler feed pump.

ivery

er feed pump delivery may be expressed as (L/h) in project documents. The pumps are, how..., tested in the manufacturer's test shops with cold water. In European practice pump delivery is given in (L/min) and/or (L/s).

The relation between these deliveries will be:

$$Q_{(L/h)} = Q_{(L/min)} \times \frac{\gamma \times 60}{1000}$$

Similarly, for Q in gallons/minute

$$Q_{(gal/h)} = Q_{(gal/m)} \times 60$$

Shaft input power required is given by:

$$N = \frac{Q \times (p_V - p_S)}{\gamma \times \eta} \times 0.02723 \text{ (kW)}$$

where

Q = pump delivery in (L/h)

p_V = discharge branch pressure (bar)

p_S = inlet branch pressure (bar)

γ = feed water specific weight (kg/dm^3)

η = boiler feed pump efficiency at a calculated point (%)

In English units:

$$N = \frac{Q_1 \times (p_V - p_S)}{\gamma \times \eta} \times 0.009325 \text{ (kW)}$$

$$= \frac{Q_1 \times (p_V - p_S)}{\gamma \times \eta} \times 0.0125 \text{ (hp)}$$

$$= \frac{Q_2 \times (p_V - p_S)}{\gamma \times \eta} \times 0.00777 \text{ (kW)}$$

$$= \frac{Q_2 \times (p_V - p_S)}{\gamma \times \eta} \times 0.010 \text{ (hp)}$$

where

Q_1 is in UK gal/min

Q_2 is in US gal/min

p_V and p_S are in lb/in^2

γ is in lb/ft^3

The pump shaft input of a boiler feed pump should be given by the manufacturer in the consistent units given later and calculated by the following equation:

$$N = \frac{Q \times H \times \gamma}{102 \times 60 \times \eta} = \frac{N_u}{\eta} \quad \text{(kW)}$$

where

Q = pump delivery (L/min)

H = head (m wg)

N_u = pump useful capacity (kW)

γ = feed water specific gravity (kg/m^3)

In English units (head H is in wg)

$$N = \frac{Q_1 \times H \times \gamma}{51500 \times \eta} \text{ (kW)}$$

$$= \frac{Q_1 \times H \times \gamma}{38420 \times \eta} \text{ (hp)}$$

$$N = \frac{Q_2 \times H \times \gamma}{42920 \times \eta} \text{ (kW)}$$

$$= \frac{Q_2 \times H \times \gamma}{3200 \times \eta} \text{(hp)}$$

For the selection of the driving machine a planning margin in its output should be considered, due to the inaccuracies in calculations of the whole system and unpredictable conditions. This is why the driving machine output (N_M) will be obtained by the following relation:

$$N_M = (1.08 \div 1.2)N$$

When speed increasing gear boxes (speed reducing gear boxes for booster pumps) and hydraulic couplings are to be set, then their efficiency should also be taken into account for calculating the pump shaft input so that:

$$N = \frac{Q \times H \times \gamma}{102 \times 60 \times \eta \times \eta_p \times \eta_{sp}}$$

where

η_p = gear box efficiency

η_{sp} = hydraulic coupling efficiency

If the pump was tested with cold water supply then the following recalculation of the pump efficiency should be carried out if pumping hot water.

$$\eta_p = \eta_{ip} = \eta_m$$

where

$$\eta_{ip} = 1 - (1 - \eta_i) \times \frac{\gamma_p}{\gamma} \times 0.1$$

$\eta_i = \dfrac{\eta}{\eta_m}$ – efficiency after subtracting bearing losses

η_m = mechanical efficiency

γ = kinematic viscositv of water at the test shop

γ_p = kinematic viscosity of warm water

H-V-HD horizontal multi-stage boiler feed pump. (Intersigma)

Turbine driven boiler feed pump 80-CHV-210-10/10 (Intersigma)

Leak-Off

The whole of the pump shaft input N is not utilized in the pump for increasing the energy of the liquid, as a part of the input that corresponds to all losses within the pump will be converted into heat. This means that the water temperature in the boiler feed pump increases proportionally to the losses within the pump.

The temperature rise will then be:

$$\Delta t = \frac{860\ N\ (1 - \eta)}{1000\ (Q + Q_k)} \quad \text{(degC)}$$

where

Q = delivery (L/h)

Q_k = amount of water flow through the balancing device (L/h)

N = pump shaft input (kW)

If the amount of the feed water flow through the balancing device is introduced before the boiler feed pump, the temperature of water entering the boiler feed pump increases to the value:

$$t_1 = \frac{(Q_{to} + Q_k)\ t_k}{Q + Q_k}$$

where

t_o = water temperature in the inlet branch less the effect of water from the balancing device (degC)

t_k = water temperature after the balancing device (degC)

A temperature rise of approximately 10 deg C is permissible in small and medium sized boiler feed pumps. At this point a device discharging this leak-off delivery (from the piping placed after the boiler feed pump) should come into operation automatically. In high pressure boiler feed pumps more detailed analysis should be made of the leak-off amount (*ie* minimum delivery) at which the leak-off device must start to open automatically.

Strength of Pipes

IN THE case of homogeneous tubes (*eg* drawn tubes), a suitable pressure rating can be determined directly in terms of the D/t ratio, by assuming a uniform distribution of stress through the tube wall, when

$$P = \frac{2St}{D}$$

where

D = outside diameter
t = wall thickness
P = internal pressure
S = material stress

This can be rendered in terms of a working pressure rating p_w where

$$p_w = \frac{2S_{max}t}{D}$$

where

S_{max} = the maximum permissible material stress for the material used. This is generally taken as one third the ultimate material stress – see Table I.

Written as a solution for tube wall thickness:

$$t = \frac{p_w D}{2S_{max}}$$

Provided the maximum material stress figure is taken within the limit of proportionality of the material, this simple formula is valid. It does not hold true for higher stress values, and thus will not accurately predict bursting pressures, *eg* using S_{ult} in place of S_{max}. It is also not valid where

TABLE I – MAXIMUM PERMISSIBLE STRESS FOR TUBE CALCULATIONS
(minimum UTS divided by 3)

Material	Condition	P_{max}	
		lb/in²	bar
Low carbon steel	as drawn	18300	1280
	drawn and polished*	28000	2000
20-ton steel	as drawn	15000	1000
Stainless (304) steel	annealed	33300	2350
	half hard	40000	2800
	hard	50000	3500
Light alloy 61S-T6	as drawn	15000	1000
Copper	annealed	6800†	480
	half hard	9000†	630
	hard	11300†	800
Tungum	annealed	22000	1550
	precipitation hardened	22000	1550
Titanium 115/125		18500	1300
Titanium 150/160		30000	2100

*cylinder tubes   † up to 150°F (65.5°C) only

the ratio D/t is 16:1 or less, as stress is then no longer uniformly distributed through the wall thicknesses, but ranges from a maximum at the inner surface, to a minimum at the outer surface. The simple formula is thus restricted to thin-walled tubes (D/t greater than 16:1). It will over-estimate the pressure rating for thick-walled tubes (D/t 16:1 or less), and in such cases an alternative formula must be used. An alternative formula which can be applied in the case of thick walled tubes and homogeneous metal pipes is:-

$$S_{max} = P \cdot \frac{R_i^2}{R_O^2 - R_i^2} \cdot \left(\frac{R_O^2}{R_i^2} + 1 \right)$$

where

R_i = inner radius of tube
R_O = outer radius of tube

Alternative formulas, written as a solution for wall thickness required are:-

$$t = \frac{D}{2} \left(\sqrt{\frac{S + P}{S - P}} - 1 \right)$$

$$t = \frac{D}{2} \left(\sqrt{\frac{3S + P}{3S - 4P}} - 1 \right)$$

For

$S = S_{max}$ the corresponding value of P is p_w

$S = S_{ult}$ the corresponding value of P is the bursting pressure

Modified formulas are used in the case of non-homogeneous tubes; and also for non-metallic tubes.

(i) In the case of welded tubes, a correction factor may be introduced; or alternatively, a higher divisor may be used to establish P_{max} from the ultimate tensile strength of the material. This is not necessarily the invariable rule, as welded tubes can have the same working strength as drawn tubes. Corrections may be applied to tubes with welded connections on a similar basis, however.

(ii) In the case of cast tubes, a nominal (and substantially lower) value may be adopted for P_{max}. Cast tubes are associated with older hydraulic systems and large pipe sizes, where pressure rating is established on empirical lines, permitting fairly large tolerances in wall thicknesses.

(iii) In the case of copper pipes and tubes intended for brazed or soldered connections, the standard thin-walled formula is derated

$$p_w = \frac{2S_{max}t}{D - 0.8t}$$

(iv) In the case of metallic tubes intended for threaded connections, an allowance is made for the reduction in tube strength due to threading. The following formula can then be used for thin-walled tubes

$$p_w = \frac{2S_{max}(t - C)}{D - 0.8t(t - C)}$$

where

C is taken as equal to the depth of thread cut, with a minimum value of 0.05 in (1.25 mm)

(v) In the case of plastic tubes, an allowance is made for the higher elastic moduli of such materials, when a suitable formula is –

$$p_w = \frac{2S_{max}t}{D - t}$$

Pipe Wall Thickness According to British Standards

Two British Standards standards are applicable for the calculation of pipe wall thickness:-

BS806 : 1975 – Specification for ferrous pipes and piping installations for and in connection with land boilers.

BS3351: 1971 – Specification for piping systems for petroleum refineries and petrochemical plants.

Design to BS806

Minimum pipe wall thickness where the outside diameter (D) is used as a basis for calculation,

$$t_{min} = \frac{PD}{2fe+P}$$

Minimum pipe wall thickness where the inside diameter (d) is used as the basis for calculation:

$$t_{min} = \frac{Pd}{2fe-P}$$

where

t_{min} = the minimum thickness (mm);

P = the internal design pressure (N/mm^2);

D = the outside diameter (mm);

d = the inside diameter (mm);

f = the maximum permissible design stress as specified in Table II (N/mm^2);

e = 1.0 for seamless and for electric resistance welded steel pipes and for electric fusion welded steel pipes complying with the requirements of BS3602 in which the weld is fully radiographed or ultrasonically tested;

e = 0.95 for electric fusion welded steel pipes complying with the requirements of BS3602;

e = 0.90 for pipes complying with the requirements of BS3601 other than electric resistance welded pipes;

The value of t_{min} is the minimum thickness for straight pipes and provision shall be made for any minus tolerance. Manufacturing considerations may make it necessary for pipes thicker than this minimum to be used.

Design to BS3351

where

t is less than D/4

$$t = \frac{Pd}{20S + P}$$

where

t = internal pressure design thickness in millimetres;

P = internal design pressure in bar;

D = outside diameter of pipe in millimetres;

S = design stress in N/mm^2 (shall be obtained from and shall not exceed those given in Table II for the design temperatures indicated therein).
Linear interpolation for intermediate design temperatures is permitted.

Pipes with t equal to or greater than D/4 require special consideration.

TABLE II – DESIGN STRESSES FOR FERROUS PIPES (BS3351)

Material	Notes	Values of S for metal temperatures in °C not exceeding															
		50	100	150	200	250	300	350	400	425	450	475	500	525	550	575	600
		N/mm²															
BS3601 HFS, CDS, Steel 22	1	125.5	114.5	103.9	99.3	94.5	89.7	85.0	75.0	65.7	56.9						
BS3601 HFS, CDS, Steel 27	1	154.5	141.3	127.5	121.7	114.2	110.0	104.0	87.7	75.8	62.0						
API 5L Grade A, seamless; steels open hearth, electric furnace and basic oxygen		122.5	111.9	101.3	95.8	91.3	86.5	82.0	73.1	64.8	55.8						
API 5L Grade B, seamless, submerged arc, spiral weld; steels open hearth, electric furnace and basic oxygen	2	153.0	139.8	126.5	119.0	113.8	108.5	102.5	89.3	75.5	62.0						
BS3602 CDS, HFS and ERW, steel 23	2	130.0	125.5	120.6	117.3	111.5	96.5	88.9	73.2	67.9	57.7	47.5	35.8				
BS3602 CDS, HFS and ERW, steel 27	2	154.5	143.0	132.0	127.0	119.0	110.0	104.2	89.7	75.9	62.3	48.6	35.8				
BS3602 EFW Grade 28 B	2.3	123.5	116.5	109.6	104.0	95.2	90.4	87.0	73.8	62.0	50.3	39.3	24.8				
BS3063 HFS, CDS, steel 27 LT-30	4.5	154.5	143.3	132.0	126.5	119.0											
BS3603 HFS, CDS, steel 27 LT-50	4.5	154.5	143.3	132.0	126.5	119.0											
BS3603 HFS, CDS, steel 503 LT-100	4.5	159.0	147.8	135.3	129.3	121.9											
BS3604 620 HFS, CDS, steel 27		166.5	144.2	144.0	139.0	124.5	114.0	109.5	104.0	101.5	98.5	92.3	81.0	62.8	43.8	28.9	17.2
BS3604 621 HFS, CDS, steel 27		144.5	144.2	144.0	139.0	124.5	115.5	110.5	106.0	103.8	100.0	92.8	81.2	64.3	49.3	34.2	24.4
BS3604 622 HFS, CDS, steel 27		144.5	134.2	129.6	125.0	120.5	115.5	111.0	106.0	103.8	101.0	92.5	82.8	63.8	47.5	36.2	26.8
BS3604 622 HFS, CDS, steel 31		164.0	164.0	164.0	164.0	156.5	150.5	144.0	137.2	134.0	130.2	107.5	82.8	63.8	47.5	36.2	26.8
BS3604 660 CDS. HFS		172.0	172.0	167.0	164.0	156.5	150.5	144.0	137.2	134.0	130.2	127.5	115.6	79.9	53.7	36.9	
BS3604 625 HFS		137.0	131.0	125.5	119.5	114.0	108.5	103.0	97.0	92.8	86.5	81.0	72.4	59.0	43.8	31.3	21.1

cont...

TABLE II – DESIGN STRESSES FOR

Material	Notes	Values of S for metal										
		–200 to + 50	100	150	200	250	300	350	400	425	450	475
BS3605 Grade 801 (304)	6	126.0	113.5	103.5	95.0	87.5	81.9	76.3	71.7	68.9	66.8	64.8
BS3605 Grade 811 (304H)	6											
BS3605 Grade 801L (304L)	6	106.8	102.8	90.3	76.5	68.2	63.3	60.0	57.2	55.8		
BS3605 Grade 822T (321)	6	129.3	127.5	117.0	108.5	105.5	102.8	102.0	101.3	100.5	99.0	97.5
BS3605 Grade 832T (321H)												
BS3605 Grade 845 (316) and Grade 845T and Grade 846 (317)	6	129.3	128.2	123.5	120.6	118.7	118.0	117.5	116.5	115.5	114.0	111.5
BS3605 Grade 845L (316L)	6	107.5	106.2	100.0	83.4	79.3	74.3	66.9	62.3	60.7	58.6	

*Intermediate values may be obtained by linear interpolation

Note 1 : Limited application (see Section 3 of BS3351).

Note 2 : The design stress values for pipe with a longitudinal or spiral weld seam incorporate weld joint factors as follows:
API 5L submerged arc welded : 1.0; API 5LS Grade B : 1.0; BS3602 ERW with Appendix A : 1.0; BS3602 EFW : 0.8

Note 3 : The values for BS3602 EFW are based on material to BS1501, 151 or 161. Grade 28B (certified or hot tested).
For temperatures over 400°C 161 material should be used.

Note 4 : The BS3603 pipes listed are impact tested and are intended for use in low temperature service

Note 5 : Up to 250°C to cover 'steaming out' and flexibility calculations.

Note 6 : The figures in parenthesis alongside the grades are equivalent AISI types. There is no AISI equivalent of 845T

FERROUS PIPES (BS3351) (contd)

temperatures in °C not exceeding*													
500 N/mm^2	525	550	575	600	625	650	675	700	725	750	775	800	825
63.3	61.3	59.7											
			55.8	48.0	37.6	30.7	22.4	17.6	13.4	10.3	7.9	6.2	4.5
96.0	94.0												
		91.7	88.5	75.8	51.3	34.5	25.5	19.6	15.2	11.8	9.3	7.6	6.5
106.0	99.3	90.3	79.3	67.6	56.2	46.6	36.8	28.3	22.0	17.6	14.4	11.8	9.6

Collapsing Pressure for Pipes and Tubes

IN THE general case of pipes where the length is eight or more times the diameter the uniformly applied pressure to produce collapse is given by

$$P = \frac{2E}{1 - \sigma^2} \left(\frac{t}{D}\right)^3$$

where

P = external pressure

E = modulus of elasticity of pipe material

σ = Poisson's ratio for pipe material

t = pipe wall thickness

D = outside diameter

The upper limit for the collapsing pressure is given when the compressive stress produced is equal to the compressive yield strength of the pipe material.

Buried Pipes

Buried pipes are subject to both internal and external loading, the general theory stating that the magnitude of internal pressure which a rigid pipe can withstand varies inversely with the magnitude of simultaneously applied external loading. The net effect on the combined load bearing strength of the pipe can be determined mathematically from the Schlick formula, as under:

$$P_1 = P_2 \left[1 - \left(\frac{F_1}{W_T}\right)^2\right]$$

which may also be written as

$$F_1 = W_T \sqrt{1 - \frac{P_1}{P_2}}$$

where

P_1 = internal hydrostatic pressure (kN/m^2) which will fracture the pipe when acting in combination with some external load F, applied in two-edge bearing.

P_2 = internal hydrostatic pressure (kN/m^2) which will burst the pipe in the absence of any external load.

F_1 = external load (kN/m) applied in two-edge bearing which will fracture the pipe when combined with some internal pressure P_1.

W_T = two-edge bearing load (kN/m) which will crush the pipe in the absence of any internal pressure.

Fig 1 shows a 'combined loading' curve for a pipe which would burst at some internal pressure P_2 or crush under some external load W_T, if either were acting alone. If, however, some lesser internal pressure P_1, is acting on the pipe in combination with an external loading F_1, the magnitude of F_1 at which fracture will occur can be determined by means of ordinates drawn to intersect at a point X on the curve.

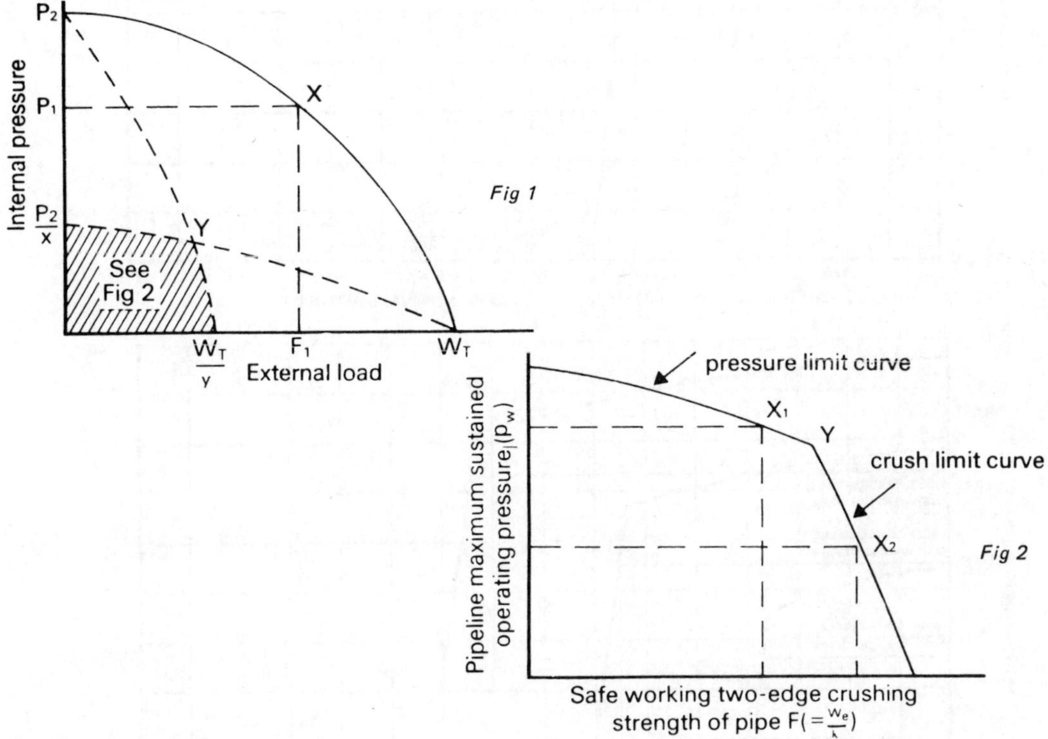

Combined Loading

The potential working envelope of a buried pipe is represented by a curve representing maximum internal pressure combined with a curve representing maximum external load (crushing pressure). Thus, because of the interdependence of the two parameters, boundaries are curves rather than straight lines and take the form shown in Fig 2.

Practical curves of this form, known as *combined loading charts,* can be devised for each particular size and class of pipe and incorporate suitable safety factors against bursting or crushing – Such a chart will indicate maximum working limits – *eg* for any given pressure limit a corresponding safe working crushing strength, or safe working pressure limit.

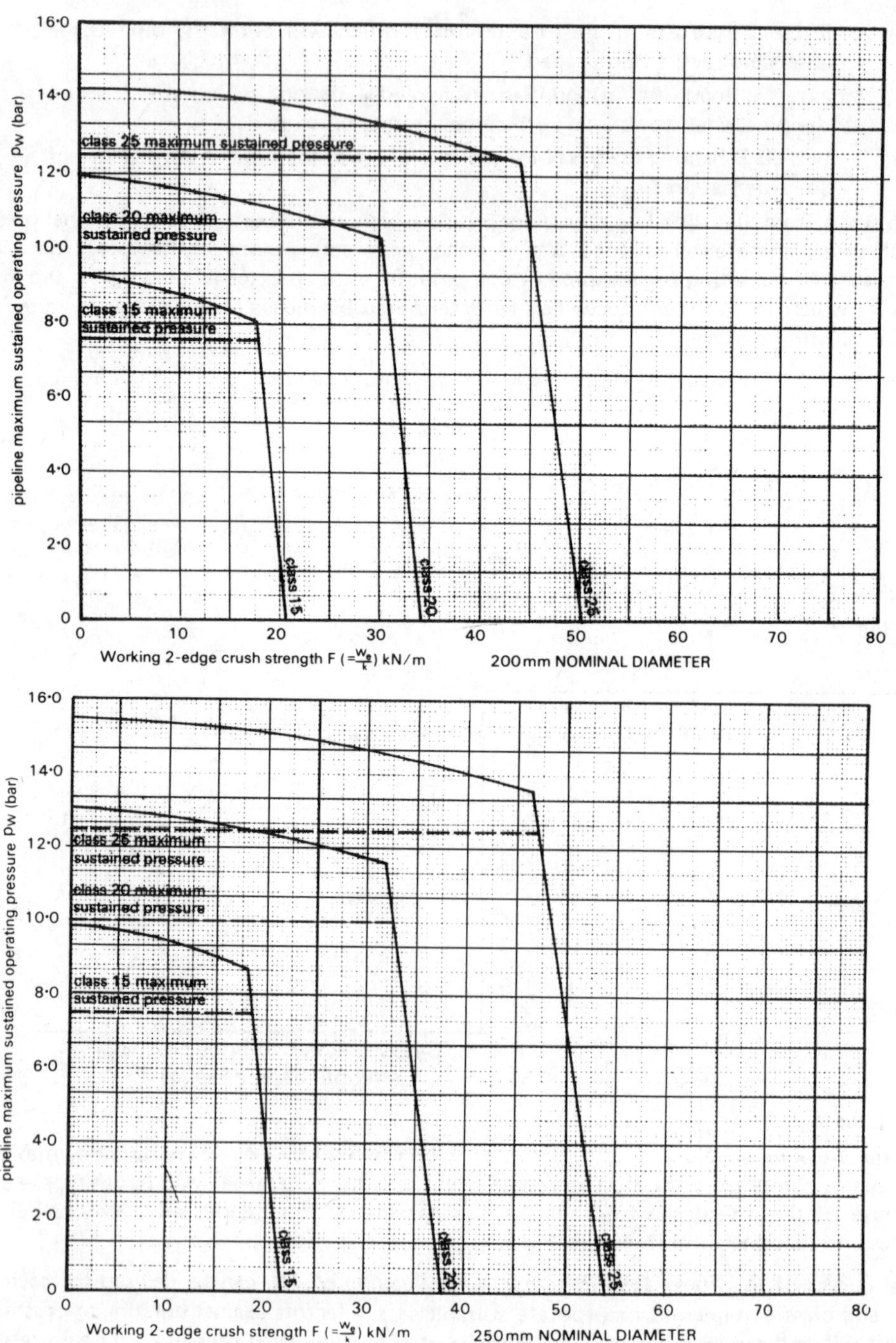

Examples of combined loading charts. TAC 'Everite' asbestos-cement pressure pipes. ...contd

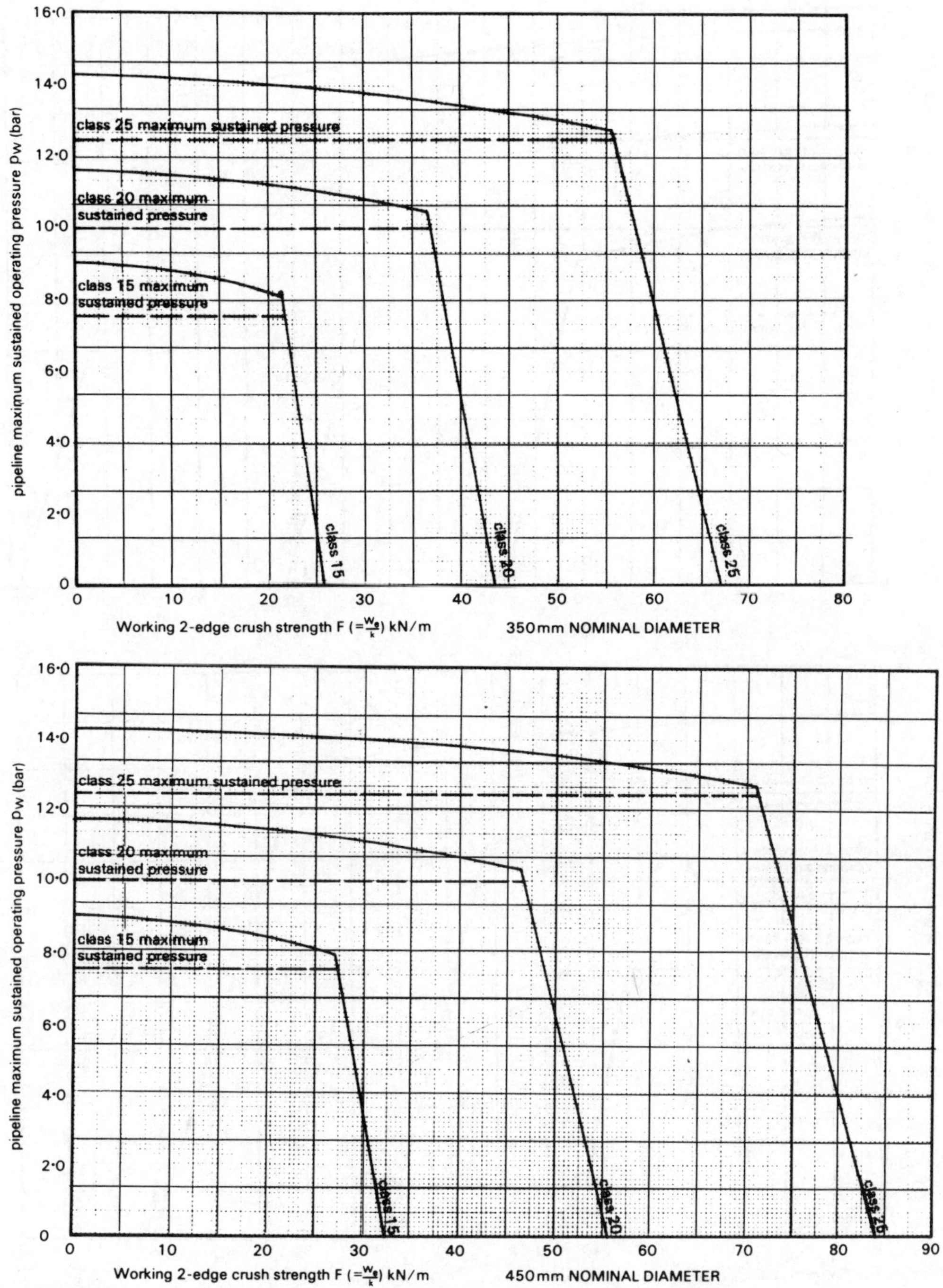

Examples of combined loading charts. TAC 'Everite' asbestos-cement pressure pipes. ...contd

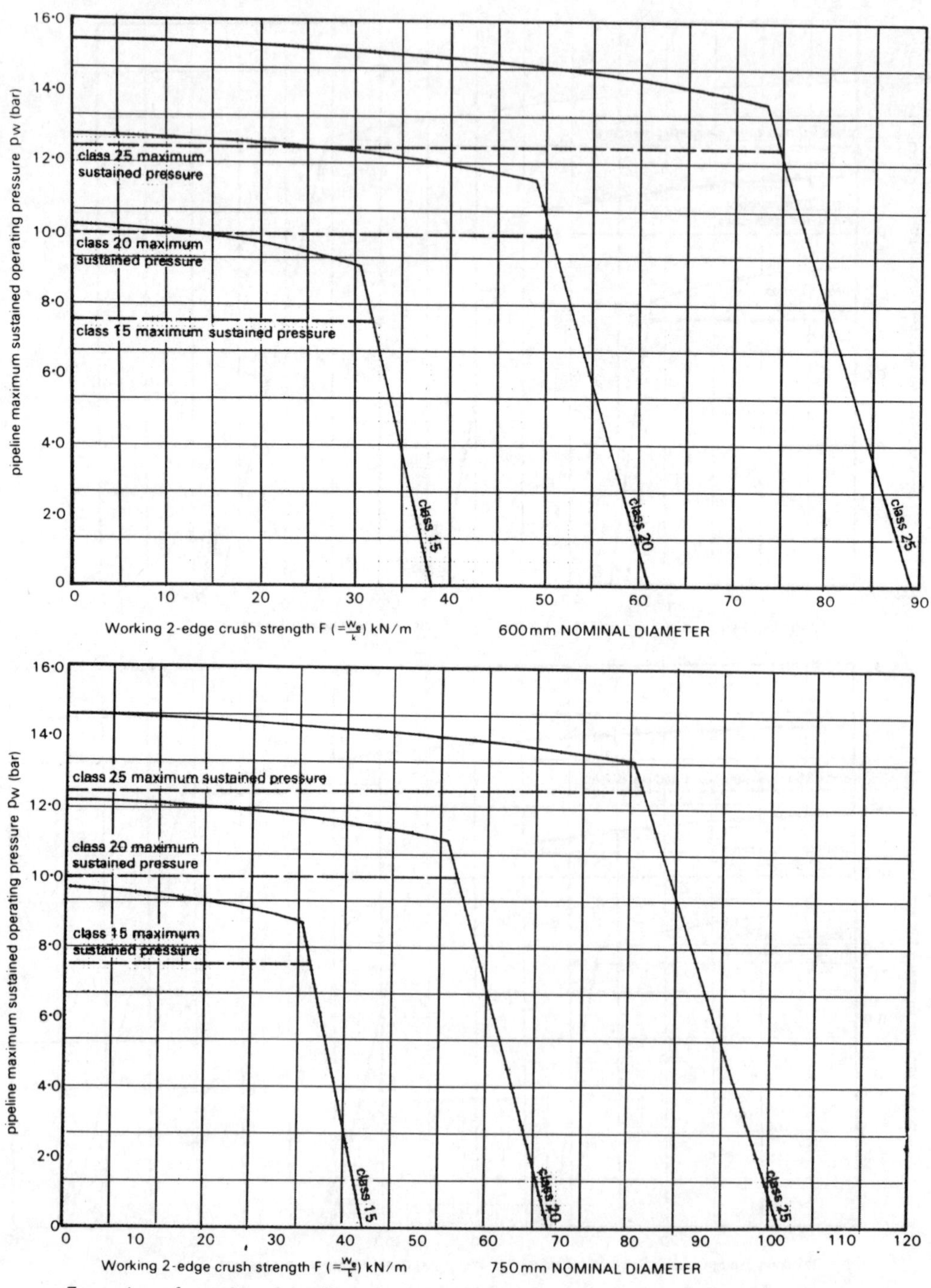

Examples of combined loading charts. TAC 'Everite' asbestos-cement pressure pipes.

Basically such loading charts are modified Schlick curves, taking into account suitable safety factors, *eg*:-

The equations for the modified Schlick curves which take the chosen factors of safety into account are:-

$$p_W = \frac{P_2}{X}\left[1 - \left(\frac{F_1}{W_T}\right)^2\right]$$

$$F = \frac{W_T}{Y}\sqrt{1 - \frac{P_1}{P_2}}$$

where

p_W = maximum sustained operating (or static) pressure (kN/m^2).

X = factor of safety against bursting when an internal pressure p_W is applied together with an external load. (See Table I).

F_1 = external load (kN/m) applied in two-edge bearing which will fracture the pipe when combined with some internal pressure p_W.

F = safe working two-edge crush strength of pipe (kN/m).

Y = factor of safety against crushing when an external load W_e is applied together with an internal pressure.

P_1 = internal pressure (kN/m^2) which will fracture the pipe when acting in combination with some external load F_1, applied in two-edge bearing.

TABLE I – FACTORS OF SAFETY (COMBINED LOADING)

Nominal diameter of pipe			
inches	**mm**	**X**	**Y**
7.9 and 8.9	200 and 225	3.5	2.5
9.8 to 19.7	250 to 500	3.0	2.5
23.6 to 29.5	600 to 750	2.5	2.5

External Loading

From preceding considerations it is evident that the safe working two-edge crushing strength, F, corresponding to a particular maximum sustained operating (or static) pressure p_W must be compared with the total external load acting on the pipe due to backfill and any other superimposed loading.

Since F is defined in terms of a two-edge strength, the following relationship is applicable:

$$kF = W_e \quad \textit{or} \quad F = \frac{W_e}{k}$$

where

W_e = total external load acting on the buried pipe (kN/m)

k = bedding factor (see Tables II and III).

TABLE II – BEDDING FACTORS – CLASS C BEDDING

Bedding angle ∝	Bedding factors (k) in different laying and backfill conditions		
	Trench and negative projection		Positive projection
	Ordinary backfill compacted between XX and YY	Ordinary backfill non-compacted between XX and YY	Ordinary compaction
30°	1.3	1.1	1.4
60°	1.5	1.2	1.7
90°	1.7	1.3	1.9
120°	1.7	1.3	1.9

TABLE III – BEDDING FACTOR – CLASS B BEDDING

Bedding angle ∝	Bedding factors (k) in different laying and backfill conditions		
	Trench and negative projection		Positive projection
	High compaction*	Ordinary compaction	Ordinary compaction
90°	2.6	1.9	2.3
120°	3.0	2.2	2.5

*At least 90% of maximum possible at the optimum moisture content (90% standard Proctor).

Notes:

(i) The above factors of safety include allowance for surge provided that the maximum sustained operating, or static, pressure plus surge (*ie* pipeline design pressure) does not exceed the maximum allowable sustained pressure for the class of pipe by more than 10%.

(ii) For pipe sizes up to 6 in (150 mm) diameter combined loading may not need to be considered as in this range, pressure pipes are designed on the basis of beam strength and consequently have bursting and crushing strengths in excess of practical needs.

The magnitude of W_e can be calculated from the appropriate Marston formulas and coefficient according to the type of soil and whether laid in trench ('narrow' or 'wide') or embankment conditions.

Trench Beddings

Wherever soil and other related conditions permit, it has been widely adopted practice over many years to lay asbestos-cement pressure pipes on the well levelled and prepared natural bottom of the trench. Selected backfill is introduced in layers not exceeding 12 in (300 mm) and properly compacted up to a level of 12 in (300 mm) approximately above the crown of the pipe. In the International Standard, beddings of this type are designated 'Class C'. This bedding embraces the Class 'C' and Class 'D' beddings described in National Building Studies Special Report No 35. See also chapter on *Buried Pipes*.

Buried Pipes

IN DESIGNING rigid buried pipelines the determination of the external load due to backfill and surface loadings is conventionally based on the methods and formulas established by Marston and Spangler. These involve lengthy and tedious calculations but are readily adapted to CAD (computer aided design). For general calculation simplified tables are available, notably those by Young and Smith (Building Research Station Report, 1970), corresponding to normal practice with concrete pipes and incorporating corrections for differences in external diameter. The latter can be a significant feature in the case of asbestos-cement pipes because of the smaller external diameter reducing the load which it has to carry. Further simplified tables have been computed on this basis.

Rigid metal pipes (*eg* cast iron) and pipes in flexible materials (*eg* reinforced glassfibre) need somewhat different treatment. The former can be laid at any depth with 3 inches (75 mm) to 24 inches (600 mm) of cover under buildings; and with not less than 12 inches (300 mm) under roads and yards subject to normal usage. Elsewhere in good ground such pipes will only need extra protection where subject to special loadings or abuse. In the latter case design is usually based on traditional empirical data; recommendations based on experience or derived from experimental data evaluating specific structural protection requirements. With flexible pipe materials due allowance must be given to the diametrical deflection produced by soil load, *eg* using Spangler's formula.

Asbestos-Cement Buried Pipelines

The following covers the use of simplified tables for the design of asbestos-cement buried pipelines. Metric units are employed throughout, consistent with current British and European practice. General assumptions are:

(i) Backfill density of 2000 kg/m^3 (125 lb/ft^3)

(ii) Frictional values $K\mu$ between 0.13 and 0.19

(iii) $r_{sd}\rho$ values of 0.5 and 0.7 as indicated.

Appropriate conversions are:

1 mm = 0.0394 in　　　1 m = 3.2008 ft

1 m^2 = 10.76 ft^2

1 kg = 2.2046 lb

$1\ kg/m^3 = 0.0624\ lb/ft^3$
$1\ N = 0.2248\ lb$
$1\ N/mm^2 = 1\ MPa = 145\ lb/in^2$
$1\ kN/m = 68.52\ lb/ft$

Notation

- B_c Outside diameter of pipe barrel.
- B_d Trench width, measured at the level of the top of the pipe.
- F_m Marston bedding factor, the value of which depends on the bedding method employed.
- F_s Design factor of safety (1.25).
- H The height of soil cover, measured from the top of the pipe barrel to the ground or road surface.
- γ Soil density in kg/m^3 (2000).
- $K\mu$ Product of the Rankine coefficient for the soil (K) and coefficient of internal friction of the soil (μ).
- $r_{sd}\rho$ Product of ρ, the proportion of the pipe diameter that projects above the bedding, and r_{sd} the 'settlement ratio'.
- W_e Total vertical external load imposed on the buried pipe, kgf/m.
- W_T Minimum ultimate crush load of asbestos cement pipe, kgf/m.

Application of the Tables

Tables I, II and III are based on assumptions that will be safe for a wide range of site conditions. The equivalent water loads are included in these Tables. By separating the concentrated surcharge loads, Table IV enables the designed to vary the calculation of the backfill load to suit individual circumstances, but here the water load must be added from Table V for pipes of 600 mm and over.

Pipes laid under verges should be designed for the full vehicle loads. Buried pipes must not be exposed to excessive loads from heavy equipment during construction. Choice of the appropriate loading category for a given location is a matter for the engineer's judgment, with due regard to possible future changes. The Tables are not appropriate for flexible pipes (pitch fibre, plastics, steel, *etc*), nor for rigid pipes supported on piles.

Design Method

For safe design, the minimum ultimate crush load (W_r) which the pipe is designed to withstand, must be greater than or equal to the computed external load (W_e) multiplied by a factor of safety of 1.25 and divided by the bedding factor F_m.

$$W_T \geqslant \frac{1.25 W_e}{F_m}$$

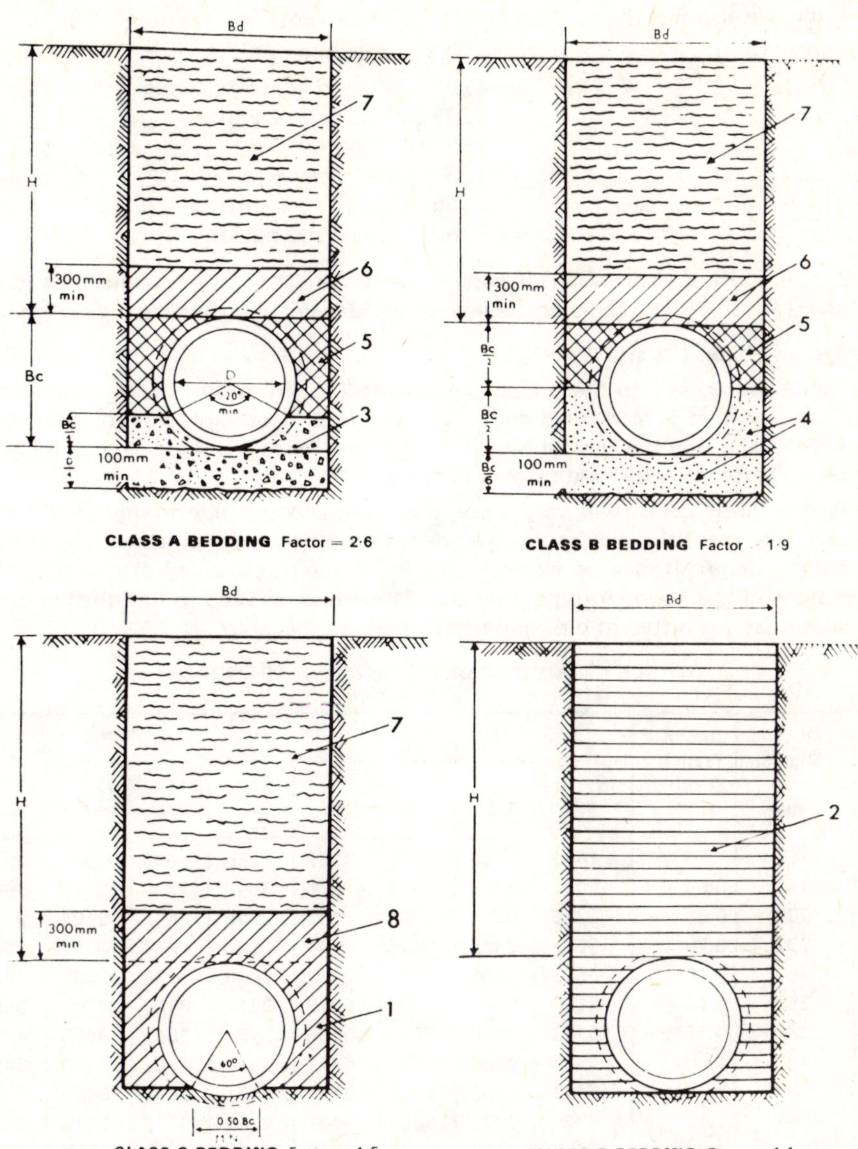

1. Lightly compacted fill.
2. Loose fill.
3. 20.5 N/mm^2 at 28 days concrete well packed under pipe.
4. Selected granular material well tamped under and alongside the pipe.
5. Selected material well tamped by hand in 75 mm to 150 mm layers.
6. Selected material lightly tamped by hand in 150 mm layers.
7. Normal fill.
8. Lightly compacted by hand.

Bd = Width of trench at crown of pipe
H = Depth of cover over crown of pipe
Bc = Outside diameter of pipe.

Fig 1 Bedding factors for pipes in trenches.

Bedding Factor (see also Fig 1)

Bedding Class	Type of Bedding	F_m
A (RC)	120° reinforced concrete cradle	3.4
A (Plain)	120° plain concrete cradle	2.6
B	360° Granular bedding	2.2
B	120° Granular bedding	1.9
C	Hand shaped trench bottom	1.5
D	Hand trimmed flat bottom	1.1

True values of F_m are considerably dependent upon standards of workmanship and good compaction of the side fill. Those tabulated assume properly maintained and supervised standards.

Method of Use of Tables I, II and III

Select the Table appropriate to the type of surface loading. From the pipe diameter and the cover depth, the external load is read off directly. Use the given bedding factor to calculate the minimum value of W_r, and find the class of pipe required from Table VI. If a pipe of sufficient strength is not available a better class of bedding may be specified.

In general the cover depth will vary along the main, and the pipe strength and bedding class must be selected to meet the maximum cover depth condition. With pipelines having considerable variation in cover depths it may be worth specifying different classes of pipe and/or bedding for different sections of the main. Alternatively the Tables may be used to find the limits of permissible cover depths for the different pipe and bedding classes available.

TABLE I – (METRIC UNITS, kgf/m) : MAIN ROADS*

Nominal Diameter mm	Outside Diameter mm	Assumed Trench Width m	W_e								
			$H =$ 0.9	1.2	1.5	1.8	2.4	3.0	4.6	6.1	7.6
100	125		1432	1352	1326	1332	1408	1535	2003	2202	2212
150	182	0.60	2047	1946	1913	1926	2040	2228	2743	2829	2881
175	208	0,60	2328	2216	2181	2197	2328	2543	2784	2856	2899
200	232	0.70	2587	2465	2428	2446	2593	2834	3342	3500	3565
225	260	0.70	2888	2756	2715	2737	2903	3174	3385	3528	3585
250	288	0.70	3190	3046	3003	3028	3212	3288	3429	3556	3604
300	339	0.75	3738	3574	3526	3556	3754	3841	4058	4257	4369
375	423	1.00	4585	4368	4292	4312	4547	4951	5930	6451	6797
450	504	1.05	5443	5190	5101	5127	5591	5894	6666	7272	7733
525	587	1.15	6320	6031	5930	5961	6291	6672	7416	8126	8680
600	671	1.20	7415	7089	6977	7012	7273	7546	8367	9190	9848
675	756	1.35	8372	8009	7882	7924	8352	8743	9819	10935	11797
750	837	1.45	9282	8883	8745	8793	9082	9479	10676	11876	12913
825	921	1.50	10230	9796	9646	9650	9826	10223	11507	12831	14041
900	1007	1.60	11220	10747	10526	10424	10586	11013	12371	13907	15190
975	1075	1.85	12000	11498	11324	11387	12000	12613	14547	16545	18307
1050	1157	1.90	12950	12405	12220	12290	12750	13380	15365	17520	19450

*TAC Construction Materials Ltd

TABLE II – (METRIC UNITS, kgf/m) : LIGHT ROADS

Nominal Diameter mm	Outside Diameter mm	Assumed Trench Width m	W_e								
			H = 0.9	1.2	1.5	1.8	2.4	3.0	4.6	6.1	7.6
100	125		1335	1122	1053	1052	1152	1316	1863	2105	2144
150	182	0.60	1850	1585	1503	1512	1667	1909	2540	2690	2782
175	208	0.60	2085	1796	1708	1721	1901	2180	2552	2697	2785
200	232	0.70	2302	1991	1897	1914	2117	2429	3083	3324	3440
225	260	0.70	2554	2217	2117	2139	2369	2719	3096	3331	3445
250	288	0.70	2806	2444	2337	2364	2621	2784	3109	3338	3449
300	339	0.75	3265	2816	2737	2772	3056	3248	3683	4002	4188
375	423	1.00	3966	3458	3301	3330	3677	4213	5462	6134	6572
450	504	1.05	4681	4095	3915	3955	4552	5014	6110	6895	7465
525	587	1.15	5412	4745	4544	4593	5082	5647	6768	7687	8370
600	671	1.20	6360	5612	5388	5447	5890	6375	7628	8689	9493
675	756	1.35	7166	6837	6088	6158	6793	7424	8987	10370	11400
750	837	1.45	7934	7026	6756	6836	7357	8019	9755	11250	12470
825	921	1.50	8737	7746	7455	7495	7927	8617	10494	12150	13550
900	1007	1.60	9576	8502	8128	8067	8510	9257	11264	13160	14660
975	1057	1.85	10236	9097	8762	8870	9782	10738	13366	15750	17740
1050	1157	1.90	11037	9818	9460	9579	10352	11364	14094	16660	18840

*TAC Construction Materials Ltd

TABLE III – (METRIC UNITS, kgf/m) : FIELDS ETC

Nominal Diameter mm	Outside Diameter mm	Assumed Trench Width m	W_e								
			H = 0.9	1.2	1.5	1.8	2.4	3.0	4.6	6.1	7.6
100	125		809	792	825	886	1054	1253	1837	2091	2133
150	182	0.60	1174	1150	1198	1288	1532	1821	2502	2670	2769
175	208	0.60	1341	1313	1368	1470	1750	2080	2509	2674	2771
200	232	0.70	1494	1463	1525	1639	1951	2319	3035	3297	3424
225	260	0.70	1673	1638	1707	1835	2185	2598	3043	3302	3427
250	288	0.70	1852	1813	1889	2031	2418	2650	3051	3306	3429
300	339	0.75	2176	2131	2221	2388	2820	3090	3615	3964	4164
375	423	1.00	2655	2578	2671	2859	3389	4021	5378	6087	6543
450	504	1.05	3157	3066	3177	3401	4212	4786	6011	6840	7431
525	587	1.15	3670	3564	3693	3954	4689	5384	6654	7623	8329
600	671	1.20	4397	4276	4423	4722	5442	6076	7497	8616	9448
675	756	1.35	4979	4843	5009	5345	6292	7089	8841	10290	11346
750	837	1.45	5535	5384	5567	5940	6803	7649	9593	11163	12414
825	921	1.50	6117	5950	6152	6513	7320	8210	10316	12046	13492
900	1007	1.60	6729	6547	6709	6998	7848	8814	11070	13049	14591
975	1075	1.85	7211	7016	7252	7731	9077	10266	13159	15630	17668
1050	1157	1.90	7795	7586	7839	8355	9605	10856	13872	16533	18766

*TAC Construction Materials Ltd

TABLE IV

Nom. Internal Diam. mm	Outside Dia. B_c mm	Trench Width B_d m	Type of Load	Total design load W_e in Kilograms per metre									
				$H=0{\cdot}6$	0·9	1·2	1·5	1·8	2·1	2·4	2·7	3·0	3·4
100	125	—	Narrow										
			Wide (0·7)	233	350	468	585	703	821	938	1056	1173	1330
			Main road		1083	887	743	631	543	471	413	364	311
			Light road		985	656	468	350	271	215	174	144	114
			Fields etc.	689	459	324	239	183	144	116	95	79	63
150	182	0·60	Narrow				1370	1550	1710	1840	1960	2080	2170
			Wide (0·7)	337	508	680	851	1022	1193	1364	1536	1707	1935
			Main road		1541	1268	1065	906	780	677	593	523	446
			Light road		1343	907	653	491	381	303	246	203	161
			Fields etc.	1002	666	470	347	265	209	168	137	114	91
175	208	0·60	Narrow				1370	1550	1710	1840	1960	2080	2170
			Wide (0·7)	384	580	776	971	1167	1363	1559	1754	1950	2211
			Main road		1750	1442	1211	1031	888	771	675	595	508
			Light road		1506	1021	737	555	431	343	279	231	183
			Fields etc.	1145	761	537	397	303	238	191	157	130	104
200	232	0·70	Narrow				1590	1810	2010	2190	2340	2470	2600
			Wide (0·7)	428	646	864	1083	1301	1519	1737	1956	2174	2465
			Main road		1942	1603	1347	1147	987	858	751	662	564
			Light road		1656	1127	815	614	478	381	309	256	203
			Fields etc.	1276	848	599	442	338	265	213	174	145	116
225	260	0·70	Narrow				1590	1810	2010	2190	2340	2470	2600
			Wide (0·7)	478	723	968	1212	1457	1701	1946	2191	2435	2761
			Main road		2167	1790	1505	1282	1104	959	840	740	632
			Light road		1832	1251	906	683	532	424	345	285	226
			Fields etc.	1430	950	671	495	378	297	239	195	162	130
250	288	0·70	Narrow				1590	1810	2010	2190	2340	2470	2600
			Wide (0·7)	529	799	1070	1341	1612	1883	2154	2425	2696	3058
			Main road		2392	1977	1663	1417	1220	1060	928	818	699
			Light road		2008	1374	997	752	586	467	380	314	249
			Fields etc.	1584	1052	743	548	419	329	264	216	180	143
300	339	0·75	Narrow				1810	2060	2300	2510	2700	2880	3030
			Wide (0·7)	619	938	1257	1576	1895	2214	2533	2852	3171	3596
			Main road		2801	2319	1952	1663	1432	1244	1089	961	820
			Light road		2328	1559	1162	878	684	546	445	368	292
			Fields etc.	1863	1238	873	645	492	387	310	254	211	168
375	423	1·00	Narrow				2480	2870	3230	3570	3870	4150	4420
			Wide (0·5)	733	1111	1489	1867	2246	2624	3002	3380	3758	4262
			Main road		3476	2881	2426	2068	1781	1547	1355	1195	1020
			Light road		2855	1969	1434	1086	847	676	551	456	362
			Fields etc.	2324	1544	1089	804	614	482	387	316	263	210
450	504	1·05	Narrow				2710	3140	3540	3910	4260	4580	4880
			Wide (0·5)	868	1318	1769	2220	2670	3120	3751	4022	4473	5073
			Main road		4126	3423	2884	2459	2118	1840	1611	1421	1213
			Light road		3363	2327	1697	1286	1003	801	653	541	429
			Fields etc	2769	1839	1297	957	731	574	461	377	313	249
525	587	1·15	Narrow				2940	3420	3860	4260	4650	5020	5350
			Wide (0·5)	1004	1529	2054	2579	3104	3628	4153	4678	5203	5902
			Main road		4792	3979	3353	2859	2463	2140	1874	1652	1410
			Light road		3883	2692	1966	1490	1164	930	758	627	498
			Fields etc.	3224	2141	1510	1114	851	668	536	438	364	290
600	671	1·20	Narrow				3170	3690	4180	4620	5050	5450	5830
			Wide (0·5)	1141	1740	2340	2940	3540	4140	4740	5340	5940	6740
			Main road		5467	4540	3827	3264	2812	2443	2139	1886	1610
			Light road		4410	3063	2238	1698	1326	1060	864	715	568
			Fields etc.	3685	2447	1726	1273	972	763	612	501	416	332
675	756	1·35	Narrow				3630	4240	4800	5350	5850	6350	6800
			Wide (0·5)	1277	1953	2629	3304	3980	4656	5332	6008	6684	7585
			Main road		6149	5110	4308	3674	3165	2750	2408	2123	1812
			Light road		4943	3438	2514	1908	1490	1191	971	804	639
			Fields etc.	4151	2757	1944	1434	1095	859	690	564	469	374
750	837	1·45	Narrow				3850	4500	5120	5710	6270	6800	7280
			Wide (0·5)	1405	2153	2901	3650	4398	5146	5895	6642	7391	8389
			Main road		6799	5652	4765	4065	3501	3042	2664	2349	2005
			Light road		5451	3795	2776	2108	1647	1317	1073	889	706
			Fields etc.	4596	3052	2152	1587	1212	951	763	624	519	413

of pipe length for cover depth H in metres

3·7	4·0	4·3	4·6	4·9	5·2	5·5	5·8	6·1	6·4	6·7	7·0	7·3	7·6
1448	1566	1683	1801	1918	2036	2154	2271	2389	2506	2624	2742	2859	2977
278	250	226	205	186	170	156	143	132	122	113	105	98	92
97	83	72	63	56	49	44	39	35	32	29	27	24	22
54	47	41	36	32	28	25	23	21	19	17	16	14	13
2260	2340	2390	2450	2500	2540	2580	2620	2640	2660	2690	2710	2730	2750
2107	2278	2449	2620	2791	2963	3134	3305	3476	3648	3819	3990	4161	4333
399	359	324	293	267	244	224	206	189	175	162	151	140	131
137	118	102	90	79	70	62	56	50	46	41	38	35	32
78	67	59	52	46	41	36	33	30	27	24	22	20	19
2260	2340	2390	2450	2500	2540	2580	2620	2640	2660	2690	2710	2730	2750
2407	2602	2798	2994	3189	3385	3581	3776	3972	4168	4364	4559	4755	4951
454	408	369	334	304	278	254	234	216	199	185	171	160	149
156	134	116	102	90	80	71	64	57	52	47	43	39	36
89	77	67	59	52	46	41	37	34	31	28	25	23	21
2710	2810	2900	2970	3040	3100	3160	3210	3260	3300	3330	3360	3390	3400
2683	2902	3120	3338	3556	3775	3993	4211	4430	4648	4866	5084	5303	5521
505	454	410	372	338	309	283	260	240	221	205	191	177	165
173	149	129	113	100	88	79	71	64	58	52	48	44	40
99	86	75	65	58	52	46	41	37	34	31	28	26	24
2710	2810	2900	2970	3040	3100	3160	3210	3260	3300	3330	3360	3390	3400
3006	3251	3495	3740	3985	4229	4474	4718	4963	5208	5452	5697	5942	6186
565	508	458	415	378	345	316	290	268	247	229	213	198	185
193	166	144	126	111	99	88	79	71	64	58	53	49	45
111	96	83	73	65	58	52	46	42	38	35	32	29	27
2710	2810	2900	2970	3040	3100	3160	3210	3260	3300	3330	3360	3390	3400
3328	3599	3879	4141	4412	4583	4954	5225	5496	5767	6038	6309	6580	6851
624	561	506	459	418	381	349	321	296	273	253	235	219	204
213	183	159	139	123	109	97	87	78	71	64	59	54	49
123	106	92	81	72	64	57	51	46	42	38	35	32	29
3180	3300	3420	3520	3610	3700	3780	3830	3910	3950	4010	4050	4090	4130
3915	4234	4553	4872	5191	5510	5829	6148	6467	6786	7105	7424	7743	8061
733	658	594	538	490	447	410	376	347	321	297	276	257	239
249	214	186	163	144	127	114	102	92	83	75	69	63	58
144	124	108	95	84	75	67	60	54	49	45	41	38	34
4650	4870	5080	5260	5440	5600	5750	5900	6020	6140	6250	6340	6440	6500
4641	5019	5397	5775	6153	6532	6910	7288	7666	8044	8422	8801	9179	9575
912	819	739	670	609	556	509	468	431	398	369	343	319	297
308	266	231	202	178	158	141	126	114	103	94	85	78	72
179	155	135	118	105	93	83	75	67	61	56	51	47	43
5160	5410	5650	5870	6080	6270	6440	6610	6760	6900	7040	7150	7260	7380
5524	5975	6425	6876	7326	7777	8228	8678	9129	9579	10030	10480	10931	11382
1084	974	878	796	724	661	605	556	512	473	438	407	379	353
366	315	274	240	211	188	167	150	135	122	111	101	93	85
213	184	160	141	124	111	99	89	80	73	66	60	55	51
5660	5950	6230	6490	6720	6950	7150	7340	7530	7700	7850	8000	8140	8270
6427	6952	7477	8002	8526	9051	9576	10101	10626	11150	11675	12200	12725	13250
1260	1132	1021	926	842	768	704	647	596	550	510	473	440	410
425	366	318	278	246	218	194	174	157	142	129	118	108	99
248	214	187	164	145	129	115	103	93	85	77	70	64	59
6190	6500	6820	7100	7390	7630	7860	8090	8300	8500	8680	8870	9020	9170
7340	7940	8539	9139	9739	10339	10939	11539	12139	12739	13339	13938	14538	15138
1439	1292	1166	1057	961	877	803	738	680	628	582	540	502	468
485	417	363	318	280	248	222	199	179	162	147	134	123	113
284	245	213	187	165	147	131	118	106	96	88	80	73	68
7230	7630	8010	8360	8720	9040	9340	9640	9900	10200	10400	10600	10800	11000
8261	8937	9613	10289	10965	11540	12316	12992	13668	14344	15020	15696	16372	17048
1620	1455	1313	1189	1082	987	904	831	765	707	654	607	565	527
545	469	408	357	315	279	249	224	202	182	166	151	138	127
319	276	240	211	186	165	148	133	120	109	99	90	83	76
7750	8190	8620	9030	9400	9760	10100	10400	10700	11000	11300	11500	11800	12000
9137	9885	10634	11382	12130	12879	13627	14375	15124	15872	16620	17369	18117	18865
1792	1609	1452	1316	1197	1092	1000	919	846	782	724	672	625	583
603	519	451	395	348	309	276	247	223	202	183	167	153	141
353	305	266	233	206	183	163	147	133	120	109	100	91	84

Method of Use for Table IV

Where the trench conditions and backfill density deviate from the norm, Table IV may be used to obtain the most economical design. The method of use is:

Step 1 – Knowing the pipe diameter and cover depth, obtain the vehicle load and the wide trench load.

Step 2 – Knowing the trench width, read off also the narrow trench load. Adopt the lesser of the two backfill loads.

Step 3 – If the soil density γ differs from 2000 kg/m^3 (125 lb/ft^3) correct the backfill load by $\gamma/2000$ ($\gamma/125$).

Step 4 – Add the backfill, vehicle loads and equivalent water load for pipes over 600 mm to obtain W_e.

With W_e determined the class of pipe and bedding required may be worked out as in Method of Use for Tables I, II and III, previously.

Example

Determine the strength classes and bedding required throughout a length of 600 mm nominal diameter asbestos-cement pipeline laid under fields at cover depth ranging from 1.2 to 6.4 m (narrow trench conditions).

Consider the use of Class B ($F_m = 1.9$) bedding or if ground conditions permit a Class D ($F_m = 1.1$) bedding. The maximum permissible external load can now be found by using the equation shown under Design Method.

ie $$W_e \geqslant \frac{1.1 \times W_T}{1.25} = 0.88\ W_T$$

or $$W_e \geqslant \frac{1.9 \times W_T}{1.25} = 1.52\ W_T$$

The values of W_T for the different classes of pipe can be found in Table VI.

W_T for 600 mm diameter Class L pipes = 4300 kgf/m

W_T for 600 mm diameter Class M pipes = 5800 kgf/m

The permissible depths of cover can now be found from Table IV.

	W_e (kgf/m)		**Depth of cover (m)**	
Class	**$F_m = 1.1$**	**$F_m = 1.9$**	**$F_m = 1.1$**	**$F_m = 1.9$**
L	3784	6536		0.6–3.7
M	5104	8816	0.6–2.1	0.6–7.0

A 600 mm Class L pipe, on a Class B bed can be laid from 0.6 m to 3.4 m. From 3.4 to 6.4 m a Class M pipe laid on a Class B bed would be needed.

If ground conditions permit, then a 600 mm Class M pipe can be flat bedded from 0.6 m to 2.1 m, and laid on a Class B bed from 2.1 to 6.4 m.

TABLE V*

Nominal diameter mm	Equivalent load kgf/m
600	210
675	270
750	330
825	400
900	490
975	560
1050	650

*When a pipe is running full, its contents are equivalent to an external load of 75% of the weight of water in the pipe.

TABLE VI – TURNALL PIPES – SIZE RANGE AND CLASSIFICATION

Nominal diameter	Class L		Class M		Class H	
	Minimum ultimate crushing load (W_T)		Minimum ultimate crushing load (W_T)		Minimum ultimate crushing load (W_T)	
mm	kN/m	kgf/m	kN/m	kgf/m	kN/m	kgf/m
100	–	–	–	–	37.95	3870
150	–	–	–	–	37.95	3870
175	–	–	–	–	37.95	3870
200	–	–	–	–	37.95	3870
225	–	–	–	–	37.95	3870
250	–	–	–	–	37.95	3870
300	–	–	35.00	3570	46.68	4760
375	–	–	38.63	3940	52.56	5360
450	–	–	43.73	4460	58.35	5950
525	37.95	3870	47.46	4840	65.70	6700
600	42.16	4300	56.87	5800	77.47	7900
675	48.05	4900	64.23	6550	86.10	8780
750	51.00	5200	68.64	7000	91.88	9370
825	55.40	5650	74.53	7600	100.71	10270
900	58.35	5950	85.57	8930	107.87	11000
975	62.76	6400	94.82	9670	116.70	11900
1050	65.70	6700	99.24	10120	124.05	12650

TABLE VII – DEPTH h_1 (mm) CORRESPONDING TO BEDDING ANGLE α

Nominal diameter (mm)	200	225	250	300	350	400	450	500	600	700	750
Bedding angle α	h_1 (mm)										
60°	← 25 →				← 50 →					← 75 →	
90°	← 50 →				← 100 →					← 150 →	
120°	← 100 →				← 200 →					← 250 →	

Buried Flexible Pipes

The following are guidelines for calculations involving buried flexible pipes.

Notation

Symbol	Definition	Unit
D_1	deflection lag factor	
d	mean diameter of pipe	mm
δ	reduction in vertical diameter	mm
E_s	elastic modulus for soil as determined at overburden pressure in tri-axial test	N/mm^2
EI/d^3	stiffness factor for pipe	N/m^2
e	subgrade modulus	$N/mm^2/mm$
Eb	bending strain	
Et	tensile strain	
H	height of cover above pipe spring line	m
k	deflection coefficient dependent upon bedding angle	
ks	Meyerhof's constant for grandular materials (taken as 1.63 n/mm 1.63 N6 1.63 $N/mm^2/m$ depth	$N/mm^2/m$ depth
Po	total external pressure on pipe	N/mm^2
Pob	external pressure on pipe used in buckling strength calculations	N/mm^2
Pod	external pressure on pipe used in deflection calculations	N/mm^2
Pc	critical buckling pressure	N/mm^2
Pe	external pressure on pipe due to backfill	N/mm^2
Pi	internal pressure	N/mm^2
Ps	external pressure on pipe due to surcharge loading	N/mm^2
Pt	external pressure on pipe due to traffic loading	N/mm^2
Pv	internal vacuum pressure	N/mm^2
t	pipe wall thickness	mm

Loads

Traffic Loads (Pt)

Traffic loading may be taken from the appropriate charts *eg* as given in N.B.S. Special Report No. 37.

For main road traffic loading normally use either HA or 45 units of HB loading in accordance with BS153 : Part 3A depending upon the quantity and type of vehicles expected.

Surcharge Loads (Ps):-

Include here long term loads.

Backfill (Pe):-

Normally use the full weight of the soil above the crown of the pipe. For buckling resistance, where ground water level is above the pipe, make allowance for bouyancy and add water pressure at crown (at invert if pipe can be empty).

Vacuum (Pv):-

Include if there is a possibility of full or partial vacuum inside the pipe.

Internal Pressure (Pi):-

Used in strength calculations to assess the quantity of glass reinforcement necessary to resist bursting.

Normally use maximum working pressure including an allowance for surges and check using test pressure with a reduced factor of safety.

Total External Pressure on Pipe (Po):-

For deflection calculations Pod = ½Pt + Ps + Pe

For strength calculations Pob = Pt + Ps + Pe + Pv

Note: Transient loads *ie* traffic loads have less effect in deflecting a flexible pipe than do permanent soil loads and surcharges. Therefore when considering deflection only, a reduction factor of 0.5 may be applied to Pt.

Deflection

Spangler's formula for the deflection of flexible pipes can be written thus:-

$$\frac{\delta}{d} = D_1 \frac{K \times Pod}{\dfrac{8\,EI}{10^6 \cdot d^3} + 0.061\,e.d.}$$

where

δ = reduction in vertical diameter

d = mean diameter of pipe

e = subgrade modulus

$\frac{EI}{d^3}$ = pipe stiffness factor

D_1 = Deflection lag factor

This is introduced to make allowance for the slow increase in deformation of some soils under sustained lateral pressure. Typical values are in the range 1.0 to 1.5 for non-pressure pipes.

For pressure pipes where internal pressure just balances external load use D_1 = 1, for pipes where high internal pressure is likely to cause re-rounding use values between 0.25 and 1; typically 0.5.

k = 0.100 for 60° bedding angle

Meyerhof and Fisher, based on Terzaghi, have shown that:-

$$e = \frac{E_s}{0.75\,.d}$$

where

E_s = modulus of elasticity of the soil at overburden pressure, as determined in a triaxial test. If no test results are available use 4000 — 10000 kN/mm² in good ground above water table, according to compaction; 2000 — 5000 kN/m² in poor ground, or in good ground below water table.

Alternatively in good ground E_S may be assessed by the following relationships:-

$E_S = k_S H.$ if backfill is dry granular material

$E_S = k_S H. \frac{1.7}{2.7}$ if backfill is saturated granular material

Where k_S (Meyerhof) is taken in the range 1.09 to 3.26 N/mm^2/m depth, a value of 1.63 N/mm^2/m depth being commonly used.

H = Height of cover above pipe spring line.

With flexible pipes the effect of the pipe stiffness on the deformation of the composite pipe/soil system is small and can usually be neglected. The Spangler formula then reduces to:

$$\frac{\delta}{d} = D_1 \frac{2.4 \text{ Pod}}{E_S}$$

The long term deflection calculated as previously should not exceed 5% of the diameter. If necessary, better quality surround material, and/or higher compaction should be specified. Allowable initial deflection will depend upon the engineer's assessment of D_1 but as a guide this should be limited to a maximum of 3%.

Buckling Strength

Based on Meyerhof and Baikie, the critical buckling pressure on a buried pipe (Pc) is given by the following equation:-

$$Pc = 4.6 \sqrt{E_S \cdot \frac{EI}{d^3} \frac{1}{10^6}}$$

Therefore permissible buckling pressure (Pob) is as follows:-

$$Pob = \sqrt{\frac{1}{F.S.} \cdot 4.6 \quad E_S \cdot \frac{EI}{d^3} \cdot \frac{1}{Cf. 10^6}}$$

where

F. S. = Factor of safety against buckling for which a value of 2 is used.

Cf = Creep Factor typically in the range 2 to 3 depending upon the long term elastic properties of the pipe wall material

$\frac{EI}{d^3}$ = Pipe stiffness factor (initial)

Bursting Strength

Bursting is resisted by glass reinforced layers acting in hoop tension. The thickness of glass reinforced layers provided is such that at working pressure there is an initial factor of safety against bursting of between 6 and 7. At test pressure a factor of safety of at least 4 is normally available.

Crushing Strength

Armaflo pipes have a crushing strength in excess of 80 MN/m^2. Crushing stress is not normally critical.

Longitudinal Strength

Armaflow pipes have longitudinal strength in excess of the requirements of British Standards 5480 Part 1 : 1977.

Strain

Bending Strain Eb

If the pipe deformed as an ellipse, the strain in the pipe wall due to bending would be

$$Eb = \frac{3\delta}{d} \cdot \frac{t}{d}$$

where

t = thickness of pipe wall

Tests by Molin showed that bedding irregularities caused deviations from this theoretical value which were dependent upon pipe stiff ness, *ie*

$$Eb = d \cdot \frac{\delta}{d} \cdot \frac{t}{d}$$

Where $3 \leqslant d \leqslant 6$: d tends towards 3 for stiffer pipes which are less affected by bedding irregularities and hence for a particular deflection are subject to correspondingly lower levels of bending strain. In non-pressure Armaflow pipes bending strain should be limited to 0.35%.

For pressure pipes bending strain is generally limited to 0.2%.

Circumferential Tensile Strain Et

Circumferential tensile strain in pressure pipes should normally be limited to an initial maximum value of 0.2% at working pressure.

Note: In pressure pipe design the effects of bending and internal pressure are not additive and may, therefore, be considered separately.

Summary of Design Criteria

Long term deflection shall be limited by the allowable bending strains listed or to 5% whichever is the lesser.

Bending Strain, 0.35% (non-pressure)
0.20% (pressure)

Circumferential tensile strain, 0.2%

Trenching

The preparation of the trench bottom to give an even bedding for the barrel of the pipe, and proper alignment of pipes, is of primary importance. In rocky ground the trench should be extended to at least 100 mm deeper than required and then made up to the required level by introduction of well rammed compactible material of a type which will not be washed away; alternatively the pipe may be embedded in a layer of freshly mixed concrete. The trench should not normally be opened up more than a few pipe lengths ahead of the point where laying is taking place.

Width of Trench

The trench width will depend to some extent on the ground conditions and depth of laying but should be kept to the practical minimum. Minimum widths for normal conditions, as used in the preparation of external loading charts are based on standard bucket sizes wherever possible and give at least 150 mm and 20 mm clearances on each side of the pipe for diameters in the ranges 200–350 mm and 400–750 mm respectively.

In narrow trench conditions the backfill load is a function of the depth of cover and width of trench (B) as measured at the level of the crown of the pipe. Thus the backfill loads will be the same for any of the trench sections shown in Fig 2.

Where a particular width has been specified it should not be significantly exceeded without reference to the pipeline designer.

The effective trench width (B) in the three examples shown is as measured at the level corresponding to the crown of the pipe.

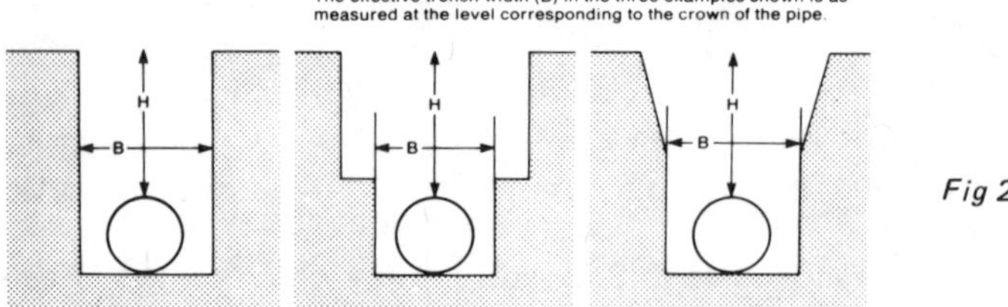

Fig 2

Depth of Trench

The depths at which a pipeline is to be laid will normally have been determined by the pipeline designer to whom reference should be made if for any reason the depths specified for a particular application require to be exceeded by a significant amount. The minimum depth of cover will depend on considerations such as frost, traffic loadings, size and class of pipeline and, as regards adequacy of anchorage, the type and compactibility of the fill material.

Preparation of Trench Bedding

Class C Bedding

In Class C bedding where the pipe barrel is to be in more or less continuous contact with the foundation soil, the initial excavation should be marginally less than the required depth in order that final levelling and preparation of the bedding may be carried out manually. Any high spots should be removed. If overdigging accidentally occurs, the correct level must be regained by introduction of selected material, well compacted. This preparatory work can be aided by use of a wooden straight-edge, of length not less than that of the pipes. One end of the straight edge should be notched. Allowance can be made for slight initial settlement due to the weight of the pipe.

Class B Bedding

In Class B bedding it is necessary to overdig the trench by an amount equivalent to the required depth of granular overlay between the pipe and the foundation ground. The straight-edge technique can again be used to obtain the gradient and approximate level of the granular bedding.

For both Class B and Class C bedding it is necessary to excavate a joint hole of sufficient length and depth to give clearance for the coupling to ensure that the pipes rest on the trench bed and not on the coupling.

Backfilling

Backfilling should be carried out in accordance with the requirements specified for the particular pipeline and can take place as soon as the joints have been made. If it is desired to leave joints uncovered until completion of pressure testing, the trench should be backfilled only over the barrel portion of each pipe. In such case it is necessary to ensure that an adequate depth of compacted backfill is applied over the crown of the pipe to prevent the pipeline lifting when the test pressure is applied.

For Class C bedding selected backfill is compacted to a suitable depth h_1, which corresponds to the bedding angles as selected for the particular nominal diameter of pipe.

The backfill between levels is ordinary soil, free from lumps and large stones. Compaction will normally be required and should be done in stages with layers of from 150 mm to 300 mm in thickness. The remainder of the excavated material can be used for the top level of backfill, the extent of compaction required depending on local factors.

For Class B bedding the pipe will have been laid on a prepared granular bed, which should be not less than 100 mm for all sizes of pipe where laid in uniform soils. Where laid in rock or mixed soils containing rock bands, boulders, large stones or other irregular hard spots, dimension h should be not less than 200 mm.

To complete the granular bedding, a further layer of suitable depth (according to pipe diameter and bedding angle α as selected) should be added and compacted.

Selected backfill, free from lumps or large stones is then compacted in layers of 150 mm to 300 mm to a height of 300 mm above the crown of the pipe. Ordinary backfill material can be used for the remainder, the extent of compaction depending on local requirements.

Joint holes should be backfilled with selected material and properly compacted.

Large stones should be kept well away from the sides of the trench to avoid the possibility of their being accidentally dropped on the pipes which have been laid.

Cavitation

THE PHENOMENON of *cavitation* is associated with a reduction of pressure occurring in a liquid system, reducing the liquid pressure at a localized area down to the vapour pressure of the liquid concerned. As a consequence vapour and small gas filled bubbles form in the liquid at this point and are entrained by the flow. As soon as they reach a region of higher pressure they suddenly collapse at extremely high velocities, with vapour condensing into liquid again. Very high implosion pressures are generated, depending on the bubble size, and may even reach 140 000–150 000 lb/in^2 (10 000 bar). Such high-velocity impacting pressures can show up as:-

(i) Noise

(ii) Vibration (critical oscillations)

(iii) Mechanical damage to construction materials.

Acrylic model of the Masoneilan variable resistance trim (VRT) showing the solid integral unit construction and alternating plate segments. This new concept helps overcome the problem of cavitation caused by high pressure drops in process control valves.

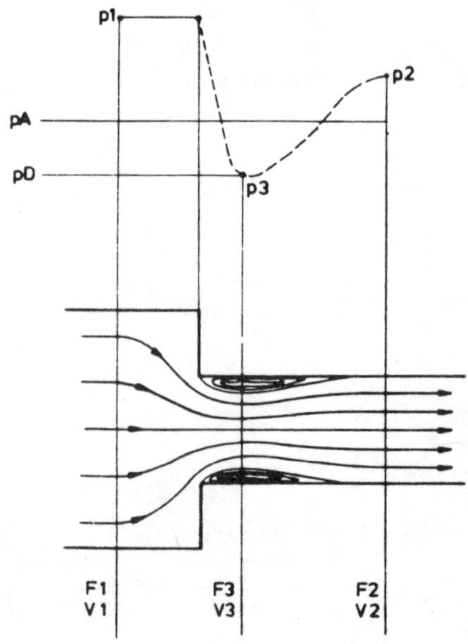

Fig 1

Critical regions for the development of cavitation in a piping system are sudden cross section enlargements or contractions, changes in flow direction and sudden changes in flow (such as at throttling gaps). The sudden contraction, illustrated in Fig 1, is a typical example. This shows the pressure distribution at the throttling point where p_1 is the upstream fluid pressure, p_2 the downstream fluid pressure (lower because of the head loss after throttling), p_A is the atmospheric pressure and p_0 the vapour pressure of the fluid. At point p_3 the pressure is reduced to the vapour pressure, generating cavitation, with subsequent pressure recovery to p_2. Somewhere between p_3 and p_2 the cavitation bubbles will collapse suddenly.

Cavitation does not necessarily lead to damage, even if it does generate noise and vibration. It depends on the intensity of cavitation, or specifically the lifespan of the bubbles from formation to implosion. Cavitation intensity decreases with an increase in the life of the bubbles.

The pressure travel gradient (p_1, p_3, p_0) is thus significant and related to the shape of the flow passage. As a general rule the pressure drop can be influenced by streamlining the flow, although this can be optimized for one predetermined flow condition only. Thus it is rather more important to try to extend the pressure travel gradient. Geometric shapes can be found which, despite cavitation, do not lead to damage, *ie* the bubble implosion occurs away from the possibility of contact with the material surfaces. From this it can be seen that only bubble implosions near the wall are destructive. If the bubbles contact material surfaces, the destruction mechanism conforms with that of liquid droplet erosion.

From the point of view of metal physics, what happens is a high-velocity deformation of the metal as a result of the bubble implosion. In many cases the mechanical erosive influence is coupled with an electrochemical corrosive influence, cavitation and corrosion occurring together. It has been shown that in the case of industrial water, damage to carbon steel and Armco (ingot) iron can be reduced by cathodic protection, *ie* the corrosive influence can be removed. Among construction materials which have proved to be less prone to cavitation, austenitic steel, single-phase copper alloys (bronzes), stainless steels and stellite armouring have been most successful. These materials are largely resistant to corrosion so that they are not subject to this additional attack.

Typical potential points of cavitation damage are:

a) Suction pipes of pumps
b) Narrow flow spaces, leakage and bearing gaps
c) Sudden changes in flow area
d) Changes in flow direction as in bends and pipe tees
e) Changes which lead to turbulence
f) Downstream of throttle and control valves; components built into the flow stream

From the following list, measures can be chosen by which disadvantageous cavitation consequences can be avoided:

1. Avoidance of turbulence by proper streamlining of flow, *eg* by means of a vaned ring in a needle valve (Fig 2).
2. Prevention of wall contact after regions of pressure drop by sudden enlargement of the pipeline (Fig 3). Developing cavitation bubbles implode in the water space. Cavitation arises in a space not endangering the material.
3. Letting pressure drop occur over sharp edges.

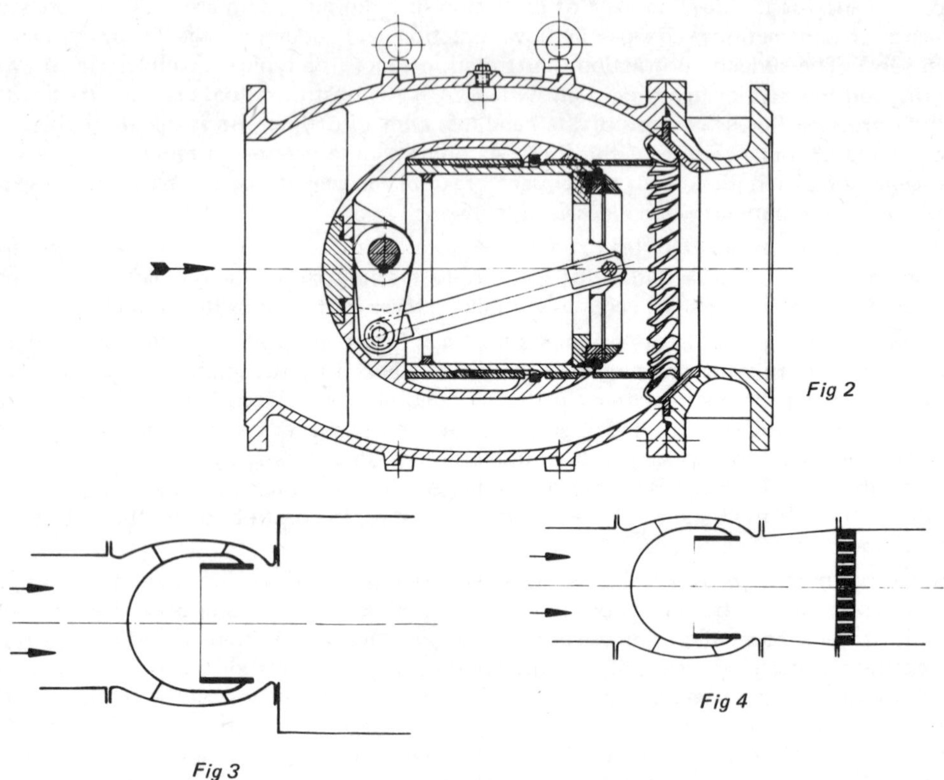

Fig 2

Fig 3

Fig 4

4. Dissipation of the kinetic energy not solely through turbulent mixing but
   a) through built-in resistance, *ie* by increasing the friction-causing wall surfaces. This causes an increase in the back pressure after a point of throttling. A pressure drop below the atmospheric pressure can be avoided, (Fig 4).
   b) Partitioning of several resisting bodies in series (multistage pressure drop).

As an approximation the number of stages required can be calculated as follows:

$$n = -6.45 \times \lg \frac{P_2}{P_1}$$

where P_1 = upstream pressure
P_2 = downstream pressure

5. Principle of flow partition into small single cross sections, *eg* hollow cylinders (Figs 5 and 6). The division into single jets beneficially influences the dynamic behaviour of the medium flowing off downstream of the throttling devices. In addition, the partition into small cross sections achieves a more uniform downstream flow in the following pipe.

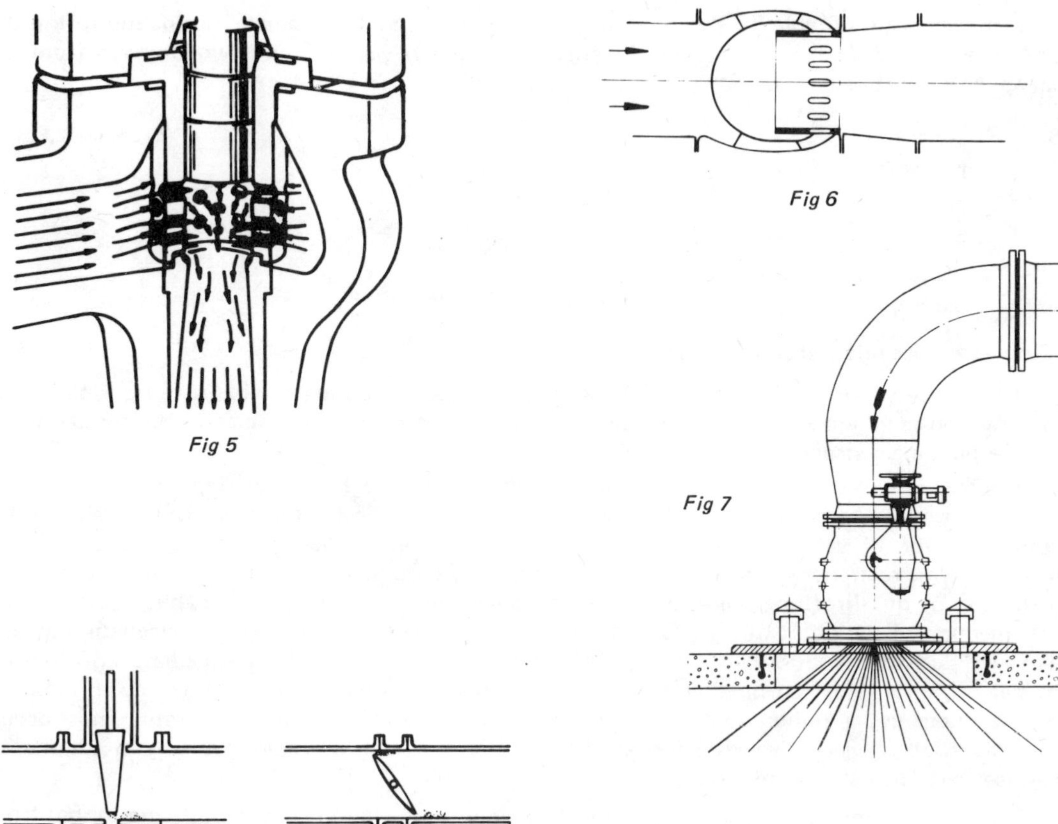

Fig 5

Fig 6

Fig 7

Fig 8

6. Change of flow contractions by the introduction of air:
   a) In free exit – similar to the effect of perlators in water taps with the well-known soft jet (Fig 7).
   b) Intensive ventilation in the flow passage produces the hydraulic character of an end closure. The arrangement of the valve on the air side is from the hydraulic point of view more favourable than that within a pipeline with ventilation.

From the above it can be seen that for throttling and control duties special valve types are required which are designed with special seat and exit configurations. In general, valves are limited only according to the nominal pressure rating. Within the standard design requirements, the special dynamic loadings in the flow passage of the various types of valves are not considered. In the case of mere shut-off valves such as gate and butterfly valves, the necessary adaptation cannot be achieved by means of design, (Fig 8). They are therefore not suitable for pronounced throttling and control duty, but because of the low permanent pressure drop are ideal for on-off duty.

Such valves are suitable for short-period throttling duties as, for example, during shut-off in the case of a burst pipe. However, when dimensioning these valves, the limits which result from the energy head must be considered.

Butterfly valves:

PN 25	7.5 m/s	(25 ft/sec)
PN 16	5 m/s	(16.5 ft/sec)
PN 10	4 m/s	(13 ft/sec)
PN 4	2.5 m/s	(8 ft/sec)
PN 2.5	2 m/s	(6.5 ft/sec)

(Flow velocities referred to valve nominal diameter).

If butterfly valves are used as safety devices in the case of a burst pipe, the responsible manufacturer considers the stresses in these circumstances and dimensions valve and operator designs in a correspondingly strong manner.

In water works and water power plants needle valves, also known as ring piston slide valves (Fig 9), have proved in more than 40 years of practice to be excellent as flow control valves, since with this type of valve individual adaptation to the given operating conditions and duties is possible. With reference to cavitation this means that by specific configuration of the outlet shape, formation of the throttling point, design of the downstream piping, and — last but not least — by the selection of the point of installation, the hydraulic conditions can be influenced directly and damages due to cavitation be avoided. With these control valves all intermediate positions, *ie* partial openings, must be possible for continuous duty to achieve variable flows or an effective change in energy, *eg* reduction of pressure. No ill effect due to cavitation or vibration must occur. The design of the needle valves offers all the advantages. The flow is guided through a ring shaped passage around a ball shaped inside body.

The outside body is so designed that the free flow cross section continuously diminishes from the inlet to the sealing and throttling point so that flow velocity increases. Shortly before the narrowest cross section a vaned ring is provided, as in the case of the Erhard needle valve, which

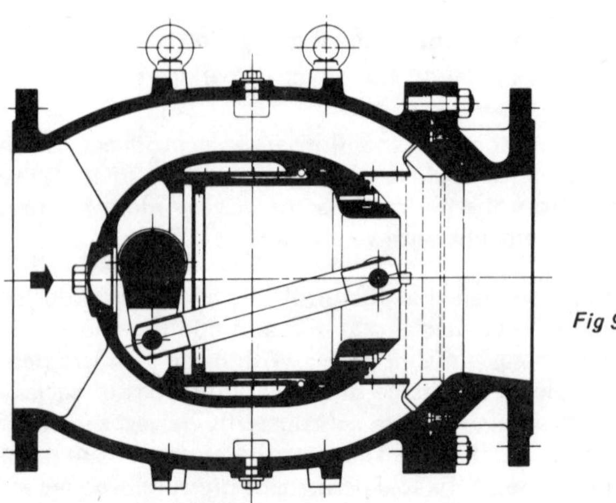

Fig 9

swirls the outer flow filaments in such a way that the fluid is forced against the wall of the downstream flow section so that detachments are avoided and cavitation bubble implosions are kept away from the wall. The shut-off piston in the spherical inside body moves in the opening and closing passage, *ie* in or against the direction of flow, and produces therefore a linear change in cross section without causing the flow direction to change. The downstream shape of the piston is sharp-edged. As against former designs with pointed piston end, the hydraulic exit flow angle can develop, depending on piston position and velocity in the water area, without touching the metallic valve parts.

To enable a control valve to fulfil its duty, *eg* continuous throttling of the rate of flow, the valve must be properly dimensioned. By dimensioning, not only the sizing of the valve is meant but also the adaptation to the duty prescribed, taking into account the specific operational conditions including assessment of the cavitational behaviour. The cavitational behaviour of a valve can be observed in a model test from which the behaviour in the actual installation can be deduced.

As a means of comparison, the cavitation coefficient sigma, also known as the Thomas cavitation number, has been introduced. This value indicates the commencement of cavitation. The cavitation number δ is calculated from:

$$\delta = \frac{P_2 + P_A - P_D}{P_1 - P_2 - \dfrac{V^2}{2g}}$$

where P_A = Atmospheric pressure
P_2 = Pressure downstream of disturbance
P_D = Vapour pressure of water
P_1 = Pressure in undisturbed upstream side
V = Velocity in undisturbed upstream side
g = Acceleration due to gravity

Responsible valve makers determine the behaviour of their valves in all-embracing tests and therefore possess comparison values for all common operating conditions. From these data the required design shapes can be deduced. In extreme cases for which test data are not yet available, the particular case is reconstructed in model tests and the required design determined. The main precondition is the precise knowledge of the operating and installation conditions already at the project stage.

Only with these data is it possible to determine optimum design of the valves and piping run with the aim of avoiding damaging cavitation effects, Among the necessary data are:

(i) Static and dynamic pressure upstream and downstream of the valve referred to the desired range of rates of flow.

(ii) Knowledge of the installation situation within the pipeline.

(iii) Knowledge of the piping run and built-in components downstream of the valve.

(iv) Knowledge of maximum permissible head loss, *etc.*

A close co-operation between designer, user and manufacturer of valves has in the past always proved to be very beneficial — and will doubtless lead to equally satisfactory solutions in the future.

Noise Control

NOISE PRODUCED in pipelines may be pump generated (changes in power and pressure, or varying amplitudes of pressure pulsations) or fluid generated (flow instability, turbulence or simple fluid friction).

Fluid-generated noise in small bore pipes with low to moderate flow rates is generally negligible, unless pressure pulsations are present, (*eg* due to valve cavitation). Thus pipe vibration, and consequent radiation of airborne noise, is usually due to the higher level of noise generated by fittings; pipe resonance is due to mechanical vibration or resonant noise generated in supporting systems.

Specifically, noise control (noise abatement) falls into two distinct categories:

(i) *Source treatment* – *ie* design of components to ensure operation at minimum noise levels.

(ii) *Path treatment* – to reduce source-generated noise to acceptable levels.

Noise due to the operation of valves, regulators and control elements is transient and related to the degree of turbulence or cavitation produced, although in specific designs and certain circumstances individual elements may be subject to vibration and generate a continuous noise. So much depends on the design and finish of the flow passages involved that no general analysis can be attempted. The noise level of such devices is dependent on the design and the localized flow velocities produced and also on the response time, where applicable. The latter effect can be minimized by arranging that the response time is not shorter than that required by the system. This will result in minimum 'hammer'. 'Water hammer' in fact, depends on the switching velocity of the valve – *ie* on the spool-switching velocity in the case of spool valves. Valves operated by dry solenoids have, in fact, uncontrolled response and so often produce 'hammer'. Wet solenoids are cushioned by the fluid so move more smoothly and open the valve passages more gradually (at the expense of some loss of solenoid power).

As a general recommendation, simple undamped ball-and-spring non-return and relief valves should not be used. On the design side, every effort should be made to ensure that the flow passages of valves are swept and free from sharp edges and corners as far as possible. Directional control valves must also be carefully designed to prevent flow instability occurring.

Source Treatment

Source treatment is difficult to describe in general terms since it is mainly concerned with the design of optimum flow paths through valves to reduce or eliminate noise that would otherwise be generated. In this respect, quite different design parameters are involved in dealing with aerodynamic noise resulting from liquid flow.

Aerodynamic Noise Reduction

Two examples of aerodynamic noise treatment are shown in Figs 1 and 2, applicable to globe and angle valve bodies. Both are cage-style valves, one using a cage with multiple slotted orifices of special shape, size and spacing; and the other a cage with multiple hole orifices. Claimed performance for the former is an 18 dB reduction compared with a conventional valve of similar type; and 30 dB reduction for the multi-hole orifice cage. The latter is also particularly effective for applications involving high differential pressures (pressure drop across the valve). A common feature of both these valves is an expanded outlet design to minimize regeneration of valve noise.

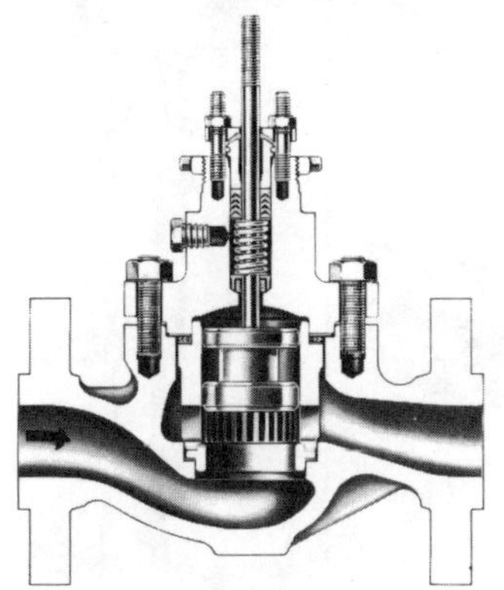

Whisper trim I cage Fisher ES valve body assembly.

Whisper trim I with slotted orifices.

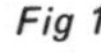

Fig 1

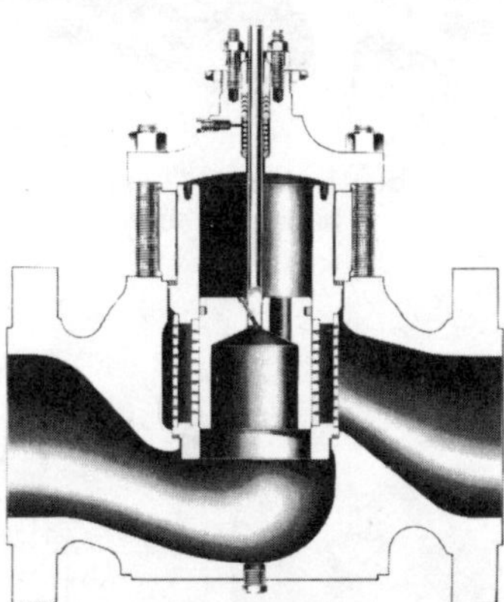

Whisper trim III cage in Fisher EWD control valve body assembly.

Fig 2

Whisper trim III cage.

Fig 3

Cavitrol I cage.

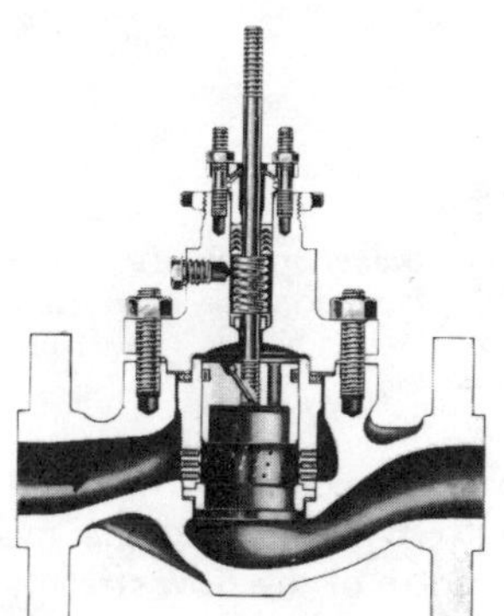

Cavitrol I cage in Fisher ED valve body assembly.

Hydrodynamic Noise Reduction

Examples of cage-type valve trims for hydrodynamic noise treatment are shown in Figs 3–6. Here the immediate aim is to eliminate or minimize cavitation. The cage design of Fig 3 uses one stage of diametrically-opposed flow holes through the cage wall to reduce both cavitation noise and damage. Each specially shaped hole directs a jet of cavitating liquid which impacts with the jet

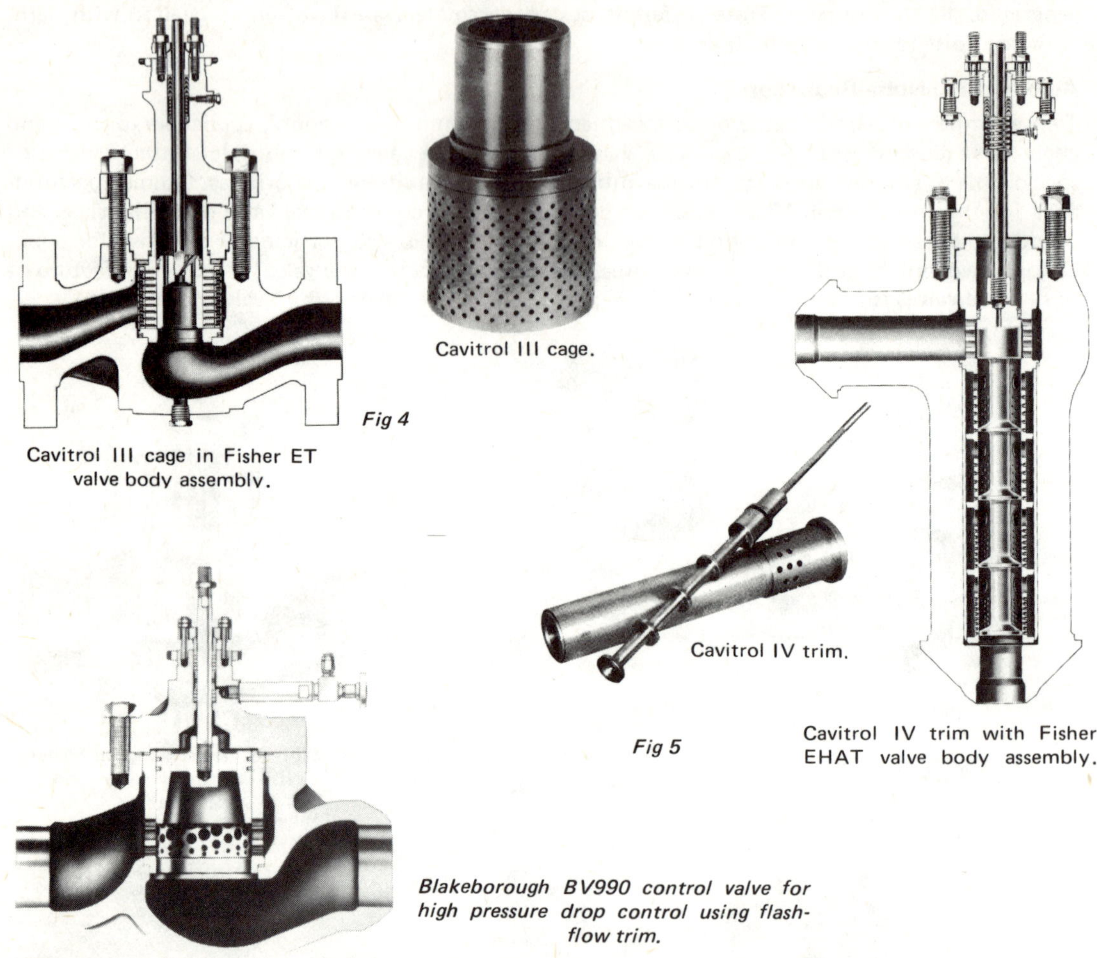

Cavitrol III cage.

Fig 4

Cavitrol III cage in Fisher ET valve body assembly.

Cavitrol IV trim.

Fig 5

Cavitrol IV trim with Fisher EHAT valve body assembly.

Blakeborough BV990 control valve for high pressure drop control using flash-flow trim.

admitted from the opposing hole at the centre of the cage. Thus, a continuous cushion is formed which prevents cavitating liquid from contacting the metal surfaces and ensures that vapour bubble collapse takes place in the centre of the flow stream.

The cage design in Fig 4 consists of one or more concentric cylindrical sections referred to as stages. The number of stages required depends on the inlet pressure and the pressure drop. In operation, the liquid undergoes a portion of the total pressure drop in each stage of the cage. This prevents the liquid in any one stage of the cage from falling to or below its vapour pressure. Therefore, formation of vapour bubbles and their subsequent collapse is eliminated.

Fig 5 shows a further trim design employing a (patented) pressure staging for elimination of cavitation with differential pressures above 3 000 lb/in^2 (200 bar). The expanding flow area design takes advantage of the ability of the liquid to undergo a greater pressure drop in the initial stages without cavitating. This results in a much lower inlet pressure to the final stage. This design also separates the shutoff and throttling locations to prevent clearance-flow erosion.

Cavitrol V trim.

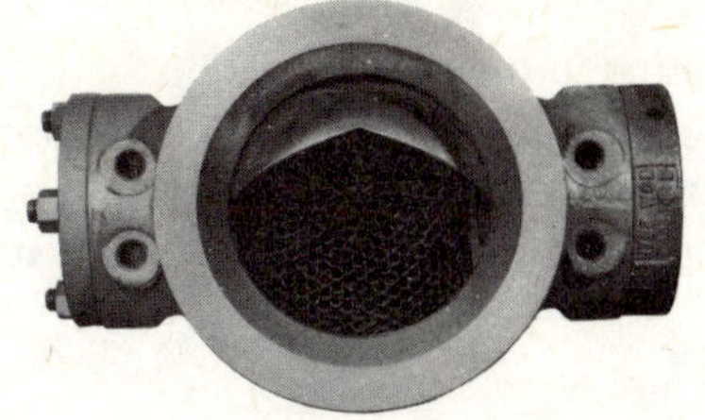

Inlet of Fisher Vee-Ball valve body assembly with Cavitrol V trim.

Fig 6

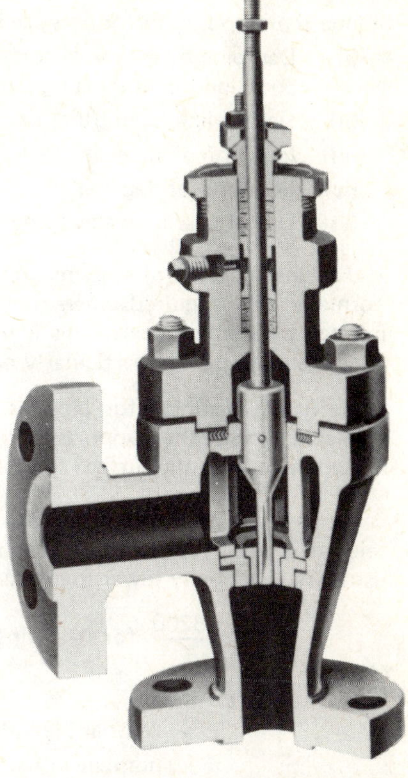

Blakeborough cage trim control valve with microflow trim.

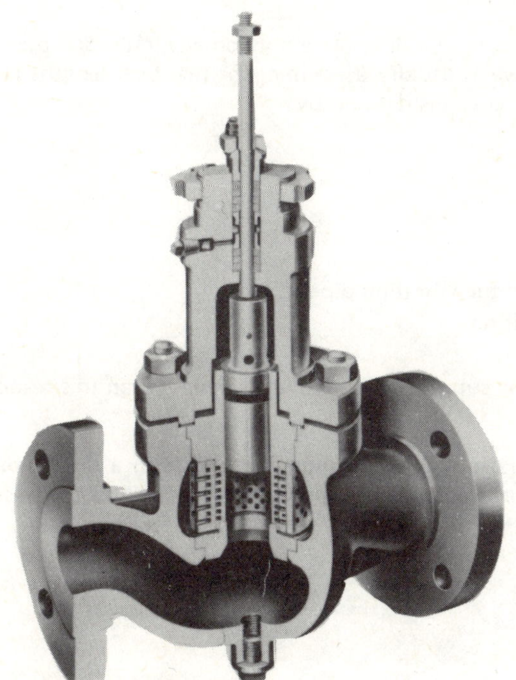

Blakeborough cage trim type control valve with the cascade trim for the safe handling of high energy conditions and noise control.

A further design is shown in Fig 6 where the trim consists of a carefully designed bundle of tubes which minimizes cavitation noise and damage by controlling the formation of cavitation bubbles. The tubes serve three functions: they prevent the flow stream from reaching its potential minimum area, they maintain maximum pressure head to reduce cavitation bubble formation, and they limit the size and number of cavitation bubbles that do form.

Path Treatment

Standard methods used for noise reduction in piping are:

(i) Damping by means of suitable isolating pipe supports. This also provides decoupling for supporting structures.

(ii) Decoupling from other sources of noise or vibration in the system.

(iii) Insertion of silencers.

(iv) Soundproof 'lagging'.

(v) Use of bearing-walled pipe.

For the majority of systems only (i), and to a lesser extent (ii), should be necessary. 'Lagging' is normally only required when there are pulsation vibrations present which cannot be damped or isolated by simple means. This is most likely to occur in pumped systems employing thin walled, large diameter piping, particularly on the suction side.

Sufficient damping for pipes is usually provided by suitable supports, or pipe clips spaced at regular intervals, the supports having resilient linings so that vibration in the pipe is not transmitted directly to the surface to which the supports are fixed.

Optimum pipe spacing can be analyzed in terms of standing wave phenomena, although this is seldom necessary. The case of axial standing wave is usually academic, for practical lengths are usually substantially lower than the critical length, which is defined by:-

$$L_a = \frac{8200}{f} \text{ for steel pipes}$$

where

L_a = resonant length of pipe in feet

= number of half wavelengths in the vibrating pipe

f = frequency of any strong vibration

Theoretically, at least, the distance between pipe supports should always be less than this resonant or critical length.

It may, however, be necessary to analyze the various possible sources of noise in a fluid pipework system in more detail in order to arrive at satisfactory noise treatment. In this case the possible sources of noise generation, in decreasing order of significance:-

(i) Pump noise (where applicable).

(ii) Appliance noise.

(iii) Control element noise.

(iv) Water hammer.

(v) Chatter.

(vi) Cavitation.

(vii) Resonance.

(viii) Pipework noise.

(ix) Thermal effects.

Bellows

Bellows – particularly rubber bellows – can be very effective in preventing pump noise from being transmitted along pipelines. Plain elastomeric bellows provide a complete isolation joint and give the best possible sound absorption as all pipe borne noise must pass through the bellows material. For best results bellows should be placed as close to the pump as possible, and the pipe to which it is connected securely anchored as near as possible to the other end of the bellows. The pump also needs to be solidly mounted to withstand both pressure forces and flexibility forces arising out of the bellows' stiffness.

Where this is not possible (*eg* the pump is flexibly mounted), or other factors (such as high pressure) mitigate against the use of plain bellows, tied or axially-restrained bellows must be used. Such bellows are pressure-balanced units. The flanges and tie-bars, however, now form a transmission path for vibration unless isolation treatment is incorporated. The simplest form of treatment is by the use of resilient bushes and/or rubber washers to prevent metal-to-metal contact between the tie-bars and the backs of the restraining flanges. Even single rubber washers can be effective, if correctly selected in terms of hardness.

Some noise reduction data obtained with representative designs are shown in Fig 7.

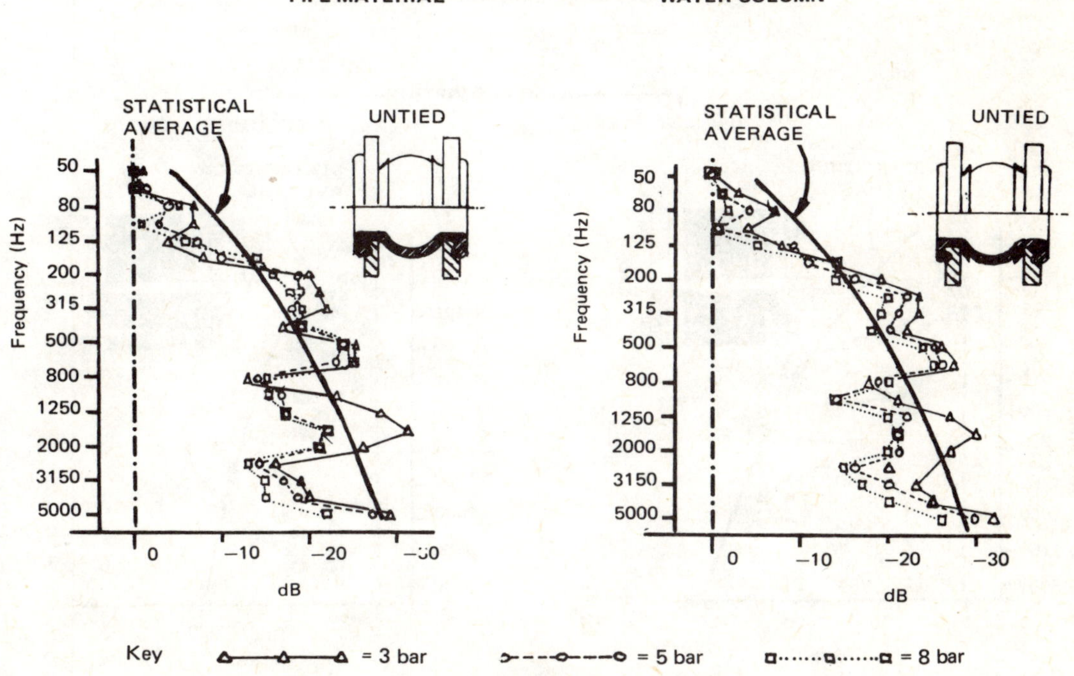

Fig 7 Typical noise reduction with simple rubber bellows (Engineering Appliances Ltd).

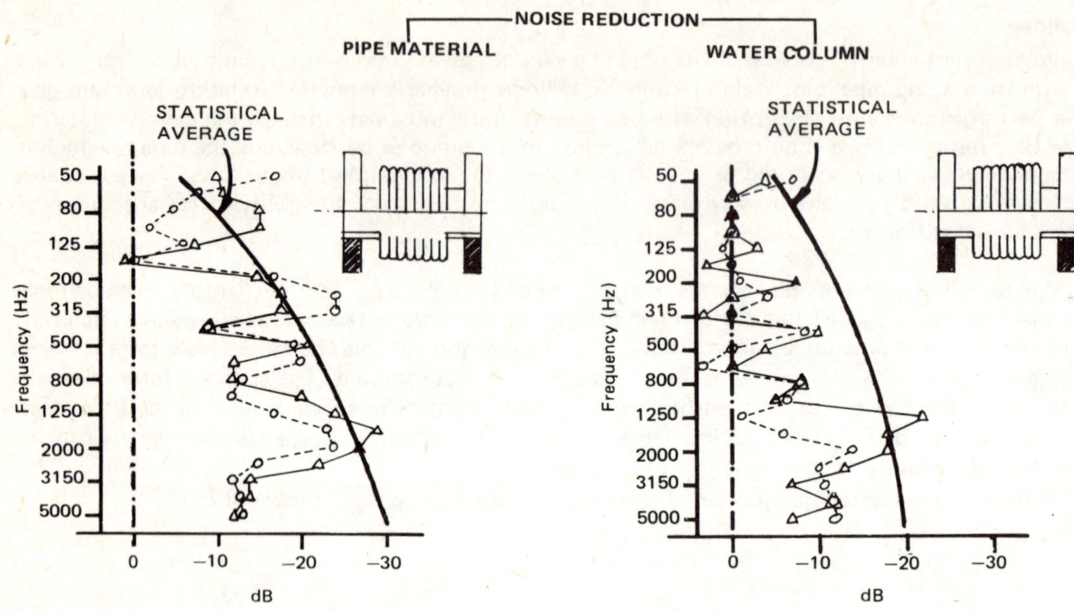

Fig 8 Typical noise reduction with corrugated steel bellows. (Engineering Appliances Ltd)

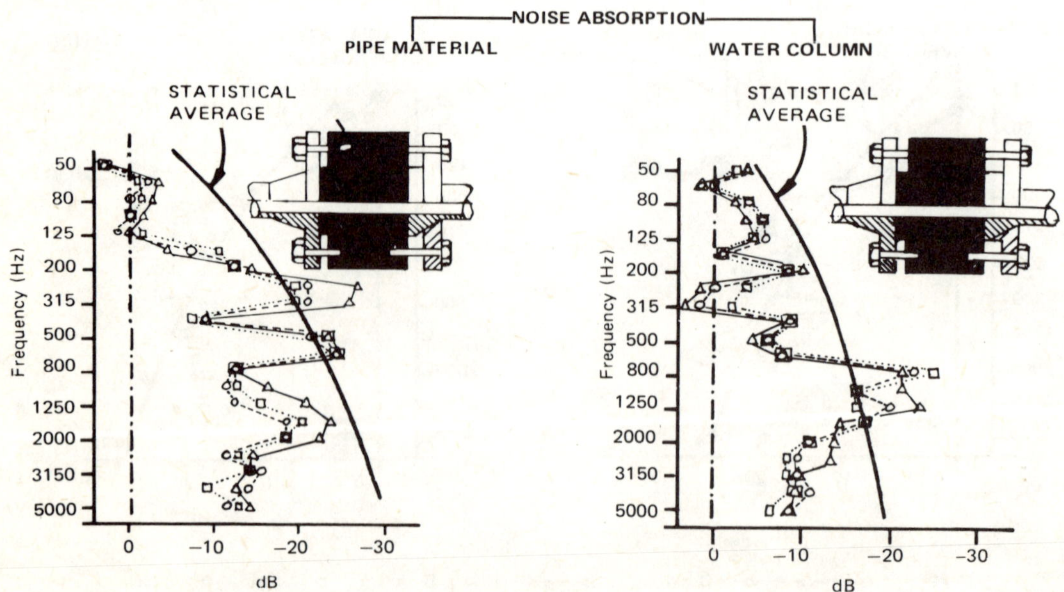

Fig 9 Solid metal/solid rubber isolating unit and typical noise reduction data. (Engineering Appliances Ltd).

Metal Bellows

Metal bellows can give inconsistent results in terms of noise and vibration isolation. Generally their performance is much below that of rubber bellows. Again some test data are shown in Fig 8.

Isolating Flanges

An alternative type of isolator is shown in Fig 9. Basically this consists of a solid rubber 'washer' of appreciable thickness, into which are bonded steel flanges. These flanges are tapped to accommodate bolts for assembling the isolator between conventional flanges without metal-to-metal contact through the joint.

Theoretically, such a form of isolator should prove better at higher frequencies than lower frequencies, although actual performance would depend on the hardness of the elastomer used. It is not generally as effective as rubber bellows, and definitely inferior for isolating lower frequencies. It does, however, have the advantage of providing a 'solid' coupling and so can be used with flexibly mounted pumps.

Acoustic Filters

Acoustic filters can be fitted to systems where pressure ripple is high. These are essentially tuned silencers which are critical in design and are usually effective over only very narrow frequency bands, although the attenuation achieved can be quite high. Untuned silencers simply comprise an expansion chamber with broader coverage but reduced attenuation. An accumulator is, in effect, an untuned hydraulic acoustic silencer and is most effective at lower frequencies. Dissipative-type silencers provide for dissipation of energy through viscous flow losses and as a consequence, consume some fluid energy. They may be combined with an untuned silencer, although the attenuation will still be appreciably lower than that of the tuned type.

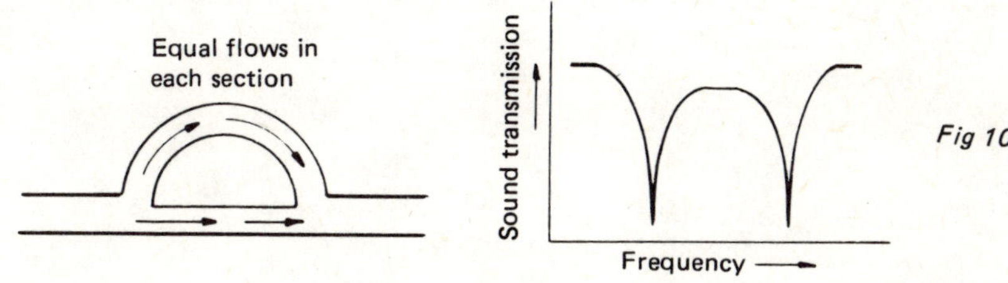

Fig 10

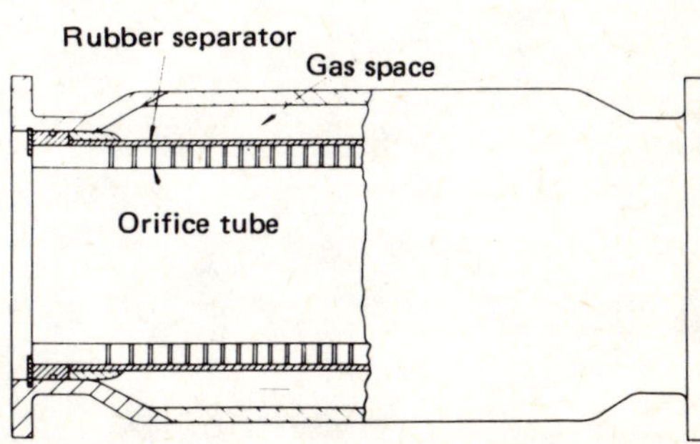

Fig 11 Pressure-release filter

In general, wave cancelling filters are to be preferred since the frequencies involved are low. If the pressure transients are narrow band, a Quinke Tube and expansion chamber can be effective – Fig 10. A major disadvantage of this and other types of simple wave-cancelling filters, however, is the relatively high pressure drop produced. The more usual form of hydraulic silencer is the pressure-release type shown in Fig 11. This gives minimum pressure drop and broad band filtering, but is pressure sensitive and needs regular routine maintenance.

Shock Preventers

Shock preventers are pulsation dampers (or accumulators) characterized by having very large flow inlet apertures which are partially closed off by liquid trying to flow back out of them. They are not shock absorbers, as they prevent shock or surge occurring. For the same reason, they do not attenuate shock.

Shock Removers

These are sensitive hydro-pneumatic devices which prevent a standing wave from passing farther down a system or from bouncing back through them. They are normally of tubular or sleeve form with a flexible membrane. Because of their length it is possible to open a membrane so that it is exposed to the increased pressure of a wave and to close it behind the wave, thus shutting it in.

See also chapters on *Cavitation* and *Flow of Liquids Through Pipes.*

SECTION 4

Valves (Services)

Air Relief Valves

AIR OR GAS trapped in a pipeline carrying a liquid can cause problems, *eg* reduce the effective flow, aggravate the effects of surge, and cause pump cavitation. Possible causes of air (or gas) entrainment are:-

(i) The pipeline was fully charged with air/gas when empty.

(ii) Air is entrained at pump suction.

(iii) Air is drawn in through faulty joints or glands.

(iv) Air/gas is trapped in pockets during pipeline filling.

(v) Air/gas in solution is released due to changes in pressure and temperature.

The problem of dealing with air/gas entrainment is not usually a demanding one. Entrained air/gas will tend to collect at high points in the system. It can then be removed by introducing *air release valves* at these points. These may be simple, manually operated valves (bleed valves), or fully automatic. In the latter case the valve should perform the following functions:-

(i) Release of air/gas accumulating in pipeline during normal pressurized operation, to prevent restriction to fluid flow.

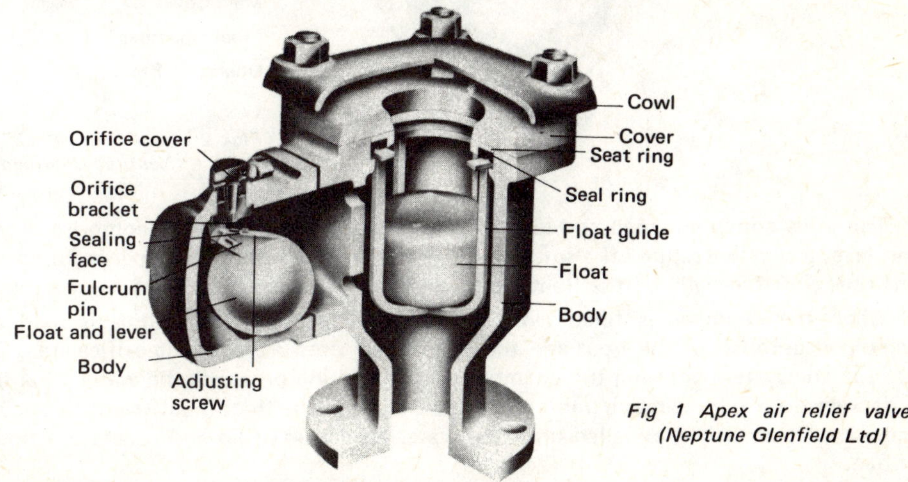

Fig 1 Apex air relief valve. (Neptune Glenfield Ltd)

(ii) Retention of the fluid in the pipeline without loss under all operating conditions.

(iii) Release of air/gas during pipeline filling at a volume rate sufficient to prevent back pressure restricting the filling rate.

(iv) Admission of air to the pipeline during emptying at a rate sufficient to prevent excessive vacuum pressure in the pipe.

Single orifice valves are capable of performing functions (i) and (ii). They are normally used where only relatively small volumes of air/gas are to be released, or where it is desirable to provide additional ventilation at operating pressures. Dual orifice valves are capable of performing all four functions. They can normally provide complete protection against air/gas entrainment under all system operating conditions.

The type of fluid product being handled also affects the design requirements of the air release valve (especially automatic valves). With sewage or industrial effluent, for example, the solids content may block the release passage(s) periodically, causing unreliable operation. This can be overcome by using large volume auxiliary float chambers to contain the fluid under all operating conditions so that it can never come into contact with the air valve elements.

An example of a dual orifice air relief valve suitable for water systems is shown in Fig 2.

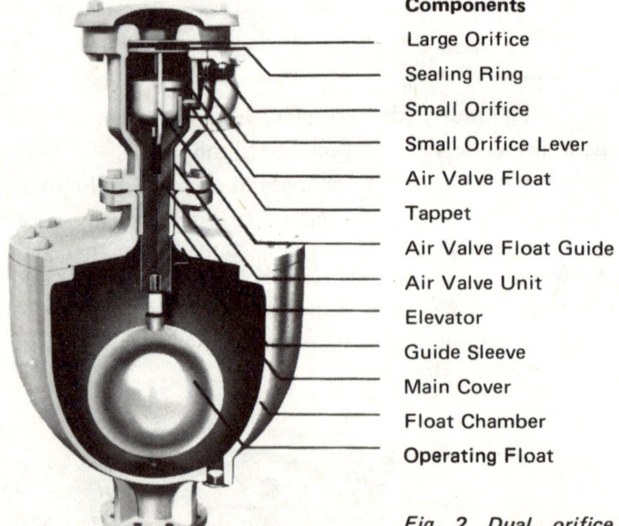

Fig 2 Dual orifice air relief valve. (Neptune Glenfield Ltd)

The valve combines small and large orifices. The small orifice valve comprises a composite float and lever assembly sealing off a small orifice vent. When the float chamber is filling with water, the orifice is closed initially by the float working through a lever ratio of 5:1.

When the chamber is filled with water under pressure, the orifice is held closed by the combined upthrust of the float and the differential pressure over the orifice area. On air accumulated in the system entering the chamber under working pressure, the water level in the chamber is depressed until it reaches a point when the weight of the float is sufficient to uncover the orifice and exhaust air. Air is expelled until the water level rises again and causes the float to close the orifice.

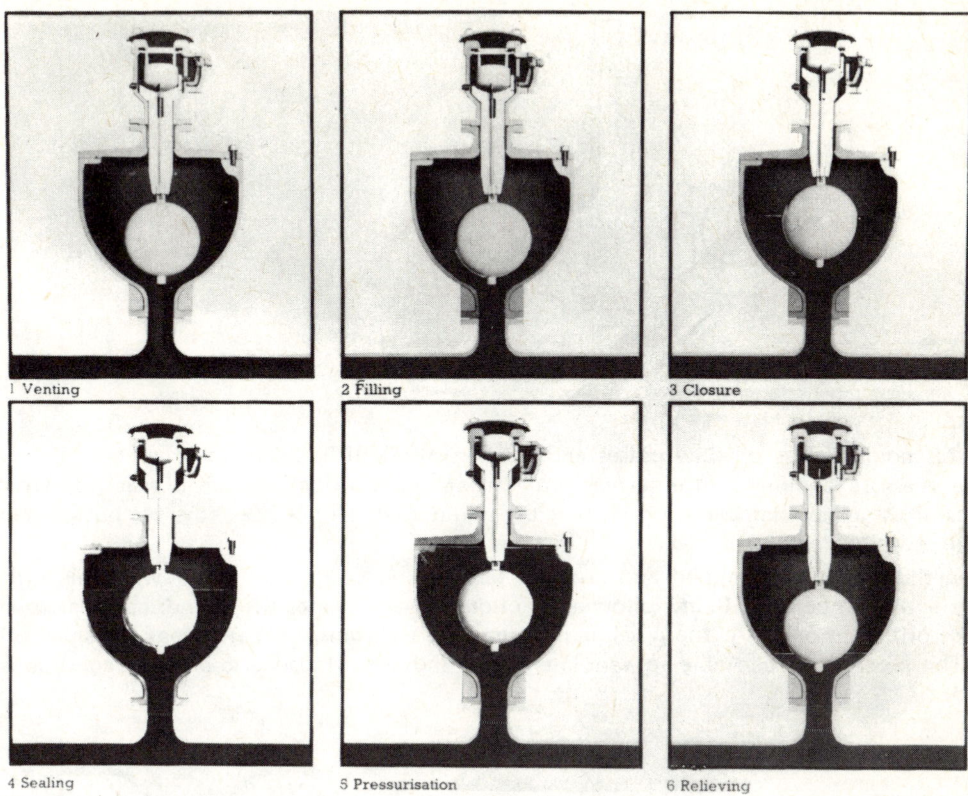

Fig 3 Operational sequence of a typical air-relief valve.

The large orifice valve consists of a float sealing off a large orifice vent to the atmosphere. The float is held at a predetermined height in its casing by a ribbed cage which also guides the float onto the seat. During the pipeline filling or emptying, the 'aerokinetic' feature holds the float off the seat and keeps it completely stable under all air outflow or inflow conditions. The valve cannot close prematurely during outflow. It closes only when water enters the casing and raises the float onto the seat.

An example of the working of dual orifice air relief valves designed for handling sewage and similar effluent is shown in Fig 3. When the pipeline is empty, the spherical operating float is suspended from the elevator within the main chamber and the cylindrical float element of the air valve is supported by the guide cage. The operating lever of the small orifice valve is held open by a tappet on the elevator. Air/gas having been inhaled or expelled from the pipelines is able to flow freely through both orifices, the design of the valve being such that the air flow creates a positive down force to hold the cylindrical float element stable within the guide cage.

As the air/gas is exhausted from the pipeline, liquid enters the main chamber and the operating float then rises with the liquid. The elevator, raised by the float, releases the small orifice valve and engages the base of the cylindrical element, which rises until seated on the rubber face of the large orifice.

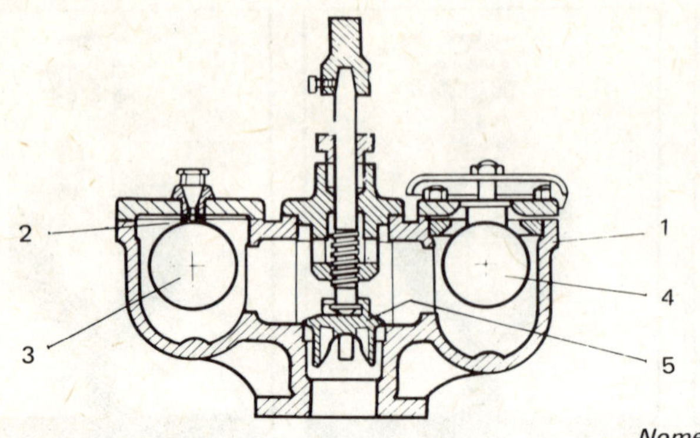

Nomenclature: Air relief valves (double orifice).

At this point air/gas outflow ceases and further inflow of liquid to the main chamber under pipeline pressure compresses the air/gas until maximum working pressure is reached. The proportions of the main chamber are such that the liquid level will not rise above the bottom face of the main chamber cover.

When the pipeline is emptied and pressure falls, the valve main chamber will drain into the pipeline and the operating float, following the liquid level, releases the cylindrical float to allow the large orifice to open. As the pipeline pressure falls to atmospheric, it opens the small orifice valve. The pipeline is then able to ventilate freely and sub atmospheric pressure conditions are avoided.

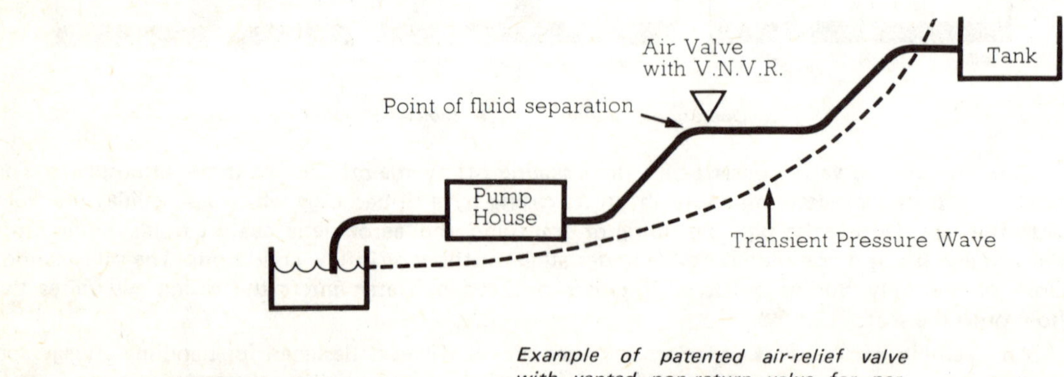

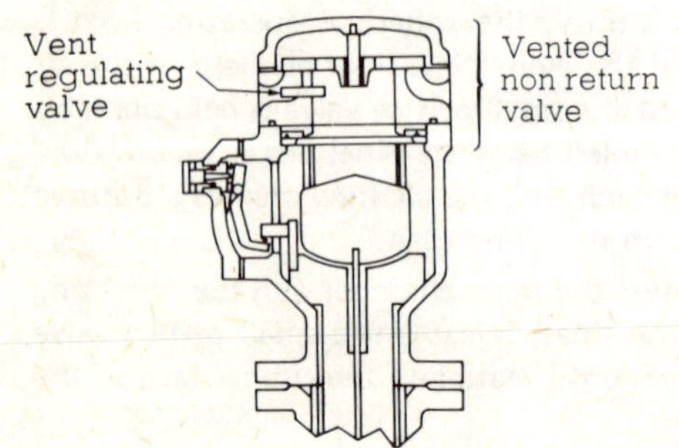

Example of patented air-relief valve with vented non-return valve for performing additional function of surge suppression.
(Neptune Glenfield Ltd)

During normal operating conditions air/gas will be released from the liquid and will collect under pressure in the main chamber, depressing the liquid level. The operating float falls with the liquid but system pressure will hold the cylindrical valve element on the large orifice seat. As the operating float approaches the limit of its travel, the tappet on the elevator opens the small orifice valve, releasing the accumulated air/gas under pressure. This in turn allows the liquid level to rise again and the small orifice valve to close, thus completing a cycle.

Positioning of Air Relief Valves

In systems handling water, air relief valves would normally be placed at all high points – *ie* where a rising section changes to a falling section. In systems handling sewage or industrial effluent rather more extensive treatment is necessary, as illustrated in Fig 4.

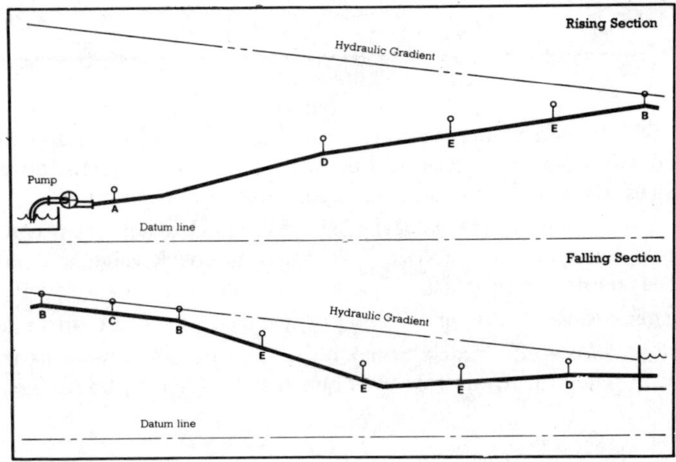

Fig 4 Typical sewage air valve location points.

Where the fluid is pumped through the pipeline it is desirable that a dual orifice valve (valve A) be located just downstream of the pump delivery valves.

Dual orifice type valves are also required at all peak points which are defined relative to the hydraulic gradient and not necessarily to the horizontal. In practice a peak may be considered as any pipe section which slopes up towards the hydraulic gradient or runs parallel to it. In the latter case the minimum requirement is a dual air valve at each end of the section (valve B); any additional valves may be of single orifice type.

Positions where an increase in down slope occurs will require ventilation by a small orifice valve which should also be installed at points of decrease in up slopes (valve D).

Pipeline sections of uniform profile also require ventilation and dual orifice units should be installed at about 2 500 feet (800 metre) intervals on these sections (valves E).

See also chapter on *Traps and Drainers*.

Check Valves

CHECK VALVES are intended to prevent a reverse flow in a line, *eg* after a pump has stopped. They are also known as non-return valves, reflux valves, flap valves, retention valves, foot valves, *etc*, in different services. Basically they can be categorized as follows.

(i) Swing-type valves (swing check valves) where the check mechanism is a hinged flap or disc — see chapter on *Flap Valves*. The butterfly check valve is a variant on this principle — see chapter on *Butterfly Valves.*

(ii) Tilting disc check valves — similar to swing-type check valves but with a profiled disc.

(iii) Lift-type valves where the check mechanism incorporates an element which lifts along an axis in line with the axis of the body seat. These may be further sub-divided into:

(a) disc check valves

(b) piston check valves

(c) ball check valves

(iv) Foot valves — specifically check valves fitted to the bottom of a suction pipe.

(v) Spring-loaded check valves.

(vi) Check and surge-suppressor valves — including multi-door check valves for larger pipelines, and electrically and pneumatically operated surge suppressor valves.

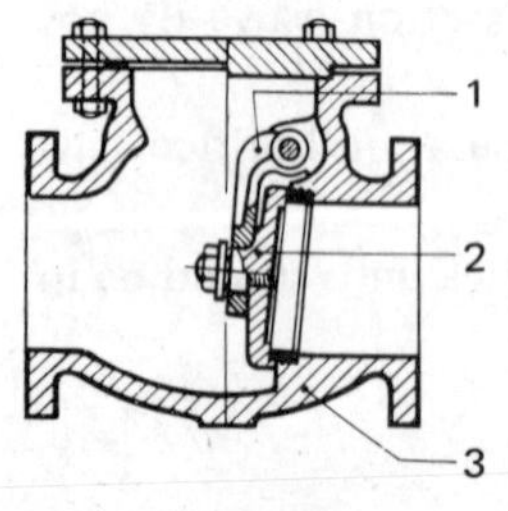

Fig 1

Swing check valve
1 — Body
2 — Cover
3 — Hinge

Piston check valve
1 — Body
2 — Disc
3 — Disc holder
4 — Cover

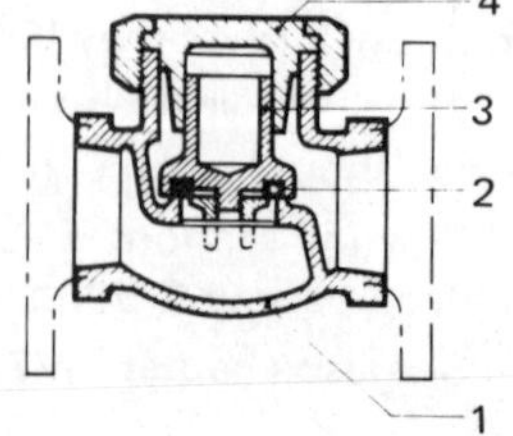

Fig 2

Check valve nomenclature.

Tilting Disc Check Valves

Basis of the tilting disc check valve is a 'lifting' section disc, pivoted in front of its centre of pressure and counterweighted and/or spring loaded to assume a normal closed position. With flow in one direction the disc lifts and 'floats' in the stream, offering minimum resistance to flow. Balance of the disc is such that as flow decreases the disc will pivot towards its closed position, reaching this before flow has actually ceased, sealing before reverse flow commences. With reverse flow, reverse flow pressure and the counterweight system hold the disc closed – Fig 3. Operation is smooth and silent under all conditions.

Valves of this type normally have resilient sealing rings mounted on a metal face. Metal seals may be used for high temperature applications.

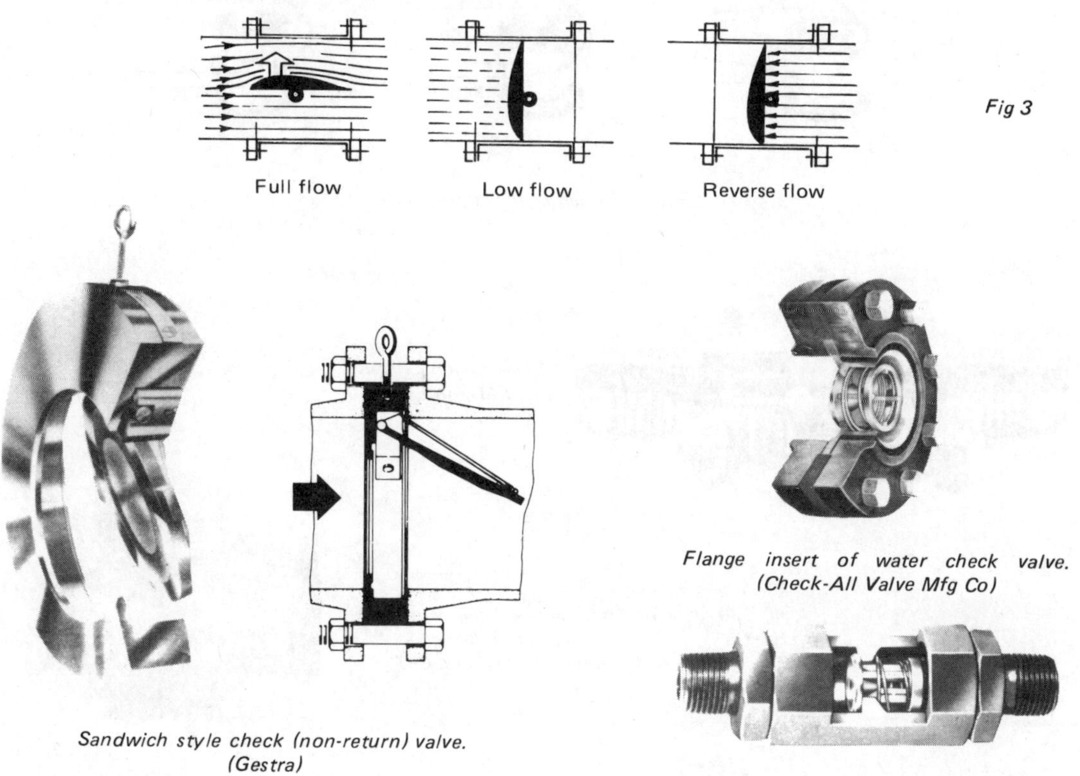

Fig 3

Sandwich style check (non-return) valve. (Gestra)

Flange insert of water check valve. (Check-All Valve Mfg Co)

3-piece miniature check valve.

Lift-type Disc Valves

Lift-type disc valves are similar in configuration to globe valves except that the disc or plug is automatically operated – *ie* is capable of floating in its seat. The disc or plug is lifted by flow in one direction, permitting through flow. With reverse flow the disc or plug is held on its seat by reverse flow pressure, giving shut-off.

Valves of this type are further categorized by geometric configuration – *ie* horizontal, angle (oblique) and vertical.

Piston Check Valves

The piston-type lift check valve incorporates a dashpot applied to the check mechanism – Fig 2 – otherwise it is basically similar to a lift-type disc valve. The advantage of the dashpot is that it provides a damping effect during operation.

Lift-type piston check valves are commonly used in conjunction with globe and angle valves on piping systems subject to surge pressures or frequent changes in flow direction.

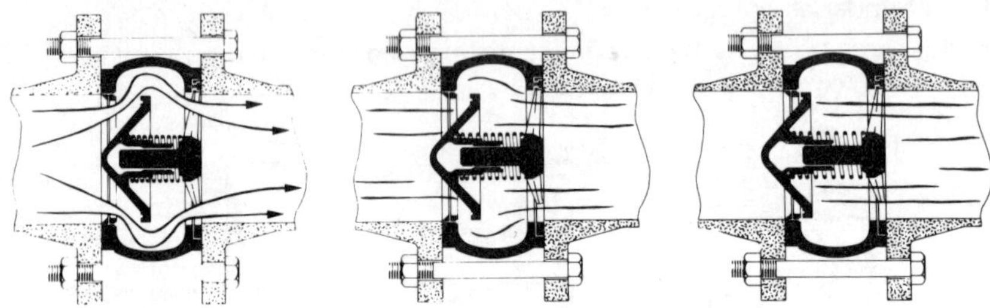

Gestra RK series non-return valve.

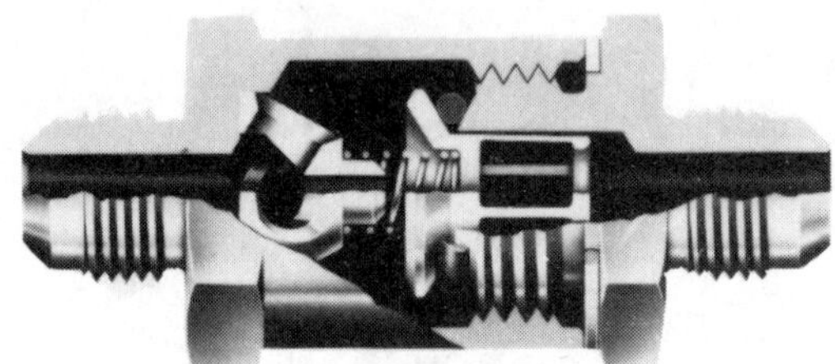

Circle seal check valve. (Tamo Ltd)

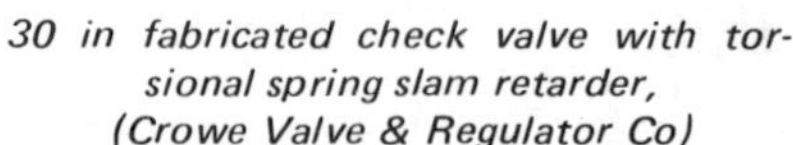

30 in fabricated check valve with torsional spring slam retarder, (Crowe Valve & Regulator Co)

Patent 'move-in' non-return valves. (Ocean)

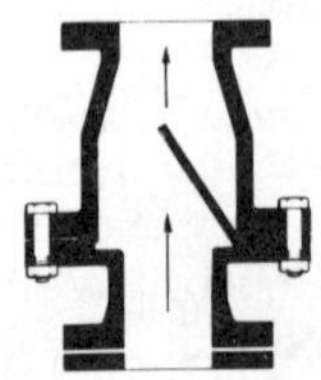

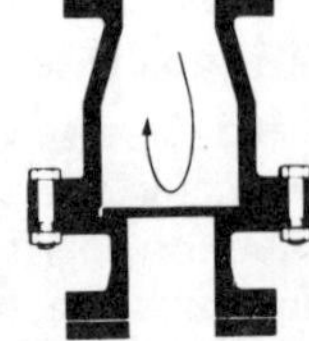

Saunders 'N' type check valve. (Saunders Valve Co Ltd)

Ball Check Valves

The check element in a lift-type ball valve is a spherical ball, suitably restrained but capable of floating off and onto a seat. With forward flow the ball is forced away from the seat, opening the valve. With reverse flow the ball is forced onto the seat to produce a seal and shut-off. A particular advantage of ball check valves is that they can prove more suitable for use with viscous fluids than other types.

Ball check valves may be of all metal construction, metal ball with resilient seat, mixed construction (metal or plastic ball), or all plastic construction.

Foot Valves

Foot valves, which often include a strainer, are fitted to the end of a suction pipe and prevent the pump emptying when it stops and so needing priming when restarting. They should have a minimum resistance to flow, with the actual valve element or flap as light as possible if the risk of cavitation is to be avoided. The valves may be of the single flap (Fig 4) or multiple flap type (Fig 5).

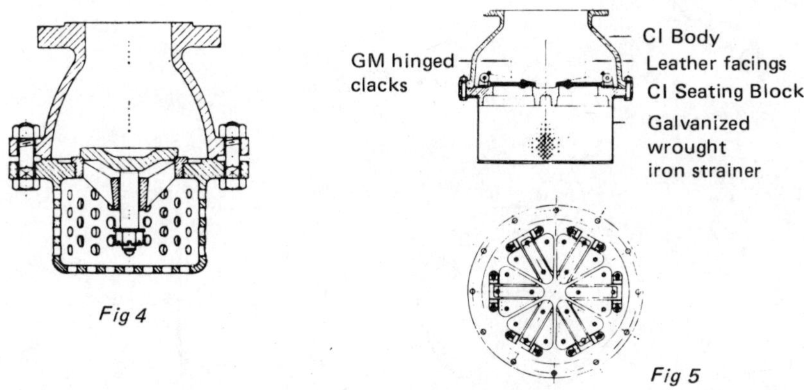

Fig 4

Fig 5

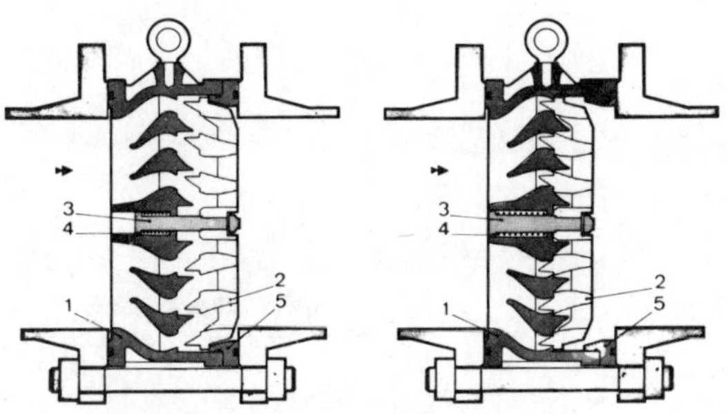

'Clasar' anti-surge non-return valve.

1 – One-piece cast body, in the form of profiled concentric rings supported by spacers.
2 – Sealing assembly, free to move longitudinally, incorporating a similar arrangement of concentric rings to those in the body.
3 – Central guide pin, engaging in the hub of the main body.
4 – Return spring to counteract the weight of the sealing assembly (eg when mounted on vertical pipe-lines).
5 – Mounting ring, incorporated with body.

Spring-loaded Check Valves

Lift-type check valves may be spring loaded for more positive shut-off action, particularly as regards more rapid response cessation of flow, *ie* they can be adjusted to close before flow has fully ceased rather than having to rely on reverse flow pressure. They can be of disc, plug or ball type and can work in any position — *ie* horizontal, inclined, upward or downward flow.

Spring-loaded check valves can be made in the widest variety of materials with stainless steel or high duty alloy springs as necessary. Opening characteristics are governed by the spring rate.

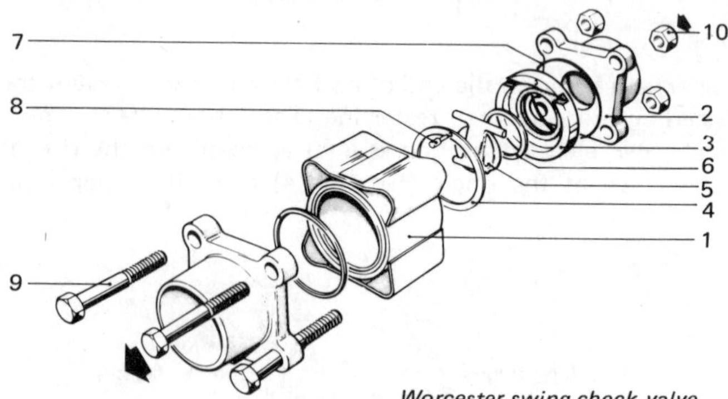

1 — Body
2 — Body connector
3 — Seat housing
4 — Retaining plate
5 — Clack
6 — Seat
7 — Body seal
8 — Orientation pin
9 — Body connector bolt
10 — Body connector nut

Worcester swing check valve.

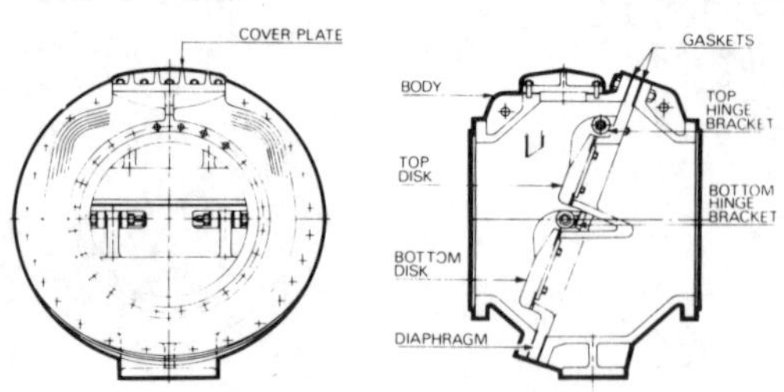

Multi-disc check valve.
(Ham Baker)

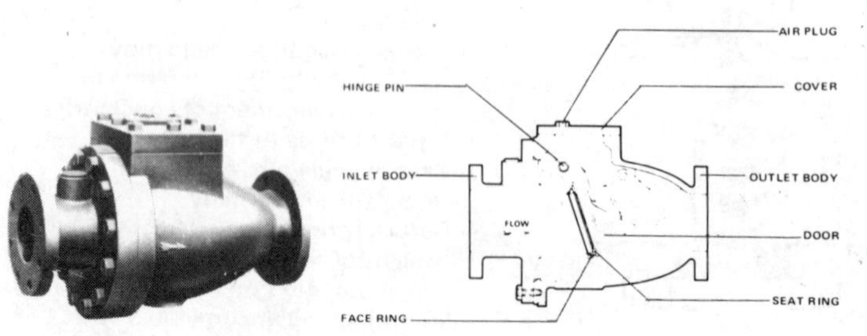

Recoil check valve.
(Neptune Glenfield Ltd)

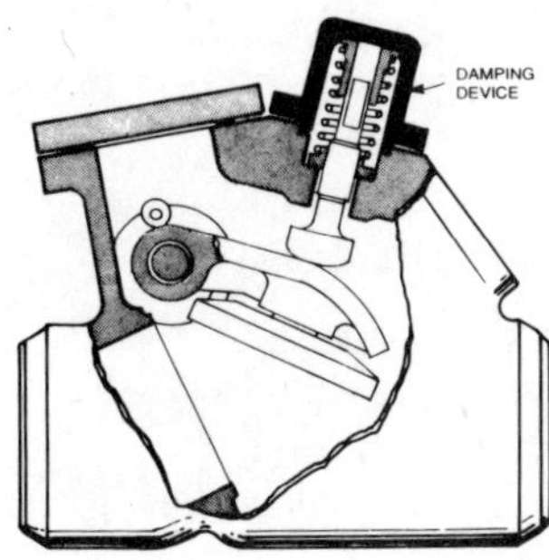

Spring-damped check valve.

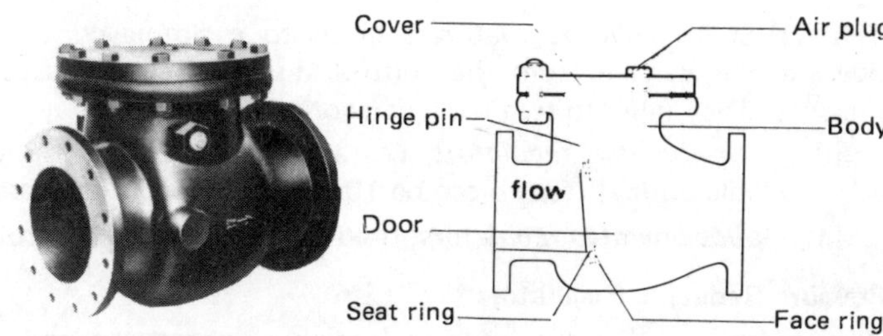

Conventional swing check valve. (Neptune Glenfield Ltd)

Check valves are commonly used in combination with flow control valves, the type and operating characteristics of which can influence the choice of check valve type.

Suitable combinations are:

Swing check valve — used with ball, plug, gate or diaphragm control valves.

Lift check valve — used with globe or angle valves.

Piston check valve — used with globe or angle valves.

Butterfly check valve — used with ball, plug, butterfly, diaphragm or pinch valves.

Tilting disc check valve — used with ball, plug, diaphragm, gate or pinch valves.

Spring loaded check valves — used with globe or angle valves.

The exception is the foot valve, normally associated with a pump (*ie* there is no other valve positioned between the foot valve and the pump).

See also chapter on *Flap Valves* and section on Non-Return Valves in the chapter, *Water Services.*

Regulators

THE DESCRIPTION *regulator* is given to a simple control device in which all the energy to operate it is derived from the control system. Regulators can be used to control flow, level or pressure. They fall into two basic categories:-

(i) *self-operated regulators,* characterized by narrow-range control (*eg* when allowable outlet pressure can be 10—20% of the outlet pressure setting).

(ii) *pilot-operated regulators*, used for broad-range control.

Pressure Reducing Regulators

A pressure reducing regulator maintains a desired reduced outlet pressure while providing the required fluid flow to satisfy a variable downstream demand. The value at which the reduced pressure is maintained is the outlet pressure setting of the regulator.

Self-Operated Regulator

As shown in Fig 1, a self-operated pressure-reducing regulator senses the downstream pressure through either an internal pressure tap or an external control line. This downstream pressure opposes a spring, which moves the diaphragm and valve plug to change the size of the flow path through the regulator.

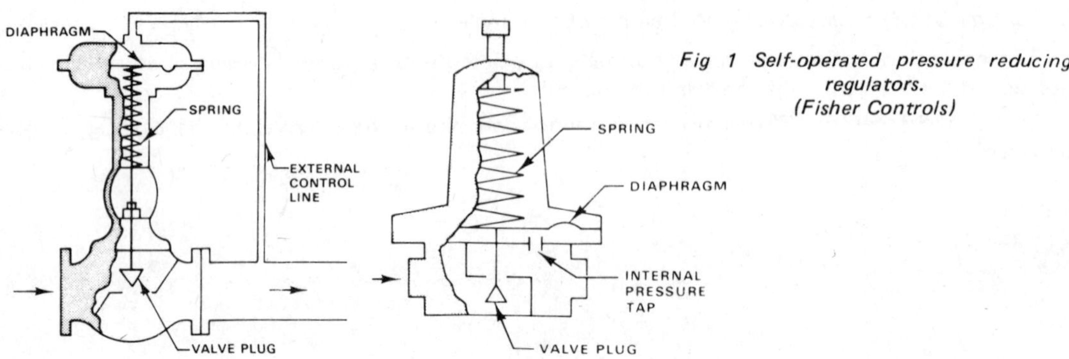

Fig 1 Self-operated pressure reducing regulators. (Fisher Controls)

Pilot-Operated Regulator (Fig 2)

The addition of a pilot to a regulator provides a two-path control system. The main valve diaphragm responds quickly to downstream pressure changes, causing an immediate correction in the main valve plug position. The pilot diaphragm responds simultaneously, diverting some of the reduced inlet pressure to the other side of the main valve diaphragm to control the final positioning of the main valve plug.

Some further examples of pressure-reducing regulators are shown in Fig 3.

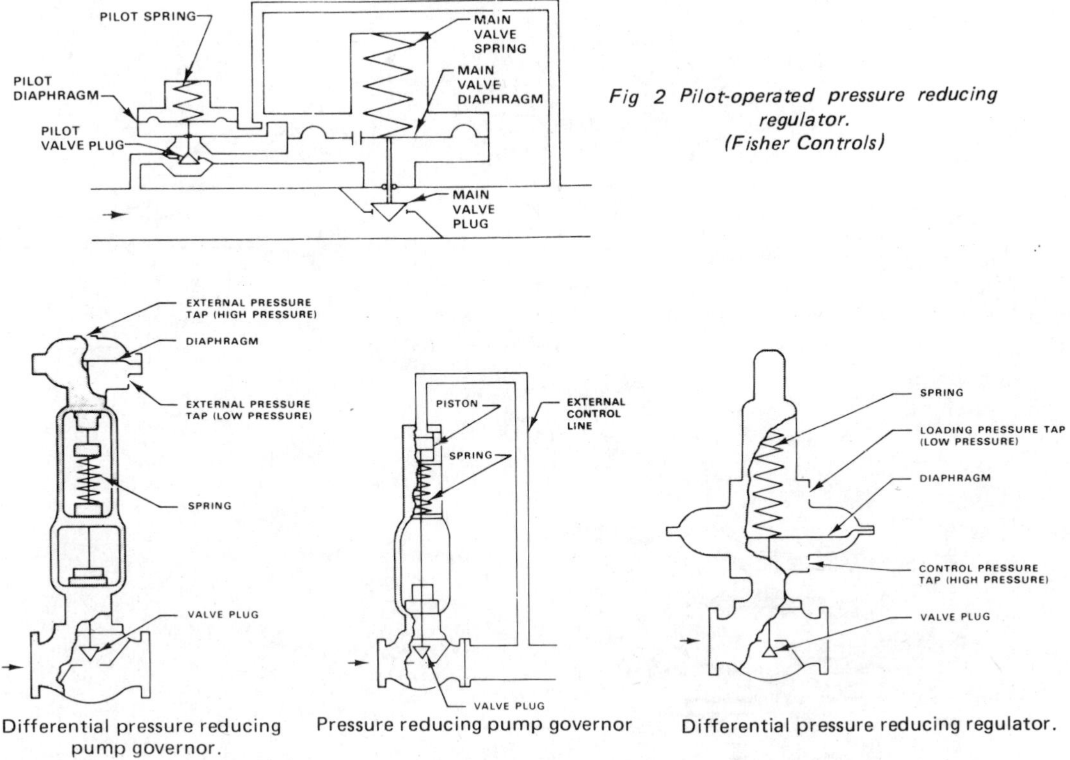

Fig 2 Pilot-operated pressure reducing regulator. (Fisher Controls)

Differential pressure reducing pump governor.

Pressure reducing pump governor

Differential pressure reducing regulator.

Fig 3

Flow Regulators

A flow regulator is used to provide a contant flow despite changes in upstream and downstream pressure.

In this case a meter valve is adjusted to the desired flow rate and functions as a fixed orifice. A diaphragm and spring operated differential valve keeps a constant differential across the fixed orifice. The constant differential pressure across the fixed orifice results in a constant flow rate for each meter valve setting.

Level regulators are used to maintain liquid level within a tank. A complete level regulator is composed of a ball float as a sensing element, an actuator, and a valve body. A level regulator may be an integral unit; it may be assembled from a ball float actuator and a valve; or assembled from a ball float sensor, a lever actuator, and a valve.

Fig 4 shows two different types of level regulators. Different combinations of level regulators are available using various ball float sensors, lever or ball float actuators, and valve bodies.

See also chapters on *Float Controls* and *Ball Float Valves.*

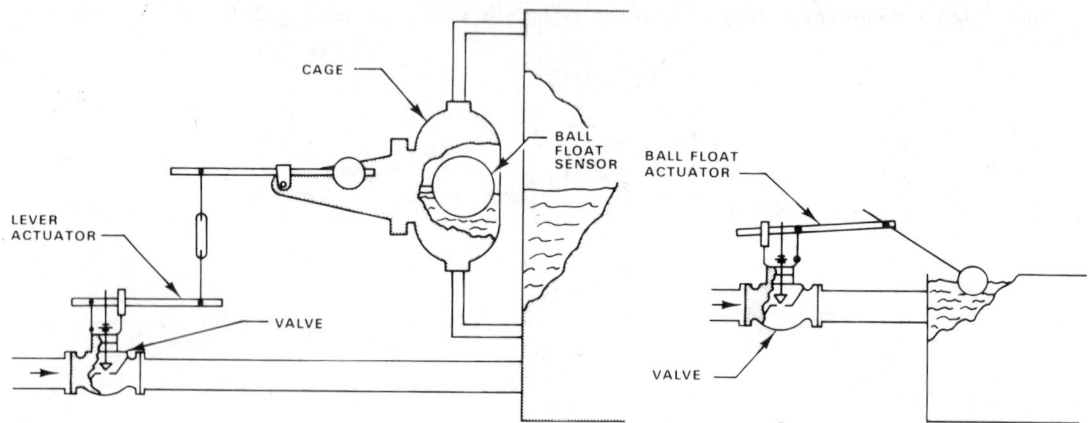

Fig 4 Level regulators.

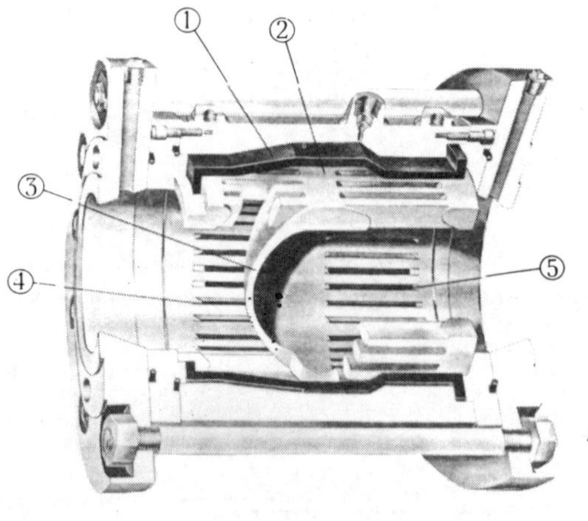

1. Expansible tube
2. Jacket space
3. Core
4. Inlet slots
5. Relief flow slots.

Fig 5 Grove flexflo gas pressure regulator.

Surge Relievers

A *surge reliever* is another form of regulator, sometimes referred to as a *surge relief valve.*

An example is shown in Fig 5, comprising a flexible tube fitted over a slotted core. The core barrier directs incoming fluid through the inlet slots in the core and down through the outlet slots, at the same time expanding the tube. The system automatically adjusts itself to accommodate pressure variations via the flexing of the tube, which is the only moving part.

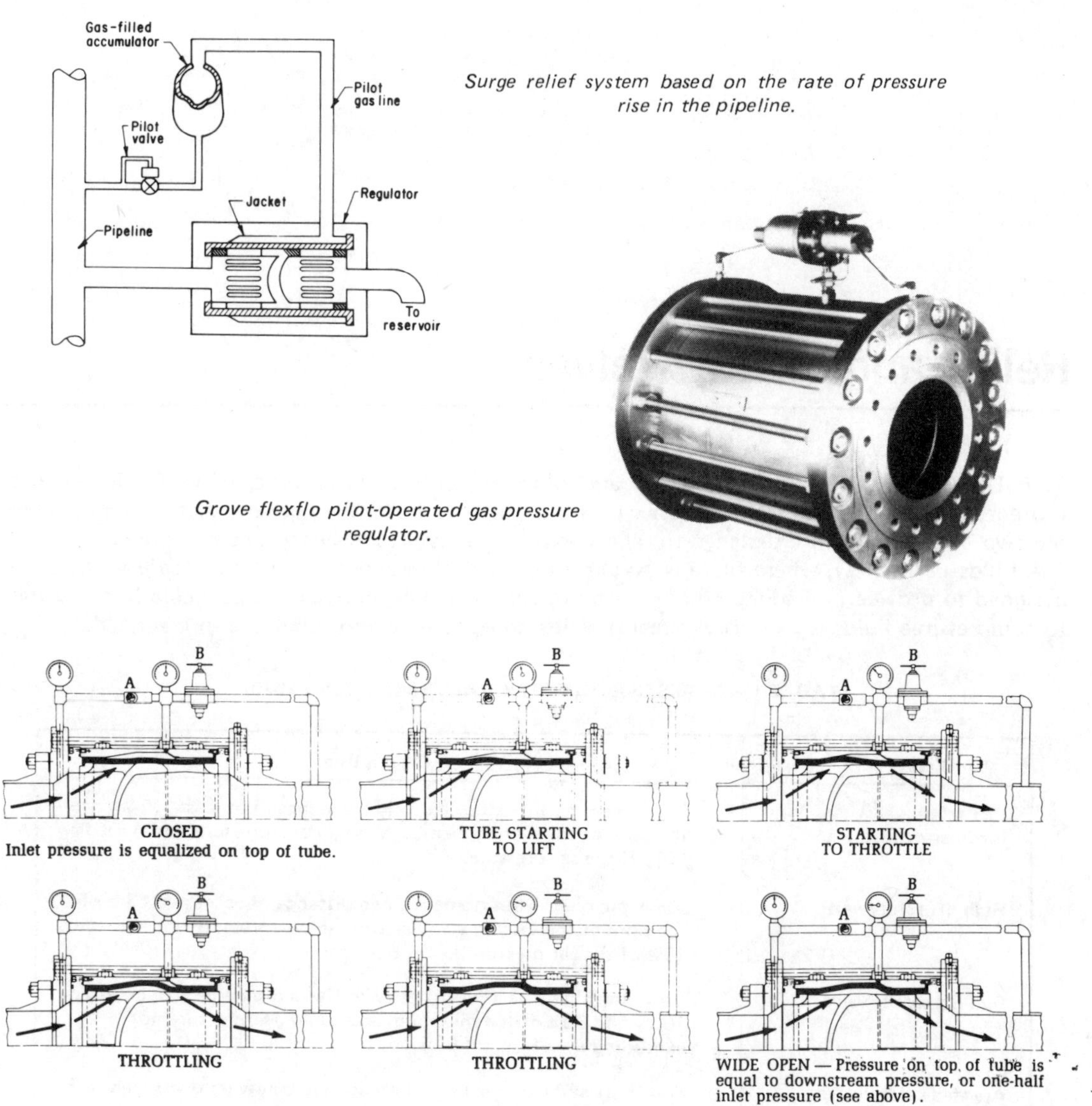

Surge relief system based on the rate of pressure rise in the pipeline.

Grove flexflo pilot-operated gas pressure regulator.

Fig 6

The working action can be followed more clearly through the diagrams in Fig 6. Pressure differential controls tube travel, which in turn is controlled by flow through the fixed metering orifice, A and the variable orifice pressure regulator, B. Pressure is controlled automatically and smoothly, with positive shut-off. When pressure restriction is critical, wide open flow occurs when the pressure on top of the tube is reduced to one half of the inlet pressure. When pressure reduction is sub-critical, pressure on top of the tube must be reduced to outlet pressure for wide open flow.

Relief and Safety Valves

THE DESCRIPTIONS *relief* and *safety* are commonly applied to all types of valves designed to protect systems or vessels from excessive pressure. There is, however, a general distinction between the two. *Relief valves* are designed to relieve excessive pressures in systems containing incompressible fluids (*ie* liquids), where there is no chance of explosion under overpressure. *Safety valves* are designed to provide immediate relief of overpressure with any fluid and are particularly applicable to compressible fluids (*eg* air, gases, steam) which could cause explosion if over-pressurized.

TABLE I – CLASSES OF SAFETY VALVES (BS5759 : 1979)*

Class	Description
Lift safety valve (ordinary class)	Valve member lifts automatically a distance of at least 1/24th of the bore of the seating member with an overpressure not exceeding 10% of the set pressure.
High lift safety valve	Valve member lifts automatically a distance of at least 1/12th of the bore of the seating member with an overpressure not exceeding 10% of the set pressure.
Full lift safety valve	Valve member lifts automatically to give a discharge area between 100% and 80% of the minimum area at an overpressure not exceeding 5% of the set pressure.
Assisted safety valve	Power-assisted valve lifting automatically below its unassisted set pressure, with fail-safe requirements.
Supplementary loaded safety valve	Valve with sealing increased by supplementary load, relieved at set pressure. Without load relief, obtains its rated capacity at not more than 115% design pressure.
Pilot operated safety valve	Valve operation via two independent pilot devices with the lift of the main valve achieved with an overpressure not exceeding 5% of the set pressure.

*A corresponding American Standard is Section VIII of the ASME code which specifies capacities, overpressure and blowdown for safety valves.

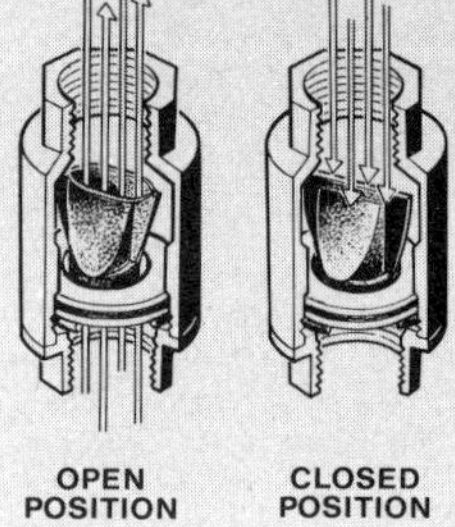

Further detailed distinctions are also drawn, leading to the following categorization:-

(i) *Safety valves* – characterized by rapid full opening or pop action, used on steam, gas or vapour services.

(ii) *Relief valves* – used primarily on liquid services for pressure relief.

(iii) *Safety relief valves* – are full opening valves capable of being used either as a safety valve or a relief valve, depending on the application.

Classes of safety valves are summarized in Table I.

Other devices and systems employed for similar duties include regulators, instrument-operated control valves for relief services and relief devices incorporating pilot- or sophisticated instrument operated sensory and control systems. The latter are available with variable set points and are more costly systems, generally only employed in more severe services.

Whilst operating characteristics may differ in detail, both safety valves and relief valves are of similar basic design, normally direct action spring type. Lever and weight or deadweight types are obsolete but may still be found in use. Alternative designs, such as torsion bar types, have limited application.

Spring loaded safety valves normally have their set pressure adjustable by means of a screw in the top of the bonnet and are normally fitted with an external lever for checking their action (or such a lever is available as an option). Relief valves may be similarly adjusted for set pressure, but do not usually have a relief-checking lever.

Types of Construction

The simplest type of pressure/relief valve is the spring loaded ball, although this has several limitations (notably a proneness to chatter due to oscillation of the spring). Poppet valves are generally preferred since their operating characteristics can be closely controlled both in design and manufacture. Plunger type valves may also be used, although again these are subject to chatter unless

TABLE II – SAFETY VALVE OPERATING PARAMETERS

Parameter	Definition
Set pressure (also known as crack pressure)	(i) *Liquid services* – inlet pressure at which valve starts to discharge under service conditions. (ii) *Gas or steam services* – inlet pressure at which the valve pops under service conditions.
Differential set pressure	Difference between set pressure and backpressure. (where present).
Overpressure	Pressure increase over the set pressure of the valve or relief device.
Blowdown	Difference between the set pressure and resetting pressure expressed either as a specific pressure or percentage of the set pressure.
Backpressure	Any pressure on the discharge side of a pressure relief valve.
Accumulation	Pressure increase over maximum allowable working pressure of a vessel or system during discharge through the pressure relief valve.
Operating pressure	Normal allowable working pressure of the system or vessel.

suitable damping is present in the fluid circuit. One advantage offered by the plunger type is that a two-diameter plunger can be used to provide a bigger discharge area for the same size of spring required, whilst at the same time incorporating a degree of self-damping. Differential plunger valves of this type are mostly used in small bore high pressure services as relief valves (*ie* in its fully open position (see also Table II).

Design Differences

A basic difference between the design of spring loaded safety valves and relief valves is that in safety valves the poppet or disc overhangs the seat to promote faster lift, whereas in a relief valve the area exposed to overpressure is the same whether the valve is open or closed. As a consequence a safety valve pops open, whilst a relief valve lifts gradually with increasing pressure until it reaches its fully open position.

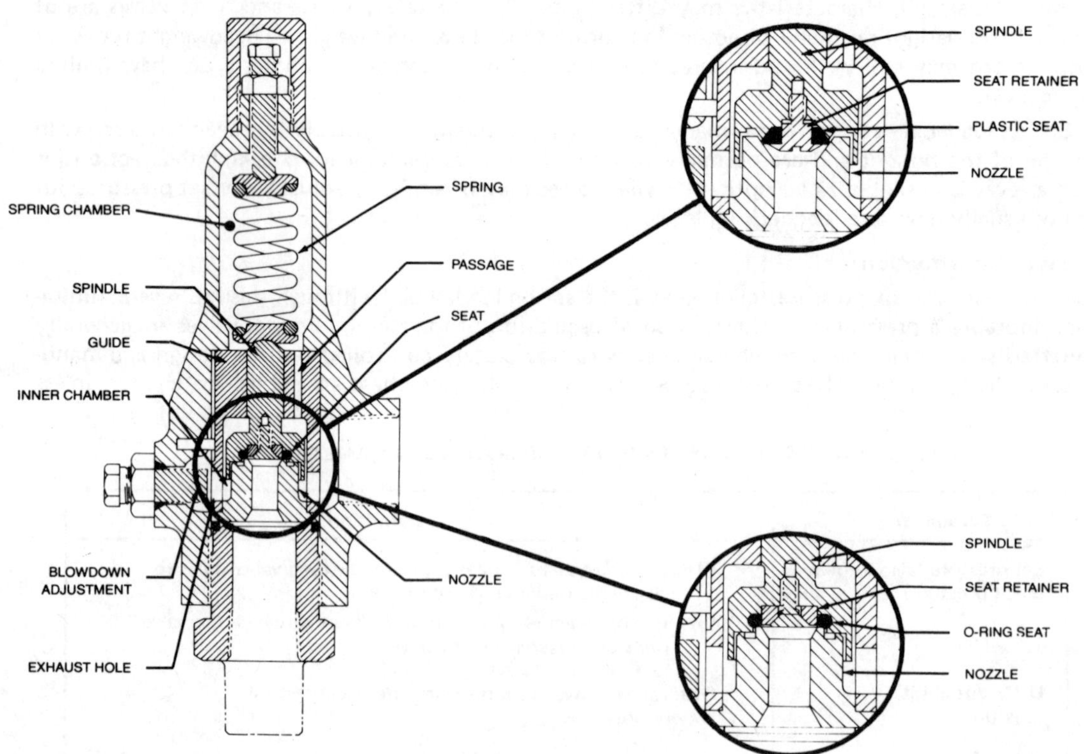

Fig 1 Typical safety-relief valve (AGFO)

Typical Design

A typical design of modern safety-relief valve is shown in Fig 1. In operation, inlet pressure is applied to the seat area and exerts a lifting force on the spindle. This lifting force is counteracted by the downward force of the spring. Under these static conditions the pressure in the spring chamber, the inner chamber and the outlet is atmospheric. If the inlet pressure is increased to the point where the total upward force overcomes the spring force the valve opens. At this opening

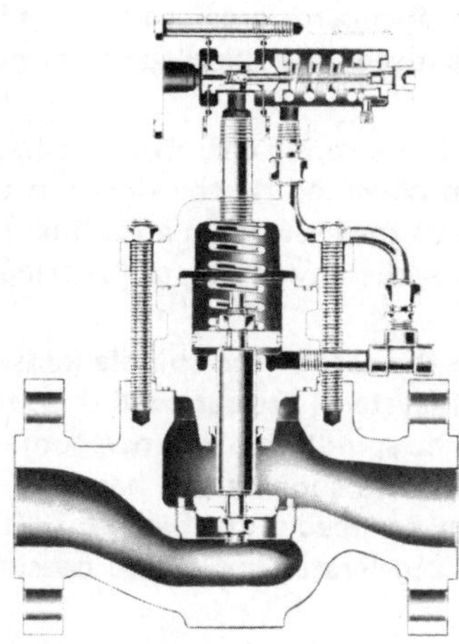

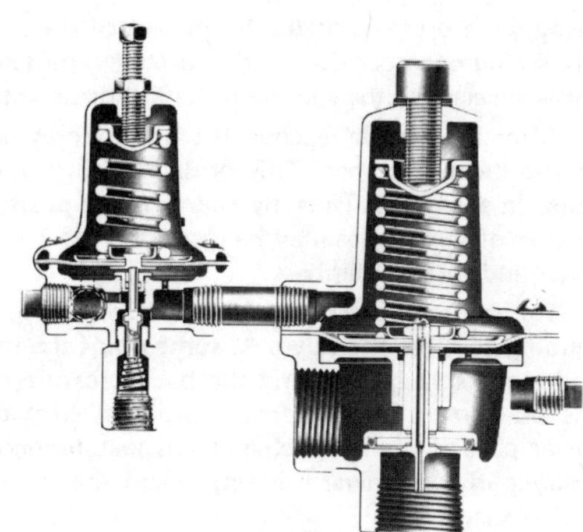

Fisher pressure relief valves.

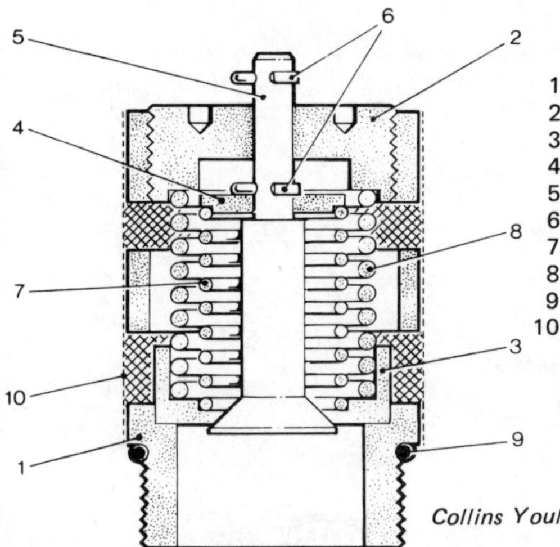

1. Body
2. Valve cap
3. Valve plate
4. Spring retaining disc
5. Vacuum poppet valve
6. Split pins
7. Vacuum spray
8. Pressure spray
9. O-ring
10. Flameproof gauge

Collins Youldon relief valve.

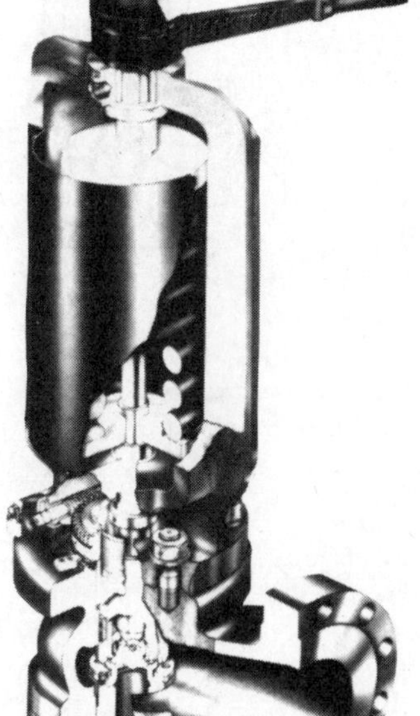

Hopkinsons full lift safety valve for design pressures 1400 to 3000 lb/in² nominal (96 to 207 bar nominal)

instant the pressure in the lower part of the inner chamber is greater than it is in the upper part or the spring chamber due to the restricted passage between the spindle and the guide. This increased force accelerates the spindle to its full open position.

After the spindle reaches its stop, the pressure in the spring chamber equalizes with the pressure in the inner chamber. This pressure, in turn, can be varied by the flow out of the four exhaust holes in the guide. Thus, by changing the position of the blowdown adjustment screw the flow out of one of these holes may be restricted more or less, thereby raising or lowering the pressure in the inner and spring chambers.

The pressure in the spring chamber acting on the top of the spindle adds to the spring force tending to close the valve. At some point the inlet (system) pressure will drop to a point where the combined spring force and the back pressure on the spindle top will overcome the lifting force on the seat area and cause the spindle to start down. As soon as this happens, the pressure in the lower part of the inner chamber is instantaneously reduced a small amount. This increases the unbalance in a downward direction and the spindle accelerates downward causing the valve to close with a snap action.

The point at which the spindle starts down is controlled by the amount of pressure in the spring chamber which is, in turn, controlled by the position of the blowdown adjustment screw in relation to one guide exhaust hole.

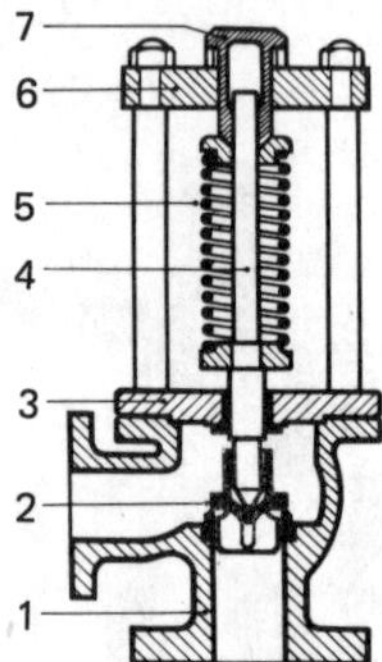

1. Body
2. Valve head
3. Cover
4. Spindle
5. Spring
6. Bridge
7. Adjusting screw

Nomenclature: Lift-type safety valve.

1. Body
2. Main valve
3. Spindle
4. Piston
5. Pilot valve spindle
6. Cap

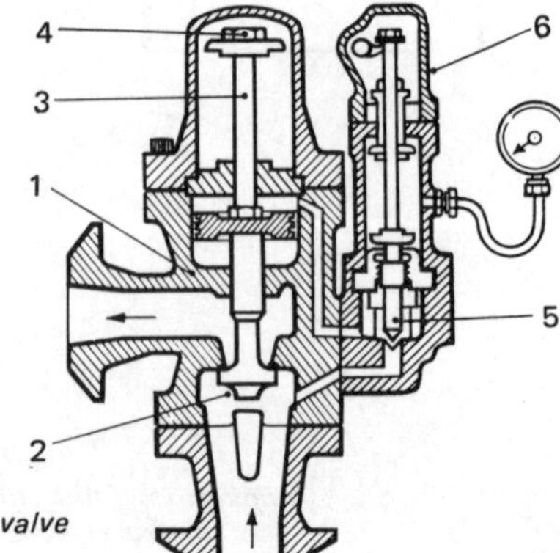

Nomenclature: Pilot-operated relief valve

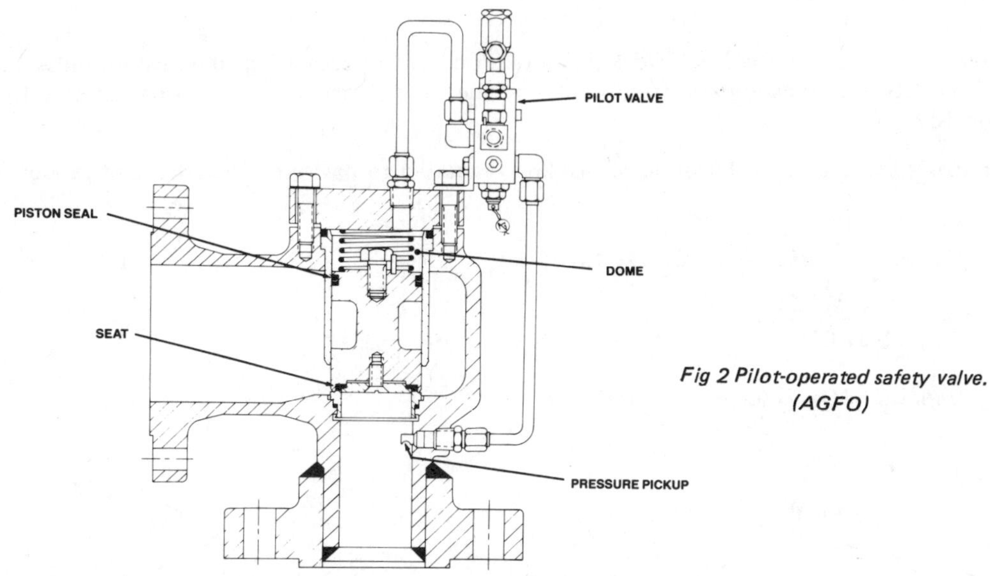

Fig 2 Pilot-operated safety valve. (AGFO)

Pilot-operated Safety-Relief Valves

An example of a pilot-operated safety relief valve is given in Fig 2. Under normal operation, the system pressure acts on the main valve seat at the bottom of the free-floating differential area piston and, by means of the pilot supply line, system pressure is also applied to the top of the piston and the pilot valve seat. Since the top of the piston is larger than the bottom (seat area) there is a large net force holding the piston down. Under static conditions the net downward sealing force increases as the line pressure increases and approaches the set point. This is exactly opposite to the conventional spring-loaded valve which reduces the net force on the seat thus normally causing the valve to 'simmer' as the set point is approached.

When the set pressure is reached, the pilot opens and partially or totally (according to pilot type) depressurizes the dome, thus reducing the load on the top of the piston to the point where the upward force on the main valve seat can overcome the downward loading. This causes a lifting of the piston and the resulting flow through the main valve.

When the predetermined system blowdown pressure is reached the pilot valve closes, the full system pressure is diverted to the dome, and the piston moves downward to close the main valve.

On higher set pressures with pop action pilots, blowdowns as short as 1% to 2% are possible, while on very low set pressures, higher blowdown percentages are necessary. Since system design also has a marked affect on blowdown, it is definitely recommended that the valve manufacturer be consulted on all applications where pop action and blowdowns shorter than 5% are desired. Unless specified, pop action valves are normally set with a 5% to 7% blowdown. Valves with modulating action will have 0% blowdown. Their action is very similar to that of a backpressure regulator.

The standard total pressure pickup is in the main valve inlet neck and ensures that the pilot relief valve senses the pressure at the valve inlet whether the main valve is closed or open and flowing.

Sizing

Safety-relief valves can be sized by calculating a required orifice area using empirical formulas; or derived from tables or capacity charts published by the valve manufacturer for a particular valve and service(s).

Formulas applicable for calculating orifice areas required to pass given quantities of vapour or gas are:-

Vapours or Gases (capacity in ft^3/min)

$$A = \frac{V\sqrt{MTZ}}{6.32\ CKP}$$

Vapours or Gases (capacity in lb/hr)

$$A = \frac{W\sqrt{TZ}}{CKR\sqrt{M}}$$

where:

A = orifice or equivalent flow area in sq ins
V = required capacity in ft^3/min
W = required capacity in lb/hr
C = gas constant based on ratio of specific heats at standard conditions
T = inlet temperature in degrees absolute ($^\circ F + 460$)
M = molecular weight of gas
P = pressure at valve inlet during flow, (lb/in^2 absolute).
(set pressure plus overpressure + 14.7)
K = valve coefficient of discharge (0.809).
This coefficient is equal to 90% of the actual coefficient and must be used for ASME code applications.
For non-code applications the actual coefficient (0.899) may be used.
Z = compressibility factor for gas or vapour; if unknown: Z = 1.0

Note:

(i) Where specific gravity of the gas or vapour is known, molecular weight can be determined for use in the aforementioned sizing calculations from the formula: M = 29 x Sp. Gr.

(ii) P_1 is the pressure at the valve inlet during relief. It may be different from process pressure due to pressure losses in the inlet piping. These losses should be determined for accurate sizing and to ensure proper valve action. A pressure drop in the inlet piping which exceeds the reseat pressure of the safety valve will cause the valve to rapid cycle. Such cycling can adversely affect the endurance life of the valve.

For STEAM (capacity in lb/hr)

$$A = \frac{W}{51.5\ KPK_bK_{sh}}$$

where

K_b = 1 when backpressure is less than 55% of relieving pressure
K_{sh} = 1 for saturated steam

For LIQUIDS (capacity in US gal/min)

$$A = \frac{V_L\sqrt{G}}{27.2P_dK_pK_v}$$

where
K_p = 1 at 25% overpressure
K_v = 1 at normal viscosities

For AIR (capacity in ft^3/min)

$$A = \frac{V_a\sqrt{T}}{418KP_b}$$

where
K_b = 1 when backpressure is less than 55% of the absolute relieving pressure.

Reactive Force

In the cases of gases or vapours it is assessed that critical flow occurs at the outlet of a safety-relief valve, when the reactive force generated on the inlet of the valve can be determined from

$$F = \frac{W\sqrt{\frac{KT}{(K+1)\,M}}}{366}$$

where

F = horizontal reactive force (lb) at centreline of valve
W = gas or vapour flow, lb/hr
K = ratio of specific heats of gas or vapour
M = molecular weight of gas or vapour
T = absolute temperature at inlet ($^\circ$F + 460)

In practical cases the actual flow may be sub-critical at the outlet, when the reactive force generated will be lower. The formula thus provides a means of estimating maximum reactive force likely to be generated.

Testing Safety Valves

Safety valves should be tested regularly to ensure that they have retained their capability of operating at design lift-off pressure. The two basic methods of testing are:-

(i) On-line testing by deliberate over-pressure of the system to determine the actual pressure at which the valve lifts off its seat.

(ii) Off-line testing by removal of the valve from its line (or position) and determination of active lift-off by hydraulic test.

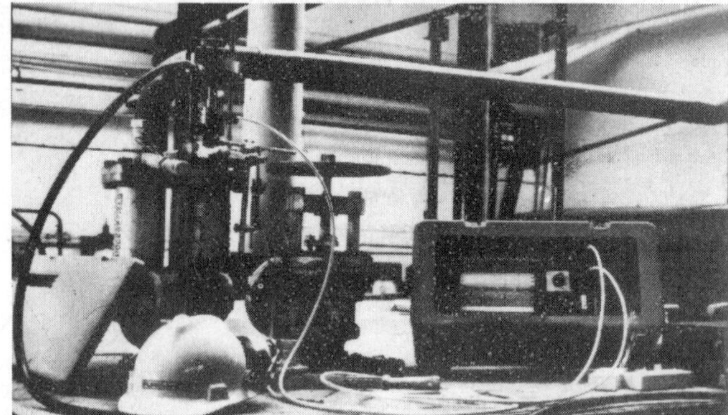

Fig 3 The Trevitest package of valve gear, recorder and hydraulic supply in position on a relief valve.

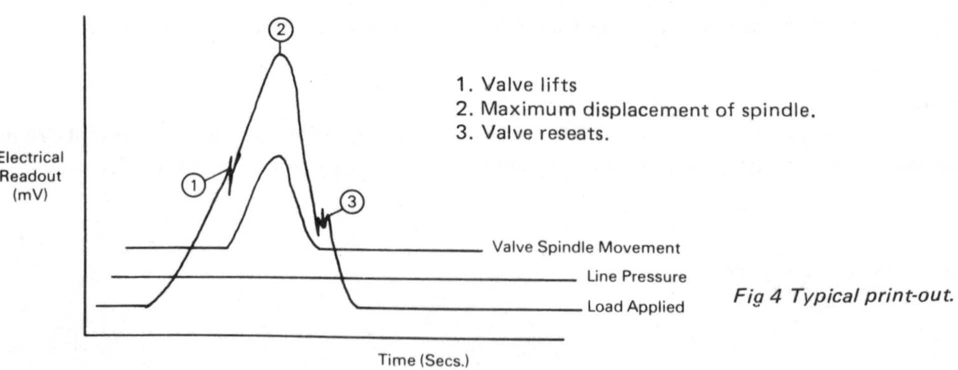

Fig 4 Typical print-out.

It is also possible to apply a hydraulic test for on-line testing using a portable hydraulic test pack – see Fig 3. This consists of a hand-pumped hydraulic generator which can be mounted on the cover of the valve to be tested, together with an adjustable jig so that a hydraulic load can be applied to the valve stem. This load is measured by a load cell, and simple development by a second transducer, the time-history of each being pivoted out by a two-pen recorder – Fig 4. From a knowledge of the valve seat and line pressure at the valve, the data obtained then gives the exact lift and reseat pressures. At the same time the actual open time of the valve is kept extremely short, minimizing waste of product.

Foot Valves

A FOOT VALVE is basically a check valve fitted to the end of a suction pipe leading to a pump. Its purpose is to keep fluid trapped in the suction pipe when the pump stops, thus maintaining a suitable prime for the pump. When the pump restarts, the suction created opens the valve, giving full flow to the pump inlet. (Foot valves are unnecessary on self-priming pumps).

Foot valves may be of simple flap type, or more usually lift check or ball check valves. They are commonly combined with an integral strainer. Some examples are:

Poppet Lift Check Valve

In the example shown in Fig 1 the poppet assembly consists of a plastic tripod which can be displaced along a bore above the valve seat. The travel of the poppet is controlled by a stop on the end of the poppet legs acting as supports for the return spring shouldered onto a washer. This spring ensures that the valve will work in any position. Main characteristics of this design are low head losses with good sealing provided by a nitrile rubber O-ring.

Fig 1

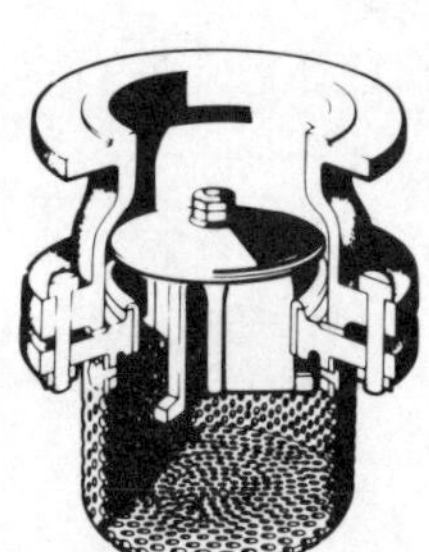

Fig 2

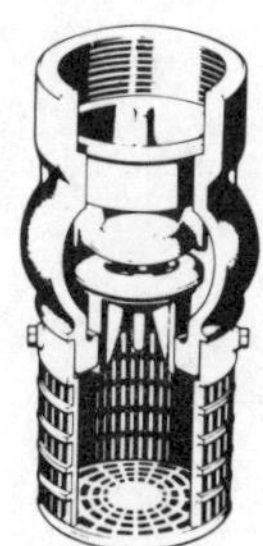

Fig 3

Fig 2 shows a design with the tripod in cast iron and with a cast iron poppet head with streamlined tripod hub. Sealing is provided by a flat gasket shouldered by the poppet head and placed on a collar-type seat. This is a simple and robust design suitable for general applications.

Fig 3 shows a further design where the all-metal poppet with profiled head is guided by three legs and restrained by a downstream stop. Sealing is by a flat seal on a flat bearing surface. Valve travel is limited by the stop. A spring can be added to ensure that the valve will operate in any position.

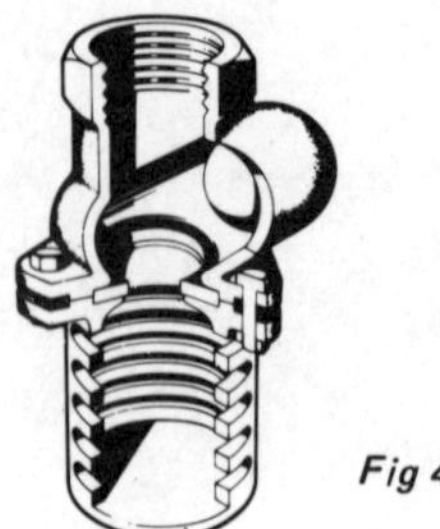

Fig 4

Ball Foot Valve

An example of this type is shown in Fig 4. This is a simple ball valve guided by an inclined cylindrical chamber and seating on an O-ring. Note that the ball is displaced laterally along its chamber with inward flow; but runs down the chamber onto its seat when the flow rate decreases. It is particularly suitable for use with contaminated waters or more viscous fluids.

All examples illustrated are of the type with integral strainer.

See also chapter on *Check Valves.*

Float Controls

THE TRADITIONAL type of float control valve is the simple on—off valve controlled by a float – see chapter on *Ball Float Valves.* The float principle is also widely used for other forms of level control valves, either directly or indirectly.

With indirect operation the float movement may be used to operate a pilot valve, which in turn operates a remote main control valve; or merely to generate air pressure on a piston-cylinder basis proportional to the rise in liquid level. At a predetermined point this pressure becomes sufficient to operate the level control valve.

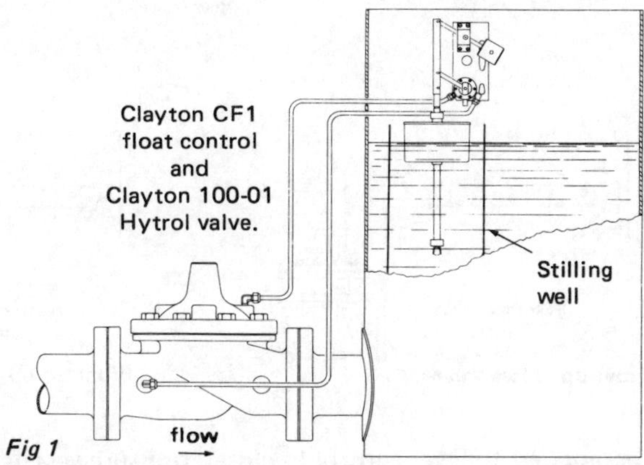

Fig 1

An example of indirect valve operation is shown in Fig 1. The float control is mounted above the high water level in the tank. The valve is installed in the pipe leading to the tank and is connected to the float control pilot by tubing.

When line pressure is used to operate the valve, tubing connections are made from the float control pilot to the valve cover, and also to the inlet side of the valve. The control may be installed at any elevation provided the flowing line pressure in lb/in^2 is greater than the head in feet between the valve and valve control.

An independent source of either air or water can also be used to operate the valve. In this case the pressure from this independent source must be equal to or greater than the pressure at the valve inlet. The independent source is connected to the float control pilot in place of the pipe running from the inlet side of the valve.

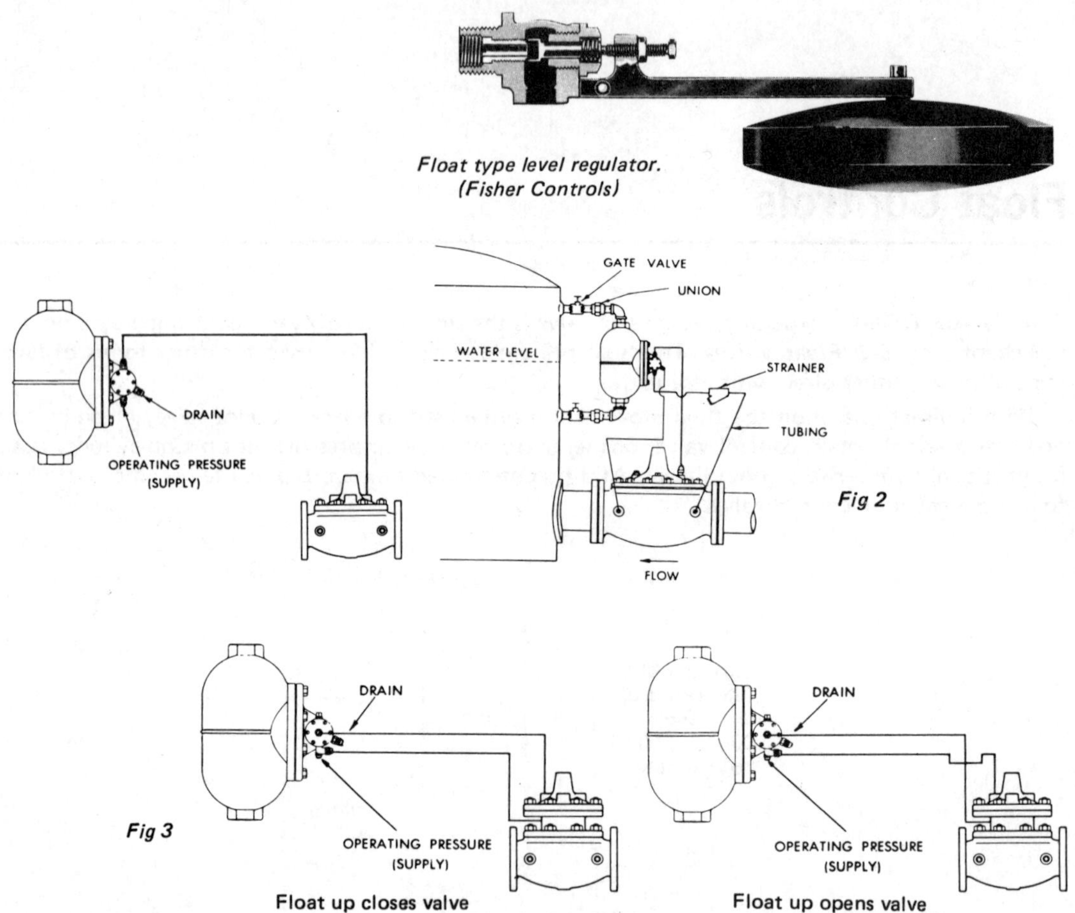

Float type level regulator.
(Fisher Controls)

Float up closes valve

Float up opens valve

A similar principle can be applied to level control in closed pressure vessels – *eg* see Fig 2. This can be worked in two different ways (depending on valve design and connection to valve ports). In either case, when the float rises, valve-operating pressure is increased. This can be used either to close the valve, which is opened again by release of pressure; or to open the valve and close it again on release of pressure. The basic descriptions in this case are: (See also Fig 3).

(i) Float up closes valve; float down opens valve.

(ii) Float up opens valve; float down closes valve.

See also chapter on *Regulators.*

Marine Valves

THE MAIN requirement in marine valves is full material compatibility with the fluid being handled – *eg* gunmetal or nickel-aluminium bronze being a normal choice for sea water systems. Corrosion problems are often aggravated by the fact that many such valves can remain open or closed for long periods – *eg* sea cocks. The type of valve used is largely immaterial provided it performs the required function, but ball valves are generally preferred. The Table lists some applications typical of naval vessels where the highest standards are normally specified.

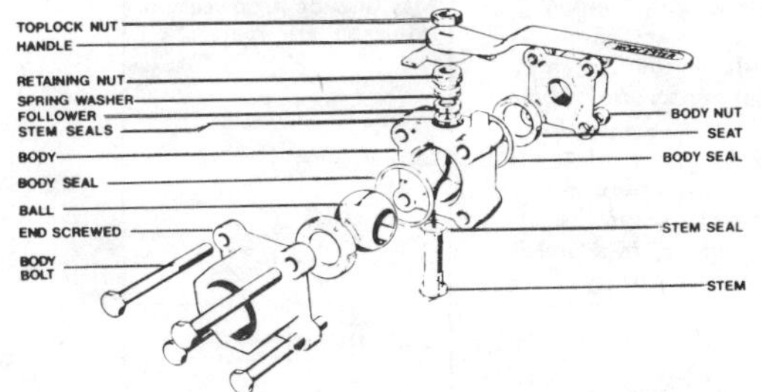

Example of marine ball valve assembly. (The Worcester Valve Co Ltd).

Non-ferrous marine ball valves (Saunders Valve Co Ltd).

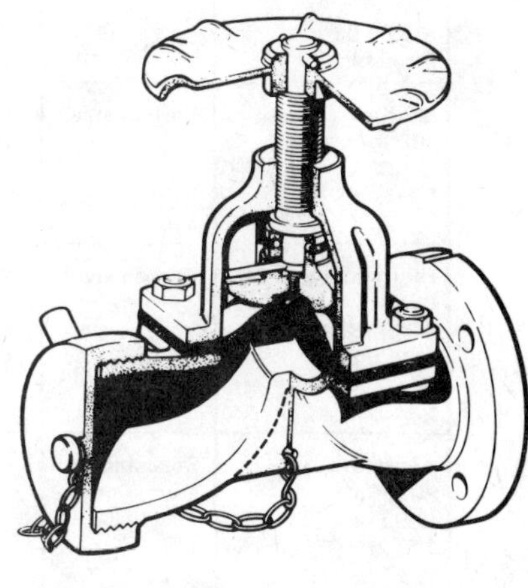

Marine tank – cleaning valve.

Application and Working Pressure	Material	Valve Type	Systems	Remarks
LP fluid services General 230 lb/in^2 (15.86 bar)	Gunmetal	Flanged	Sea water Chilled water Fresh water Air Furnace fuel oil Diesel fuel oil Lubricating oil	May also be produced in non-magnetic materials
LP fluid services General 230 lb/in^2 (15.86 bar) MP fluid services 500 lb/in^2 (34.47 bar)	Ni.Al.Bronze	Screwed Female BSP	Sea water Chilled water Fresh water Air Furnace fuel oil Diesel fuel oil Lubricating oil	
LP fluid services General 230 lb/in^2 (15.86 bar)	Aluminium	Screwed Female BSP	Some cooling systems Air systems Fuel systems Lubricating oil Where weight is at a premium	These valves are non-magnetic
LP fuel systems 230 lb/in^2 (15.86 bar) Firesafe	Carbon steel	Flanged	Furnace fuel oil Diesel fuel oil Lubricating oil	
LP fuel system 230 lb/in^2 (15.86 bar) Firesafe	Stainless steel	Flanged	Helicopter fuelling systems and some gas turbine fuel systems where scrupulous cleanliness is required. Sea water displaced fuel systems where any possible corrosion risk is secondary to a firesafe requirement	May also be produced in non-magnetic materials
HP fuel systems 900 lb/in^2 (62.05 bar) Firesafe	Carbon steel	Flanged	Furnace fuel oil or diesel fuel oil supply to certain types of steam boilers	'Firesafe' valves
MP fuel systems 400 lb/in^2 (27.58 bar) Firesafe	Carbon steel	Flanged ANSI 300	Fuel supply for certain steam atomization type of boilers. Refrigerant gas for refrigerant systems	
HP hydraulic fluids 4000 lb/in^2 (275.8 bar)	Carbon steel	Screwed Female BSP or Unified	Missile handling hydraulics HP air Gun turret general service hydraulics Propeller pitch control	'Firesafe' valves
Actuators 80 lb/in^2 (5.52 bar) 120 lb/in^2 (8.27 bar)	Steel and aluminium			

LP = Low Pressure MP = Medium Pressure HP = High Pressure

Self-Acting Reducing Valves

SELF-ACTING reducing valves generally fall into two main categories (i) *direct acting* valves and (ii) relay- or *pilot-operated* types.

An example of a direct acting type is shown in Fig 1. In this type the main valve opening is determined by deflection of a spring-loaded diaphragm. Reduced pressure which is fed back from the outlet side of the valve is applied to the underside of the diaphragm; the combined effect of this with the loading of the adjusting spring determines the opening of the main valve and hence the pressure drop across the reducing valve.

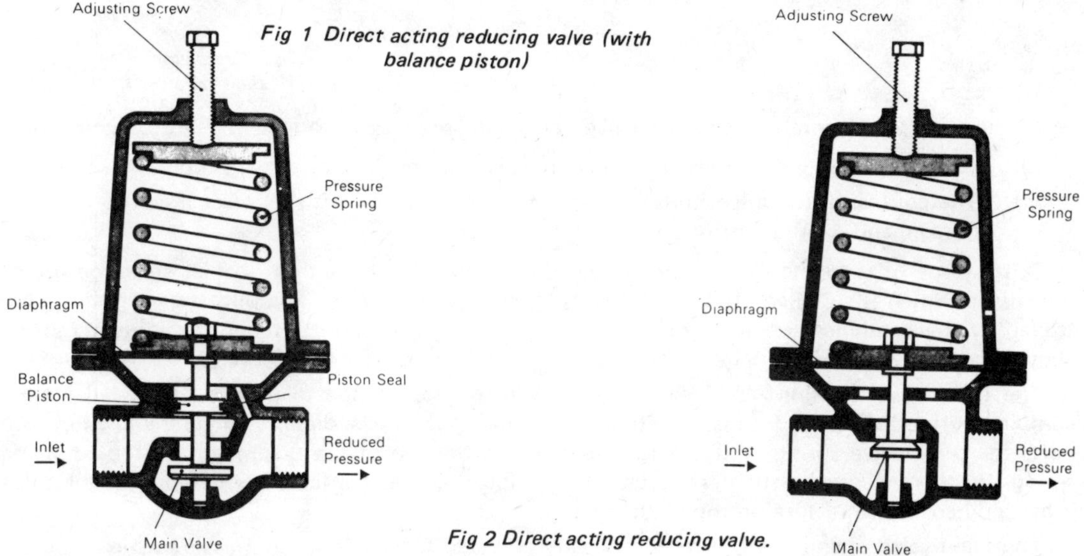

Fig 1 Direct acting reducing valve (with balance piston)

Fig 2 Direct acting reducing valve.

Materials used for the diaphragm include rubber, synthetic rubbers and stainless steel. Alternatively the diaphragm may be replaced by some form of bellows.

A direct acting reducing valve with a balanced piston is shown in Fig 2. This is basically similar to the unbalanced type but the main valve is balanced by a piston so that it remains unaffected by variations in inlet pressure.

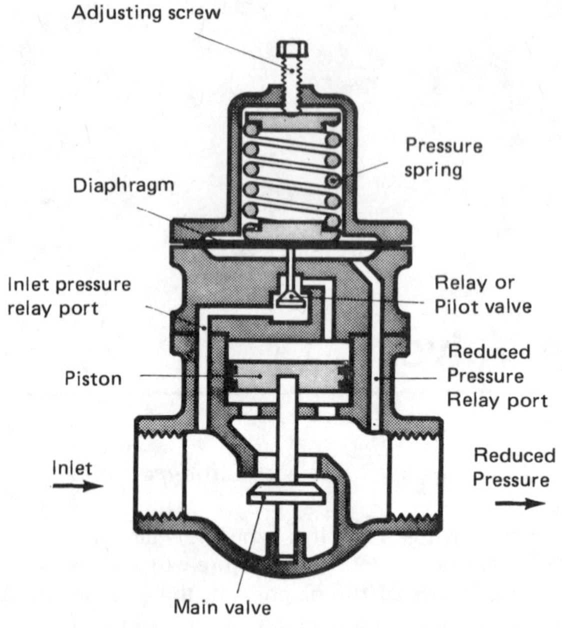

Fig 3 Relay operated reducing valve

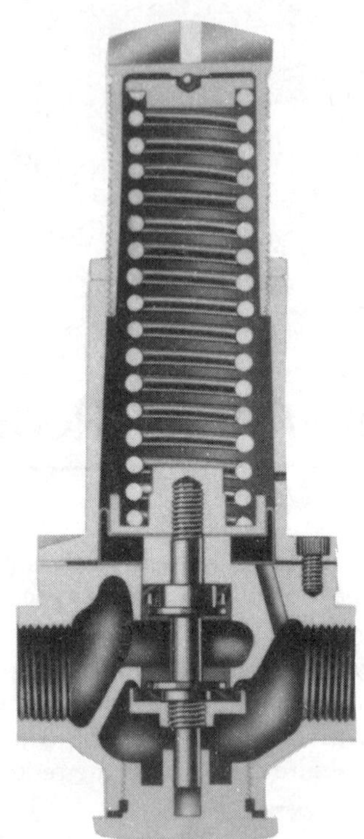

Class T reducing valve (IMI Bailey Valves Ltd)

A relay- or pilot-operated reducing valve for steam services is shown in Fig 3. It comprises:

(i) The valve body which contains the main valve and seat and piston assembly.

(ii) The control head which houses the pilot valve assembly with its associated diaphragm and main adjusting spring.

In this type of reducing valve the main valve is opened by the action of inlet steam on top of the piston which is supplied via the inlet relay port and pilot valve. The pilot valve itself is controlled by the combined action of reduced pressure under the diaphragm and the adjusting spring above it.

The pilot valve opening (and hence the pressure on top of the piston) is controlled by the combined effect of reduced pressure acting on the underside of the diaphragm via the low pressure port, the pilot valve spring, and the load exerted on the top of the diaphragm by the adjusting spring. By this means, the slightest variation in reduced pressure affects the opening of the pilot valve and hence the pressure on top of the piston.

The main valve opening is thus automatically adjusted by the reduced pressure which is accurately maintained despite variations of inlet pressure and/or capacity.

This type of valve is not suitable for use on liquids because there is a danger of a hydraulic lock, caused by liquid becoming trapped between the piston and pilot valve when the latter closes as the outlet pressure tends to rise. This would prevent the main valve from closing and, as a consequence, an excessive rise of pressure would occur.

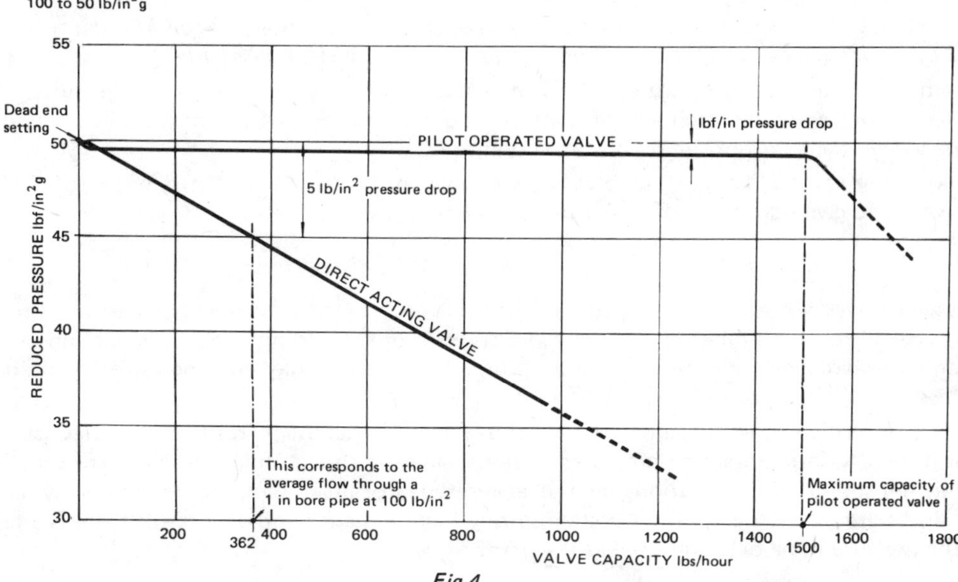

Fig 4

The characteristics of both the relay operated and direct operated reducing valves are shown in Fig 4. Both curves are shown for 1 in (25 mm) valves reducing from 150 lbf/in^2 (10 bar) to 50 lbf/in^2 (3.5 bar).

It should be noted that in the case of the direct acting valve (including those with balance pistons) the outlet pressure falls as the flow through the valve increases. Thus if the valve is set at a no-flow setting of 50 lb/in^2 (3.5 bar), the outlet pressure falls by about 5 lb/in^2 to 45 lb/in^2 (0.35 bar to 3.15 bar) when passing an average flow for this type of valve. The direct acting valve is usually made equal in size to the inlet pipe.

In the case of the pilot operated valve it will be seen that apart from an initial pressure drop of ½ lb/in^2 (0.035 bar) from the dead-end setting of 50 lb/in^2 (3.5 bar) the outlet pressure remains constant until maximum rated capacity is reached.

It should also be noted that in the examples shown, the 1 in (25 mm) pilot operated valve is capable of passing a flow of more than four times that of the direct acting valve, and with only ½ lb/in^2 (0.035 bar) pressure drop as compared with 5 lb/in^2 (0.35 bar) drop.

Applications of Reducing Valves

Reducing valves are used for reducing one pressure to another, control being via throttling of the fluid through the valve and its seat. Reducing valves should never be deliberately oversized as if the valve is too big then the lift of the valve will be small, when wire drawing or erosion of the valve and seat can result. Additionally, small variations in valve opening cause large changes in flow which at small flow demands can lead to pulsating pressure being generated in the downstream flow.

The following notes designate the main fields of application of self-acting reducing valves.

Air or Gases

This includes all compressed air systems for use with power tools, pneumatic control systems, *etc* and control valves for the storage and distribution of industrial gases, *etc.*

Both direct acting and pilot-operated reducing valves may be used on these duties and are selected according to the accuracy of control required and whether or not the valves are intended to give a dead tight shut-off under no-flow conditions.

Depending on the design, the valve may or may not need to be fitted with 'soft' pilot and main valves to give tight shut-off under no-flow conditions.

Water

Reducing valves are extensively used in industrial and domestic water distribution, fire protection systems and the limitation of water pressures in high buildings, *etc.* Direct-acting valves with piston valves are generally used for these duties. As a general rule reducing valves are used mainly as pressure limiting devices in water distribution systems.

Because of high peak demands at times of heavy industrial usage, water authorities usually have great difficulty in maintaining pressures in the systems, although high pressures are usually available at the source of distribution, such as at reservoirs or main pumping stations. Very large pressure drops are experienced in the system during high demands and as a result there is a tendency for the pressure to be below normal at the point of usage.

However, when the total demand in the system drops, much higher pressures are experienced in the distribution system and these are frequently in excess of the normal pressure ratings for the equipment being used. This can give rise to burst mains, or excessively high discharge rates from domestic fittings, such as water closets, wash basins, *etc* or storage tanks.

It is therefore common practice to fit a reducing valve in the line which under high flow conditions normally operates in the wide open position and presents only a nominal resistance to flow (such as would be experienced with an ordinary globe stop valve). However at periods of low demand and high pressure, the reducing valve becomes effective and reduces the pressure in the downstream mains to an acceptable limit. It is important in such applications that the outlet pressure is not affected by inlet pressure variations and for this reason the direct acting valves with piston balance are admirably suited to this application.

Other Liquids

In this field reducing valves are used for such applications as controlling ram pressures on hydraulic presses; bearing lubrication systems in rolling mills and heavy industrial equipment; and for pressure control in fuel oil systems. Again valves are normally used for these applications. In many applications the flow is relatively constant and the outlet pressure from the reducing valve therefore remains constant.

In fuel oil systems the flow variations are normally of the order of 50 to 100%, in which case the outlet pressure variation would probably be of the order of 2 to 3 lb/in^2 (0.14 to 0.21 bar) depending on the size of valve used. The variation would be in the order of 5 lb/in^2 (0.35 bar) between full and no flow conditions.

Steam

This particular category covers by far the majority of reducing valve applications and in general there are two broad sections.

TABLE I – STEAM PIPE CAPACITIES*
lbs/hour dry saturated

Pressure lb/in² *bar*	PIPE SIZE IN INCHES												
	½	¾	1	1¼	1½	2	2½	3	4	5	6	8	10
5 *0.35*	12 *0.4*	32 *0.4*	63 *0.4*	106 *0.4*	163 *0.4*	320 *0.4*	536 *0.4*	813 *0.4*	1560 *0.4*	2550 *0.4*	3820 *0.3*	6180 *0.3*	10000 *0.3*
10 *0.69*	15 *0.5*	40 *0.5*	79 *0.5*	133 *0.5*	206 *0.5*	404 *0.5*	676 *0.5*	1055 *0.5*	1962 *0.5*	3220 *0.5*	4810 *0.5*	8000 *0.4*	12950 *0.4*
20 *1.38*	18 *0.6*	54 *0.7*	107 *0.7*	182 *0.7*	300 *0.8*	586 *0.8*	980 *0.8*	1400 *0.7*	2680 *0.7*	4390 *0.7*	6550 *0.7*	11300 *0.6*	15650 *0.5*
30 *2.07*	24 *0.8*	69 *0.9*	137 *0.9*	245 *0.9*	377 *1.0*	740 *1.0*	1240 *1.0*	1880 *0.9*	2600 *0.9*	5900 *0.9*	8810 *0.9*	14700 *0.8*	22300 *0.7*
50 *3.45*	35 *1.2*	98 *1.3*	203 *1.4*	357 *1.5*	582 *1.6*	1140 *1.6*	1910 *1.6*	2640 *1.4*	4880 *1.3*	8000 *1.3*	12000 *1.3*	21450 *1.2*	33400 *1.1*
75 *5.17*	47 *1.6*	136 *1.8*	284 *2.0*	495 *2.1*	780 *2.2*	1493 *2.1*	2500 *2.1*	3800 *2.1*	7100 *2.0*	11600 *2.0*	16900 *1.9*	30600 *1.8*	48400 *1.7*
90 *6.21*	53 *1.8*	157 *2.1*	332 *2.4*	581 *2.5*	896 *2.5*	1755 *2.5*	2940 *2.5*	4460 *2.5*	8280 *2.4*	13570 *2.4*	19450 *2.2*	35400 *2.1*	54500 *1.9*
100 *6.90*	59 *2.0*	173 *2.3*	362 *2.6*	615 *2.6*	984 *2.8*	1928 *2.8*	3220 *2.8*	4840 *2.7*	9040 *2.6*	14820 *2.6*	21800 *2.5*	38900 *2.3*	61500 *2.2*
120 *8.27*	68 *2.3*	204 *2.8*	431 *3.2*	755 *3.4*	1182 *3.5*	2320 *3.5*	3900 *3.5*	5710 *3.3*	10920 *3.3*	17910 *3.3*	26000 *3.1*	46900 *2.9*	73000 *2.7*
150 *10.34*	83 *2.8*	252 *3.5*	539 *4.1*	945 *4.4*	1420 *4.5*	2900 *4.4*	4770 *4.2*	7100 *4.2*	13600 *4.2*	22300 *4.2*	32500 *4.0*	59100 *3.8*	93000 *3.6*
180 *12.41*	96 *3.3*	296 *4.2*	642 *5.0*	1131 *5.3*	1748 *5.4*	3440 *5.5*	5890 *5.7*	8750 *5.5*	16680 *5.4*	27300 *5.4*	40000 *5.2*	71500 *4.8*	112000 *4.5*
200 *13.79*	107 *3.7*	324 *4.6*	708 *5.6*	1238 *5.9*	1942 *6.1*	3880 *6.3*	6530 *6.4*	9750 *6.4*	18880 *6.3*	31000 *6.3*	44300 *5.8*	80000 *5.5*	126000 *5.2*
220 *15.17*	116 *4.0*	354 *5.0*	770 *6.0*	1350 *6.4*	2120 *6.6*	4200 *6.8*	7150 *7.0*	10850 *7.0*	20800 *7.0*	34000 *7.0*	49000 *6.5*	90000 *6.3*	141000 *5.9*
250 *17.24*	133 *4.5*	408 *5.7*	871 *6.7*	1525 *7.1*	2385 *7.3*	4760 *7.6*	8100 *7.8*	12360 *8.0*	23800 *8.0*	39000 *8.0*	57000 *7.6*	105000 *7.8*	168000 *7.0*
300 *20.69*	157 *5.3*	496 *7.2*	1025 *7.8*	1798 *8.3*	2800 *8.5*	5610 *8.9*	9550 *9.2*	14700 *9.6*	28400 *9.6*	46500 *9.6*	68900 *9.4*	125900 *9.1*	202000 *8.9*

*IMI Bailey Valves Limited

Note: Figures in italic show pressure drops (lb/in²) for equivalent lengths equal to 360 pipe diameters. When using this table, allowance should be made for the effects of bends and fittings in the pipe line.

(a) *Power*

Reducing valves are only occasionally used on power installations involving steam, *ie* direct steam supply, steam engines and turbines, *etc.* In these cases the general principles of application still apply, although special problems do sometimes arise in the case of reciprocating machinery which may give rise to pulsations in the pipework system, and these can be amplified by the reducing valve itself. This is normally overcome by providing adequate pipe volume both upstream and downstream of the reducing valve to act as a 'steam accumulator'.

(b) *Process*

With saturated steam, temperatures and pressures are strictly related, and because of this it is frequently found convenient to control temperature by controlling the steam pressure. Application in the process field includes space heating, kitchen equipment, sterilizing equipment, curing process in the rubber and plastics industries, *etc,* industrial cooking equipment *etc.* In fact anywhere steam is used as a heat transfer medium reducing valves will invariably be installed.

In general only low pressure steam, usually below 50 lb/in^2 (3.5 bar), is used for process purposes. At such low pressures the latent heat content of the steam is relatively high and is easily transferred from the steam to the product being processed.

Sizes of Pipes and Fittings

The inlet and outlet pipes should be sized to suit the maximum steam demands of the system (*eg* see Table I). Pipe sizes should always be determined in terms of pressure drop and not by such rules as arbitrary steam velocities.

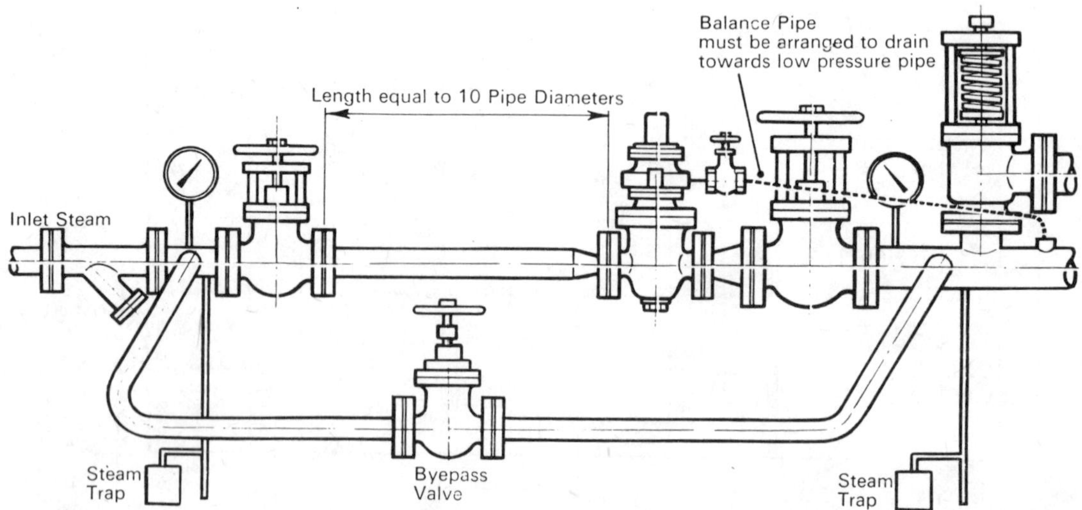

Fig 5 Typical reducing valve installation using globe stop valves.

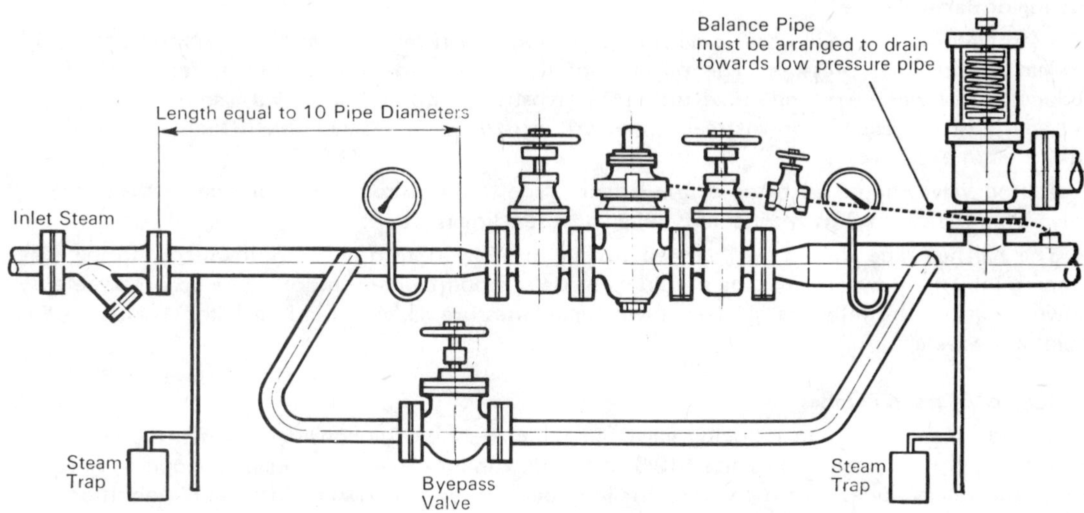

Fig 6 Typical reducing valve installation using parallel slide valves.

Correct sizing of pipework and fittings associated with all valves is extremely important in order to obtain the best possible operation. Specifically:

(i) Strainers should always be equal in size to the inlet pipe.

(ii) When globe valves are used as inlet and outlet stop valves these should also be equal in size to the respective pipe into which they are fitted.

(iii) When parallel slide valves are used as stop valves these can be fitted equal in size to the reducing valve for reduced pressures between 30% and 70% of the inlet pressure. They should be equal in size to the respective pipe when the pressure difference between the inlet and outlet is 30 lb/in^2 (2 bar) or less, or in other conditions. When they are connected directly to the reducing valve the length of distance pieces between should not exceed three pipe diameters.

(iv) In order to provide a streamlined flow at the approach to the reducing valve, a straight length of pipe equal to 10 pipe diameters should be provided between the fitting and the reducing valve (this does not apply to parallel slide valves). Typical reducing valve layouts are shown in Figs 5 and 6.

Steam Traps

Whenever possible pilot-operated reducing valves should be sited at some point in the pipeline where they cannot become flooded with condensate during periods of low flow or prolonged shutdown. If this is not possible then steam traps (and if possible dirt pockets) should be fitted to both the inlet and outlet pipework to remove any condensate which may accumulate in the vicinity of the reducing valve.

Condensate can be trapped between the piston and pilot valve when the steam flow is resumed and this prevents the main valve from closing as the reduced pressure may continue to rise above the setting and eventually cause the relief valve to blow.

Fitting of Balance Pipes

It is strongly recommended that a balance pipe should be fitted when the reduced pressure is 10% or less of the inlet pressure. The purpose of this pipe is to improve the performance of the reducing valve when working under difficult downstream conditions. It will also help to counteract any pressure drops in downstream pipework caused by undersized pipe fittings, *etc,* providing they are not excessive.

A stop valve should be fitted in the balance pipe to allow complete isolation of the reducing valve from the steam flow (particularly when a bypass line is fitted).

The balance pipe should be arranged to fall to allow it to drain into downstream pipe. The tapping into the downstream pipe should be made at a point where smooth flow occurs preferably downstream of the relief valve. The downstream pressure gauge should be fitted as near to this point as possible.

Reducing Valves in Parallel

As already mentioned two reducing valves in parallel should be considered when the minimum flow through the system is less than 10% of the maximum capacity of a single reducing valve, or when the valves are expected to work for long periods on 'no-flow' or 'dead-end' conditions, or working in partially completed plant systems. A typical layout is shown in Fig 7.

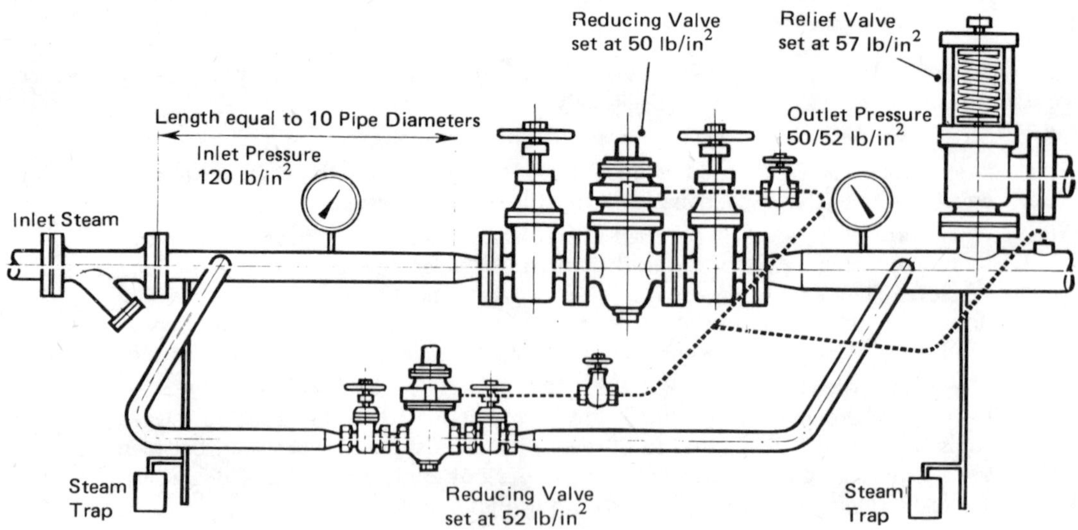

Fig 7 Parallel arrangement of reducing valves.

In order that the valves can deal effectively with minimum capacity variations of less than 10% two unequally sized reducing valves, having a maximum capacity equal to the required capacity, should be connected in parallel with the outlet pressure of the smaller valve set 2 to 3 lb/in^2 (0.14 to 0.21 bar) higher than that of the large valve. In this way the larger valve would shut at low demands leaving the smaller valve to handle the low flows. As the demand increases the larger valve will open automatically as the reduced pressure falls, and share the load with the small valve. By this method capacity ratios of up to 100:1 can be obtained.

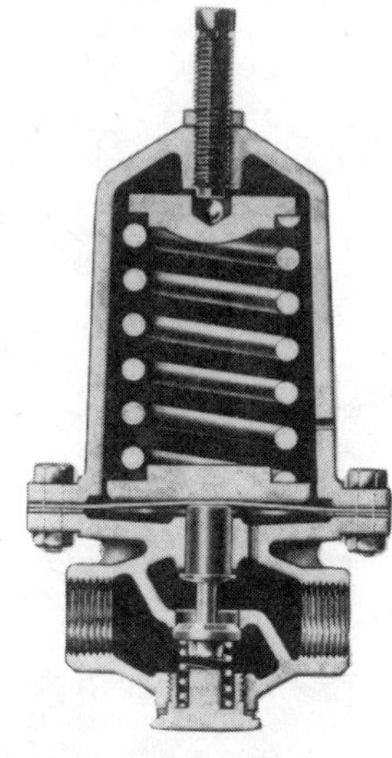

Class H2 reducing valve.
(IMI Bailey Valves Ltd)

Superheated Steam

Superheated steam is less dense than saturated steam and, therefore, for the same pressure drop the reducing valve will have slightly smaller capacity. The reduction in capacity is dependent on the amount of superheat. Capacity figures quoted in manufacturers' catalogues are normally for dry saturated steam. When steam is superheated above 50°F (28°C) before it enters the reducing valve, dry steam capacities should be multiplied by the following factors.

100 to 150°F (56 to 83°C) of superheat – 0.89
150 to 200°F (83 to 111°C) of superheat – 0.86
200 to 300°F (111 to 167°C) of superheat – 0.82

Fire Resistant Valves

THE API (American Petroleum Institute) fire tests were designed specifically for valves in oil and gas production plants. API 607 was designed to cover lower pressure refinery valves. API RP6F is the equivalent for higher pressure valves. Other test specifications include OCMA FSV.1, EXXON BP3-14-1 and FM 6033 (see Table I).

The basis of all fire tests is that a pressurized valve must operate after being burned at a specified high temperature for a specified period, and leakage after burning (which will destroy soft seals) must remain within specified limits.

The OCMA FSV.1 test has been in use for some years but is now widely superseded by the more realistic API tests – *eg* RP6F for pipeline valves and API 607 for refinery valves. Unlike the OCMA test, API RP6F limits the size, range and pressure that one test can cover. Approval applies to valves up to twice the basic size of the valve tested and pressure ratings not less than 50% nor more than 200% of the test valve.

The broad object of a fire test is to establish acceptable levels of leakage after exposure to a fire for either a 15 minute, or a 30 minute, time period. This allows the user to select a valve best suited for a particular service.

A maximum 30 minute test duration has been established on the basis that this represents the maximum time during which a fire should normally be extinguished. Fires of greater duration are considered to be of major magnitude with consequences greater than those anticipated in this test.

Specific API test requirements are

(a) The valve shall be tested in the closed position with water, with the stem and bore in the horizontal position. Check valves will be tested in their normal operating position.

(b) The valve will be uniformly enveloped in flame having a temperature of 1400–1600°F (761 to 871°C) average of two thermocouples, one located one inch (25 mm) below the valve and the other one inch (25 mm) from the upper stem packing box on the horizontal centreline. No reading shall be below 1300°F (704°C). Piping upstream of the test valve larger than one inch (25 mm) nominal pipe size or one-half of valve nominal pipe size (whichever is smaller) must be enveloped in flame for a distance of at least 6 inches (152 mm).

TABLE I – SUMMARY OF TEST SPECIFICATIONS

Test Specification	OCMA FSV.1	EXXON BP3-14-1	API 607 First Edition	FM 6033	API RP6F
Stem position	Vertical	Vertical	Horizontal	Not specified	Horizontal
Bore position	Horizontal	Horizontal	Horizontal	Not specified	Horizontal
Valve open or shut	Open	Open	Shut	Shut	Shut
Test pressure during burn	30 lb/in^2	25 lb/in^2	1000°F class pressure rating (7% of full W.P.)	125 lb/in^2	75% of full W.P.
Test medium	Kerosene or diesel fuel	Liquid hydrocarbon	Water	Not specified	Water
Valve body temperature	Sufficient to destroy soft seat	1200°F minimum	1100°F	Not specified	1400°–1600°F
Burn duration	15 minutes	15 minutes	30 minutes	15 minutes	15/30 minutes
When seat leakage measured	After test	After test	During test	During test	During test
Maximum external W leakage	No appreciable leakage	Leakage shall be negligible	20 ml/min/ inch dia.	0.1 qt/min (94.6 cm^3/min)	100 cu^3/min/ inch dia. (valve closed) 400 cm^3/min/ inch dia. (valve open)
Maximum seat leakage	10 ml/min/ inch dia.*	10 ml/min/ inch dia.*	40 ml/min/ inch dia.	Individual drops	400 cm^3/min/ inch dia.
Operability	3 cycles open to shut	3 cycles open to shut	1 cycle open to shut	Must be operable	1 turn shut to open

*In no case shall leakage rate exceed 100 ml/min

(c) The end connection piping-to-valve joint leakage (flanged, threaded, or welded) is *not* considered a part of this test and is not included in the allowable external leakage. For the test, it may be necessary to modify this joint to eliminate leakage.

Suggested systems for fire testing to API specifications are shown in Figs 1 and 2. Fig 2 is a schematic outline for systems using compressed gas as the pressure source. Test procedure is as follows:

(i) Open valve(s) (Items 5 and 6) at water source, and any necessary vent valves (Item 17) to flood the system and purge the air. The test valve may have to be placed in the partially open position in order to completely flood the valve body.

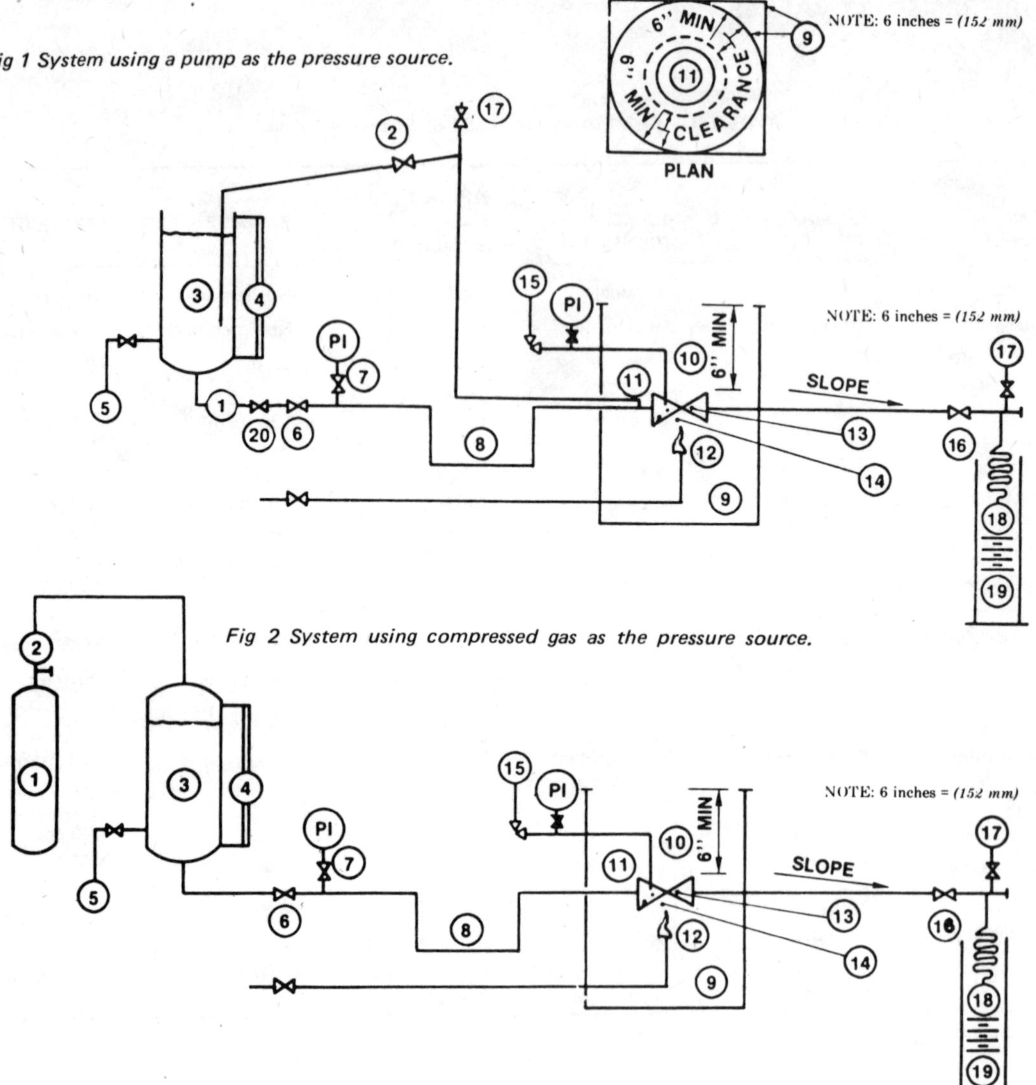

Fig 1 System using a pump as the pressure source.

Fig 2 System using compressed gas as the pressure source.

1. Pressure source.
2. Pressure regulator and relief.
3. Vessel for Water.
4. Calibrated sight gauge.
5. Water supply.
6. Shutoff valve.
7. Pressure gauge.
8. Piping arranged to provide vapour trap.
9. Enclosure for test – horizontal clearance between any part of the valve and the enclosure shall be a minimum of 6 inches (152 mm).
10. Minimum height of enclosure shall be 6 inches (152 mm) above the top of the valve.
11. Test valve mounted horizontally with stem in horizontal position.
12. Fuel gas supply with minimum of three (3) burners located at 120°.
13. Flame temperature thermocouple – located 1 inch (25 mm) from upper stem packing box on horizontal centreline.
14. Flame temperature thermocouple – to be located 1 inch (25 mm) below the centre of the valve body.
15. Pressure gauge and relief valve (if required) connected to centre cavity of valve.
16. Shutoff valve.
17. Vent valve.
18. Condenser.
19. Calibrated container.
20. Check valve.

(ii) Close fill valve (Item 5) and vent valves (Item 17), and close the test valve (Item 11). The system upstream of the test valve should be completely water filled and the system downstream shall be drained.

(iii) Pressurize the system to the appropriate pressure from Table I. Maintain this pressure during all testing. Record the reading on the calibrated sight gauge (Item 4). Empty the graduated downstream container (Item 19).

(iv) Open fuel supply, establish a fire, monitor the flame temperature, and when the average of the two thermocouples (Items 13 and 14) reaches 1400°F (761°C) start the test. Maintain the average temperature between 1400–1600°F (761–871°C) for the test duration. No reading shall be less than 1300°F (704°C).

(v) Record instrument readings (Items 7, 13, 14 and 15) every two minutes for the test duration.

(vi) At the end of the test duration (15 minutes or 30 minutes), shut off the fuel.

(vii) Immediately determine the amount of water collected in calibrated container (Item 19) to establish total through valve seat leakage. Continue recording the amount of water collected for use in establishing the external leakage rate. If the test valve is of the upstream sealing type, the volume of water that is trapped between the upstream seat seal and the downstream seat seal, when the valve is closed, shall be determined before the test is started and identified in the test report. It is assumed that during the test this volume of water would move

TABLE II – TEST PRESSURE DURING FIRE TEST

	Valve Rating (PN)*	Test Pressure lb/in^2	(bar)
Spec 6D Valves	150 (20)	210 ± 10%	(44.5 ± 10%)
	300 (50)	540 ± 10%	(37.2 ± 10%)
	400 (64)	720 ± 10%	(49.6 ± 10%)
	600 (100)	1080 ± 10%	(74.5 ± 10%)
	900 (150)	1620 ± 10%	(111.7 ± 10%)
	1500 (250)	2700 ± 10%	(186.2 ± 10%)
	2500 (420)	4500 ± 10%	(310.3 ± 10%)
	(bar)		
Spec 6A Valves	2000 (138)	1500 ± 10%	(103.4 ± 10%)
	3000 (207)	2250 ± 10%	(155.1 ± 10%)
	5000 (345)	3750 ± 10%	(258.6 ± 10%)
	10000 (690)	7500 ± 10%	(517.1 ± 10%)
	15000 (1034)	11250 ± 10%	(775.7 ± 10%)
	20000 (1379)	15000 ± 10%	(1034.2 ± 10%)

*(PN) is the pressure class designation utilized in ISO (International Standards Organization) documents

through the valve, past the downstream seat seal and be collected in the calibrated container. Since this volume has not actally leaked past the upstream seat seal, it may be deducted from the total volume measured in the downstream calibrated container when determining the through valve leakage.

(viii) Allow the test valve to cool to 200°F (93°C) or less. Use temperature sensitive crayons or other suitable means to indicate valve body temperature near the thermocouples (Items 13 and 14). Record the level in the sight gauge (Item 4). Use the initial and final readings to determine total leakage during the test.

(ix) Close the shutoff valve (Item 16) and operate the test valve against test pressure differential (Table II) to the full open position.

(x) Measure and record external leakage for a minimum of five minutes after valve is in the full open position at test pressure. Divide the total external leakage by the duration of the test in minutes to obtain external leakage rate. The test system, excluding the test valve, may be adjusted during the test period to keep the test within the limits specified herein.

Fire Hydrant Valves

FIRE HYDRANT valves are commonly screw-down gate or globe valves purpose-designed to meet the requirements of International Standards – *eg* BS750 which calls for a delivery of 34 lit/sec (450 gallons/min) at a constant running pressure of 1.7 bar (25 lb/in^2). Ease of operation is also a most important feature of all valves and fittings used in firefighting.

An example of a modern design with a 'captive' internal valve is shown in Fig 1. Particular features are:-

(i) *Spindle and Loose Nut Arrangement* – in preference to an internal valve assembly having the screwed portion integral with the internal valve.

(ii) *Wearing Parts* – reduced to a minimum and can be replaced individually without renewing the complete internal valve assembly.

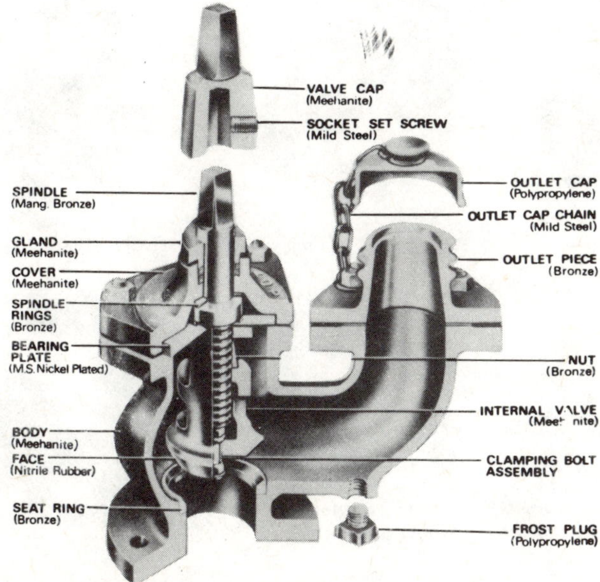

Fig 1 Sectioned view of typical Glenfield screwdown fire hydrant to BS750 with 'captive' internal valve.

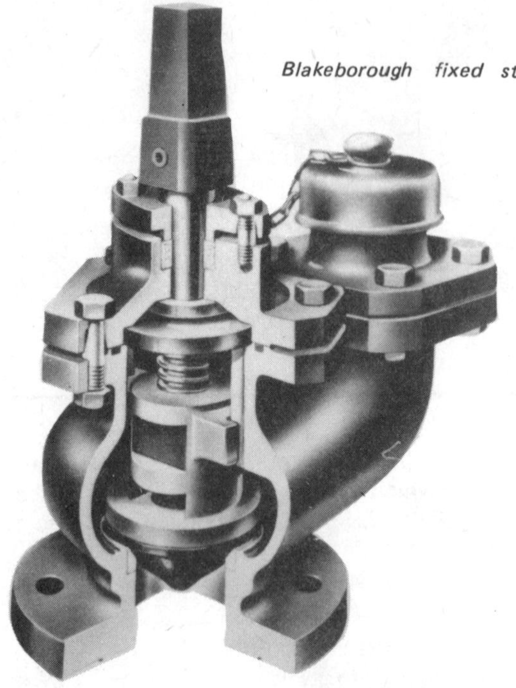

Blakeborough fixed stopper hydrant.

Gate valve hydrant.
(Neptune Glenfield Ltd).

Horizontal Pattern

Bib Nozzle Pattern

Right Angle Pattern

Double Outlet Valve

Globe valves for firefighting equipment.
(John Morris & Sons Ltd).

Oblique Pattern

(iii) *Spindle Thread* — the stronger stub Acme form in place of square thread.

(iv) *Face* — synthetic rubber instead of leather, thus obviating any possibility of bacterial growth.

(v) *Guides in Body* — simplified arrangement

Materials

Body, Cover, Internal Valve, Gland and Spindle Cap — high duty Meehanite cast iron.

Spindle — manganese bronze to BS2872 GZ114.

Bearing Plate — nickel plated mild steel to BS950 Grade EN3.

Spindle Rings and Seat Ring and Valve Nut bronze — to BS218.

Outlet Piece — to BS1400.

Valve Face — nitrile rubber, retained by brass face washer, clamping bolt and self-locking nut.

Frost Plug and Outlet Cap — polypropylene.

Outlet Cap Chain — black enamelled mild steel.

Packing — greasy hemp.

Discharge

Captive Internal Valve — 37.6 lit/sec (496 gallons/min) with running pressure of 1.7 bar (25 lb/in^2) at hydrant inlet valve fully open (32 mm (1¼ in) travel).

$$Q = 28.6\sqrt{h} \text{ or } 99.2\sqrt{h} \text{ for valve fully open}$$

where

Q = lit/sec or gallons/min
h = bar or lb/in^2 running pressure

Loose Internal Valve — 36.4 lit/sec (481 gallons/min) with running pressure of 1.7 bar (25 lb/in^2) at hydrant inlet and valve fully open (32 mm (1¼ in) travel).

$$Q = 27.7\sqrt{h} \text{ or } 96.2\sqrt{h} \text{ for valve fully open}$$

Temperature Control Valves

TEMPERATURE CONTROL valves fall into broad categories:

(i) *Mixing valves* for adjusting the flow rates of two different fluids (*eg* hot and cold water) joining in a common stream. The degree of 'mixing' thus determines the temperature of the common stream.

(ii) *Thermostatic valves* which open or close at a predetermined fluid temperature as sensed by an integral thermostat unit.

(iii) *Thermostatically controlled valves* where the thermostat is remote from the valve and merely generates an electric signal controlling the valve actuator on an on–off basis.

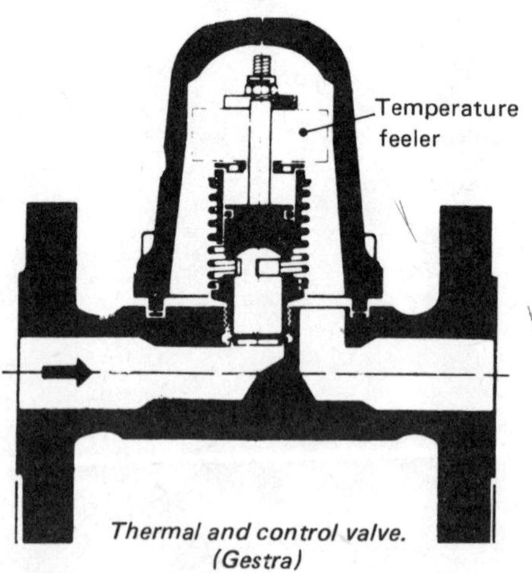

Thermal and control valve. (Gestra)

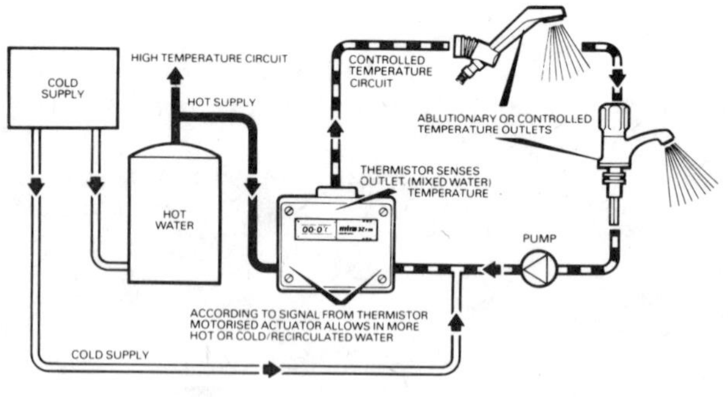

Fig 1
Schematic layout of a recirculating blended water system controlled by a Mira 32 rm.

Mixing valves may range from simple manually operated taps connected to a common outlet, to semi-automatic and fully automatic systems. An example of a modern system applicable to domestic and other blended water recirculating systems is shown in Fig 1. The temperature controller is an all-electronic unit, incorporating a thermistor positioned in the outlet and monitoring

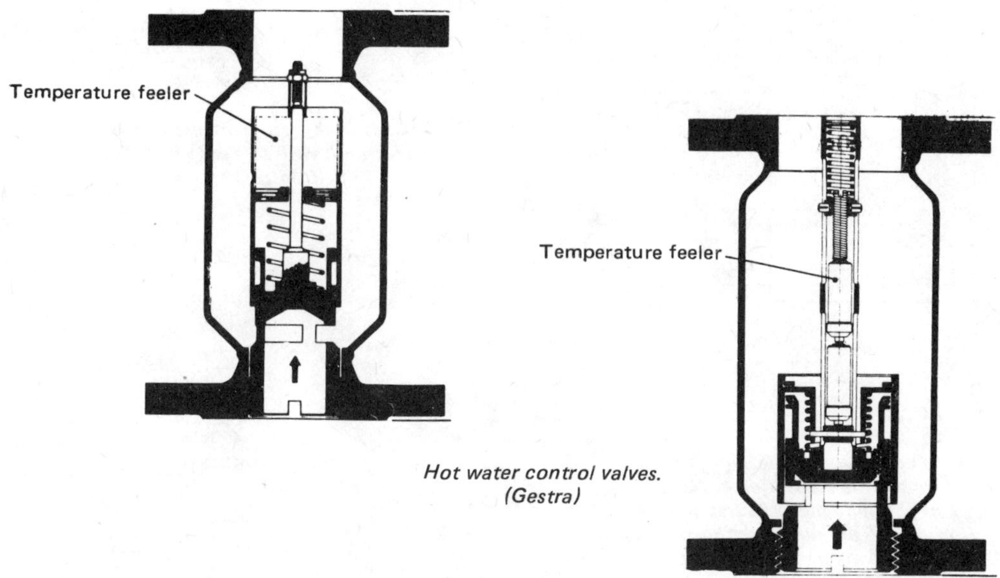

Hot water control valves.
(Gestra)

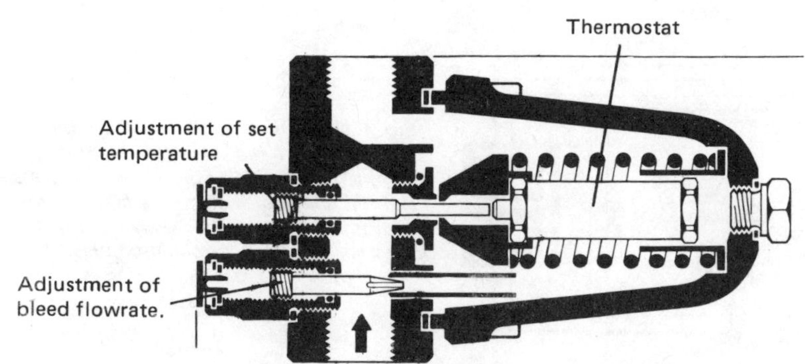

Example of cooling water control valve.
(Gestra)

the recirculating temperature. This is continuously compared with the set temperature. Any 'error signal' present starts the motorized actuator in the unit to adjust the amounts of hot and cold water until the outlet temperature corresponds to the set temperature, when the error signal becomes zero.

Thermostatic valves are designed either to open at an increase in temperature; or to close with an increase in temperature. The former are used in cooling circuits and the latter in heating circuits.

Some examples of their applications are given in Figs 2, 3, 4 and 5.

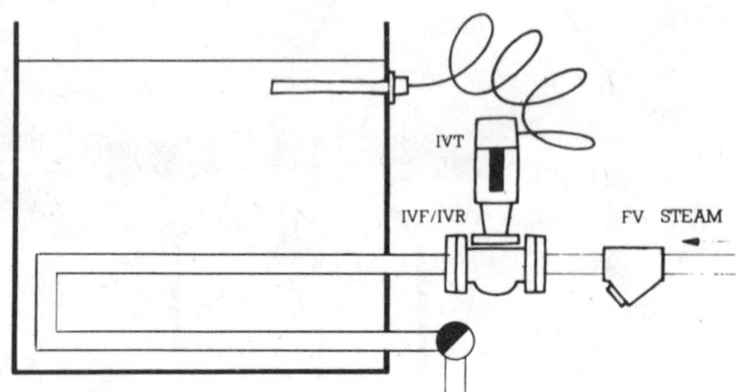

Fig 2 Controlling the temperature in a galvanic bath with IVT + IVF/IVR. (Danfoss)

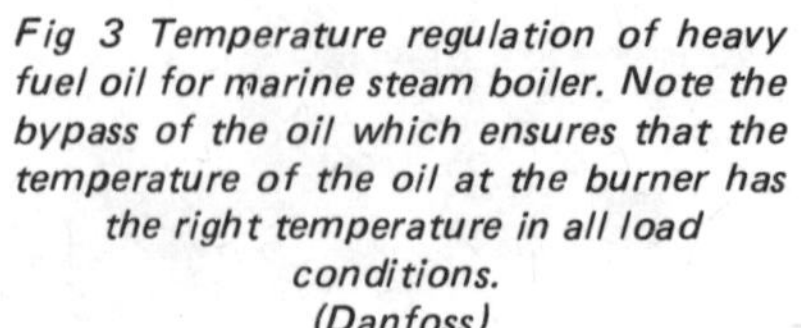

Fig 3 Temperature regulation of heavy fuel oil for marine steam boiler. Note the bypass of the oil which ensures that the temperature of the oil at the burner has the right temperature in all load conditions. (Danfoss)

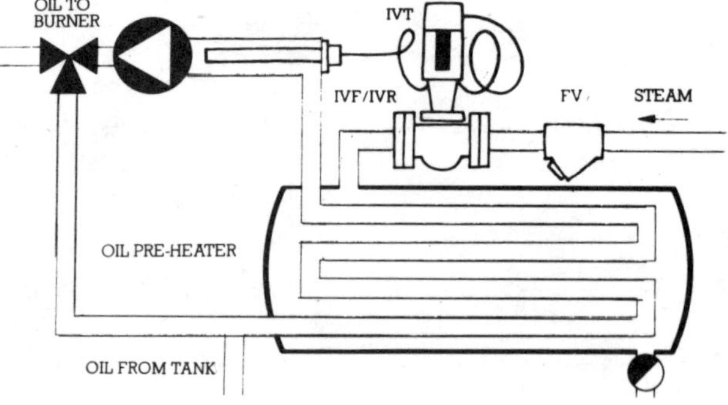

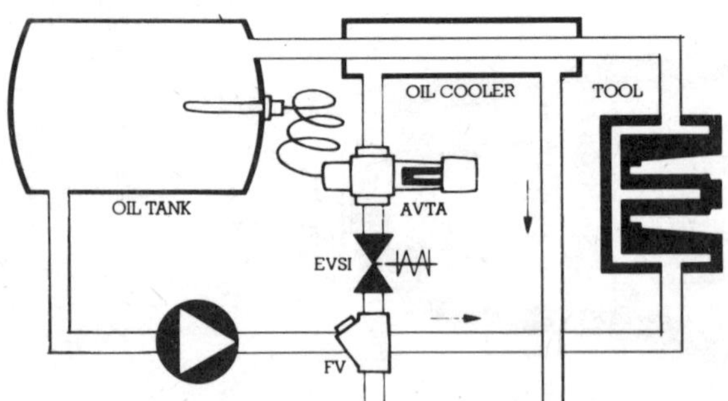

Fig 4 Cooling of tools. AVTA controls the cooling of the oil in the plant where compact flow oil coolers are used. As soon as the tool/machine is taken out of operation the solenoid valve stops the cooling water. Application plastic moulding machines, presses. (Danfoss).

Fig 5 Cooling of compressors. Thermostatic cooling water valve on compressor. The valve must in this case be equipped with a bypass since the sensor will otherwise never come into contact with the heated cooling water. (Danfoss).

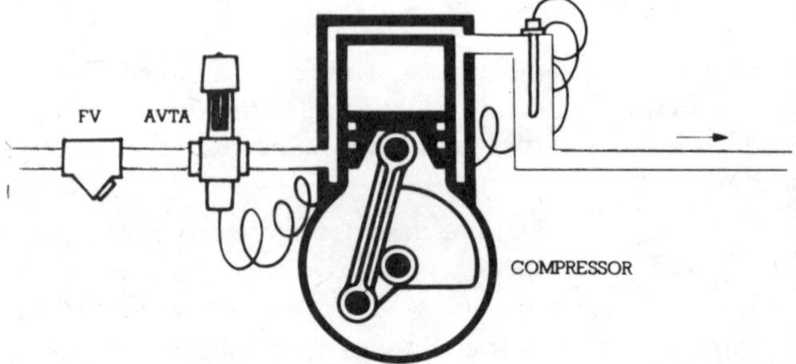

SECTION 5

Ancillaries

Valve Actuators

DEVICES FOR the actuation of valves range from simple manual gearboxes to highly sophisticated motorization of valves with automatic control and programming by computer. There is also a distinction between actual requirements. The vast majority of applications are concerned with straightforward opening/closing of valves. Other systems may call for continuous modulating control which can set limits on the usefulness of both mechanical and energy systems.

It is probably most realistic to deal with the overall subject under three main headings:

(i) Main types of valve actuators available

(ii) Choice of energy systems

(iii) Electric/electronic controls

Main Types of Valve Actuators

The main types of actuators used are

(a) manual operators

(b) piston (pneumatic or hydraulic)

(c) electric solenoid

(d) diaphragm

(e) motor actuators (electric, pneumatic or hydraulic)

All have their particular virtues and uses.

MANUAL OPERATORS

Apart from the obvious job of providing a means whereby a man can open or close a valve, the requirements of a manual valve actuator may encompass any or all of the following:-

(i) *Convert motion from linear to rotary*

Valves of the gate, globe, diaphragm and pinch type utilize linear motion of the obturating members to achieve a seal. Handwheel operation implies rotary motion, and conversion is generally by nut and screw, which may be part of the valve or of a separate manual actuator.

(ii) *Withstand thrust*

When the threaded nut forms part of the manual actuator, it will have to withstand the operating thrust developed and incorporate suitable thrust bearings.

(iii) *Lock the valve in position between operations*

For valves operated by linear screw thread, position locking is achieved by the use of an irreversible, (*ie* low-efficiency) thread. Butterfly valves must also be restrained against self-operation as a result of dynamic flow, and again this is customarily ensured by using irreversible, low-efficiency, gearing such as worm and wheel between valve stem and handwheel.

A Mastergear, quarter-turn manual actuator — with cover removed to show worm and wheel construction. A spur box on the input side reduces the manual input effort required to operate the valve.

Section view of a typical manual, bevel-gear operator for multi-turn valves. (Torkmatic 'B' type)

In both cases the efficiency must be low enough so that not only does the valve not move from an intermediate position if pressure is applied to the stationary valve, but that it will also stay in position if manually operated with flow present. There have been many occasions when valve slamming has occurred under high-flow conditions with a gear ratio which was thought to be irreversible, but proved not to be when the handwheel was started under heavy flow conditions, and the valve took over.

This dynamic effect is less of a problem with plug and ball valves which have a higher ratio of static friction to dynamic torque, and in any case, are not suited in standard form to flow regulation.

Whatever manual means is adopted for the operation of valves, the main constraint is always the power available from man himself. It is difficult for a man to exert more than about 1/10 horse-power continuously for any length of time by hand.

otork
ork Controls Limited
ssmill Lane
h, Avon. BA1 3JQ
phone (0225) 28451
ctuator manufacturers for
e International valvemaking
dustries

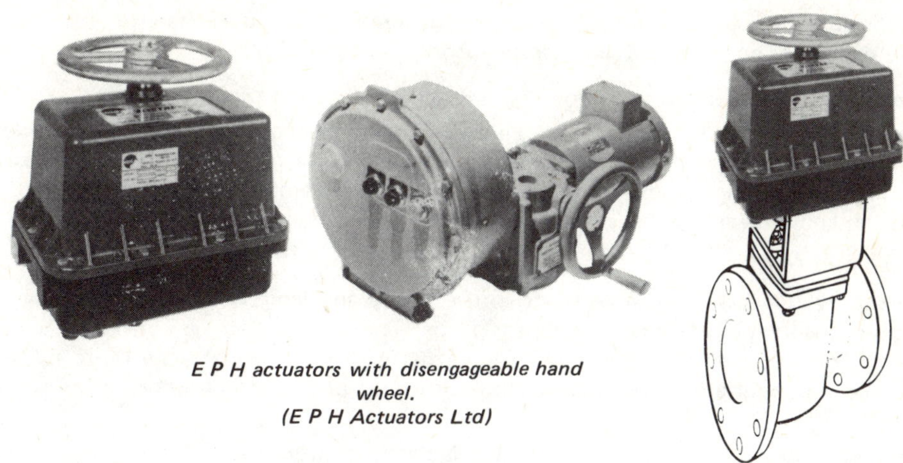

E P H actuators with disengageable hand wheel.
(E P H Actuators Ltd)

Size and type of valve, line pressure and other factors will determine the power required. For larger valves this will nearly always mean the introduction of an intermediate gearbox with a handwheel capable of operating the valve comfortably with a man's strength — *ie* with a rim pull below 60 lbs (27 kg).

Man can exert up to three times this force momentarily, by pulling his weight on a handwheel, so that the extra force required to seat or unseat a valve is not a problem if the gearing is adequate. However, the fundamental principle of gearing to reduce input torque inevitably means that the number of turns required of the handwheel increases. For large valves and pressures the number of turns is such that manual operating time may be measured in hours rather than minutes.

There is no way, therefore, that emergency manual operation can be quick. The alternative philosophy is to accept that large valves are difficult to operate, and require at least two men, or a team in an emergency, operating a larger handwheel with far less turns. There is no ready solution to this dilemma, other than power operation.

Another reason for utilizing reduction gearing is to reduce the torque to be transmitted via shafts and couplings to a remote operating point, *eg* spur gearing of a gate valve to a pedestal, say, on a floor above.

A side handwheel is generally more convenient than one with a vertical shaft. Bevel gear actuators are therefore widely used to reduce the operating torque of gate valves and bring their handwheel to a particular side for access.

The use of worm gearboxes and equivalent nut and screw scotchyoke actuators for quarter-turn valves automatically turns the drive through a right-angle, which may therefore be oriented in the right direction. Additional gearing for torque reduction is then usually provided by spur gears, although a further change of direction via a bevel gear may sometimes be needed.

Signalling

A problem with manually operated valves — especially in plants with centralized control — is the absence, usually, of any standard means of signalling the valve status, or of confirming that a required operation has been carried out. If this is necessary, special provision has to be made.

Often this is done by attaching some form of external limit-switching mechanism, but this may not be reliable enough or environment- and tamper-proof.

Handwheel Drives for Powered Actuators

All the aforementioned points apply to auxiliary manual drives for power actuators. However, the fact that the drive is auxiliary only and not the main means of operation may make design compromises in interests of economy more acceptable than they would be if manual operation were the regular and only means.

PISTON ACTUATORS

A piston and cylinder, which can be made in practically any length or diameter readily converts hydraulic or pneumatic pressure into linear force. This can further be converted to part-turn operation by rack and pinion or linkage. The virtue for the 90° valve is the finite stroke of the piston, which has resulted in the main volume of use of piston actuators being for quarter-turn valves.

Another advantage of the piston and cylinder is the capability of maintaining pressure to hold a valve position, whereas electric motor actuators must be de-energized at end of stroke, otherwise they burn out, *and* be mechanically self-locking against line pressure or other forces acting on the valve. The pneumatic piston actuator may also be useful for certain modulating valve control requirements.

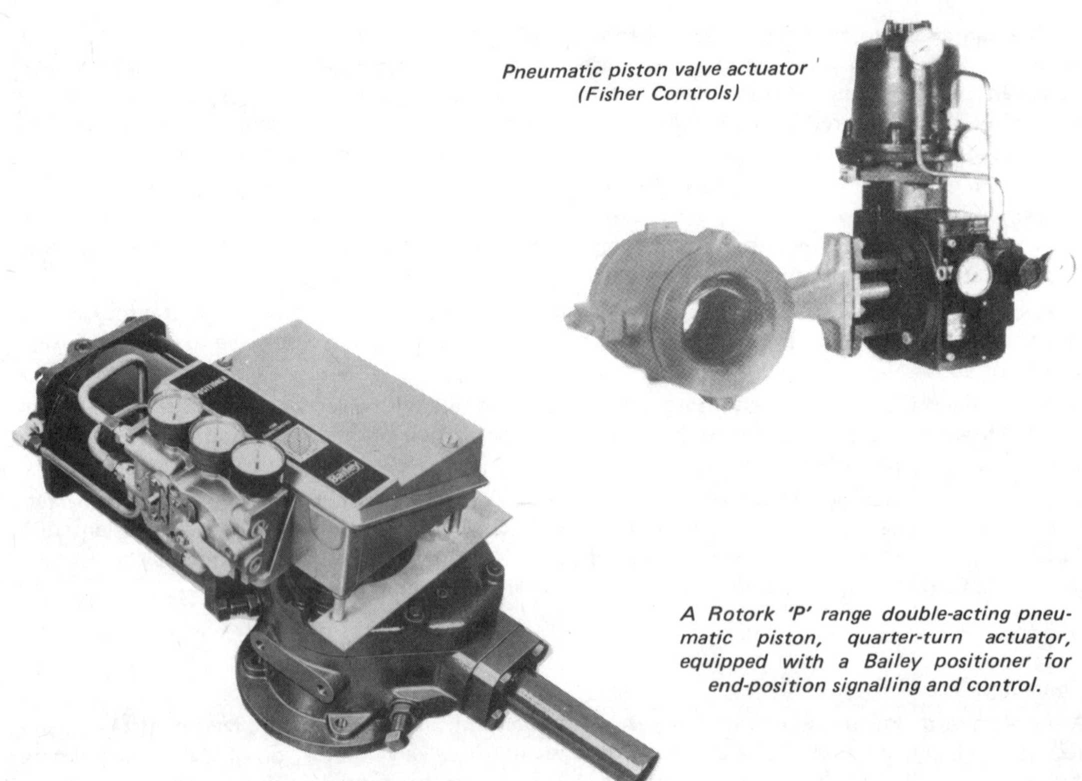

Pneumatic piston valve actuator (Fisher Controls)

A Rotork 'P' range double-acting pneumatic piston, quarter-turn actuator, equipped with a Bailey positioner for end-position signalling and control.

ELECTRIC SOLENOID

The electrical solenoid is limited to very small powers indeed, *ie* to pilot duty rather than actuation duties for other than the smallest valves. However, a development, known as Quadrak, is now available which fills the gap that exists between the capabilities of ordinary solenoid type actuators and the larger motor driven type, particularly for quarter-turn ball valves up to 50 mm (2 in) bore.

DIAPHRAGM ACTUATORS

The diaphragm actuator is a variation of the piston and cylinder, and most economical to manufacture, but is even more limited in stroke. It therefore finds its application primarily on short-stroke control valves, where its absence of sliding parts makes it ideal for continuous modulating duty.

MOTOR ACTUATORS

The motor actuator, which can be pneumatic, electric or hydraulic, is particularly suited where stroke is long, because a motor with gearbox has unlimited stroke. However, there are different mechanical virtues between the motor and piston and cylinder. As mentioned earlier, the piston can be kept continuously pressurized, whereas the electric motor cannot; a piston can maintain its position, but the drive of a motor must be mechanically self-locking in order to maintain its position when switched off. The motor does, however, have the virtue of kinetic energy – the ability to take a 'running jump'. Kinetic energy, however, places mechanical constraints on the use of motor actuators as the energy must be controlled and not misapplied.

In many circumstances motor actuators are not very efficient. For example, when driving through a worm gear and nut and screw, say on a gate valve, up to 90% of the motive power is trying to wear itself out and only 10% is useful. This is acceptable for intermittent valve operation, but it is one of the reasons why geared motor actuators are not well suited to continuous modulating or positioning control via stem nuts or self-locking gears. This applies to all forms of geared motor actuators whether pneumatic, hydraulic, or electric; the last may also be limited in operating frequency by motor heating and switching considerations.

CHOICE OF ENERGY SYSTEM

Clearly the selection of the energy system for a particular valve-operating duty is not something that can be made in isolation. Overall design considerations, safety requirements, availability of supplies and total installed initial cost and subsequent maintenance costs all need to be considered.

Sometimes practical valve or environmental needs dictate the use of a specific type of actuator, in which case the energy source is predetermined. Sometimes the non-availability of an energy source – or the prohibitive cost of providing it – will rule out certain actuator options. What follows, therefore, is only a brief resume of the basic systems available, and the practical and economic pros and cons of each for various duties. This must, of course, be related to the foregoing section on actuator types and *their* performance capabilities.

Advice is readily available from most actuator manufacturers on the choice, sizing, adaptation, installation and commissioning of valve control systems. For the unwary or inexperienced specifier there may be some risk of bias in such advice from manufacturers of particular types of actuator. This is rare among leading suppliers whose own reputation must stand or fall on the soundness of the advice in terms of performance and reliability.

Pneumatic Systems

If the operation is to be within the confines of a plant where compressed air is available, the cheapest method of valve actuation is a pneumatic piston and cylinder. Pneumatic operation over a distance is, however, limited entirely by the high cost of making adequate compressed air available over long distances. Ease of storage of compressed air facilitates fail-safe operation under electric power failure conditions. The major limitation of pneumatics is available force due to practical pressure limitation – 6–8 bar (80–100 lb/in^2) being the normal maximum.

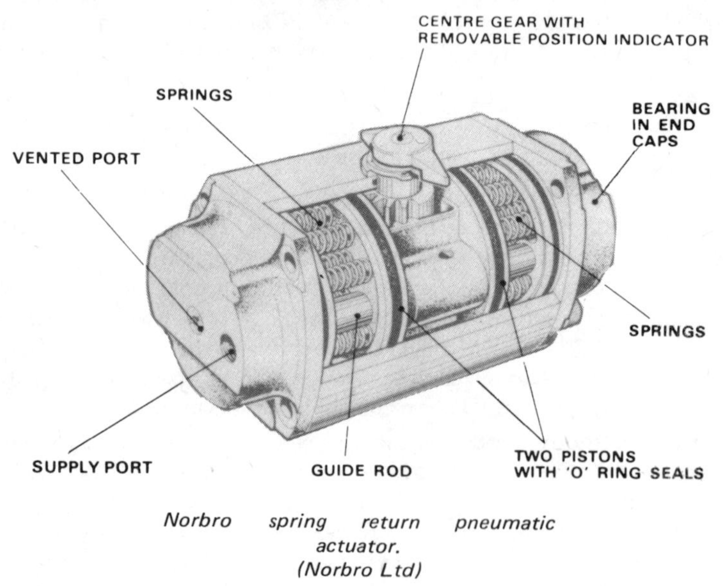

Norbro spring return pneumatic actuator.
(Norbro Ltd)

The main limitation of pneumatic *motor* actuators is one of overall efficiency – as mentioned earlier – because pneumatic energy in large quantities is expensive. However, it is important to acknowledge the need or preference for pneumatic motor actuators in certain circumstances – where perhaps electricity supplies are not available or not allowed, or where the certainty of operation, regardless of any contingency, from a stored energy source (gas or air bottle) is required. In this context it should be noted that recent developments include an all-pneumatic actuator, with all the control, remote signalling and other attributes normally only available on the more sophisticated electric motor actuators.

Hydraulic Systems

Hydraulic systems are the natural choice if an exceptional duty requires a large amount of stored power. The forces available, speed and controllability of the hydraulic actuator are virtually unlimited, *ie* the only method of operating a large valve in seconds is to pump up a hydraulic accumulator first. Unless these particular virtues are needed, the system is uneconomical and, for this reason, is not commonplace.

Furthermore, the hydraulic system is very expensive for operation over long distances. A pneumatically-operated valve is vented to atmosphere, but with the hydraulic system, the hydraulic fluid has to be returned through a line of greater section to avoid pressure drop. For

Rotork all-pneumatic valve actuator with built-in limit controls, weather proofing and other features normally only associated with sophisticated all-electric actuators. It can be powered from a regular compressed air supply or from pressurized air/gas cylinders.

some special situations, such as tanker cargo valves, however, hydraulics provide the only acceptable means of centralized control.

Combined Pneumatic/Hydraulic Systems (Air-Oil Cylinder)

There are some specialized applications where the combination of pneumatics and hydraulics can meet particular operational, fail-safe and other needs. Usually the principle involved is to use a plant's existing compressed-air supply to drive a hydraulic pump which, in turn, is used as the regular valve actuator energy source, but also tops-up an accumulator for stand-by emergency use. Such systems can be devised to cater for virtually any supply failure conditions. They are, however, usually specially custom-engineered. A very typical application is for self-powered actuators in gas pipelines from which the live gas can pressurize a hydraulic accumulator system for emergency isolating valves.

Electric Systems

There are two main factors to be considered here: an electric motor as the power source, and electricity for the control system. Undoubtedly, full electric operation and control offers the greatest flexibility, especially where operation over distances in excess of, say, 100 metres (100 yards) is required. Remote indication by electrical position feedback and provision for stand-by manual operation is inherent if the system is totally electric.

ELECTRIC CONTROL

Because much of the development in industry today is in the direction of centralized control – which in turn involves remote electrical control of valves – a closer look at electric valve actuators and their control systems is justified.

Before going into the detail, a general historical point needs to be made, regarding the cost-effectiveness of electric valve actuation. This technology has evolved with astonishing speed. Among the leaders in the field, this evolutionary phase has resulted in very advanced, sophisticated designs suitable for instant adaptation to practically any kind of valve, in any environment in any part of the world. That this can lead to over-simplification and unnecessarily high costs for more

FX550 electrically-driven trolley-mounted actuator. (Flexible Drives (Gilmans) Ltd)

mundane valve actuation needs has been acknowledged by some manufacturers. They are now applying the benefits of their design/applications experience to the question of greater cost-effectiveness for certain classes of duty.

Generally the choice between the more sophisticated design and the more economical approach depends upon whether or not the actuator is to be installed in a hazardous zone and therefore needs to be flameproof/explosion-proof and/or if it has to interface with computer control systems. If not, then the later designs, aimed at industries such as the water, sewage and electricity generating, offer most of the inherent advantages of the more sophisticated type, but at less cost. They also offer the option of separate thrust bases — again an economy for use where thrust is absorbed by the valve itself and torque-only actuators are needed, such as ball, butterfly and certain gate and globe valves.

The following observations on electric valve actuator design and control are generally applicable to most types.

Sealing

The weakness of electrical equipment is that electricity and water do not mix. However, electric actuators are required to operate in a wide variety of environments: high temperatures, steamy atmospheres, adverse weather conditions, mud, sand, flooding and so on. Thus the primary problem for the designer is not the mechanical duty — that is negligible — but to keep the environment out.

For example, a plant operating five days a week, fifty weeks a year, with a motorized valve of one minute stroke time, opening in the morning and closing in the evening, would demand a total actuator working life of only seven days in twenty years. Thus proper environmental sealing to enable an actuator to do nothing with complete reliability is paramount. A minimum requirement is that it should be watertight and flood-proof.

As a further safeguard to exclude the environment, the terminal area — which has to be exposed for wiring on site — should be separately sealed from the rest of the equipment, to prevent the ingress of dirt or water during installation — the most vulnerable period in an actuator's life.

Control

The actuator motor has, of course, to be switched by contactors, solid-state power switches being not yet cost-effective for such intermittent duty. These can be separate from the actuator, but having produced a good, sealed enclosure, the system is now considerably simplified if all the control gear is built into the actuator housing itself.

One British manufacturer has pioneered this concept of integral control gear; they dramatically make the point that with the starter and all control gear built into the actuator, the number of cables required to wire up the starter, actuator, control station and power supply is reduced by half (three instead of six) and the number of terminations cut from 45 to 19. This is obviously significant in terms of installation cost, ease of commissioning and long-term reliability, quite apart from the benefits of a complete 'package' which is totally sealed from the environment.

This sealed package can also accommodate interposing relays to facilitate the control of valves by low voltages over considerable distances. The integral control concept simplifies the whole process of motorized valve engineering procurement, testing and installation with undivided contractual responsibility and quality control of the essential motive elements.

Valve Positioning

There are two basic requirements in the opening and closing of valves: when closing to be sure that the valve is properly and tightly seated, yet without excessive force being applied; when opening, to be certain that the valve is fully opened without excessive overrun or strain on the backseating.

These requirements are met by a combination of torque and limit switches. Normally a torque switch is set to disconnect the motor when the closing torque reaches a preset level; when the valve opens, a travel limit switch disconnects the motor after a preset number of turns. In the opening direction, however, there is back-up protection by another torque switch, thus providing in each direction, against mid-stroke seizure or obstruction. Additional switch contacts can be incorporated to operate at each end of the valve stroke, for remote signalling and/or interlocking control circuits.

Varipas actuator-positioner. (Worcester Controls Ltd)

'Hammerblow' Effect

All motor actuators inherently develop kinetic energy and its associated flywheel effect, which are valuable for intermittent valve operation because, upon a reversal of direction, backlash allows the motor to reach full speed before taking up the drive. However, this same kinetic energy must be kept down to a reasonable level, to avoid excessive overrun on switch-off, and to limit the damaging forces available in the event of malfunction, *eg* due to incorrect wiring.

Efficiency

It has been said already that many valve motor drives are mechanically inefficient – only 10% useful energy when, for instance, they are driving through a worm gear and the nut and screw of, say, a gate valve, and not much higher when driving a butterfly valve through self-locking gears. Wearing of the inefficient gears or stem nuts is the primary limitation on frequency and continuity of operation of such drives, and also limits the practical speed of operation of large screw-operated valves and penstocks. Nevertheless, the many advantages of control, power source, interfacing with control instrumentation and so on, compensate handsomely for the mechanical inefficiency, provided the actuator duty is properly considered.

Fail-Safe Facilities

A limitation of the electric motor actuator is, of course, that it will not work without an electricity supply. To cater for emergency conditions without electric power, it is possible to add a standby pneumatic motor to an electric actuator for a single, fail-safe, operation using power from a nitrogen or air bottle instead of an expensive compressor plant.

Supervisory and Computer Control Facilities

Because the evolution of environmentally sealed electric motor actuators with packaged control gear has eliminated the problem of providing reliable hardware in exposed situations, the latest developments aim to improve system reliability by providing all the facilities required for supervisory and computer control. For example, if the local power supply to an actuator fails, it might not be discovered until someone needed to operate the valve possibly weeks later, when it could be too late to correct in time for an emergency function.

Suppose someone has forgotten to reset the local/remote selector switch for remote control – the actuator then remains inoperable from the control room. Suppose a motorized valve in a sequence control system fails to complete the stroke required, or is manually operated away from the position required by the control room – how does the system react to and cope with these problems?

The very latest actuator designs are able to detect these and other problems as soon as they arise, and can give an immediate alarm to enable corrective action to be taken before further operation is required.

The design of such actuators enables virtually any depth of automatic supervision or fault analysis to be achieved. At the same time the basic local and remote control facilities remain unchanged, so that straightforward valve control requirements are met as previously. By using the 'information' available from a limit switch at each end of travel, a monitoring relay and a pair of contactor contacts, every phase of operation of the actuator can be monitored, up to full fault analysis and trouble-shooting by digital computer. Thus, although an installation may not require the facility immediately, subsequent expansion and addition of supervisory and computer facilities are catered for by this standard actuator.

Continuous Modulating Control

Electric motor actuators have a limited usefulness for this duty, because of the problems of mechanical inefficiency and overshoot and some considerations of motor heating and switching frequency. Some manufacturers therefore offer an alternative method of dealing with continuous modulation by the use of a packaged electro-hydraulic actuator which works directly from electric power. These utilize the inherent high mechanical efficiency of fluid-power systems to avoid the wear and tear problems.

The pneumatic diaphragm or piston operator is particularly suitable for continuous modulation control. The only limitations are those of compressed air itself. These include maximum force available, because of pressure limitation, and also the fact that rigid positioning cannot be achieved using a compressible medium like air. Reliability of the system is also dependent on the absence of dirt and moisture, which can be readily transmitted with the air through piping. The pressure drop in the pipe also limits the distance over which a pneumatic control signal can be transmitted satisfactorily.

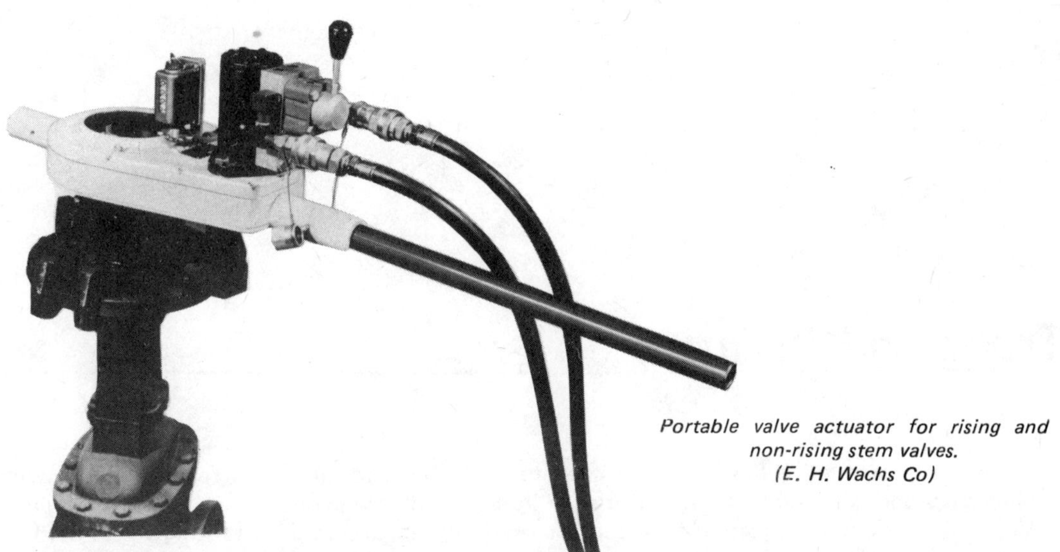

Portable valve actuator for rising and non-rising stem valves. (E. H. Wachs Co)

As plants get bigger and the valves increase in size and pressure, the limitations of air become more apparent. A large plant incorporating electronic instrumentation and pneumatic control valves will require installation of both electric and pneumatic systems, and suitable signal conversion equipment from one to the other.

PORTABLE VALVE ACTUATORS

Portable valve actuators or valve wrenches can be used to provide portable turning power to replace manual effort in opening or closing valves on pipeline systems not fitted with permanent valve actuators. For example, water distribution piping systems generally have no permanent actuators installed, opening and closing of valves,when required, being by manpower. Equally, probably the majority of process valves do not have permanent actuators installed and, in fact, are not frequently closed. However, during turn-around time, or in emergencies, these valves must be closed and this can be difficult and time consuming using manpower alone.

A further advantage of portable valve actuators is that they permit implementation of valve exercising programmes designed to service and operate valves in order to keep them in good condition, *eg* eliminate 'valve seizure' problems which can arise when valves are left in one position for long periods. This applies particularly in water systems.

Portable valve actuators and valve wrenches may be designed for operation by electricity (*ie* powered by electric motors), compressed air or hydraulic power, and are normally adaptable for fitting a wide range of standard valves. The best designs provide operation 'feel', as well as speed control, reversibility and the ability to make instant stops and starts and reverses needed to free up sluggish gate valves, hydrants and sluice gates, *etc.* Typically speeds of up to 20–25 rev/min may be provided, with operating torques up to 1360 Nm (1000 lb/ft) for handling valves in the size range 152–18288 mm (6–60 inches).

The majority of such models are readily portable (*eg* weigh less than 18–23 kg (40–50 lb)). Larger, heavier models may be trolley or truck mounted.

Traps and Drainers

Traps or vent traps are components designed to vent collected air or gas from liquid systems automatically and continuously without loss of liquid. In this respect they are similar in function to air relief valves, but the description trap is specifically applied to such venting valves used on heat exchangers, surface condensers and hot water heating systems.

Drainers are the complement of traps. They are valves used to drain oil, water or condensate automatically and continuously from pressurized gas or vapour systems, and also for interface control where continuous draining of the heavier liquid is required.

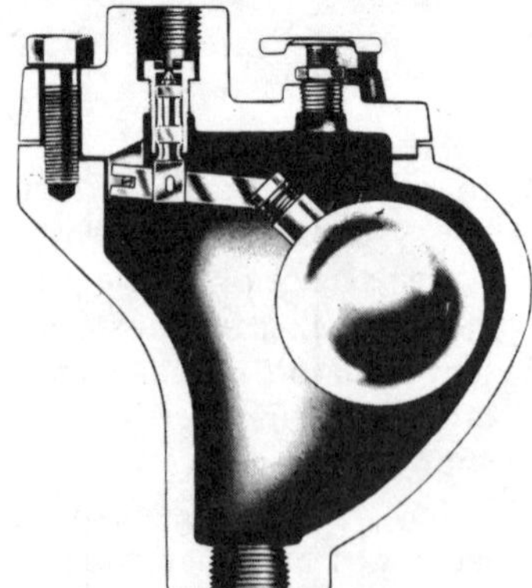

Vent trap (sectioned). (Fisher Control Co)

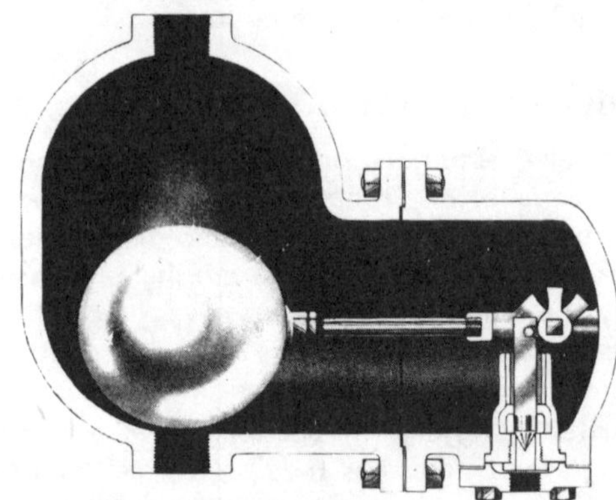

Automatic drainer (sectioned). (Fisher Control Co)

Fig 1

Both traps and drainers are normally float-operated valves, comprising a float chamber, float assembly and cover with a vent cock – see Fig 1. Typical materials of construction are:

Float chamber and cover – cast iron or steel

Float – copper or stainless steel

Plug and liner assembly – stainless steel

Vent cock – brass

See also chapter on *Air Relief Valves*.

S

SE ... ategories:

... ernal flow leakage when the valve is

... of fluid.

... welded).

... han valve stem seals, but can be subject to more
... ic flow when the valve is in the open position.
... tight, and remain so. Some general services can
... delaying the onset of necessary maintenance. At
... hat maintain sealing integrity even after being

... and at the same time maintain a clean environ-
... quirements can range between extremes. Where
... eepage may be tolerable, and the effect in the
... r hazardous product, absolute seal integrity is

... e, gate, disc, *etc*, closing against a seat (category
... utterfly valve) rotated to present a blank area to
... be self-sealing by nature of their flexible con-
... tegory 3).

... s may be used, viz:

... ng element and its face.

... esilient face.

... metal face in which is inset a resilient seal

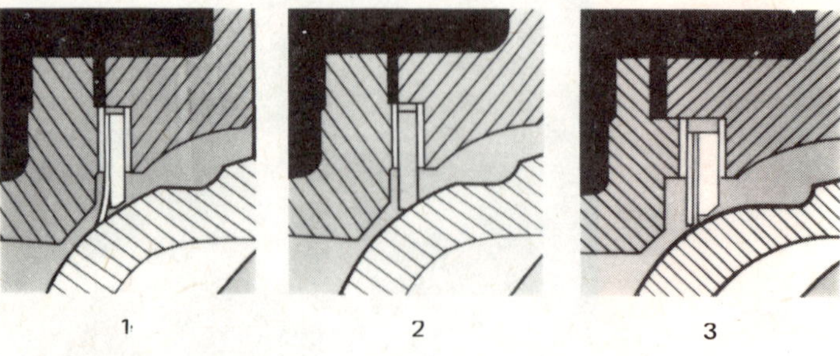

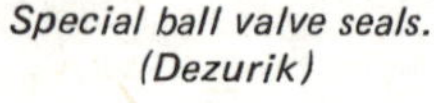

Special ball valve seals.
(Dezurik)

1. Laminated seat
2. Rigid seat
3. Laminated steel/PTFE seat

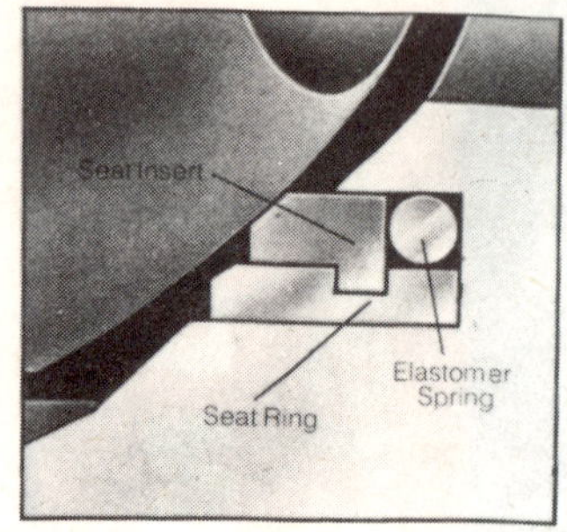

Example of ball valve seal with protected seat.
(W-K-M Division, ACF Industries Inc).

Metal-to-metal sealing provides the greatest strength but (usually) the greatest susceptibility to wear, seizure, or degradation of performance through entrapped particles when handling fluids with a solids content. *Wire-drawing* or premature erosion of the valve seat caused by high localized fluid velocities being generated between the valve at its seat in a partially open position is also a possibility. This condition can be generated by back pressure causing a valve to lift slightly from its closed position; or through differential expansion when hot fluids are involved.

Metal sealing against a resilient seat provides a more positive and tighter shut-off without the necessity for fine finishing of the two surfaces. The resilient seat may also be able to accommodate solid particles and still provide complete sealing. The chief limitations of this arrangement are that the *maximum pressure* which can be accommodated is limited to one below that which will cause the resilient seat to extrude; and the *maximum service temperature* is that of the resilient material (elastomer or plastomer). This will inevitably be lower than that of a metal-to-metal combination.

Metal sealing against a resilient seal inset in a metal seat or face can provide much higher working pressures, but again the maximum working temperature is limited by that of the resilient material.

Values of 'mixed' plastic and metal construction may utilize the plastic element as the resilient member without recourse to separate treatment. Problems here can be wear (nylon will wear away steel, for example), and the considerable difference in thermal expansion coefficients between metals and plastics. This particular problem does not arise in valves of all plastic construction, although these are limited in size range, pressure rating and temperature rating.

Category 2 valves may rely on lapped finishes for sealing. This can prove quite satisfactory in small sizes, particularly when the fluid being handled is a lubricant (*eg* spool valves for oil-hydraulic services invariably seal in this way). It is less satisfactory in larger sizes and when handling non-lubricating fluids, unless the valve itself is lubricated. Thus some types incorporate additional sealing elements to provide positive sealing.

Category 3 valves need no special attention to sealing requirements. The sealing potential of a diaphragm in fact, (and the positive isolation provided by the flexible tube of a pinch valve), can also eliminate the need for valve stem and bonnet seals.

Valve Stem Seals

The normal method of providing a spindle seal is by a packed gland, or in some cases by an O-ring or O-rings. In the case of *non-rising stems* the seal has only to accommodate rotary motion; also the gland can be located close above the threaded position of the stem. With *rising stem* valves the

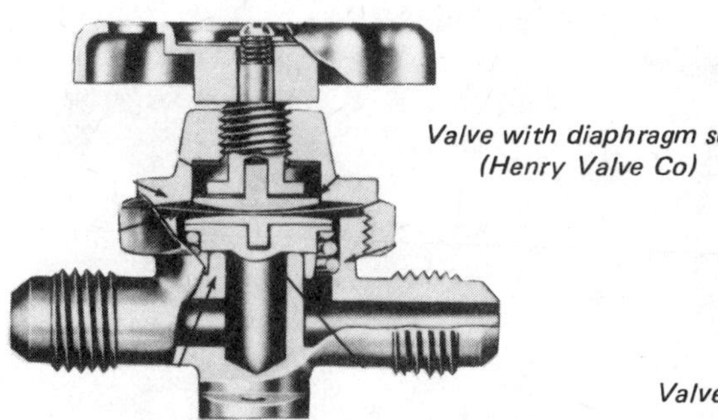

Valve with diaphragm seal (Henry Valve Co)

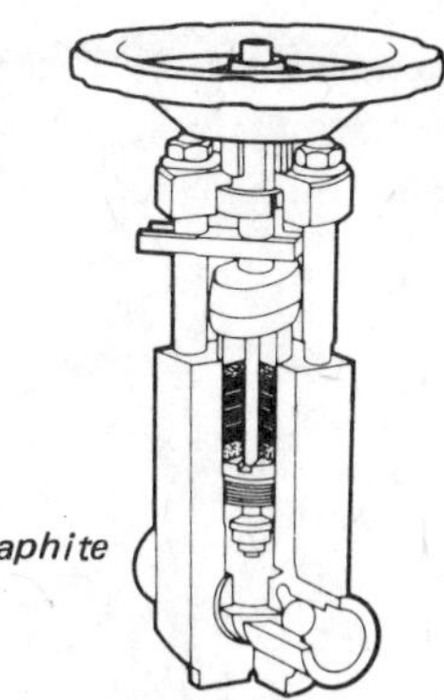

Valve stem seal with expanded graphite rings. (E.D.F.)

seal has to accommodate both rotary and linear motion and must be of sufficient height above the threaded portion of the spindle to be clear of it when the valve is fully open (*ie* stem has risen to its highest level). In the case of sliding-stem valves the seal has only to accommodate linear motion.

To some extent this will dictate the type of seal used – *eg* O-rings and most other types of gland rings readily accommodate rotary motions, but can prove less satisfactory with linear motions. Packed glands, on the other hand, are capable of accommodating both rotary and linear motions.

Packings used include asbestos, asbestos/graphite, PTFE, PTFE/graphite, PTFE treated fibres in plaited and 'superplaited' forms; synthetic rubbers and rubber/fabric composites in the form of moulded gland rings; wrapped metallic or metal foil packings; graphited fibre rings and die-formed expanded graphite rings. Loose fills of granulated PTFE and other compositions, or asbestos fibres may also be used. Loose packings should normally be used with solid or split end rings to prevent extrusion of the loose material, although this may not be necessary if the gland and neck bush are in first class condition.

Bonnet Seals

The valve bonnet may be case or formed integral with the body, in which case it will have a separate top cover or cap which may be bolted or screwed in place. In either case the top cover usually forms both the upper bearing for the valve stem and the gland follower for tightening the gland. The spindle seal can then also act as the gland. With screw-in or screw-on bonnets the threaded length screwing into the body acts as the bonnet seal. A tight thread is obviously desirable for maximum sealing efficiency. This will normally result in some distortion of the bonnet when assembled, and a looser fit when the bonnet is unscrewed and replaced at a maintenance period. Also there is the possibility of a screwed bonnet unscrewing slightly under vibration or through some other cause, again reducing the efficiency of the seal. To a large extent these limitations can be overcome by assembly with a thread sealant, which must be of non-hardening type to permit unscrewing for maintenance. An alternative solution is to incorporate an additional static seal which is compressed into position on assembly. The majority of screw-in or screw-on bonnets, however, do not have separate seals.

With bolted-on bonnets a gasket type static seal is normally used to ensure leak tightness. This should present no particular problems provided the choice of gasket type and material is suitable for the service, sufficient gasket area is present, and the assembly is tightened within specified closure pressure requirements for the material relative to the pressure to be sealed. Requirements in this latter respect may be different for gases and liquids for the same gasket material.

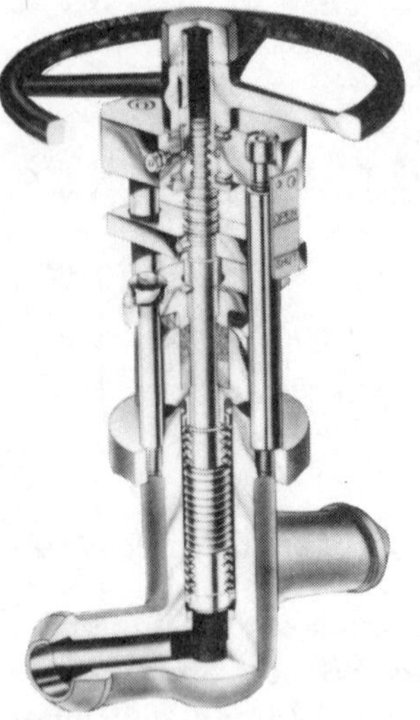

Globe valve with bellows seal. (Hopkinsons Ltd).

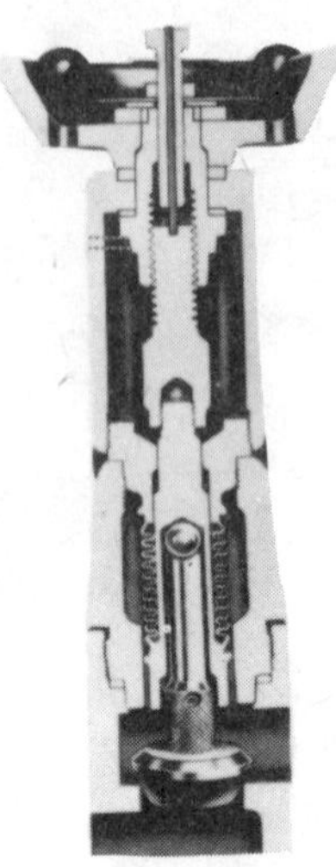

Bellows sealed globe valve (B.V.M.I. Limited)

Bellows sealed needle valve. (Ashford Controls Ltd)

'Glandless Valves'

The principle employed in glandless valve construction is to isolate the bonnet interior completely from the fluid present in the valve body. This, in effect, provides a 100% bonnet seal and eliminates the need for a stem seal entirely. The chief alternatives are:-

(i) The use of a diaphragm covering the whole of the bottom of the bonnet, thus sealing off the bonnet interior volume completely. The valve stem then extends through the diaphragm (with attention to maintaining a seal at this point), with the flexibility of the diaphragm allowing normal stem movement.

(ii) The use of a bellows to isolate all the 'working' points carried inside the bonnet. Entry of fluid into the bonnet is then restricted to the interior volume outside the bellows. Secondary seals may or may not be necessary to ensure that there is no possibility of external leakage from this volume. Equally a secondary seal may be used to inhibit entry of the fluid into the non-isolated bonnet volume to work as a first line of defence; or to prevent leakage in the event of bellows failure.

Bellows bonnet seals may employ elastomeric bellows, but more usually metallic bellows. Apart from providing higher temperature and pressure ratings, metallic bellows are also suitable for vacuum services down to high vacuum.

Valve Connection Seals

Threaded valve connections are normally used with no seals, relying on the threads to provide their own labyrinth type seal. Thread sealing compounds of the non-hardening type may be used where it is felt necessary – *eg* often for resisting the loosening effects of vibration rather than actual sealing.

TABLE I – SUMMARY OF PACKING TYPES AND PROPERTIES

Type	Material	Construction	Temp. (max. or range) °F	Temp. (max. or range) °C	Maximum Pressure lb/in²	Maximum Pressure bar	pH	Suitable for
Natural fibre	Cotton	Plaited or braided with lubricant	199	90	–	–	6 – 9	Water
	Hemp	Plaited or braided with lubricant	176	80	–	–	5 – 9	Water
	Flax	Plaited or braided with lubricant	160–250	70–120	–	–	–	Water
Synthetic fibre	Nylon, rayon, etc.	Plaited or braided, impregnated with PTFE	–	–	–	–	–	Not used for valve packings.
Asbestos	White asbestos	Plaited or braided with graphite, mica or oil lubricant	–22 to 570	–30 to +300	–	–	5 – 12	Steam, liquors, etc.
Asbestos (reinforced)	White asbestos, inconel wire reinforcement	Plaited or braided with graphite, mica or oil lubricant	–60 to + 1380	–50 to + 750	3600 to 9500	250 to 650	4 – 11	Water, steam, solvents, hydrocarbons, incorporating acids and alcohols
Asbestos graphite	Asbestos fibres mixed with graphite	Wet-spun, dust-free type preferred as smoother than dry mix	–30 to + 570	–40 to + 300	–	–	5–12	High temperature and steam services.
PTFE	PTFE yarns or tape	Plaited or braided	–328 to + 490	–200 to + 250			0–14	Water, steam, hydrocarbons, Soft packings
	PTFE yarns with lubricant	Plaited or braided with graphite	–150 to + 480	–100 to + 250	430	30	0–14	Water, steam, hydrocarbons, acids, alkalis, solvents, etc.
	PTFE yarns, treated with PTFE dispersion	Plaited or braided	–328 to + 570	–200 to + 300	1450	100	0 – 14	All media, but little used for valves.
PTFE/graphite	PTFE/graphite mix	Solid extrusion	–150 to + 480	–100 to + 250	1450	100	0 – 14	Water, oils, hydrocarbons, alkalis, acids, alcohols, etc.
PTFE/aramid	Aramid fibres treated with PTFE dispersion	Coated fibres braided or plaited and impregnated with lubricant	–364 to + 570	–220 to + 300	2900 tc 14,500	200 to 1000	1 – 14	Not particularly su'table for valve stem seals.
Expanded graphite	Pure expanded graphite	Tape form	–328 to + 1100	–200 to + 600	9350	300	0 – 14	High temperature services; sealing gases and low viscosity fluids; all services requiring superior leak tightness.
	Pure expanded graphite	Flexible plait	–328 to + 1100	–200 to + 600	4350	300	0 – 14	
Carbon fibre	Amorphous carbon yarns treated with graphite powder	Twisted or plaited	–328 to + 1100	–200 to + 600			0 – 14	High temperature services; but little used for valves.
Glass fibre	Glass fibre yarns	Braided with added lubricant	–	–	–	–	–	Corrosive media (except strong alcohols)
Alumina silicate	Alumina silicate	Plaited, with or without inconel wire reinforcement	2300	1260			0 – 9	Extremely high temperature services

Flanged end connections may be bolted up with or without gaskets (or groove-fitting static sealing rings, such as O-rings). Gaskets used may range from conventional sheet materials to spiral wound gaskets, depending on the service involved and operating conditions.

Number of Packing Rings

Tightening pressure required in a gland is independent of the number of rings. It depends solely on the quality and characteristics of the packing rings and the pressure to be sealed. The optimum number of rings will, thus, depend on the packing section, gland geometry, and the type of packing used. Experience would indicate a minimum of three and a maximum of five rings as a good general choice, lacking more specific data (*eg* with die-formed expanded graphite rings satisfactory valve stem sealing may be achieved with only one or two rings).

Gaskets

Static joints may use a variety of gasket materials. The majority of conventional multi-purpose gasket materials are made from white asbestos fibres with an elastomeric border; or agglomerated cork, for more restricted services (*eg* water, hydrocarbons and gases).

Modern alternatives to CAF (compressed asbestos fibre) gaskets have recently appeared in the form of polyaramid fibre gasket materials and also pure expanded graphite. The former are suitable for such applications involving water, steam, gases, hydrocarbons, chemicals with the range pH 1–13, and gases at temperatures up to 660°F (350°C). Expanded graphite gaskets with or without metallic reinforcement offer pressure/temperature combinations better than those of any other gasket material and are particularly suitable for valve seals between body and bonnet, and for pipe flange seals.

See also chapters on *Nuclear Valve Glands, Fire Resistant Valves* and *Valve Spindle Corrosion.*

Heated Pipes and Valves

HEATING CAN be advantageously applied to pumps, pipelines and valves in systems handling viscous fluids or working at very low ambient temperatures where the product viscosity has increased to unacceptable levels. The governing parameter is normally the pump performance. Heating can maintain pump efficiency or, in the case of reciprocating and rotary pumps, maintain the delivery of a given size of pump or eliminate the need for a larger size of pump running at a slower speed. In extreme cases, heating the fluid may be the only satisfactory method of handling solid substances or very viscous semi-solids in a fluid form with a particular or preferred type of pump, rather than with a special pump.

Except in instances where heating is essential, the economics of the operation can be assessed by comparing the cost of heating with the gains realized by increased pumping efficiency. This involves both the initial cost of the heating system and the operating or running costs, which may, however, be offset by hidden gains such as improved pump life, or the suitability (with heating) of a cheaper pump. Detailed analysis can, therefore, only be undertaken for individual installations. Systems heating, it should be emphasized, is widely used by various industries and has come more and more to the fore during the past twenty years, particularly with the advent of electric surface heaters which are very simple to install and make it possible to heat virtually any component or part of a complete pumping system.

Viscous materials which are commonly handled in a heated state include acids, asphalt, bitumen, gelatine, glucose, grease and heavy oils, paints, petroleum jelly, phenols, plastics, resin, sodium, tar, varnishes and waxes. Certain substances which are in a solid state at normal ambient temperatures may also be rendered fluid for pumping by moderate heating.

The reduction in viscosity achieved by heating will depend both on the rise in temperature and the viscosity index of the fluid. Fluids with a low viscosity index will undergo a relatively rapid decrease in viscosity with rising temperature although the viscosity index itself is not a criterion which can be employed to predetermine viscosity at any particular temperature. This is best done by empirical measurement or can be estimated on an ASTM graph in the absence of known viscosity figures for the fluid over a range of temperatures. The amount of heating required to achieve a particular fluid viscosity can then be calculated on the basis of fluid volume and specific heat of the fluid, although this is more usually done on semi-empirical lines.

The heating of solids to render them fluid represents a slightly different problem. Thus most liquids tend to absorb heat uniformly upwards due to the generation of convection currents within the liquids. Solids with poor heat transfer characteristics will melt first in the immediate vicinity of the applied heat, the rate of further melting then depending on the speed at which the heat can be transferred through the liquid to the unmolten material. This may influence the type of heater employed and its position, particularly for the heating of drums or containers.

Heating of the delivery container only is the simplest and most direct solution in many cases. The heated fluid is readily transferred to the pump and through the system and will not normally lose much heat in the process. Viscosity will thus remain substantially the same as in the heated delivery tank. However, difficulties may arise on shut-down due to 'freezing' or thickening of the fluid left in the lines, pump and valves which may make restarting difficult or even impossible, and additional heating may well have to be applied to other elements in the system.

Heating the pump only may be a satisfactory alternative, provided the fluid is not so viscous on the unheated suction side that the pump performance is seriously affected. In the latter case, the delivery container may also be heated with advantage.

The requirement for complete pipeline heating is, basically, to replace heat lost under static conditions. If the material enters the pipeline at the correct temperature and flow is not interrupted the question of extra heating should not arise. Heat losses may be reduced or minimized by lagging, but this will not solve the problem of starting a pump when the unheated pipeline is either empty or full of cold liquid. On the other hand, it is generally uneconomic to heat the pipeline purely to heat the liquid flowing through unless this is specifically demanded.

Valve heating may be required to maintain the liquid flow through the valves and to avoid 'freezing' or coagulation of the fluid within the valves, particularly under static conditions. This may be of importance in some applications and not necessary in others. To avoid the possibility of valve seizure it is better, if possible, to eliminate valves entirely from a system handling a fluid which may coagulate under static conditions.

Heating Methods

Heating is mainly applied to pumps or delivery vessels, which subjects are outside the scope of this particular Handbook. Suffice it to say that the basic methods of heating pumps are

(i) Circulation of a suitable heating fluid through a double-walled pump casing.

(ii) Electrical heating by special heating elements incorporated in the pump casing.

(iii) Heating by special jackets incorporating electrical heating elements.

Methods (i) and (ii) demand special pump constructions. Method (iii) can be applied to standard (non-heated) pumps, with certain limitations.

With method (i) hot water can be used as the heating fluid to achieve temperatures of up to 195–205°F (90–95°C), or somewhat higher in a closed pressurized system. Steam heating can provide temperatures of the order of 300–355°F (150–180°C), depending on the steam pressure available.

A wider range of working temperatures can be provided by electrical heating, with the advantage that electricity is clean, reliable, easily controlled and capable of repeating performance exactly.

A further advantage of electric heaters is that they can be rendered 'flameproof' using metal sheathed mineral insulation heating elements with flameproof glands screwing into flameproof junction boxes which are Buxton certified. It is then necessary to ensure that the stabilization temperature reached is below the ignition temperatures of flammable concentrations which could be present. This can be ensured by the use of suitable control to maintain the temperature of the equipment at the correct level and at the same time keep power consumption to the minimum.

Heated Pipelines

Heated pipelines are mainly used only in the chemical industry for processes involving the conveyance of viscous fluids which do not flow readily at ambient temperatures. These take the form of jacketed pipes in carbon steel (usually), stainless steel or other alloys, through which steam or some other heating fluid can be circulated. Jackets are normally made from the same material as the pipe to avoid difficulties which might arise from electrolytic corrosion due to two dissimilar metals being in contact. Erosion-corrosion may also be a problem and stainless steel impingement plates are often fitted at the point of entry of the heating fluid to protect the pipe from excessive wear.

Jacketed pipes are a standard production in straight lengths, using two different sizes of pipes. Bends may be constructed either by bending the product and jacket pipes by hot forming, or by the use of welded fittings, depending on size, bend radius, *etc.*

Jacketed pipe assemblies are also supplied for site assembly, completed by butt welds at the joints. In this case studs are welded to the inside of the jacket pipe to ensure concentricity.

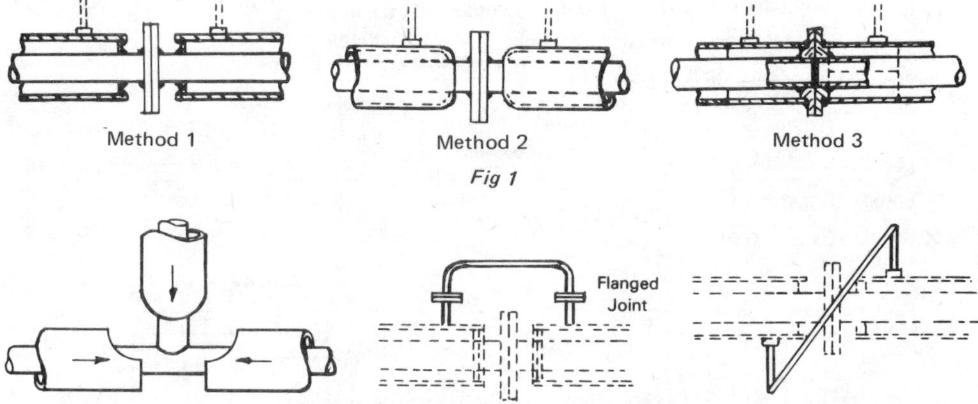

Fig 1

Fig 2 Connections for jacketed pipes.

Examples of jacketed pipe assemblies are shown in Fig 1. In method 1, a plate is welded in to seal the gap between the jacket and product pipe at coupling points. This is usually the cheapest method. In method 2 the jacket ends are hot formed and then welded to the product pipe. This is the most commonly used system. Method 3 is normally only applied to pipes where the whole run must be heated. The jacket in this case is carried the whole length of the product pipe and welded to the back of the flange. This involves the use of special flanges.

Examples of connections used with jacketed pipes are shown in Fig 2. Connecting jackets by the parallel method requires condensate drainage for each pipe, but is the more efficient method. Series connection connects the bottom of the upstream pipe to the top of the next, when condensate is carried along with the heating medium. In this case drainage need only be arranged at suitable intervals, not for each pipe. This is usually the preferred – and cheapest – method.

Heated Valves

Valves are less commonly heated than pumps or pipelines, but are adaptable to similar treatment to pumps if strictly necessary. Methods (1) and (2) are unusual, except for specialized valves of relatively large size. Electric blankets or flexible electric heating tapes are generally more practical, depending on the size of valve concerned.

Pipeline Heaters

Flexible electric heating tapes provide a simple and convenient method of heating pipe runs and various types have been developed to deal with different applications and working conditions. A loading of approximately 10 watts per foot run (33 watts per metre) is usually adequate for most general purpose pumping applications and will compensate for temperatures up to 210°F (100°C) when straight traced and applied under woven glass cloth. This is suitable for use at temperatures up to 660°F (350°C). To render the tape waterproof an additional layer of silicon rubber may be extruded over the standard insulation. Further types are covered by PVC sheeting for full weather-proofing (*eg* for frost protection of water pipes and outdoor installations).

A tape with a loading of 15 watts per foot run (49 watts per metre) will compensate for temperatures up to 300°F (149°C) when straight traced under lagging. A spiral lay for the tape will appreciably increase the temperature compensation provided.

Typical heater tape construction is shown in Fig 3. The tape itself is normally made from special glass fibre weave covering the PTFE protected heating element.

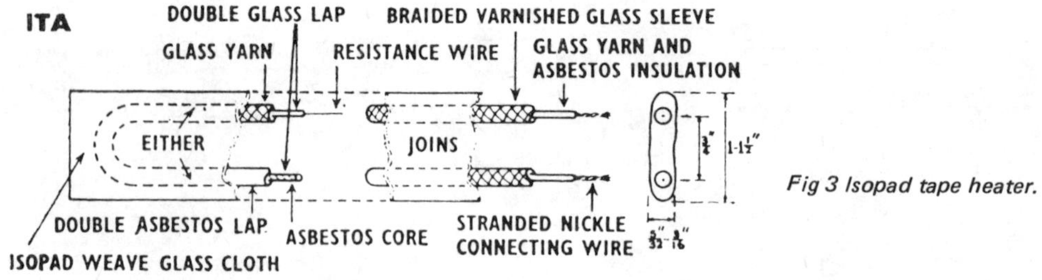

Fig 3 Isopad tape heater.

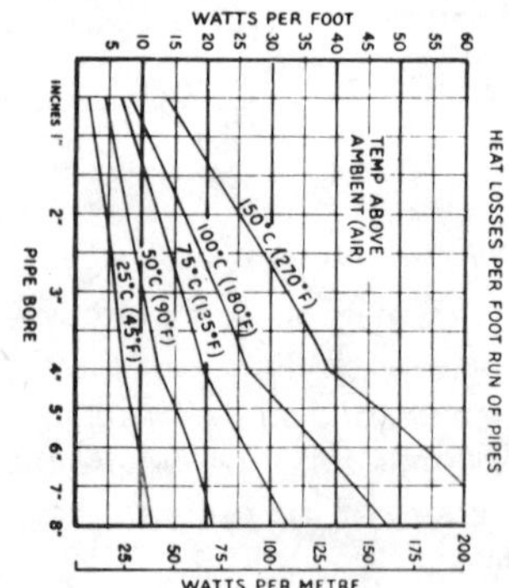

Fig 4 Pipe heat losses with 1in insulation thickness. (Isopad Ltd)

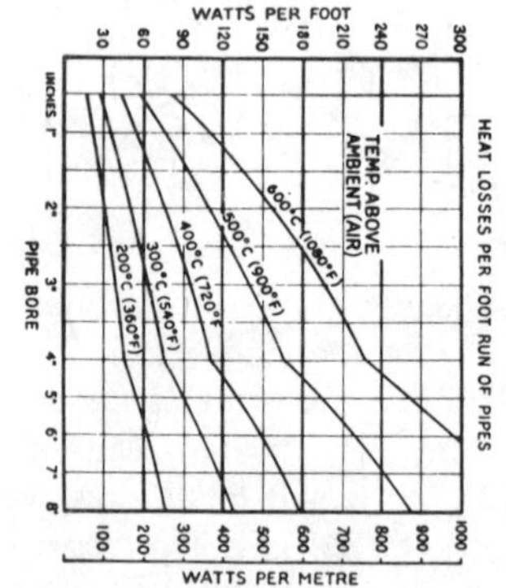

Fig 5 Pipe head lossses with 2in insulation thickness. (Isopad Ltd).

TABLE I – HEAT LOSS FACTORS FOR OTHER THICKNESSES OF LAGGING

Lagging Thickness	½ inch (12 mm)	1 inch (25 mm)	1½ inch (37 mm)	2 inch (50 mm)	3 inch (75 mm)	4 inch (100 mm)
Factor for Fig 4	1.60	1.00	0.80	0.70		
Factor for Fig 5				1.00	0.80	0.65

The heating of pipelines requires the replacement of heat lost through the thermal insulation. Using standard insulating materials heat losses can readily be analyzed, enabling an exact performance to be established. Thus Fig 4, which shows the heat losses per foot (and metre) of pipe up to 8 in (200 mm) bore under one inch (25 mm) of thermal insulation, is also a guide as to the wattage required per foot run of pipe to maintain the correct temperature.

Fig 5 shows heat losses for a typical higher temperature application with two inch (50 mm) thicknesses of thermal insulation. Correction factors for other thicknesses of insulation are given in Table I.

An extension of pipeline heating is provided by electrically heated flexible hose. This has particular application where viscous fluids have to be handled through a flexible line and where there may be a serious reduction of flow or even blockage due to cooling in low temperature conditions.

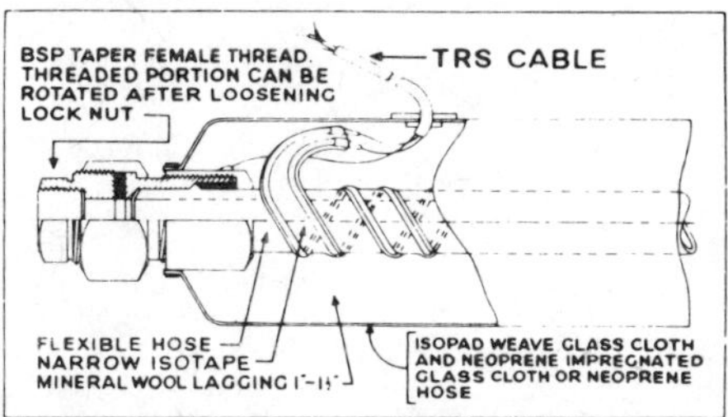

Fig 6 Heated flexible hose.

A typical hose unit is shown in Fig 6 comprising an Isotape or heating element spiralled around the hose to give uniform heating. Thermal insulation is fitted over this assembly, finished in neoprene covered glass cloth or a thin flexible rubber outer covering. Heated flexible hoses of this type are made in lengths up to 30 feet (9 metres) and bore sizes up to 4 in (100 mm). Alternatively, standard flexible tubes can be turned into heated hoses by fitting a suitable heating element and thermal insulation. In such cases it is important to ensure that the construction materials of the basic hose are suitable for withstanding the temperature generated by the heating element, which is normally of the order of 30% higher than the temperature at which it is desired to maintain the fluid. The addition of thermal insulation, necessary with a heated hose, also reduces the flexibility compared with standard hose.

1

2

Dezurik jacketed valves.
1. Partially jacketed. 2. Fully jacketed.

SECTION 6

Pipelines

Pipe Joints and Couplings

THE TYPE of pipe coupling or method of jointing used to give satisfactory service depends on a variety of factors, *eg*

(i) Pipe diameter

(ii) Pipe material

(iii) Pressure rating and other service requirements

(iv) Facilities available

In many cases specific types of points or couplings may be specified for particular classes or types of pipes. In other cases the choice may be open and any one of a variety of methods or types of coupling may be employed according to circumstance or preference.

The type of coupling or joint suitable is often dictated by the pressure rating of the system, compression or expansion fittings normally being employed for high pressure work. The pipe material is concerned mainly with compatibility – *eg* the avoidance of electrolytic corrosion under adverse conditions and the suitability of the material for the type of coupling chosen. Some tube and pipe materials, for example, cannot be flared, thus eliminating this type of coupling as a choice; thin walled pipes and tubes are not suitable for threading; and so on.

Pipe diameter affects the choice only in respect of the range of standard and proprietary couplings and fittings available. Couplings for general use are normally produced in a range of nominal overall diameter sizes and nominal bore sizes, the latter consistent with piping having outside diameter to standard specifications.

In practice there is a broad distinction between couplings and joints used on larger size pipes and general service pipelines, and those used on small bore pipework systems. The former embrace the classic methods of jointing pipes – see Table I – and proprietary joints which may be suitable for a range of pipe materials or specifically designed to suit particular types of pipes. Some examples of proprietary pipe joints are given in Table II.

With small bore pipes and tubes the range of proprietary fittings available is more numerous, but most follow similar principles to obtain a leak-tight, pressure-resistant rigid joint.

High pressure couplings, in general, are limited in maximum diameter size and thus large diameter pipes (or non-standard diameters) would generally best be made up in different fashion. For example, facilities available largely govern the practicability of applying welded or bonded jointing techniques to large diameter pipeline installations fabricated on site.

TABLE I – CLASSICAL PIPE COUPLING METHODS COMPARED

Method	Advantages	Limitations	Remarks
Threaded joints	fast assembly practical for small diameter good pressure and temperature resistance minimum skills required to install no welding slag in system	needs thick pipe no direct connection of valves and fittings	Weakens pipe at joint. Low installation cost with no special skills required.
Flanged joints	good for installing and dismantling valevs and fittings accepts high pressures and temperatures reusable	difficult to align needs space for bolt installation expensive	
Welded joints	any diameter lighter leaktight	not possible in hazardous system cannot be dismantled needs flanges for valves expensive no prefabrication allowances	Requires special skill for welding operation

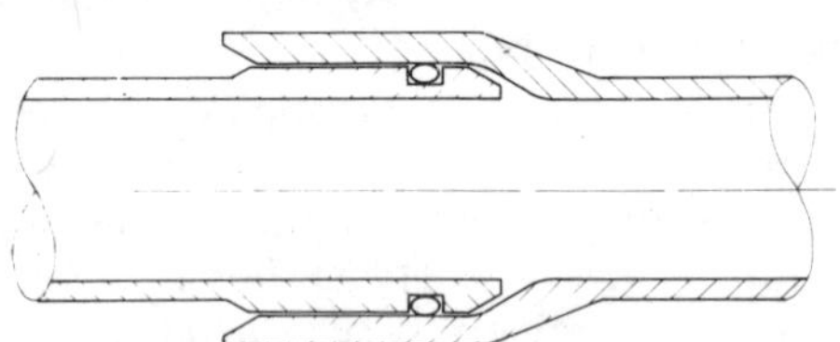

Spigot-and-socket joint with watertight packing.

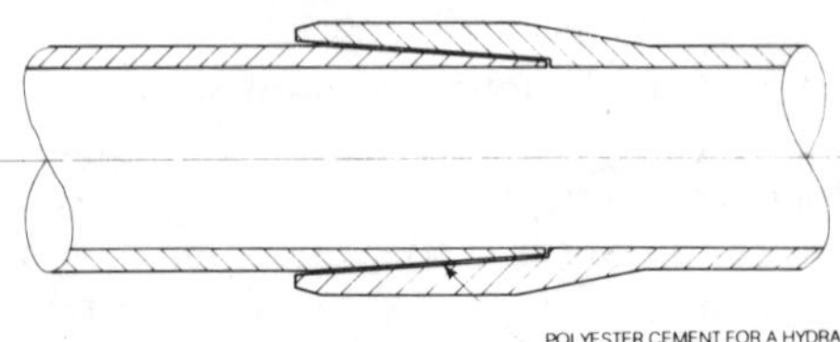

Spigot-and-socket joint with glueing

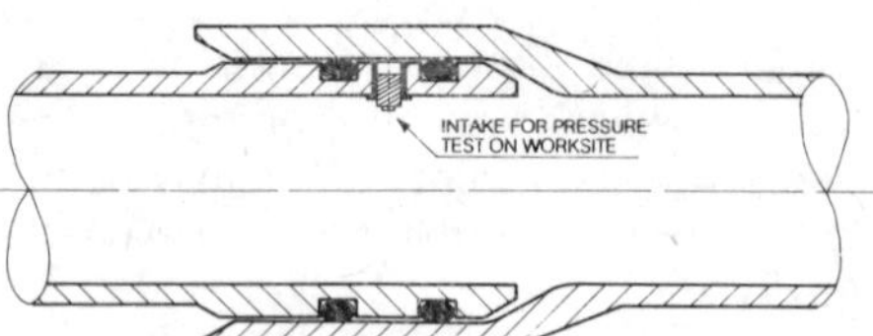

Spigot-and-socket joint with double watertight packings.

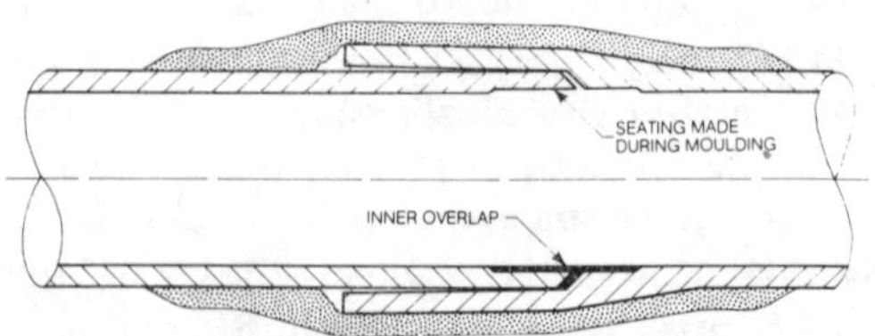

Spigot-and-socket joint with glueing and overlap.

Examples of spigot and socket joints for GRP pipes. (Vetroresina)

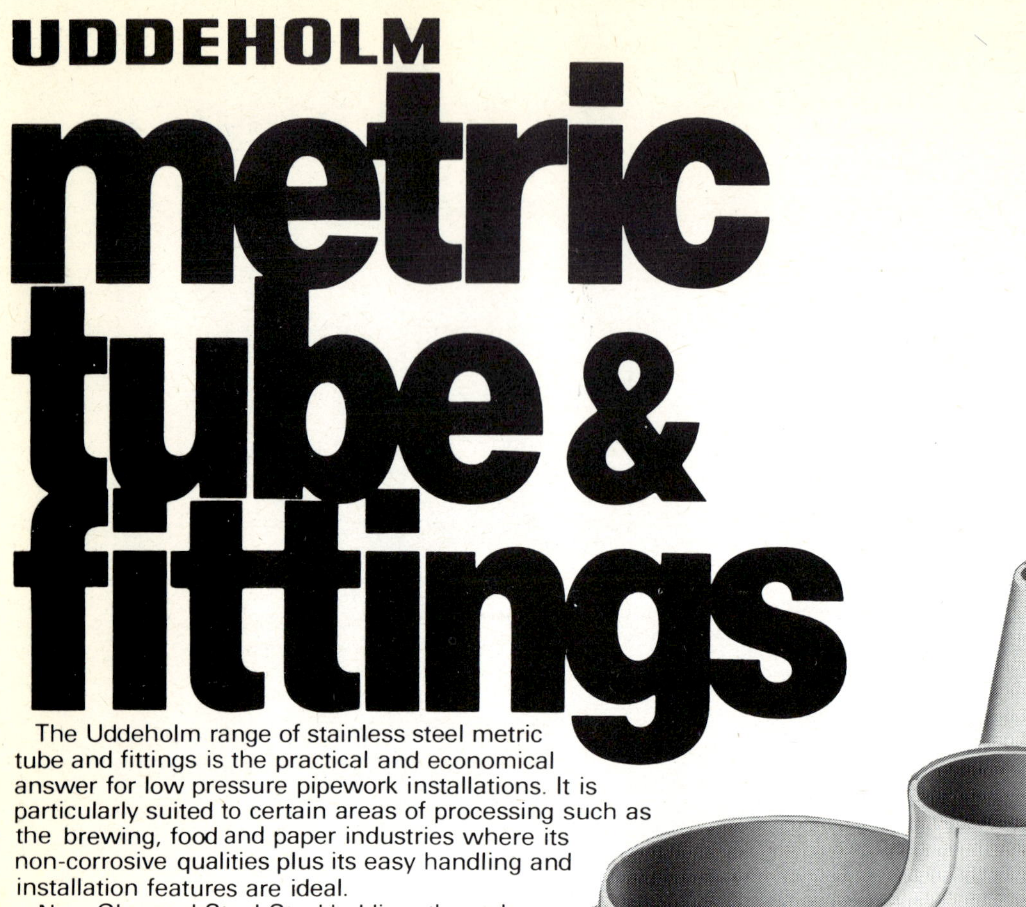
UDDEHOLM
metric
tube &
fittings

Flanged Couplings

Flanged couplings are widely employed on large diameter pipes, the flange being screwed or welded to the prepared pipe ends. In some cases the flange is fabricated integral with the length of pipe. Various methods of flange mating are employed. These include:-

(i) One flange carries a male boss which fits into a recess in the matching female flange. Soft copper or aluminium or similar gasket material may be clamped between the mating surfaces.

(ii) Both flanges may be recessed or grooved and identical, sealing with a gasket or compound.

(iii) Both faces may be flat, either with ground faces for a pressure seal or nominally flat surfaces, relying on the use of a gasket or sealing compound to seal.

(iv) One flange is recessed for an O-ring, the other flange being plain (Fig 1). On clamping up the O-ring provides the pressure seal and does not require such high mechanical compression as a flat gasket.

(v) Both flanges are grooved to accommodate an O-ring. Sealing principle is identical to (iv) but with the advantage that both flanges can be identical.

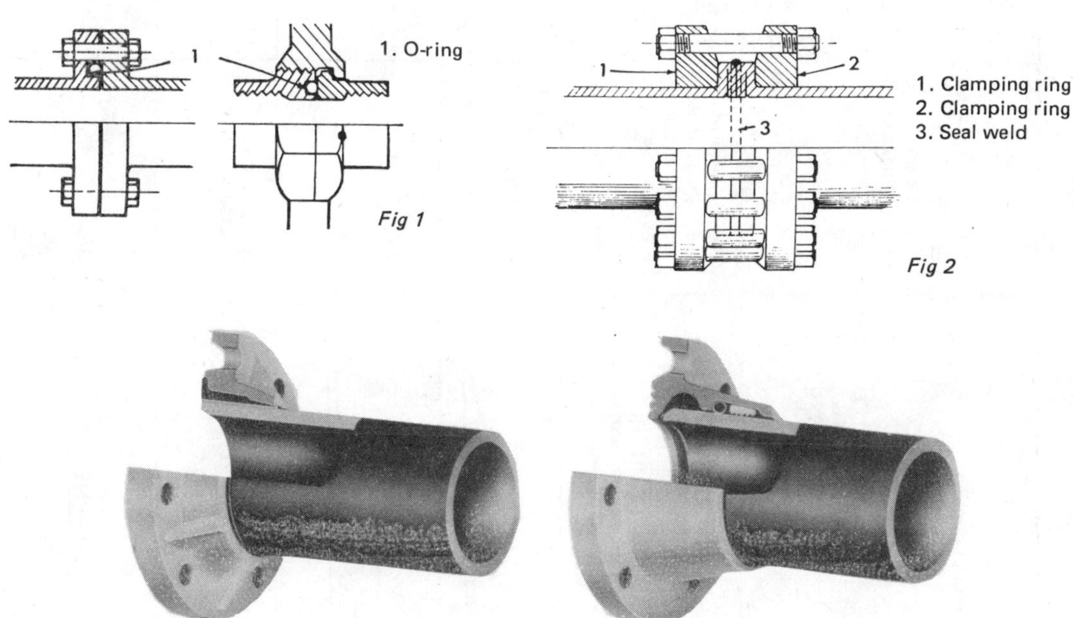

Fig 1

Fig 2

Hawle flange joints for plastic pipes.

Flange joints may be completed (i) by bolting the flanges together with through-bolts; (ii) by clamping up with external compression rings tightened by bolts; (iii) by welding the flanges; (iv) by use of an external nut engaging on a threaded flange periphery (this method being largely restricted to O-ring unions or similar couplings employing a resilient sealing ring. A method of using clamping rings in conjunction with a seal weld is shown in Fig 2 which enables the joint to be broken and re-made, if necessary, without detriment to its efficiency.)

TABLE II – SOME PROPRIETARY PIPE JOINTS

Joint	Type	Special Features	Suitable Pipes
Viking Johnson Coupling	Bolted up mechanical with wedge shaped outer sealing rings	Can accommodate expansion and contraction, limited joint movement and long radius Stepped type can accommodate different pipe diameters	Most rigid pipes including:- Steel Ductile iron UPVC Asbestos – cement Stainless steel Alluminium alloy
Aqualock	Development of the Viking Johnson coupling for polyethylene pipes		Exclusively for water industry
Tyton	Push-fit assembly with double seal	Flexible joint	For pipe sizes up to 600 mm (up to 24 in)
Stantyte	Development of Tyton joint	Accommodates limited ovality in larger pipes; protects joint against high backfill loads	Pipe sizes 700–1 200 mm (27–48 in)
Stanlock	Rolled up version of Stantyte joint		Asbestos-cement pipes 80–1 200 mm (3–48 in)
Turnall Anchor	Push-fit sleeve with rubber sealing rings	Simple assembly; provides flexible joint	Asbestos-cement 450–750 mm (18–30 in)

Tacseal	Push-fit sleeve with sealing rings	Simple assembly; provides flexible joint	Asbestos-cement pipes 100–375 mm (4–15 in)
Hepseal	Polyester moulding with O-ring seal between socket and spigot	Also in form of integral glass filament wound socket and polyester O-ring spigot	Clay pipes 300–800 mm (12–32 in)
Widnes	Machined sleeve grooved to accommodate rubber sealing rings	Provides angular deflection and 'chain'	Asbestos-cement pipes 75–750 mm (3–30 in)
Gibault	Rolled up mechanical joint comprising collar and two flanges with rubber sealing rings		Asbestos-cement pipes 75–250 mm (3–10 in)
Stanton-Cornelius	Flexible push-fit roll-ring joint with 'O' ring seal	Can accommodate slight movement in buried pipes	Cement and asbestos cement pipes
PJE	Clamped-up groove location collar with elastomeric seal	Example of a pipe coupling requiring forming of pipe ends	Thin walled metallic pipes 37.5–1000 mm (1½ – 24 in)

Screwed Connections

Screwed connections may be used on low pressure systems on thick-walled metallic tubes (both ferrous and non-ferrous), and also on thick-walled plastic tubing. Screwed BSP taper threads or American Standard (API or Briggs) are widely used for steam, air, gas, water and oil services.

Taper threads are invariably best for screwed connections, the best seal being given by having a slightly greater taper on the male than on the female part of the joint. BS21:1957 (amended July 1959) specifies BSP standard pipe threads from 1/8 inch to 6 inch nominal bore, inclusive of jointing threads, and longscrew threads. Jointing threads relate to pipe threads for joints made pressure tight by the mating of the threads for assembly with either taper or parallel internal (female) threads. Parallel external (male) pipe threads are not considered suitable as jointing threads. Longscrew threads relate to parallel external (male) pipe threads used for connectors where a pressure-tight joint is achieved by compression of a soft material onto the surface of the external thread by tightening a backnut against a socket.

The effective length of thread engagement with wrench-tightened BS pipe thread is given in Table III for the full range of diameter sizes. This figure gives the effective length of threaded end accommodated within the mating fitting or coupling when assembled – *ie* the actual pipe length required is the distance between faces of the end fittings (assuming a threaded fitting at each end), plus twice the threaded engagement length. Other screw thread data are given in Tables IVA, IVB, IVC, IVD and IVE.

TABLE III – LENGTH OF ENGAGEMENT BSP TAPER PIPE THREADS

Size (inches)	1/8	1/4	3/8	1/2	3/4	1	1.1/4	1.1/2
Engagement (inches)	1/4	3/8	3/8	1/2	9/16	5/8	3/4	3/4
Size (inches)	2	2.1/2	3	3.1/2	4	5	6	
Engagement (inches)	15/16	1	1.1/8	1.3/16	1.3/8	1.9/16	1.9/16	

TABLE IVA – AMERICAN 'DRYSEAL' TAPER PIPE THREAD (NPTF)

Nominal Size in	Threads per inch	Male o.d. at gauging point in	Gauging distance from small end in	Length of full thread in
1/8	27	0,4032	0.1615	0.2639
¼	18	0,5360	0,2278	0,4018
3/8	18	0,6714	0,2400	0,4078
½	14	0,8355	0.3200	0.5337
¾	14	1,0460	0,3390	0.5457
1	11½	1,3082	0,4000	0,6828
1¼	11½	1.6530	0.4200	0,7068
1½	11½	1,8919	0,4200	0.7235
2	11½	2,3659	0,4360	0,7565
2½	8	2.8622	0.6820	1,1375
3	8	3,4885	0,7760	1,2000
3½	8	3,9888	0,8200	1,2500
4	8	4,4871	0,8400	1,3000
5	8	5,5493	0,9400	1,4060
6	8	6,6060	0,9600	1.5130

TABLE IVB – BSP – PARALLEL

Nominal Size in	Threads per inch	DIAMETERS (INCHES)		
		Major	Effective	Minor
1/8	28	0,3830	0,3601	0.3372
¼	19	0,5180	0,4843	0,4506
3/8	19	0.6560	0,6223	0,5886
½	14	0.8250	0,7793	0,7336
5/8	14	0.9020	0.8563	0,8106
3/4	14	1,0410	0,9953	0,9496
7/8	14	1,1890	1,1433	1,0976
1	11	1,3090	1,2508	1,1926
1¼	11	1,6500	1,5918	1,5336
1½	11	1.8820	1,8238	1,7656
1¾	11	2,1160	2,0578	1,9996
2	11	2.3470	2.2888	2,2306
2¼	11	2.5870	2.5288	2.4706
2½	11	2,9600	2.9018	2,8436
2¾	11	3,2100	3.1518	3,0936
3	11	3,4600	3,4018	3,3436
3½	11	3,9500	3,8918	3,8336
4	11	4,4500	4,3918	4,3336
5	11	5,4500	5,3918	5,3336
6	11	6,4500	6,3918	6,3336

Sound pressure-tight joints are possible with properly matched threaded connections which give metal-to-metal contact over the whole length of thread in engagement. Packing, thread sealing compound or a PTFE tape wrapping may be used to provide leak-tightness.

TABLE IVC – BSP – TAPER

Nominal Size in	Threads per inch	DIAMETER (INCHES)		
		Major	Effective	Minor
1/8	28	0.3830	0.3601	0.3372
¼	19	0.5180	0.4843	0.4506
3/8	19	0.6560	0.6223	0.5886
½	14	0.8250	0.7793	0.7336
¾	14	1.0410	0.9953	0.9496
1	11	1.3090	1.2508	1.1926
1¼	11	1.6500	1.5918	1.5336
1½	11	1.8820	1.5238	1.7656
2	11	2.3470	2.2888	2.2306
2½	11	2.9600	2.9018	2.8436
3	11	3.4600	3.4018	3.3436
3½	11	3.9500	3.8918	3.8336
4	11	4.4500	4.3918	4.3336
5	11	5.4500	5.3918	5.3336
6	11	6.4500	6.3918	6.3336
7	10	7.4500	7.3860	7.3220

Note: For tubes 7 inch and upwards the ends must be specially sized before screwing to ensure ample thickness below the root of the thread.

Screw Thread Data

These tables summarize the geometric characteristics of the various types of screw threads used on hydraulic couplings and fittings of British, European and American origin or specification. It should be noted that in the case of bore size tubes, outside diameters are normally standardized in accordance with BS980.

For metric working CETOP bore sizes are recommended for use throughout Europe, when outside diameters are in accordance with CETOP RP 31H.

TABLE IVD – UNF THREADS (SAE AND JIC COUPLINGS)

Use	Nominal Size in	DIAMETER (MEAN)				Threads per inch
		MAJOR		MINOR		
		mm	in	mm	in	
SAE Couplings / JIC Couplings	7/16	10.97	0.4321	9.42	0.3710	20
	½	12.57	0.4946	11.00	0.4334	20
	9/16	14.14	0.5567	12.41	0.4886	18
	¾	18.89	0.7438	16.98	0.6670	16
	7/8	22.05	0.8682	19.82	0.7805	14
	1.1/16	26.68	1.0551	24.20	0.9527	12
	1.5/16	33.15	1.3051	30.52	1.2027	12
	1.5/8	41.05	1.6175	38.48	1.5150	12
	1.7/8	47.42	1.8675	44.82	1.7650	12
	2	50.60	1.9925	48.00	1.8899	12

Screw Threads

British Standard Pipe Threads are exactly the same as the R-thread (Germany, Sweden, *etc*), and the G-thread (Switzerland). Sizes are consistent with bore size tubing having outside diameters standardized in accordance with BS980. Widely used American threads are the NPT and NPTF ('Dryseal'), together with UNF threads for SAE and JIC couplings. All of these thread sizes are

TABLE IVE – METRIC (ISO) THREADS

Size (Major Diameter) mm	Pitch mm	Effective Diameter mm	Minor Diameter mm	Depth of Thread mm
6	1	5.350	4.700	0.650
7	1	6.350	5.700	0.650
8	1.25	7.188	6.376	0.812
9	1.25	8.188	7.376	0.812
10	1.5	9.026	8.052	0.974
11	1.5	10.026	9.052	0.974
12	1.75	10.863	9.726	1.137
14	2	12.701	11.402	1.299
16	2	14.701	13.402	1.299
18	2.5	16.376	14.752	1.624
20	2.5	18.376	16.752	1.624
22	2.5	20.376	18.752	1.624
24	3	22.051	20.102	1.949
27	3	25.051	23.102	1.949
30	3.5	27.727	25.454	2.273
33	3.5	30.727	28.454	2.273
36	4	33.402	30.804	2.598
39	4	36.402	33.804	2.598
42	4.5	39.077	36.154	2.923
45	4.5	42.077	39.154	2.923
48	5	44.752	41.504	3.248
52	5	48.752	45.504	3.248
56	5.5	52.428	48.856	3.572
60	5.5	56.428	52.856	3.572
64	6	60.103	56.206	3.897
68	6	64.103	60.206	3.897
72	6	68.103	64.206	3.897
76	6	72.103	68.206	3.897
80	6	76.103	72.200	3.897

*Subsequent sizes in steps of 5 mm

consistent with fractional inch size pipes and tubes. In the case of NPTF threads the thread crests are crushed when drawn wrench-tight and so cannot subsequently be loosened and reused without some loss of sealing efficiency. For metric working CETOP bore sizes are recommended for use throughout Europe.

Bonded Joints

Adhesive bonded joints are a proposition only where the tube or pipe material is suitable for such treatment – *eg* certain types of plastic tube may be jointed by adhesive bonding – but in general the performance achieved will be inferior to that given by alternative methods.

In the case of bonded glass plastic joints pipe ends are formed so that the male end of the pipe can be spigoted into the female end and drawn up tight with clamps. An annular clearance space between the two overlapping sections is then completely filled with resin which is then set, resulting in a permanent, highly satisfactory joint with a strength greater than that of the rest of the pipe. The resin used is basically the same as that used in the fabrication of the pipe (an epoxy resin, selected for the service temperature required), reinforced with a mineral filler (usually mica). Special equipment is needed to make such bonded joints since the joint area must be preheated and hot-setting resin is normally employed, calling for further controlled heating of the joint area to produce a cure. The necessary equipment can, however, be rendered in mobile form for working on site.

Adhesive bonding on metallic pipes offers no advantages and many disadvantages compared with conventional jointing methods.

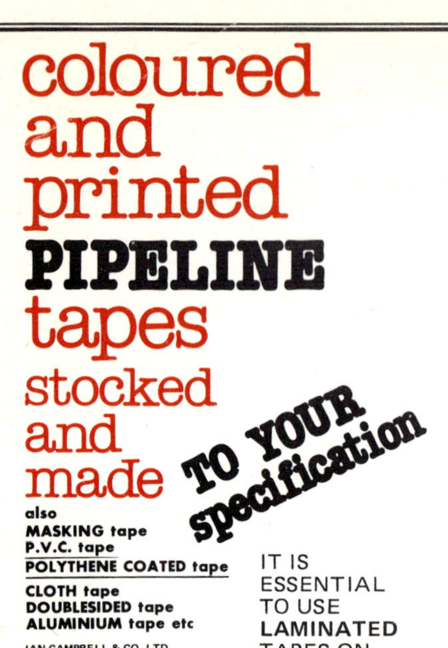
coloured
and
printed
PIPELINE
tapes
stocked
and
made
TO YOUR
specification
also
MASKING tape
P.V.C. tape
POLYTHENE COATED tape
CLOTH tape
DOUBLESIDED tape
ALUMINIUM tape etc
IAN CAMPBELL & CO. LTD.,
Britannia Works,
West Road,
London. E15 3PZ
01–472–6018/9
01–470 1223/6
IT IS
ESSENTIAL
TO USE
LAMINATED
TAPES ON
PIPES TO B.S.
SPECIFICATION
OVER 50 YEARS SERVICE TO INDUSTRY

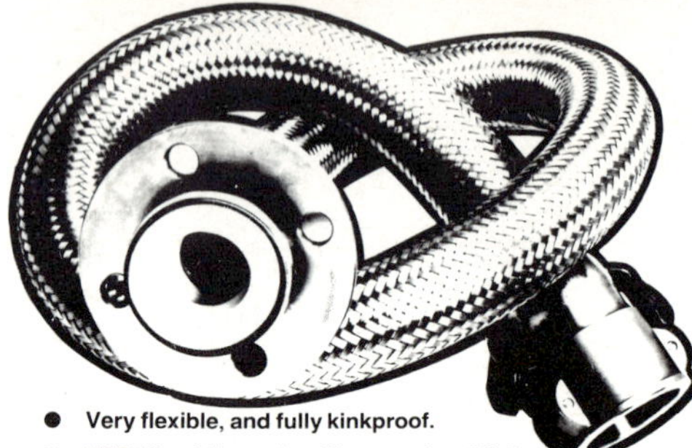

TTP
PRINTERS LTD
(Subsidiary Company of The Trade & Technical Press Limited – Est. 1932)
Graphic Art
Typesetting
Litho Platemaking
Litho Printing
Book-Binding
Technical and scientific reproduction a speciality
13/15 Creek Rd., East Molesey, Surrey. KT8 9BE. Tel., 01 941 2100

Welded Joints

Welding is one of the most completely satisfactory types of joint for high pressure high temperature steam services and similar duties. Large diameter high-duty alloy steel pipes are almost invariably fabricated by butt-welding where practical.

The most suitable methods of fusion welding are manual metal arc (MMA) using coated electrodes and gas-shielded processes such as MIG and TIG. Four types of welds are shown in Table V particularly applicable to the welding of stainless steels although most can also be used for plain steels.

TABLE V – TYPES OF WELD JOINT GROOVES

Type of Groove		Dimensions				
			t mm	α	d mm	k mm
I-groove square edge close butt	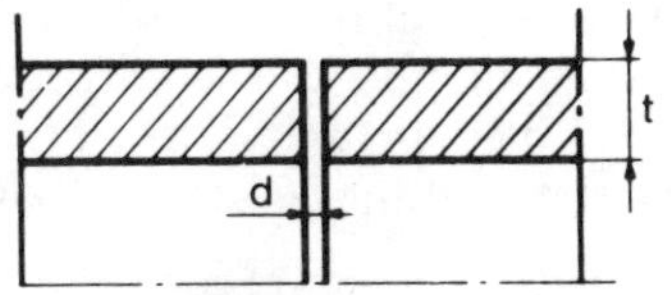	MMA	$\leqslant 3$		1-2	
		TIG man.	$\leqslant 3$		1-2	
		TIG autom.	$\leqslant 4$		0	
		MIG	$\leqslant 3$		1-2	
V-groove	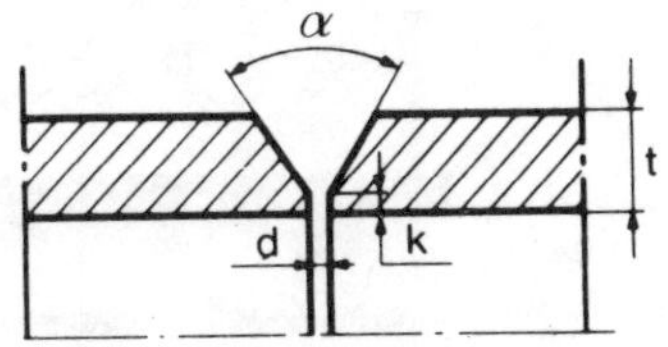	MMA	$3 < t < 15$	$\geqslant 50^\circ$	1-3	0-2
		TIG man.	$2.5 < t < 8$	$\geqslant 60^\circ$	0-1	0-2
		MIG	$3 < t < 110$	50°	0-2	0-2
U-groove	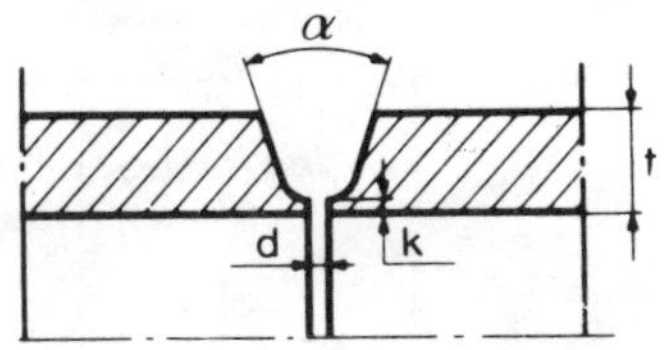	MMA	> 12	30°	1-3	2-3
		TIG	> 6	30°	0-2	1-3
		MIG	> 8	30°	1-2	2-3
Modified U-groove	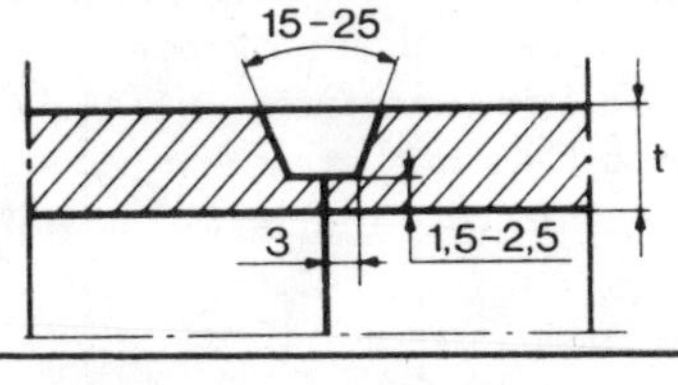	Root pass by automatic TIG welding. Other passes by MMA, TIG or MIG welding. t usually < 8 mm.				

Considerable care is required both in matching the weld metal to the type of pipe steel and also in the welding procedure and post-welding treatment for high duty systems. Radiographic or similar methods of examination for crack detection and flaws are employed whenever practicable and the most modern methods of welding are superseding the older, traditional workshop methods.

Brazing and Soldering

Brazing and soldering are acceptable for low pressure work, particularly for copper and copper alloy pipes. Lead pipes, where used, are also commonly 'plumbed'. Soft soldering is not generally considered satisfactory, whilst brazing is generally to be preferred to silver soldering on the score of strength, although silver soldering is rather easier to apply. The method may appear cheaper and more direct than the use of proprietary couplings, but this is seldom the case except in special circumstances.

High frequency brazing is ideally suited to stainless steels, whilst flame brazing or furnace brazing is also possible. Brazing of lap joints should incorporate an overlap between three and six times the thickness of the parent metal.

Pyrotechnic Jointing

For on-site work a number of patented joints have been produced capable of producing welded joints through the heat generated by a chemical fuel incorporated in the joint make-up. Typically, for example, such a joint may consist, basically, of a close fitting sleeve carrying two rings of welding alloy – Fig 3. This sleeve is surrounded by a chemical fuel and the whole encased by a refractory outer sleeve or cover. The pipe ends are squared off, cleaned and inserted into the joint. The fuel is then ignited via a fuse, the heat generated by the fuel melting the welding alloy rings and raising the pipe metal locally to sufficient heat to complete a weld. The refractory outer sleeve localizes the heat of the chemical fuel and in the process is itself fused into a brittle mass which can be chipped away on cooling, leaving just the metal sleeve of the basic joint welded to the two pipes.

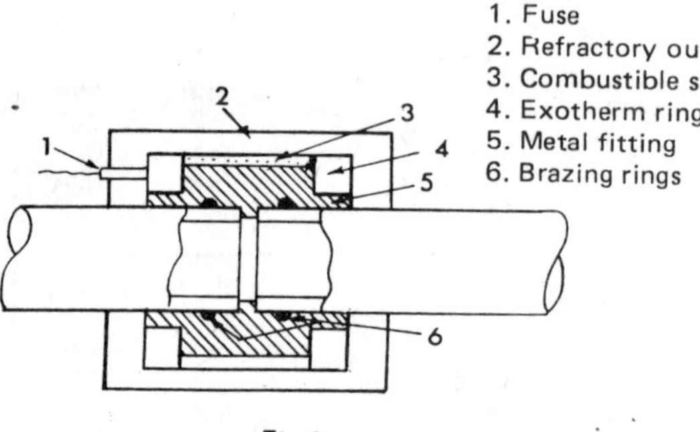

Fig 3

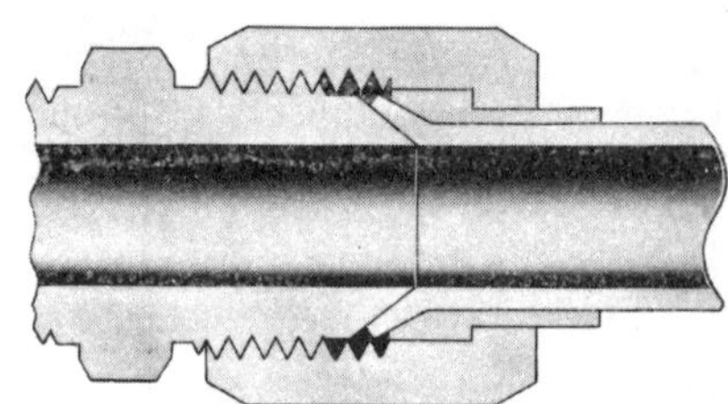

Typical flanged tube coupling. (Parker Triple-Lok)

Joints of this type rely entirely on the 'self-contained' action of the propreitary item and do not require specialized knowledge to complete. The complete process of making the joint is only a matter of a few minutes. Variations on this theme are also used for on-site welding where conventional welding equipment is not available, employing 'Thermit' or similar pyrotechnic heating. In such cases, however, the success of the joint may depend very much on the skill of the individual operator.

Tube and Small Pipe Couplings

Couplings for thin walled tubes and pipes are generally classified as *expansion* or *compression* types; or as *flared* or *flareless* couplings, respectively. There are also variations on these two basic principles for obtaining a pressure-tight joint from a coupling assembly.

The basis of an expansion coupling is that the tube end is cut square and flared, the coupling itself typically embracing a threaded nipple with centre shoulder, a nut or threaded ferrule (for each end), and a sleeve (also for each end). The flared end of the tube conforms to the sleeve and projects evenly just beyond the sleeve and the joint is completed by tightening up the nut or ferrule – Fig 4. Any flare angle between 30 degrees and 90 degrees may be used, although 30 degrees is typical of British practice and 37 degrees of American hydraulic practice. Flared joints are limited in application to the more ductile tubing, *ie* tubing which can be flared cold without cracking or undue difficulty.

The basis of a compression fitting is that a sleeve or olive is clamped between two shoulders, one on the body and one on the ferrule or nut of the coupling and so compressed on tightening the nut that it is forced to grip the pipe or tube wall securely – see Fig 5. It is essential that the end of the tube projects well through the olive and in the case of hard tubing considerable torque may have to be applied to the nut to tighten the joint effectively. Used on more ductile tube materials – *eg* copper and brass – the olive may be replaced by a shaped collet which, on tightening, produces an annular deformation in the tube wall to grip and lock the tube within the coupling, sealing being effected by the fit of the tubes or pipes in the ferrule – Fig 6. An essentially similar form of coupling, normally used with an internal sleeve, is often used for jointing plastic tubing.

Flared Couplings

The use of a split collar (or split collet) in conjunction with a flared coupling provides protection for the pipe and also compensates for variations in flare angle and pipe diameter – see Fig 7. This is the principle used on a variety of individual proprietary and patented couplings.

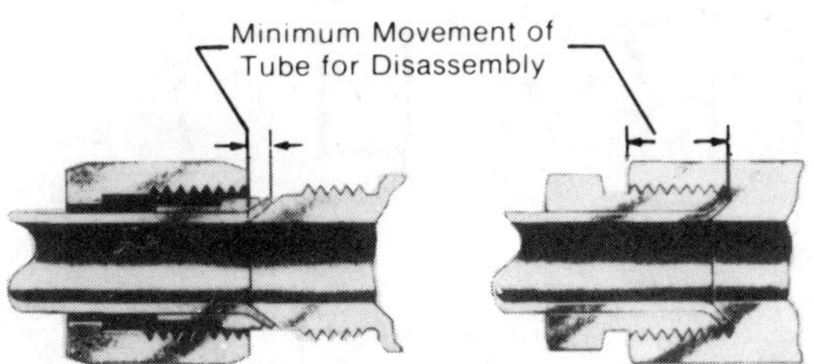

Fig 4 Flared type tube fittings.
(Parker).

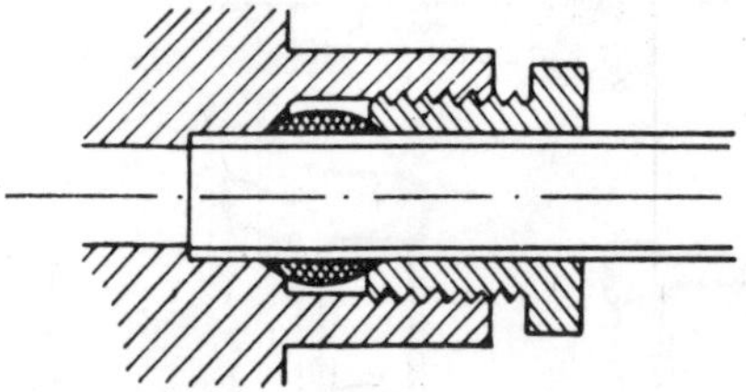

Fig 5 Simple olive compression joint.

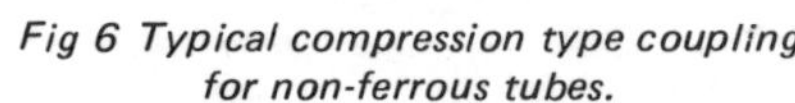

Fig 6 Typical compression type coupling for non-ferrous tubes.

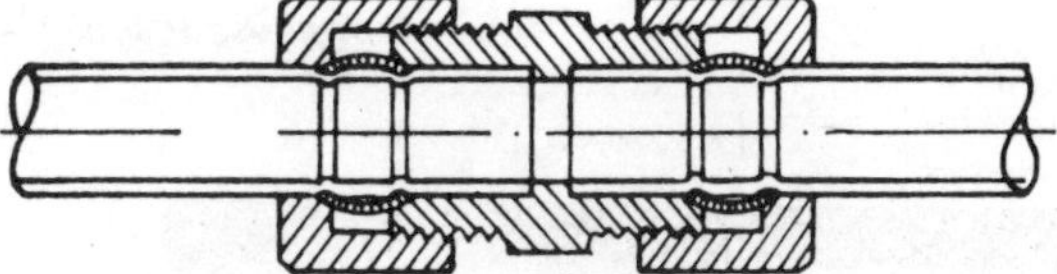

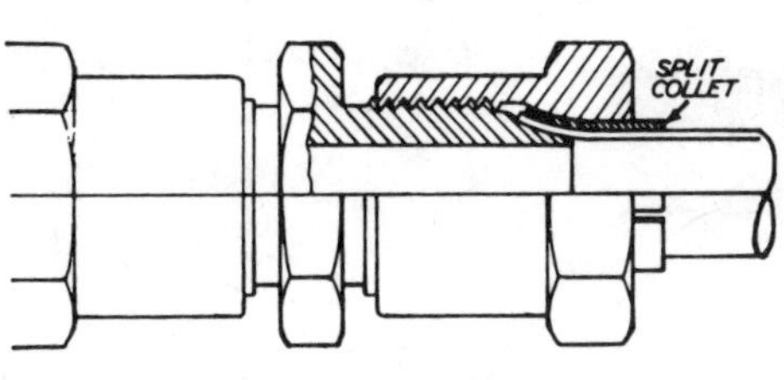

Fig 7

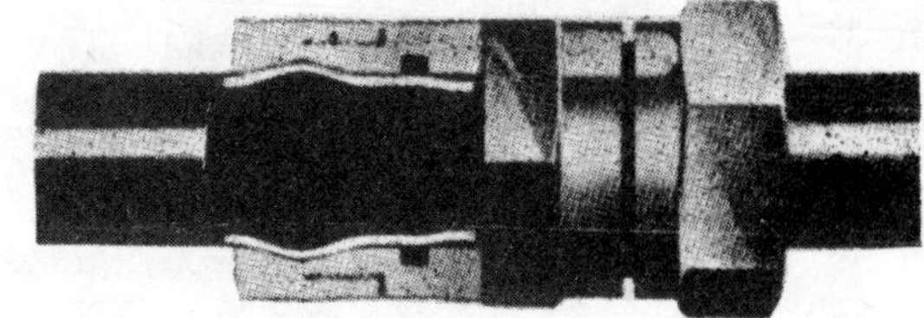

Fig 8 Section view of the Keela Tite coupling.
(Industrial Hydraulics Ltd).

Semi-Flared and Formed-Pipe Couplings

Although basically flareless type couplings, these types involve some shaping of the tube or pipe ends by special equipment. The coupling shown in Fig 8 utilizes a specially shaped pipe end formed to accurate shape by oil pressure, incorporating a short length of original diameter leading to a smooth enlargement followed by an accurate diameter which gives the O-ring the required compression in its groove in the coupling body. The remainder of the formed section introduces a conical bulge on which the coupling nut clamps to retain the tube securely in the coupling. Forming can be done by means of a portable tool with a hand-operated pump and no weakening on the tube takes place during forming.

In another type – the JR flareless coupling, Fig 9 – annular projection is rolled outwards near the end of each pipe, sufficiently far in from the end to prevent splitting during the rolling operation. A pressure-tight seal is provided by the pipe ends pressing against the shoulders of the nipple. Tightening of the nuts applies pressure to the sealing rings through the sleeves whilst the conical portions of the sleeves are also forced into contact with the tapered ends of the nipple.

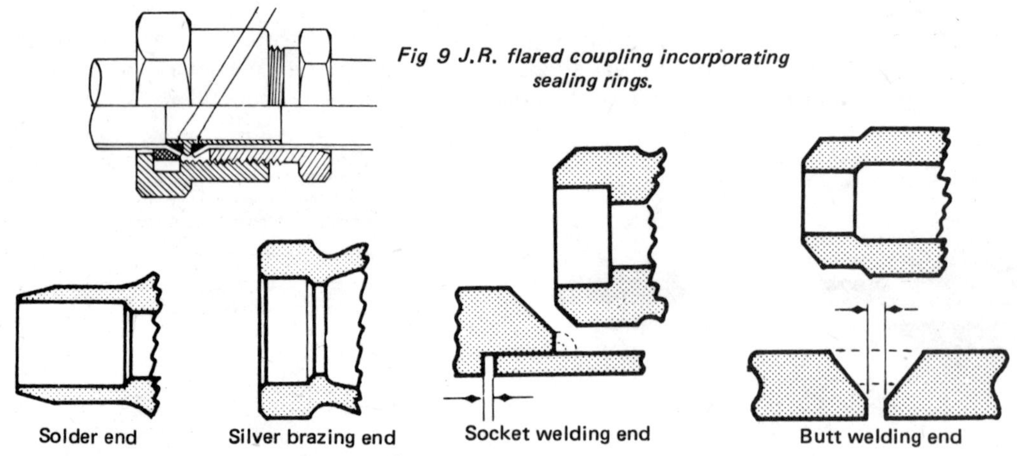

Fig 9 J.R. flared coupling incorporating sealing rings.

Pipe ends for soldering, brazing and welding.

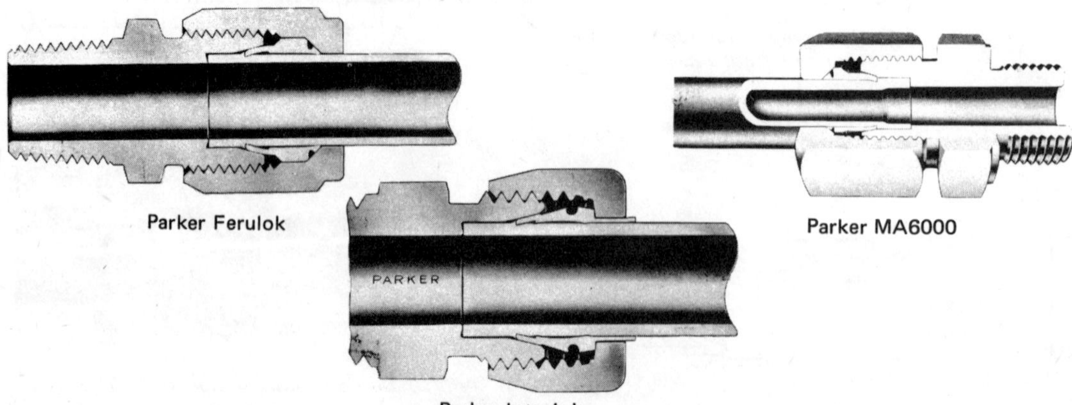

Examples of flareless tube couplings.

Flareless Couplings

A number of patent fittings of this type employ a hard sealing ring in place of an olive which, on tightening the nut, cuts into the tube or pipe wall to form a metal interlock. Although basically a compression joint in principle, the sealing ring may or may not actually be compressed onto the surface of the pipe as in ordinary joints, according to the design of sleeve. Sealing may be accomplished by end pressure, by compression, or a combination of both principles.

The Ermeto coupling – Fig 10 – utilizes a case hardened sealing ring which is initially a loose fit on the pipe or tube. On tightening the nut the leading edge of the ring is forced to move along the pipe and at the same time to cut into it, throwing up a ridge of displaced metal in front. The depth of cut is controlled so there is no danger of weakening the pipe whilst at the same time any grooves or inequalities in the pipe surface are cut through to produce a consistent seal. The sealing ring in its final tightened form is bowed and thus also acts like a spring washer to resist vibration and accidental loosening of the nut, the actual seal depending on the butt-end tightness of the completed joint against a shoulder in the coupling or fitting.

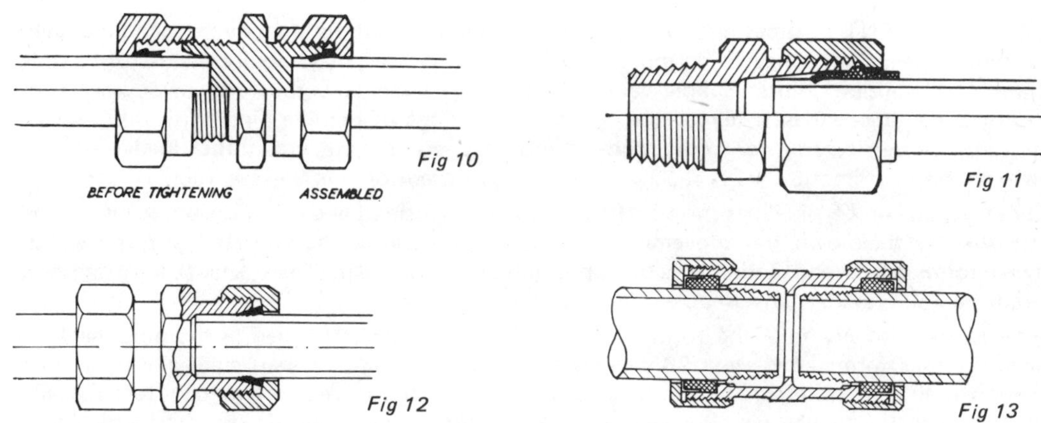

Fig 10

Fig 11

Fig 12

Fig 13

A similar type of coupling – Fig 11 – utlizes the axial pressure produced by a sleeve to form a primary seal, but the tube is forced into a gradually tapered section in the connector or coupling body, thus providing an additional sealing action. The outer end of the sleeve in this case is slotted to form spring fingers to grip the tube when the nut is tightened and provide support against vibration.

A further variant – the 'Supakone' coupling – Fig 12 – utilizes a hardened wedge as the sealing ring. This conical wedge is so shaped that when the nut is tightened the leading edge of the cone is first forced down by the taper on the bore of the annular seating, subsequently cutting into the tube surface raising a burr which eventually fills the space between the cone and the body when the joint is complete.

A low pressure type compression coupling where elastomeric rings are employed to provide the seal is shown in Fig 13. Here the elastomeric joint rings are compressed by the gland nuts screwed on to the sleeve of the coupling and the coupling can be used on pipes with either plain or screwed ends. Since there is no mechanical connection between the pipes such a coupling can accommodate a certain amount of flexing but where pressures are more than about 0.7 bar (10 lb/in^2) it may be necessary to secure the pipes against end movement or separation.

Expansion/Contraction Joints

AXIAL MOVEMENT of pipes due to thermal expansion and contraction may be accommodated by suitable pipeline geometry or, in more severe or specific cases, may require the use of flexible sections of expansion joints. Simple geometric treatment can take the form of straight runs interrupted by looped bends (expansion loops), or sections of flexible pipe. With more severe conditions, or particularly for protection of mounted equipment, *etc*, thrust loads resulting from expansion/contraction are best accommodated by *expansion joints* — see Table I.

Quite a number of pipelines have sufficient flexibility using bends to absorb expansion and contraction by their own free movement; or extra bends can be incorporated to provide this. Suitable forms are L and Z bends with approximately equal legs. These shapes have the best inherent flexibility. Alternatively *pipe loops* may be incorporated.

Pipe loops normally take the form of cold or hot bent loops formed in the pipe itself, or retangular loops formed by using 90 degreee angle bends — Fig 1. A particular advantage of a pipe loop is that it is pressure-balanced. Disadvantages are the increased frictional loss (usually negligible); size (particularly in the case of large diameter pipes); and dead weight which may have to be supported. Pipe loops, too, can be difficult to accommodate in a crowded installation, or in buried pipelines.

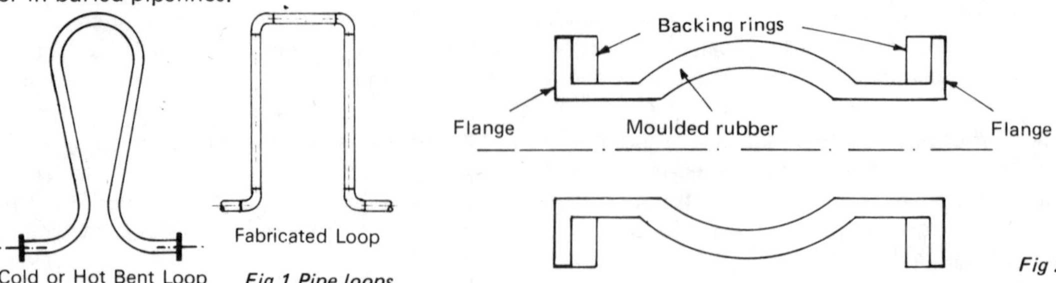

Fig 1 Pipe loops.

Fig 2

The simplest type of expansion joint is of rubber, moulded either from homogeneous rubber, or, more usually, of laminated construction with a high tensile textile reinforcement. The characteristic shape is a flexible arch with integral flanges at each end — Fig 2. These flanges are often backed by steel plates for bolting up, which may or may not be bonded in place.

TABLE I – BASIC TYPES OF EXPANSION JOINTS

Type	Movement(s) Accommodated	Durability	Shock Resistance	Corrosion Resistance	Typical Applications and/or Remarks
Rubber	Axial and lateral	Very good	Fair	As good as elastomer used	Condenser pipelines marine applications
Reinforced rubber	Axial and lateral	Excellent	Excellent		
Plastic bellows	Axial and lateral	Excellent	Very good	Excellent (PTFE)	Limited pressure rating. Widely used in contact with chemical and corrosive fluids
Metal bellows: (i) simple	Axial *	Fair to good	Fair	Depends on metal	Performance improved by use of multi-ply construction
(ii) twin	Axial and lateral †		Fair to good	Depends on metal	Greater movement. Lateral movement usually depends on length of connecting tube
(iii) tied			Very good	Depends on metal	Tie bars take pressure loads
(iv) hinged			Very good	Depends on metal	Articulated single bellows, usually in sets of two or three
(v) gimballed			Very good	Depends on metal	Gimbal mounting of bellows permits angular rotation in any plane
Compound	Axial and lateral °	Good	Very good	Depends on construction	Various designs. More expensive
Slip	Axial	Very good	Very good	Very good	Requires gland packing or seals. Larger movement than bellows. Pressures up to 500 lb/in^2 (35 bar) or more.

* May have lateral compensator for limited lateral movement † Limited ° Depending on design

The choice of elastomer depends on the fluid to be handled. If necessary, the lining can be of one rubber type, *eg* to resist oils, and the outer casing of a tough, abrasion resistant rubber to resist scuffing, or selected to be compatible with the external ambience. Corrosion resistance can thus be virtually complete. Rubber (and particularly reinforced rubber) also has excellent resistance to shock loading and erosion and exhibits general freedom from fatigue. The main limitation is the maximum service temperature for the rubber used, which is normally of the order of 175–195°F (80–90°C). Pressure rating is also somewhat limited, being of the order of 150 lb/in^2 (10.5 bar), decreasing with increasing bore sizes. High pressure expansion joints of this type can, however, be provided by special constructions.

The resilient nature of the material means that rubber expansion joints can accommodate both axial and lateral movement, although the latter tends to become more limited as the degree of reinforcement is increased and the joint becomes stiffer.

Plastic Bellows

Plastics are an alternative to rubbers for simple expansion joints, although in this case the form of the joint is normally a convoluted bellows. PTFE is the material commonly chosen because of its excellent chemical resistance. Pressure rating is low but can be increased by stiffening the bellows and also adding reinforcement, *eg* a spiral wire. The temperature range is rather better than with rubber expansion joints, for PTFE formulations can be worked up to about 390°F (200°C).

Plastic bellows expansion joints are principally used in pipelines carrying corrosive chemicals, solvents and other aggressive fluids at moderately low pressures, and in pipeline sizes from about ½ in (12.5 mm) up to 8 in (200 mm). Rubber bellows are also used for working temperatures up to 175–195°F (80–90°C).

Metallic Bellows

Metallic bellows immediately offer higher service temperature ratings, which can be extended beyond 1110°F (600°C) if necessary, by the use of heat-resistant alloys. They are also more suitable for lower temperature ratings since metals will remain flexible well below the brittle point of elastomers and plastomers. Aluminium bellows, for example, are particularly suitable for cryogenic services. Pressure rating can also be increased by a suitable choice of metal and wall thickness (at the expense of reduced flexibility). Material choice is also affected by compatibility requirements.

A typical installation is shown in Fig 3.

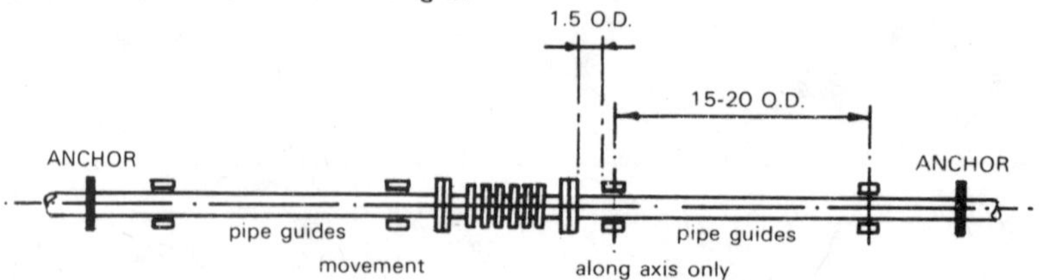

Fig 3 Axial bellows expansion joints.

The inherent disadvantages of metal bellows are fatigue and possible stress-corrosion cracking. Both limitations can be relieved to some extent by using an optimum form of convolution producing the most uniform stress distribution as the bellows are subjected to stress cycling.

In general, it is held that a rounded convolution is best in this respect. A more open V-form convolution may be adopted in specific cases, however; this form is easier to clean and also easier to inspect. The depth of convolution also affects the stress distribution, shallower convolutions reducing metal stretching to a minimum when the bellows are originally formed from plain or preform tube – Fig 4.

A further possibility is the use of multi-ply construction for the tube. This is claimed to give a substantial increase in fatigue life over solid wall tube of the same thickness and form. A further advantage offered is that the inner and outer metal layers can be of a suitably resistant metal with the intermediate layers of steel to provide high strength. Flexibility is, however, reduced with multi-ply construction and, of course, cost is increased.

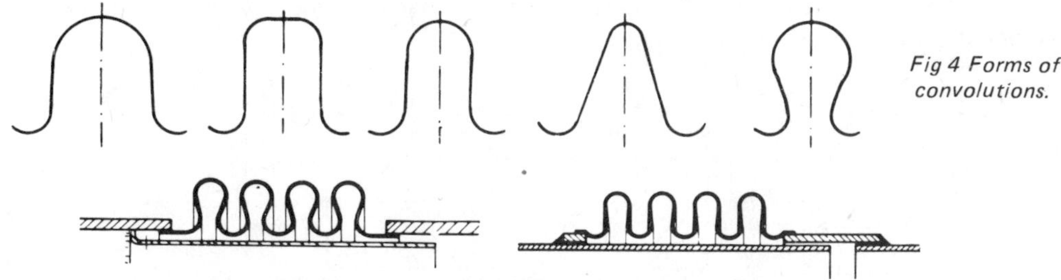

Fig 4 Forms of convolutions.

Fig 5 Internal sleeves for bellows joints. Light/medium duty (left) and heavy duty (right).

Sleeves

Convoluted bellows offer a poor flow path for fluids, tending to generate high turbulence. Not only is this undesirable from the point of view of loss of delivery head (increased flow resistance), but also from the possible scouring action of localized high fluid velocities. Convoluted bellows may, therefore, be fitted with smooth bore internal lining tubes or telescopic sleeves, (Fig 5).

External sleeves may also be fitted to convolute bellows to eliminate dirt traps and protect the bellows against accidental damage or external corrosion. Corrosion is not normally a problem since the usual material chosen for metallic bellows is stabilized stainless steel.

In general, however, sleeves should only be used with bellows where strictly necessary – *eg* internal sleeves where flow velocities are high and external sleeves where external protection is definitely required. Equally, sleeves should not be used to perform the function of pipe girders. Bellows are normally too short to give a reasonable bearing length.

Flexibility

The flexibility of simple metallic bellows is mainly dependent on the convolutions and wall thickness, strength and flexibility posing opposite requirements. Available axial movement must be limited according to the material stresses developed. Small lateral movement may also be tolerated, depending on the individual design.

Twin bellows, connected by a plain tube, can provide increased axial movement (due to the increased bellows length possible without introducing buckling-type instability). The presence of the central tube also permits a reasonable degree of lateral movement to be tolerated without high material stresses developing. As a general rule, the amount of lateral movement is dependent on the length of the central tube.

Plain bellows have the ability to absorb movement in all three planes. They cannot be used, however, where:

(i) Pressure is so high as to impose unacceptable forces on the pipe anchor (or pump).

(ii) No suitable anchor point can be provided near a mounted component (*eg* a pump).

(iii) The pump is flexibly mounted and any thrust forces imposed on the pump could produce dangerous misalignment of the system.

Where plain bellows are not suitable, axially-restrained or *tied* bellows must be used.

For use at higher pressures the bellows may be restrained by tie bars – Fig 6. These bars may be spherically seated or hinged, depending on the type of movement to be accommodated. The aim is for the tie bars to take the pressure loading, allowing the bellows to deform to accommodate lateral or offset movements.

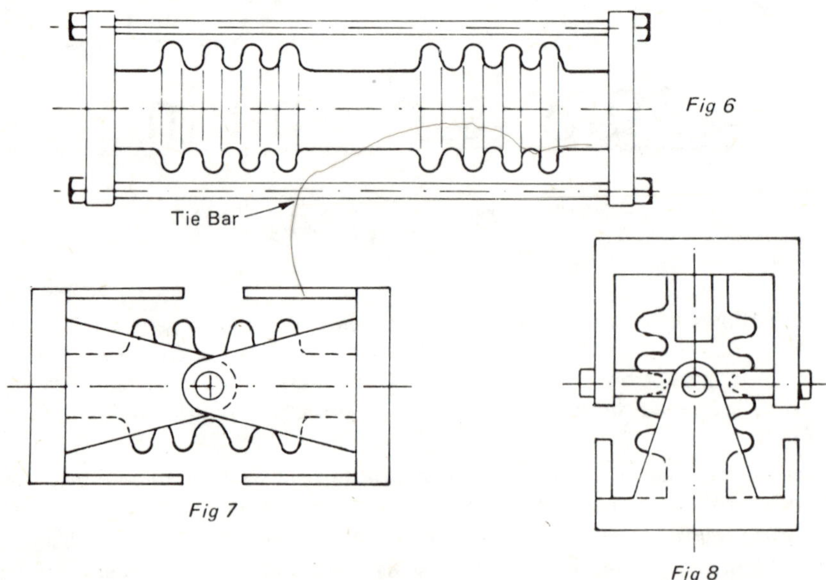

Fig 6

Fig 7

Fig 8

Angular Movement

Where angular movement has to be accommodated a fully articulated assembly may be used. Thus Fig 7 shows a simple hinged expansion joint using hinged plates for the articular movement, which also takes the pressure load, leaving the bellows to provide flexibility for angular rotation in one plane. The only real disadvantage of such a design is that movement is severely restricted to one plane. Since thermal expansion motions are likely to be present in more than one plane, simple hinged expansion joints are usually employed in sets of two or three which may or may not be set in the same plane.

Angular rotation in any plane can be provided by a single gimballed expansion joint. In this case two pairs of hinge plates are used, mounted on a common gimbal ring – Fig 8. There are many possible variations on this theme, such as the use of gimbal cups mounted on a gimbal ring with the bellows in the form of a cuff. The gimbal movement itself provides a smooth bore flexible pipe joint, the bellows in this case working as a seal. Some other standard forms of bellows are shown in Fig 9.

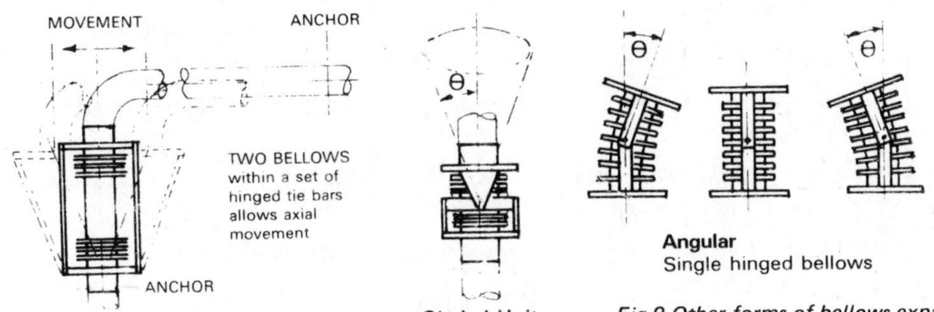

Fig 9 Other forms of bellows expansion joints.

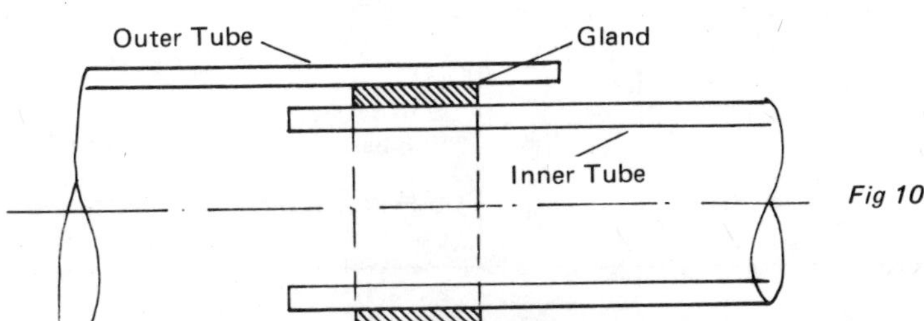

Fig 10

Slip Joints

A slip joint consists of two tubes, one sliding inside the other and with the clearance space sealed by a gland —Fig 10.Such a joint can obviously only accommodate axial movement, but the degree of movement possible is considerably greater than that which can be provided by bellows. The joint itself does not transmit thrust forces, other than those generated by the friction of the gland. Where only axial movements have to be accommodated the only real disadvantage of the slip joint is that the gland needs regular maintenance, *eg* repacking as necessary, and may also need regular lubrication. When the gland has to be repacked it may be necessary to shut down and depressurize the system, although some designs of slip joint allow for the gland to be repacked under normal system pressure.

Typical materials for slip joints are summarized in Table II. Various types of seal may be employed, both non-metallic and metallic. In the latter case, the seal may be in the form of metal piston rings. Another type of seal is the so-called fluid ring where two sets of sealing rings or piston rings are employed and the annular space between them is filled with a high viscosity fluid which acts as the seal and also provides lubrication for movement. Various proprietary designs of slip joint are shown in Fig 11.

Slip joints may need the addition of a condensate trap for specific applications, *eg* for steam lines, but with suitable design and choice of materials they are usually free from condensation and corrosion troubles. In this respect they can be better than bellows joints, which in some applications can suffer from condensation unless internally lined.

Compound Joints

The description 'compound' is sometimes given to expansion joints capable of accommodating both axial and lateral movement, but should, in fact, include all joints of 'compound'

TABLE II – SLIP JOINT MATERIALS

Body	Sleeve	Gland	Bush	Remarks/Performance
Gunmetal	Gunmetal	Packing		Steam up to 120 lb/in^2 (8.5 bar). Water and general services up to 160 lb/in^2 (11.25 bar).
Cast iron	Gunmetal	Packing		Steam up to 55 lb/in^2 (3.8 bar). Water and general services.
Cast iron	Gunmetal	Packing	Gunmetal	Steam up to 150 lb/in^2 (10.5 bar). Water and general services up to 250 lb/in^2 (17.5 bar).
Cast steel	Gunmetal	Packing	Gunmetal	Steam services up to 350 lb/in^2 (25 bar).
Forged steel	Mild steel (plated)	Piston rings		Steam, gas, oil and other services.
Steel (precision bore)	Steel (precision bore)	Piston rings		Services up to 500 lb/in^2 (35 bar).
ABS		Ring seal		Up to 150 lb/in^2 (10.5 bar) } for plastic pipelines carrying acids, alcohols, chemical liquids, etc.
PVC		Ring seal		Up to 100 lb/in^2 (7 bar) } for plastic pipelines carrying acids, alcohols, chemical liquids, etc.

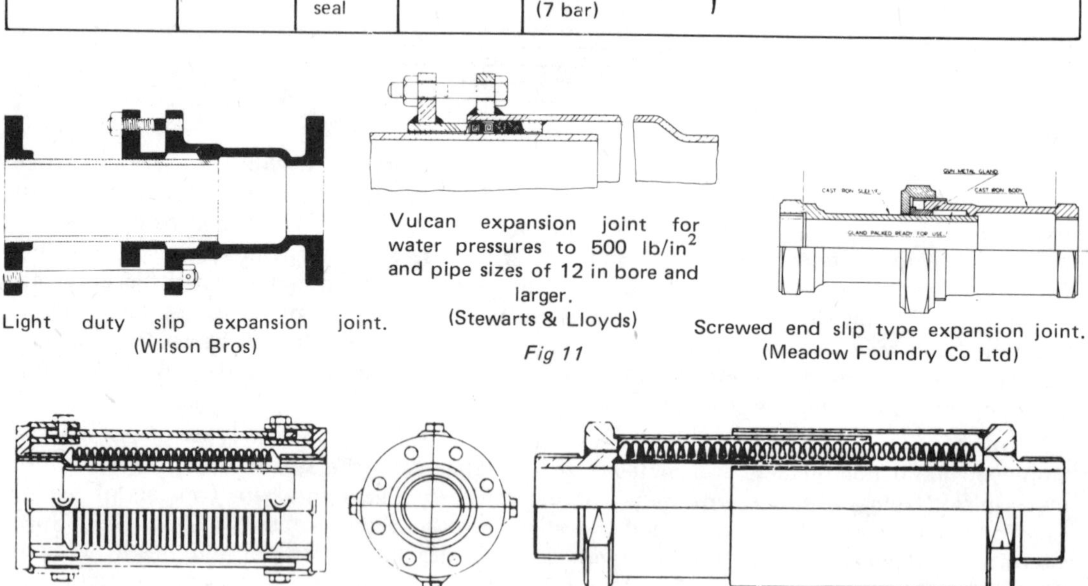

Light duty slip expansion joint. (Wilson Bros)

Vulcan expansion joint for water pressures to 500 lb/in^2 and pipe sizes of 12 in bore and larger. (Stewarts & Lloyds)

Fig 11

Screwed end slip type expansion joint. (Meadow Foundry Co Ltd)

Fig 12

construction. The smooth-bore gimbal joint previously described, for example, is really a compound joint, although designed specifically to accommodate angular movement only. Slip joints are produced in combination with bellows, the latter acting as a seal – Fig 12. Such compound joints are suitable for axial movement only, but may be further modified to accommodate angular movements, *eg* by hinging the ends. Again there is a variety of proprietary joints which can be classified as being of compound construction.

Corrosion and Cathodic Protection

THE CORROSION of metals can be classified under five separate headings:-

(i) *Simple Corrosion* of individual, isolated metal components. It may be anything but simple in its mechanism.

(ii) *Bi-metallic corrosion,* often called galvanic corrosion, the result of two or more dissimilar metals being connected together with an electrical conductor and immersed in an electrolyte – *ie* a liquid like salt water which can pass an electric current. This type is usually much simpler to understand.

(iii) *Crevice corrosion,* or the setting up of a localized corrosion cell because of 'oxygen starvation'. It is also known as differential aeration corrosion and anaerobic corrosion; it applies principally to stainless steels.

(iv) *Stray current corrosion.*

(v) *Bacterial corrosion* – occurring particularly in soils.

To complete the picture there is also stress corrosion (corrosion initiated or accelerated by stress); erosion corrosion (associated with moving water); solution cell corrosion; and various forms of erosion caused by cavitation and fatigue, or combinations of other forms of corrosion.

Simple Corrosion

Single metal fastenings or fittings are subject to simple corrosion – *eg* mild steel immersed in salt water will rust; on the other hand pure metals, *eg* good wrought iron does not. The action is electro-chemical – the salt solution acting as an electrolyte causing the formation of numerous electrolytic cells (like elementary batteries) with the result that parts of the metal (anodic areas) are eaten away (see Fig 1). These anodic areas are tiny and are of course connected to the neighbouring areas which may be cathodic due to differences in composition or degree of protection from oxides, paint or gas bubbles in the vicinity. The progressive effect is an overall roughening and depletion of the surface as well as discolouration (rusting).

Often the effect is limited to the formation of a surface coating inhibiting further attack. This gives certain metals, notably stainless steel, aluminium bronze, and aluminium alloys their corrosion resistant properties.

In a similar manner certain alloys — brass being a major case — can have one of the constituent metals dissolved away by electro-chemical attack, leaving them in a very weak state. Thus continual immersion in water can leach out the zinc content of brass leaving only a weak spongy structure mainly of copper — a process known as dezincification.

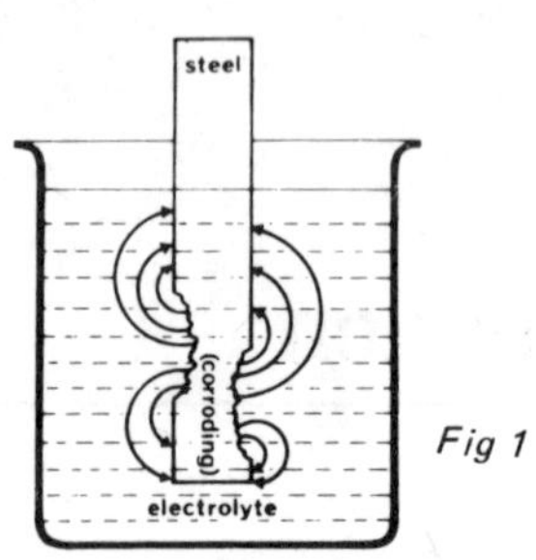

Fig 1

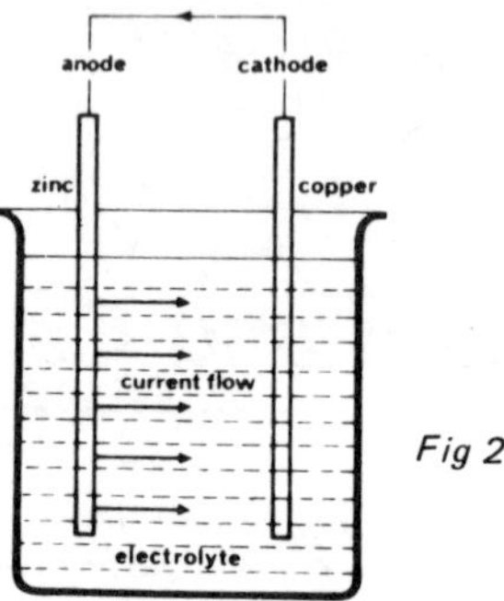

Fig 2

Galvanic Corrosion

The likelihood and rate of corrosion are increased dramatically if two dissimilar metals are in contact or connected electrically in the presence of an electrolyte (*eg* a salt solution). In this case an electrolyte cell is formed between the two metals with the less noble or anodic metal being eaten away quite rapidly, see Fig 2.

The main point is that while many metals may be relatively corrosion proof as far as simple corrosion is concerned, no metals are completely corrosion free, and can readily become one of the partners in generating bi-metallic corrosion.

The rate of attack is directly related to the position of the two metals on the galvanic scale. The farther apart they are on this scale the greater the potential of the cell, and the faster the rate of attack on the more anodic metal. Under such circumstances there will be no attack on the other (cathodic) metal until all the anodic metal has been eaten away. This principle is used in cathodic protection schemes — *ie* fitting an anode (normally high purity zinc) to the underside of the hull, connected by an electrical conductor (a bonding wire) to all other metal fittings.

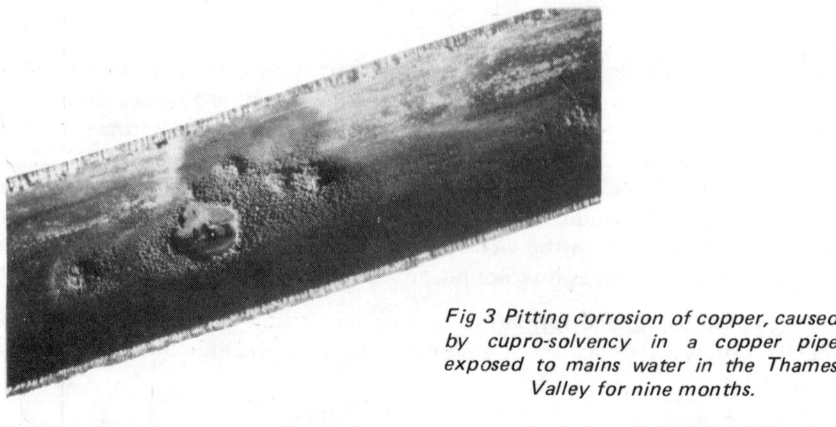

Fig 3 Pitting corrosion of copper, caused by cupro-solvency in a copper pipe exposed to mains water in the Thames Valley for nine months.

Crevice Corrosion

This is the type of corrosion that affects the corrosion-resistant metals – *eg* stainless steels and nickel-copper alloys. Equally it can be a source of even more rapid attack on any homogeneous metal – *ie* is quite distinct from galvanic corrosion which requires the presence of a dissimilar metal, (Fig 3).

The 'active' agent in crevice corrosion is a difference in the electrolyte present, or more specifically differential aeration, establishing localized corrosion cells leading to pitting of the metal surface. In extreme conditions, (*eg* a length of pipeline running from a moist to a dry environment, or from a low resistance section to a soil which is a comparatively poor conductor) quite large currents can be generated. This phenomenon is commonly associated with any length of buried pipeline running through soils with considerable variation in electrical resistivity.

Examples of soil resistivity and corrosiveness are given in Tables I and II.

TABLE I – SOIL RESISTIVITY AND CORROSIVENESS

Resistivity (ohm–cm)	Corrosiveness
0 – 1000	very corrosive
1000 – 3000	corrosive
3000 – 5000	mildly corrosive
5000–10000	slightly corrosive
over 10000	not normally corrosive

TABLE II – PARAMETERS AFFECTING CORROSIVENESS OF SOILS

Parameter	Least Corrosive	Most Corrosive
Texture	fine sandy soils	coarse clay soils
Uniformity	homogeneous	mixed soils
Organic content	low content	high content
Acidity	lowest acidity	highly acid soils
Alkalinity	lowest alkalinity	highly alkaline soils
Oxygen content	well aerated soils	compacted soils

Stray Current Corrosion

Stray electrical currents may be picked up by linked metallic systems from a variety of sources. The flow of such a current through the system establishes a 'cell' condition. The point at which the current *leaves* will then carry metallic ions into solution, with corrosion or accelerated natural corrosion occurring at this point. The severity of such corrosion is related to the value of the stray current flowing and its duration of flow. Typically a stray current of 1 amp flowing through a steel pipe for one year would result in approximately 20 lb (9 kg) of steel being dissolved away from the region where the current leaves the system. By contrast the other end of the system, where the current enters, would be *protected* from corrosion.

Only stray d.c. currents will have this effect – *ie* picked up from external d.c. current sources through leakage or accidental connection. However, stray a.c. sources can induce small d.c.

currents under certain circumstances through rectifier effect also being present. This is often noticeable on lead cable sheaths and on lead-jointed pipes in close proximity to overhead electric grid lines.

The general treatment for eliminating stray currents is bonding together of all metallic compounds with adequate earthing. With dissimilar metals in such a chain, however, the least noble (most anodic) can become subject to accelerated corrosion.

Bacterial Corrosion

Some bacteria depend on the reduction of sulphate in soils rather than oxygen and can thus live in the absence of oxygen. Their presence in soils produces sulphides which are corrosive to most metals. Such bacteria are actually killed by exposure to oxygen, so would normally be destroyed when the soil is excavated. They can survive and develop bacterial corrosion in compacted soils, particularly those plentiful in sulphates and other soluble salts.

Cathodic Protection

The simplest form of cathodic protection is by the use of a *sacrifical anode.* This confines the corrosion in a bi-metallic system to the anode and is effective in inhibiting galvanic corrosion of the main (protected) metal(s) for the life of the anode material.

To be most effective the anode material selected should be widely separated from the main metal on the galvanic scale, towards the anode end (see Table III). Logical choices are thus magnesium, zinc and aluminium. The most effective choice will also depend on the electrolyte present — *eg* zinc is most effective in salt waters; aluminium or magnesium more effective in brackish (fresh) waters. Also the anode metal needs to be of high purity. If not, it may be effectively displaced downwards on the galvanic scale (*ie* be more cathodic) with reduced protection effect; *ie* it may 'protect' the more noble constituent in its own composition rather than the main metal to which it is connected.

TABLE III — GALVANIC SCALE OF METALS

Anodic end or most 'active' electro-chemically

	Potential		Potential
Magnesium	1.6	Silicon bronze	0.26–0.29
Zinc	1.1	Gunmetal	0.3
Galvanized iron	1.0	Stainless steel (passive)	0.35–0.35
Aluminium	0.8–1.5	Tin bronze	0.24–0.31
Cadmium plating (or steel)	0.8	Copper-nickel — 90:10	0.22–0.28
Iron and mild steel	0.7–0.8	80:20	0.21–0.27
Cast iron (grey)	0.7	70:30	0.17–0.23
Stainless steel (active)		Nickel aluminium bronze	0.15–0.22
Lead	0.55	Silver solder	0.10–0.20
Tin and tinned steel	0.5	Stainless steel (passive)	0.05–0.10
18/10/3 Stainless steel (active)		Monel 400, monel K500	0.4–0.13
Chromium plating	0.25–0.45	18/10/3 Stainless steel (passive)	0.04–0.10
Brasses	0.3–0.4	Silver	
Solder (lead/tin)	0.28–0.36	Titanium	
Copper	0.30–0.36	Platinum and gold	
Manganese bronze	0.28–0.32	Carbon (graphite)	

Cathodic end (most noble) or "passive' electro-chemically

For buried pipe protection a long anode life is obviously desirable, as replacement requires excavation. An anode life of ten years is a typical target figure, the weight of anode required depending on the metal used and the corrosiveness of the soil. Another problem is that in soils with high resistivity the total resistance in the 'cell' circuit can be so high that current flow is virtually inhibited – *ie* there is little or no sacrifical solution of the anode. At the other extreme, with soils having very low resistivity it may be necessary to incorporate an additional resistance in the circuit to limit the cell current, and thus prolong the life of the anode. A generally accepted 'design' value for cell current is 100 mA.

Magnesium is the preferred metal, having a substantially higher electrode potential than zinc or aluminium. The following are typical anode weights used in various soils, which should be capable of a protective life of ten years maintaining a 'cell' current of 100 mA. (See also Table IV).

Soil resistivity ohm–cm	**weight of magnesium anode**
below 750	23 lb (10 kg)
1 500–3 000	14.5 lb (6.5 kg)
above 3 000	7.7 lb (3.5 kg)

The particular advantages of cathodic protection using a sacrifical anode are simplicity and low cost. Anodes can be attached or connected directly to tanks or similar components, where they are readily accessible for inspection and replacement. With buried systems, anode sizes are selected for longer life, the anode position(s) being designated by marker posts

TABLE IV – SACRIFICIAL ANODES

Anode	Typical Weight Loss per Year	Available Potential (volts)	Remarks
Magnesium	17 lb (7.7 kg)	1.6	Used on underground pipelines, especially where surfaces are poorly protected by coatings. Efficiency 50%.
Aluminium	8 lb (3.5 kg)	1.15	Used on offshore structures. Efficiency 80%.
Zinc	26 lb (11.8 kg)	1.1	Used on submarine pipelines Efficiency 95%

Impressed Current Systems

In this system an external anode is chosen from one of the more cathodic materials (*ie* material from the 'noble' end of the galvanic scale), with a continuous current applied to it. The anode then forms the positive side of a complete d.c. circuit embracing pipeline or system to be protected and the soil or water surrounding the pipe line. The value of the impressed current inhibits any galvanic cell current being generated and thus prevents corrosion occurring on the protected metal. Current densities required can range from as low as 0.00001 mA/ft^2 (0.0001 mA/in^2) for well coated pipes in certain soils up to 0.1 amp/ft^2 (10 amp/in^2) or more for bare steel pipes submerged in sea water.

Anode materials used include graphite, 14% silicon iron, titanium and platinized titanium. Anodes are normally placed in a special trench or ground bed of fine coke. Such systems are normally designed for a life of the order of 20 years, with minimal maintenance. Although more expensive to install initially, they are normally more cost-effective in the long term for buried or submerged pipeline protection. The main point to check is that the correct current flow is maintained.

Electrical Continuity

Both systems of protection – sacrifical anode and impressed current – rely on continuous electrical connection through the run of the pipeline (or other structure) being protected. Not all metal pipe joints do provide low resistance electrical continuity, in which case bonding must be used across each joint. This consists of a separate conductor (bonding cable) joining the two pipe lengths across the pipe coupling. The cross section of such a bonding cable should be sufficient to produce a negligible voltage drop.

Some pipes are manufactured with bonding studs moulded in or welded on to which lugged bonding cables can be bolted. Other methods of fitting bonding cables include drilling and plugging and Thermite welding. The latter technique is generally to be avoided on internally lined pipes (*eg* as used in the water industry).

Corrosion Control by Insulating Joints

Since galvanic corrosion is produced by a 'cell' current flowing along a metal pipe (or structure), another basic method of controlling corrosion is to isolate a metal pipe from the environment by using a suitable insulating material. This effect can be achieved, to a certain degree, by careful coating; and if the pipeline continuity is also broken up at specified points by isolating joints, it allows a degree of corrosion current control.

Whilst this dual approach considerably reduces the overall corrosion problem, it is unable to ensure complete inhibition of all the various types of corrosive effect. This is because technology has, so far, been unable to offer the engineer a fully isolated structure. It is thus also essential to apply 'cathodic protection' which, in conjunction with the precautions of coating and correctly located insulating joints, allows complete inhibition of corrosion by regulating all current flow off the pipeline surface.

However, to have any form of cathodic protection acceptable from a technological or economical viewpoint, it is in any case essential to ally coatings, and cathodic protection, with controlled continuity break-up by means of isolating joints, and also to use these as a means of preventing unwanted current flow onto adjacent structures.

It is possible to achieve isolation, under certain circumstances either to a greater or lesser degree, using a special type of flange gasket with insulating bolt sleeves and washers. However, these can present the following disadvantages:

(i) Gradually decreasing electrical resistance.

The principal difficulty is that these joints are generally assembled whilst installing the pipework, or even assembled some time after pipework installation by flange splitting. Consequently, the achievement of a satisfactory form of electrical installation is very much related to the care and skill of the installation fitter – often working under extremely onerous conditions. Furthermore, it is impossible to make any form of satisfactory electrical insulation in damp or high humidity conditions. Where flanged joints are utilized, it is also virtually impossible to carry out an adequate test of electrical resistance to the surrounding ground, because of the paralleling path between adjacent pipework sections and the soil itself. Consequently, field assembled flanged joints have the following weaknesses:

(ii) A low resistance to both mechanical and bending stress.

(iii) The possibility of gasket leakage, particularly when installed after pipework pressure testing, because it forms a discontinuity in the approved mechanical properties.

(iv) Unacceptable internal electrical resistance, in the case of conducting, or semiconducting, liquids.

(v) For full effectiveness, the absolute necessity of keeping the joint in a suitable chamber, to prevent parallel current flow between adjacent pipe sections, and also to permit periodic checking.

Desirable features of an insulating joint are thus:-

(i) Joints should be prefabricated — and thus provable by hydraulic and electrical tests.

(ii) Non-separable and monolithic construction specifically designed with regard to the mechanical stresses on the various sections, the location of the sealing gasket, and the precompression of this item during assembly, to enable a perfect seal to be formed against the internal and external environments.

(iii) Joints should have high combined electrical resistance due to suitably high dielectric characteristics of the insulation materials chosen, plus the coating of the internal surface by resin to prevent bridging.

(iv) *Anti-tracking* should be eliminated, with all the related metallic surfaces so arranged as to have perfect insulation between them right up to the test voltage.

(v) Closing welds should be constructed in such a way that the joint electrical characteristics remain unchanged during the life of the pipelines.

It must be appreciated that if any form of insulating joint causes an electrical short circuit or hydraulic failure, it will result in a replacement cost or true repair costs which can even be as much as the initial cost of the joint itself. Many of the failures related to cathodic protection systems, for example, can be traced to inadequate isolation at structure extremities, causing uncontrolled loss onto other apparatus nearby. To rectify these faults may necessitate taking the pipeline out of operation either involving a stopping or cutting operation, which in itself is extremely expensive.

It is for this reason when considering the type of joint to use, that engineers not only have to evaluate the mechanical and electrical properties related to the pressure requirements, but also have to take note of the experience and long term reputation of the joint supplier. Consequently, a prudent choice should be that of a company which has long term experience in manufacture of specialist insulating joints of all types, and where the prototypes have been rigorously tested, backed up by continuous production line quality control.

Fault Diagnosis and Conditioning Monitoring

PIPELINES are normally surveyed for one of the following reasons:-

(i) to investigate repeated or isolated blockages.

(ii) to check on structural integrity, for connections from new sites, or expected working life for maintenance expenditure budgets.

(iii) to determine exact location of pipelines, branches and connections to update drawings or carry out maintenance work.

(iv) to inspect installations prior to handover.

(v) to check on scale or corrosion build-up on industrial heating or process pipelines.

(vi) to detect and locate leaks in buried pipelines.

(vii) to carry out interior inspection of welded joints on long runs of metal pipe (*eg* gas or oil lines).

(viii) to test individual components, *eg* safety and relief valves on line.

Techniques used obviously vary with individual industries and installations, size of pipes, *etc.*

T.V. Surveying

T.V. surveying is a particularly versatile method of pipeline inspection generally applicable to pipes ranging in size from 2 in up to 4 feet or more in diameter (50 mm to over 1 m). Such systems can also be designed to function under widely varying conditions. They consist of:-

(i) a video camera.

(ii) umbilical cable.

(iii) control unit.

(iv) lighting leads.

(v) video signal recorder (*eg* videotape).

(vii) power supply.

(viii) ancillary equipment.

Video Camera Units

Video camera units may range in size from as small as 1 in (25 mm) diameter upwards. Small units are obviously necessary to survey small diameter pipelines and negotiate bends. A normal

1 in (25 mm) diameter camera, for example, can be expected to negotiate 90 degree bends in 4 in (100 mm) diameter pipelines.

Various lighting attachments may be used for illumination of pipes of different diameter, including the use of a rotating mirror in front of the lens at 90 degrees to the line of the pipe so that ports and branch lines can be screened circumferentially.

An umbilical cable of suitable length connects the camera to the control console. This carries lighting power, rotating mirror power and vision signal; it is stored on a 12-way slipring drum or similar device that contains a trip measuring device plus an electronic pulsing unit which relays back to the monitor the distance the camera has travelled down any one line. This distance can be zeroed or preset to any measurement at any one time.

The control unit, which is basically the power unit for the camera, controls lighting intensity, remote focussing capability, and rotating mirror positioning.

Lighting heads are normally low voltage lamps of different sizes and wattages housed in intrinsically safe glass fronted housings for attachment to the front of the camera.

Various items and relevant information need to be shown on the monitor screen during a survey such as:-

Sight address, date, pipe diameter, run number, location and description of faults, *etc.*

The equipment used to produce the characters shown on the screen can come in various shapes and sizes with differing capabilities, but is normally known as a word processor or screen writer.

Ancillary equipment carried by the unit may include: generator, winches, steel rods, drain stoppers, life line, harness, hydrant stand pipe, hoses and key, extension leads, gas detector, road cones, cable rollers, field telephone set with up to 400 yards of cable, float lines, *etc.*

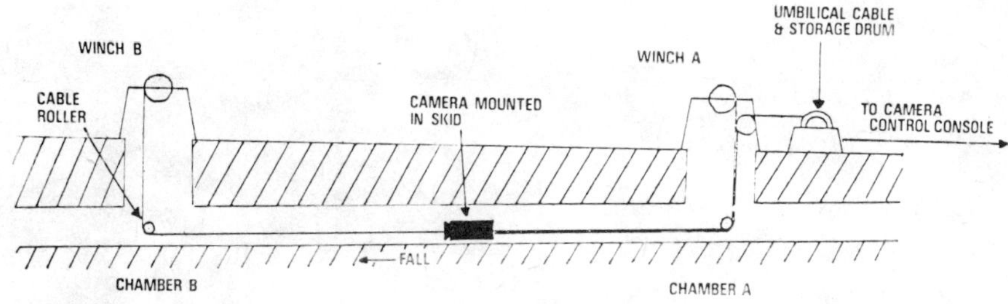

Fig 1

Survey Methods

The most common method used to carry out a T.V. survey is that shown in Fig 1, whereby a line is passed through the run and a camera is towed by a winch from chamber 'A' to chamber 'B'. Winch 'A' is purely a safeguard, so that if the run being surveyed is damaged to such a degree that camera progress is impeded, the camera can be retrieved by winching back on a steel band instead of pulling on the umbilical cable. As previously stated, as the camera and umbilical cable are pulled down the run, the measuring wheel on the drum transmits back to the monitor the distance the camera has travelled from the beginning of the run for location purposes.

Runs are normally surveyed in the direction of the fall of the drain. The reason for this is that if the camera is pulled against the flow it can create a bow wave, causing droplets of water to splash onto the camera lens, so distorting vision. Waste matter can also be deposited on the lens which could obliterate vision completely. Surveying against the fall is only possible if the run is 'dry' during the survey.

A land line telephone system may be used for communications between vision controller and winchman, so that the camera can be stopped anywhere along the run to focus on any problem area.

When the survey is completed the camera and skid are detached from the umbilical cord and safety winch cable. The two cables are then wound back to chamber 'A'. Alternative methods of transporting the T.V. camera by rodding are shown in Fig 2.

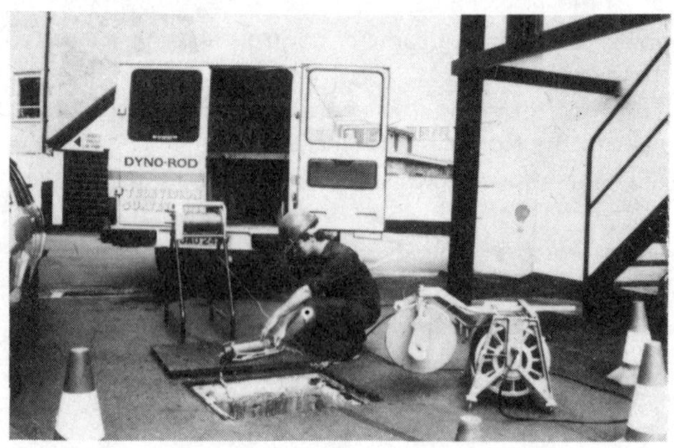

T.V. survey equipment. (Dyna Rod Ltd)

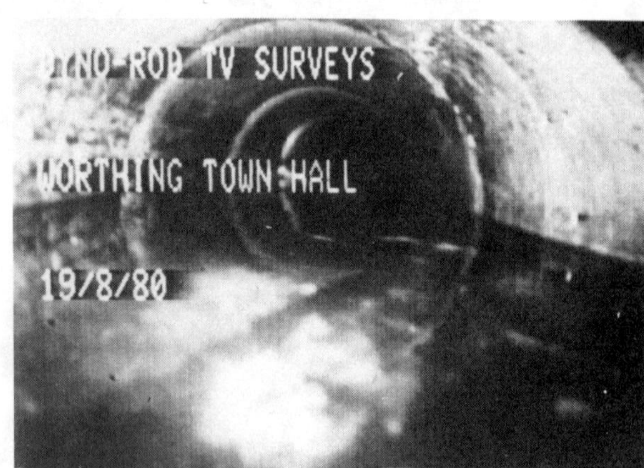

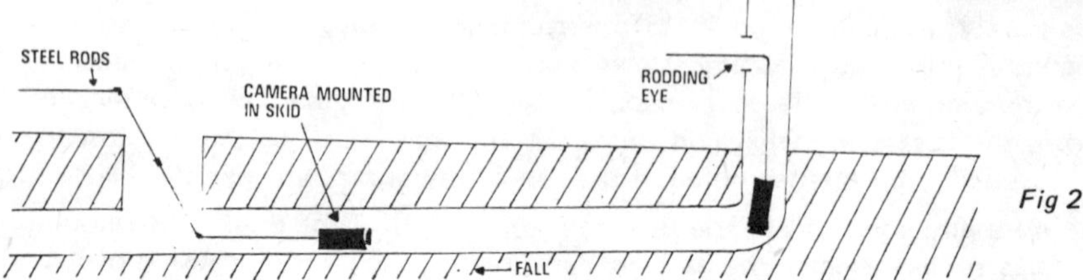

Fig 2

Position Surveys

Metal detection can be used to trace the path of buried metal pipelines, and also plastic pipelines where a metalized identification strip has been buried with the pipe. Simple hand-held units of this type are effective up to depths of 6 feet (2 metres). More powerful units may be used for detection at greater depth, and also for surveying wrapped or coated pipes. In the latter case the instrument may also be capable of detecting flows or breaks in the pipe wrap.

Such instruments may work in the inductive or conductive mode. In the inductive mode a linear or circular aerial is used as a 'direction finder' to establish a small signal which establishes the exact location of the pipeline run relative to the aerial. With conductive mode operation no aerial is used and direct contact is made with the pipe to be traced to feed the signal directly into it.

Location of Submarine Pipelines

Four different methods may be used to locate submarine pipelines — echo-sounding, magnetometers, sonar-scanning (and similar acoustic methods), and seismic profiling. Echo-sounding and acoustic methods can only be used to detect submarine pipes which are exposed above the seabed. Magnetometers and seismic systems can detect buried submarine pipelines.

The most common method of locating exposed submarine pipelines is by side-scan sonar, using a towed 'fish' with laterally-directed transducers generating acoustic pulses of frequencies of the order of 100 kHz. Various techniques are used depending on the depth of water involved, the length of pipeline, *etc* and the feasibility of employing underwater acoustic transporters. In shallow waters suitable results may be achieved more simply and economically using transit sonar, when the sonar transducer is mounted on the hull of the running vessel, or conventional high-resolution echo sounders.

The application of a magnetometer or seismic profiler for the detection of buried submarine pipelines requires the use of detectors towed behind the survey vessel which then performs a series of transverses to intersect the expected pipeline axis.

Magnetometers measure the strength of the earth's magnetic field, which is affected by the presence of a steel pipeline. Essentially, then, a magnetometer detects the presence of such a buried pipeline by its anomalous magnetic effect. This magnetic effect is usually proportional to the centre of the distance between the magnetometer sensor and the object (pipeline). For this reason it is necessary to lower the magnetometer 'fish' as close as possible to the seabed.

With a seismic profiler a piezoelectric transducer or 'pinger' emits pulses of acoustic energy with single frequencies in the range 1.25—1.4 kHz at a high pulse rate. The 'fish' is towed at a height of 15—30 feet (5—10 m) above the seabed, concentrating the downward directed energy beam over a small area of seabed. A proportion of acoustic energy penetrating the seabed is reflected back by the lined pipeline producing a characteristic deflection on a graphic recorder.

LEAK SEALING

Leaks in pipelines obviously need treatment at source, *eg* replacement of leaking fittings or glands, or repair or replacement of damaged components. Temporary measures such as binding with PTFE tape or 'bandaging' are seldom effective, largely because it is difficult to seal any pressurized system which has developed a leak. There are, however, processes which enable on-line sealing to be tackled with the plant remaining in full production and the pipeline at normal operating pressure — these rely on injecting a compound which forms a moulded seal over or around a leakage point. Such compounds are normally thermosetting materials which soften when initially warmed and then curve to form a tough moulding.

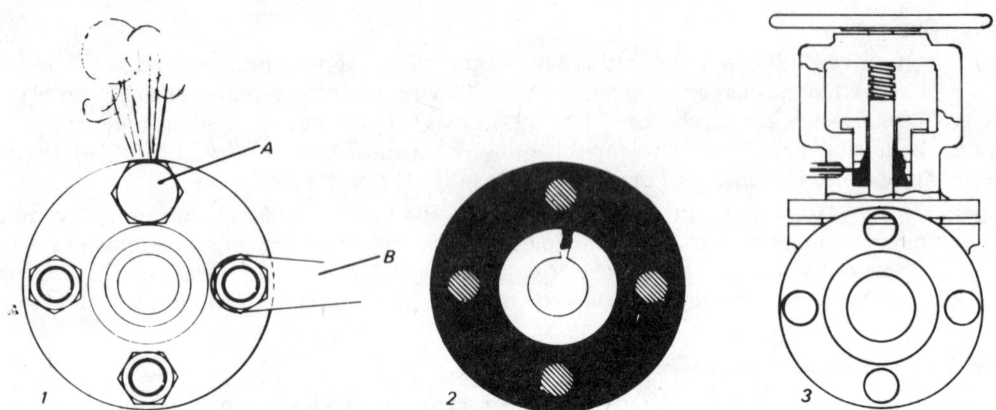

1. Preparation of a flanged joint for the Furmanite process.
   A. Furmanite device exhausting the leak to atmosphere.
   B. Injection device.
2. A new gasket is formed around the periphery of the original gasket.
3. A valve gland is repacked on line through direct injection into the stuffing box.

Fig 3

An example of treatment of a leak from a flange due to gasket failure is shown in Fig 3. The process involves forming an injection moulding around the failed gasket. This is achieved by providing a path for the leak through special adaptors fitted to the studs or bolts, or outer edge of the flange. The gap between the flanges is fitted with a peripheral seal.

Compound is then injected through the adaptors into the space between the peripheral seal and the failed gasket, starting from the side opposite to the leak. During injection this compound fills all the grooves and pits in the faces of the joint.

The compound used depends upon several factors, but is usually thermosetting, to form a tough, resilient mass quickly. The repair can be regarded as permanent but, like any gasket, its life will depend on the extent of thermo-cycling and other factors such as the nature of the contents of the pipeline.

Valve Glands

Somewhat different treatment can be applied to leaking valve glands. The compound is injected into the stuffing box and once the leak has stopped the gland follower is backed off and the gland fully repacked.

Other Applications

Leaks from heat exchanger joints, screwed couplings, pipe units, riveted joints, *etc* can also be dealt with by these methods; even a leak in the wall of a pipe or weld. In this case a special retaining box is fabricated to fit around the leaking area and once mounted, is injected with sealant. This type of treatment has been applied successfully in oil refineries, chemical plant, conventional and nuclear power stations and numerous other industries.

Pipeline Cleaning

THREE DISTINCT methods may be adopted for pipeline cleaning – electromechanical cleaning, pigging or high pressure water jetting. These are described in detail under separate headings.

Electromechanical Methods

Electromechanical methods of cleaning pipes and tubes involve the use of powered tools to cut or loosen the contaminants within the pipe. The tool is usually attached to a flexible cable or rod which enables it to be introduced to the working area even when this is at some distance from the entrance pipe, or when it is beyond a number of bends. Power is transmitted to the tool by the axial rotation of the cable or rod, which in turn is driven by a motor attached to the other end.

A variety of tools and drive systems is available, and in one form a vacuum system is combined with the cable feed to assist in the extraction of debris from the pipe. Electromechanical cleaning methods may be used on pipes in the range ½–12 inches diameter over lengths of up to 200 ft from access points to remove deposits and blockages arising from a wide variety of contaminants.

Equipment

Primary equipment comprises an electric or air motor, gear box and cable feed device. Cables or rods, and tools are exchangeable items. Versatile general purpose machines are normally rated at 0.2 to 1 hp and provide a cable rotation of a few hundred rev/min. This provides a relatively high torque enabling a variety of cleaning techniques to be used. Specialized machines having vacuum extraction, for applications such as cleaning boiler tubes up to 3 inches in diameter, are rated at 0.5–3 hp and produce higher rotational speeds, Variable speed, forward and reverse facilities are frequently incorporated in both types of machine.

A typical arrangement is shown in Fig 1 for transmitting the power from the motor to the flexible cable and for providing a variable cable length. Cable is stored inside the drum and the required length can be pulled out of the central guide tube. Rotation of the drum then imparts axial rotation to the cable.

Flexible cable is usually of metal construction in the form of a tightly coiled spring having an outer diameter of from ¼ to 1 inch depending on the duty. Performance of larger diameter cables is improved by a braided inner cable, slightly smaller than the internal diameter of the spring, that provides a core onto which the spring can tighten when transmitting power to the tool. Tools clip or screw into the end of the cable and are selected specifically for the type of cleaning action required.

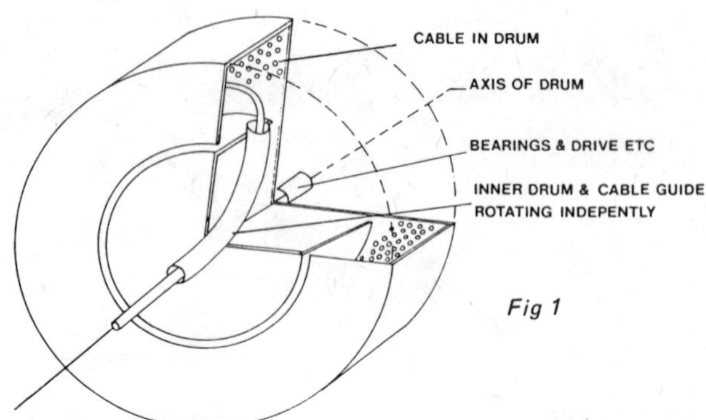

Fig 1

The following Table illustrates the range and typical uses of the many tools available. However, it is frequently the case that tools are developed specifically for individual jobs, and the list should not be considered to be exhaustive.

Type of Tool	**Principal Use**	**Example**
Brush	Removing loose deposits, burnishing.	Soot, light scale, rust
Surface scraper	Removing medium-hard deposits.	Scale, fats, detergents, chemical deposits
Serrated rotor	Removing hard deposits	Carbon scale, limescale, resin
Spear blade	Breaking through soft blockages	Sewage, aggregation of solids
Drills	Penetrating hard blockages	Concrete, resin
Carbide cutter	Penetrating hard blockages	Concrete, resin
Saw blade	Sawing through tangled fibres	Tree roots
Hook	Retrieving solid objects lodged in pipe	Masonry, tools

Optimum Cleaning Methods

Careful selection of equipment is an essential step if pipes are to be cleaned efficiently, and this depends upon possessing adequate information on the following factors:

- Internal pipe diameter
- Distance from access points
- Number and location of bends and branches
- Materials of construction
- Physical properties of material to be removed
- Distribution of material to be removed
- Health hazards

Internal diameter influences not only the size of cutting tools but the choice of flexible drive cable. Cables must be small enough to pass through the pipe, including obstructions, and flexible enough to pass round bends. They must also be sufficiently rigid to transmit the required power to the tool without coiling up inside the pipe, and a fine balance normally exists between size, stiffness and flexibility.

Distance from access points must clearly be limited to the maximum reach of the cable. This depends on a number of factors, including the angle of the pipe and the number of bends.

Internal Pipe Diameter		Typical Maximum Length of Cable feet (metres)			
(inches)	(mm)	General Purpose		With Vacuum	
½	12	75	(22)	50	(15)
2	50	100	(30)	100	(30)
4	100	150	(45)	100	(30)
6	150	200	(60)	100	(30)
12	200	200	(60)	–	–

The maxima given are only a guide and can vary greatly between applications. In favourable cases it is possible considerably to exceed these distances by joining lengths of cable.

Material of pipe construction influences the choice of cleaning tool insofar as the tool must be selected to avoid damage to the structure of the pipe. The physical properties and the distribution of the material to be removed affect the choice of cleaning tool, cable size and motor power. They may also influence the decision whether to use a complementary cleansing action such as flushing, back flushing or vacuuming.

Unfortunately it is often difficult to determine these factors because of the inaccessibility of the materials involved. Samples taken from the end of the pipe may not be representative, and it is frequently necessary to carry out trial runs before the correct choice of equipment can be determined.

Applications

Electromechanical cleaning methods have general applicability in the removal of unwanted materials from pipes and tubes of all types. With the correct choice of equipment and attachments in the hands of skilled operators, they can be successful in dealing with the following materials:-

Superficial or heavy deposits on pipe walls
- lime scale
- soot and carbon
- ochre
- rust
- fats, grease, detergents
- oil and tar
- chemical deposits

Sediments and blockages caused by
- any of the materials listed above
- cement and concrete
- aggregation of miscellaneous solid materials
- sewage
- tree roots

Typical locations in which these methods are commonly used are:-

fuel lines	chimneys and flues	smoke tubes
transportation pipeline	hot water pipe systems	water tubes
water mains	ductwork	heat exchangers
effluent pipelines		evaporators
drains and sewers		economizers
process pipework		

Pigging

Pigging is a method of cleaning pipes and pipelines by hydraulically or pneumatically propelling a tightly fitting plug, or pig, through the pipe. Contaminants are removed from the walls by the passage of the pig, and these are pushed forward and out of the pipe by the combined action of the pig and surplus propellant escaping past it. Pigs may also be used for gauging pipelines and for plugging pipes during pressure testing.

There are two classes of pig construction, foam and steel, with several variations of each for the performance of specialized functions.

Pigging may be used in pipes from ½–56 inches bore and over distances of many miles, to remove sediments and superficial deposits arising from a wide variety of contaminants.

Equipment

The propulsive force on the pig can often be provided by the normal contents of the pipeline, if this is available at a convenient pressure and flow. Otherwise air or water must be supplied by means of a suitable pump provided specially for the purpose. Hydraulic propulsion is preferred on safety grounds, but compressed air is sometimes required if it is necessary to avoid wetting the interior of the pipe. In all cases suitable pressure limiting devices and gauges must be fitted.

Pigging pressures are usually in the range 10–100 lb/in^2 and depend on the bore of the pipe, decreasing with increasing pipe diameter. The type of pig and the duty it has to perform also influence pressure requirement.

The pump is sized to give the desired propellant flow rate, which is determined by the pipe bore and the optimum pig speed for the application. Pig speed can thus be controlled by adjusting the pump delivery rate.

Pig launchers may be placed between the pump and pipe to enable the pig to be fed into the pipe without damage to either. Similarly, a trap is usually placed at the far end of the pipe to catch the pig as it leaves the pipe, which it may do at high speed.

Foam pigs are made of polyether or polyurethane foam of from 2–9 lbs/ft^3 density, chamfered on the leading edge and with a sealing disc on the rear pressure face. On some types the outside surface is coated with elastomeric polyurethane to increase the resistance to abrasion from the pipe wall. Abrasive foam pigs, in addition, have bands of abrasive grit around the circumference to improve the cleaning action. Whilst normally made with a circular cross-section, foam pigs can be fabricated with other profiles for cleaning ductwork.

Steel pigs are in comparison relatively complex, consisting of a steel frame which seals against the pipe wall by means of two or more elastomeric cup seals. Except in the case of gauging and bi-directional pigs, cleaning is achieved by a series of scrapers or brushes arranged around the outside of the frame. Gauging pigs consist simply of the frame and seals, while the seals of bi-directional steel pigs are placed in opposition to each other to enable the pig to be driven in either direction.

The following Table summarizes the principal applications of each type:

Type of Pig	Principal Uses	Examples
Foam	Swabbing; Removal of thin, soft or liquid films	Sludge or water in air lines; to avoid damaging pipe surfaces
Coated foam	Removal of films and heavy sludges	Oily deposits
Abrasive foam	Removal of heavy or hard deposits	Rust; slag; hard scale in water, effluent and hydrocarbon lines
Cleaning (brush)	Removal of miscellaneous superficial materials	Light rust and scale; precommission cleaning
Scraper	Removal of hard or strongly adhered films and deposits	Corrosion; wax in effluent and hydrocarbon lines
Gauging	Proving minimum bore prior to commissioning	
Bi-directional	Plugging pipeline during hydrostatic testing	

Optimum Cleaning Methods

Careful selection of equipment is essential if a pigging operation is to be successful, and this depends upon possessing adequate information on the following factors:

- Internal pipe diameter
- Bend radii
- Location and type of branches, valves, flow meters and other obstructions
- Location of access points
- Material of construction
- Pressure rating of pipelines
- Constraints on propellant type
- Physical properties of material to be removed
- Distribution and thickness of material to be removed
- Health hazards

Nominal pipe bore, actual bore and bend radii are of crucial significance in selecting the sizes and types of pigs to be used. Very tight bends or changes in direction can be negotiated only by foam pigs. Where there is a possibility of the pig becoming stuck it is necessary to use a bi-directional pig so that it can be withdrawn from the blockage.

Pig Type	Range of Actual Bore D, (inches)	Minimum Bend Radius	Uni-Bi-directional
foam	½ to 48	no limits	Bi
steel	2 to 56	1½D to 5D	either

The physical properties and distributions of the material to be removed influence the type of cleaning action required and hence the choice of pig. In many cases a range of different pigs will be used in succession to provide the optimum cleaning action at each stage of the operation. Unfortunately it is often difficult to determine in advance which pigs will be needed, and it is sometimes necessary to make the choice of pig types as the operation proceeds.

It is essential to identify beforehand all intersections of pipe runs, change of bore and obstructions in the bore from valves, pumps and monitoring equipment, and to plan the pigging operations carefully.

In complex process pipework it may be necessary to break the operation into separation sections. On long pipelines there is no reason, however, why a pig should not be sent many miles. If desired it can be followed electrically or magnetically by the use of tracking equipment.

Applications

Pigging as a technique is suitable for cleaning the walls of pipes and removing sediment from the invert. It is particularly suitable for long runs of pipework having many bends and few access points, which cannot easily be cleaned by any other method. Pigging is not suitable for unblocking pipes which are completely closed.

With the correct choice of pigs and equipment pigging can be successful in dealing with a wide variety of materials of which the following are examples:-

Removal of

- weld and nodular scale
- slag
- rust and corrosion
- lime scale
- ochre
- cement powder
- wax
- waste liquid
- sedimentation
- contamination

Special applications

- gauging
- hydrostatic testing
- swabbing (disinfectants, *etc)*

Typical locations in which these commonly occur are:-

Pipelines

- crude oil
- natural gas
- refined hydrocarbons
- chemical
- water
- effluent
- dewatering
- slurry

Process pipework

- refined hydrocarbons
- petrochemicals
- general chemical
- cooling water
- food and brewing

High Pressure Water Jetting

With high pressure water jetting techniques, deposits and blockages in pipes are broken up by the action of fine, high velocity jets of water directed onto them from suitably shaped nozzles, and the resulting debris is washed out of the pipe by the flow of spent water.

Pressures of up to 20000 lb/in^2, in some cases higher, and a wide range of flow rates are used, depending on the nature of the material to be removed. Water is delivered to the nozzle through lances for cleaning small bore pipes and tubes, and through flexible, self-propelling hoses for cleaning larger pipes.

With lances, straight pipes of 5/8–6 inches bore can be cleaned for a distance of up to 20 feet from access points. Hoses are capable of cleaning pipes of 2–42 inches bore (up to 60 inches with attachments) for a distance of up to 400 feet from access points, and they may also negotiate a number of bends.

Equipment

A conventional pump assembly consists of a diesel or electric power unit, high pressure water pump, header tank, control valves and instrumentation, all mounted on either a skid, trailer or vehicle. Fully flame-proofed versions are available for use in hazardous environments.

The choice of pump is determined by the pressure and flow requirements of the pipe to be cleaned. Pumps are usually of the triple plunger type, enabling a range of pressures and flow regimes to be selected by changing plungers. The alternative diaphragm type of pump, often used in low power applications, may also be used over an extended range of pressures, although this can only be achieved efficiently by a change of pump heads.

The power requirements of the pump are determined primarily by the pressure and flow rate. A useful rule-of-thumb is that, within the limits of flow and pressures for a given pump, the product of available flow rate and pressure are approximately constant. The majority of high pressure water jetting pumps are rated at between 25 and 200 hp, going up to 500 hp. A typical specification for a 100 hp pump, for example, would be to deliver 10 gal/min at 10 000 lb/in^2.

Hoses are selected to withstand the maximum operating pressure with an acceptable margin of safety, and to transmit the required flow rate with the minimum of pressure drop along lengths of up to 400 feet (120 metres).

A primary consideration in the selection of lances for use in small pipes is the bore of the pipe, as sufficient clearance must be left around the outside of the lance without restricting the flow through the bore. Maximum lance length is 20 feet (6 metres).

A range of nozzles is available to suit the application with a choice of the number and orientation of jets, and materials of construction.

High pressure water is extremely dangerous. All pressure devices must be able to withstand the operating conditions, and must be properly maintained. Similarly, all operators must be correctly equipped with protective clothing.

Optimum Cleaning Methods

Careful selection of pressures, flow rates and nozzle configurations is an essential step if pipes are to be cleaned efficiently, and this depends upon possessing adequate information on the following factors:-

- Internal pipe diameter
- Distance from access points
- Number and location of bends and branches
- Location of valves and other obstructions
- Physical properties of material to be removed
- Distribution and thickness of material to be removed
- Health hazards

The physical properties of the material to be removed is the principal factor in determining the pressure required. The following Table illustrates cutting abilities of various pressures, although there can be considerable variations in practice.

Pressure Range		Examples
Up to 4000 lb/in^2	(275 bar)	Grease, paraffin wax, crude residues. Algae, pulp, asbestos, PVA, food residues, loose masonry, clay, mud, silt.
4000 to 8000 lb/in^2	(275–550 bar)	Boiler scale, carbon, potash, asphalt, cement, plaster, mastic, PVC, unbonded paint, rust, oils.
8000 to 20000 lb/in^2	(550–1380 bar)	Polymers, bonded paint, resins, plastics, synthetic rubber, coke, concrete, silicates, mill scale

Increasing flow rate has the general effect of increasing the rate at which material is cut. When large quantities of contamination have to be removed from a pipe, and particularly when it is completely blocked, high flow rates are essential. This is also true in the case of large diameter pipes when sufficient cleansing flow must be maintained in the bottom of the pipe to keep debris in suspension.

The distribution of the material within the pipe influences the configuration of nozzle jets which are required, forward facing nozzles being used to clear blockages, and rear facing nozzles for flushing out loose material.

Applications

Water jetting is particularly suitable for removing large amounts of unwanted materials from pipes, including the cleaning of long lengths of totally blocked pipes. It offers advantages in plant down-time, cost and, very often, effectiveness of cleaning.

SECTION 7

Services

Water Services

THE USE of the plug valve for water supply systems dates back to Roman days and is still widely used in its modern forms together with various other types of rotary movement and screw-down shut-off valves. The main exception has been the ball valve. Only during the last ten years or so has this type been developed to meet the technical and economic requirements of water supply systems. Various manufacturers in the meantime have attempted to provide an alternative for drinking water systems by offering 'intermediate' ('mixed') construction types, such as the segmented gates and ball valves. In practice, however, these types of constructions generally turned out to be hydraulically unstable and required very high actuating moments.

Compared with the general class of shut-off valves, the ball valve offers the following potential advantages:

- (i) In the open-position, it provides free passage of water in the supply system with a diameter equal to the supply connections.
- (ii) It gives an unimpeded flow profile without any distortion.
- (iii) It offers the smallest resistance to flow, that is, a very small loss in pressure over the comparable supply line distances.
- (iv) It is completely adaptable.
- (v) The change in the cut-off from open position to closed position requires a minimal change in place.
- (vi) The precise shape of the cut-off guarantees a seal of great integrity.
- (vii) It offers a favourable ratio of weight to stability resulting from the design of the ball or hollow sphere which withstands the pressure.
- (viii) It has low installation costs.

Ball Valves for Water Supplies

It was not until there were ball valve designs which took into account the specific requirements of a drinking water supply (such as resistance to the formation of deposits and acceptable hydraulic performance in intermediate positions), and the cost requirements (namely, amortization of the high use of energy even at low flow rates and short periods of useful operation), that the use of ball valves for all aspects of drinking water supply systems became a subject of interest. In contrast to the design of a conventional ball valve with seat rings on both sides to form a seal, the principles of the rotary piston valve were borrowed as the basis for the design of a valve having flow around a ball section — Fig 1.

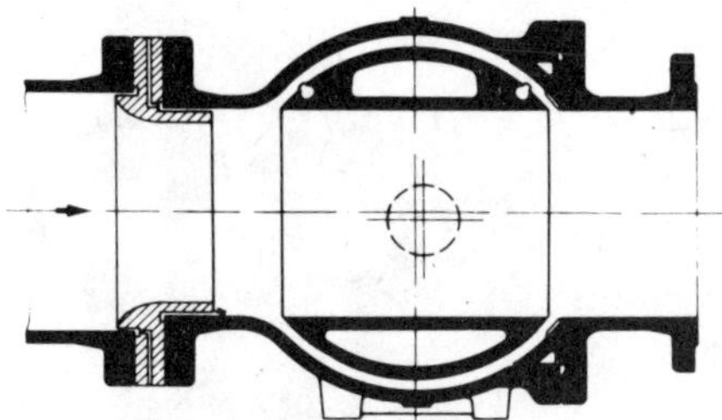

Fig 1 Concentric ball valve. Open position.

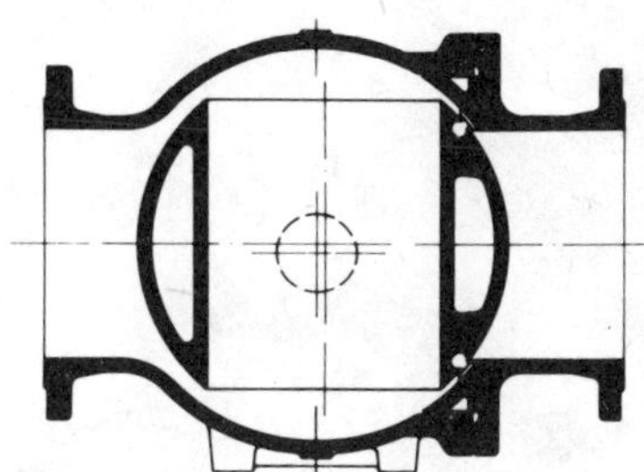

Fig 2 Ball valve, closed position.

This type of circulation stabilizes the flow through the bore of the ball and prevents the damaging effects of cavitation and disintegration which occur in the intermediate positions. The ring gap is used for the through-flow of the medium as soon as a point of constriction is attained at which the pressure loss in the central bore is greater than the pressure loss in the ring gap. The basis for the stabilization of the flow is, however, that the ring gap for circulation exhibits small transient changes in the surface, so that sudden changes in pressure or velocity can be avoided. The desired resistance of the ball valve to the formation of deposits, of the sort that might be formed from minerals and foreign particles such as sand, can be achieved in a design which makes use of a surface not requiring special finishing — Fig 2.

This type of design requires the fixed placement of the seal elements, so that a rubber or an elastic preformed seal of a conventional variety would be first and foremost. To this end, there are already well-tested seals from the area of shut-off valves which have been around for decades and can be easily adapted. The counter seat in the housing can use a metallic seal made of corrosion-resistant steel with inlet and outlet edges having especially large radial sections which then provide a useful fixing in place of the rubber-elastic preformed seal. In addition, the bearing for the turning point of the ball is placed on an eccentric, which reduces the frictional load of these seal elements to the smallest levels, which is, of course, a necessary requirement. Mineralogical deposits and foreign particles can in this fashion block the ball in its entire periphery, so that the operating forces of the drive unit are not sufficient to move the ball.

For this reason, these types of ball valves are given a relatively large gap between the outer surfaces of the ball and the inner surfaces of the housing, to try to avoid gap corrosion — the formation of hard layers of deposits and corrosion. Additionally, the ball is provided with a scraper rim extending beyond the turning radius, which provides only a line contact and otherwise gives the ball surface free room in which to turn. As a consequence, even with a large amount of deposits, the operating forces are sufficient to actuate the valve. In intermediate positions of the ball valve, this overall ring gap results in a washing action, and the medium moving through the valve produces a kind of self-cleaning effect. This provides the necessary flow characteristics.

One design of this type of valve is the ball valve with an alternative opening, which in a state of no pressure permits the exchange of the elastic ring seat of the ball without removing the valve itself from the supply line (Fig 3). This device is particularly useful with large nominal diameters. It is a further requirement that the components for transmitting the motion and the drive unit

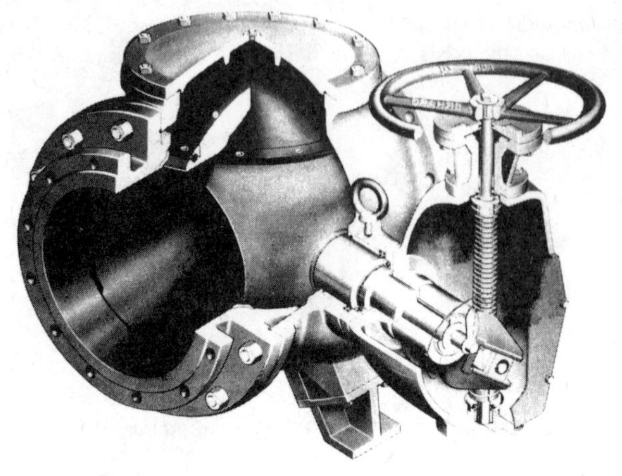

Fig 3

Blakeborough SF resilient seal wedge gate valve for water works use to BS5163. 16 bar rating. 50–300 mm bore

must be solidly built and require no maintenance. For this reason massive shaft bearings in the horizontal direction are needed. These are not exposed to the deposits of solid matter, and thus are not located at the deepest point. The resulting possible lateral arrangement of the driving mechanism results in a relatively low construction height. These construction considerations have led to a durable and maintenance-free design of ball valves for the drinking water supply using well-tested components at competitive prices.

Non-Return Valves in Water Systems

The complexity of water distribution network dynamics creates an unstable equilibrium of pressure and back-pressure, constantly modified by the user, leading to different appliances or collectors, some of which can be compared with veritable drains: retention vats in factories, sinks and their dishwater, baths and their bathwater or soapwater, washing machines, central heating circuits with anti-scale additives, *etc.*

As a result there is the possibility that water provided through the network can be polluted by waste water returning from the consumers, returning to the mains or passing from one consumer point to another without going through the mains (from one apartment to the other). This may be caused by:

(i) *Depression on the mains:*

A considerable call for water (fire hydrants for example) or intervention in the main pipe (repair, new branch, breakage) can create a depression.

(ii) *Over-Pressure at the Consumer:*

All systems of high pressure, of course, but also all appliances for hot water production, sanitary or otherwise, instantaneous or otherwise, can be the reason for this.

(iii) Simultaneous appearance of low pressure on the mains and high pressure at the consumer.

TABLE I – PROTECTION SYSTEMS

Method	Geometry	Remarks
Barometric loop	33 ft (10 mm) loop without branch	Physically very safe, but costly and not always realizable. Inoperable in case of leakage.
Overflow safety gap (i) total overflow	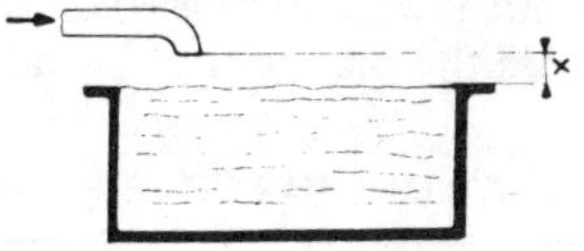	Safe primarily but ineffective if tap is extended with a sample pipe
(ii) at partial or overflow limit		Safe if outlet flow is well dimensioned in comparison with inlet flow. Ineffectual if the tap is extended with a sample pipe.
(iii) Diverted overflow	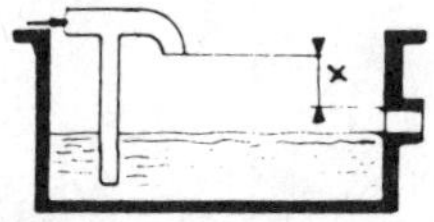	Safe principle but limited in practical application
Disconnectors (i) Without moving parts	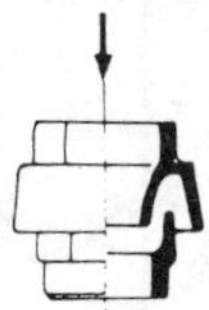	Safe principle but may involve substantial head loss.
(ii) With moving parts (a) on-line	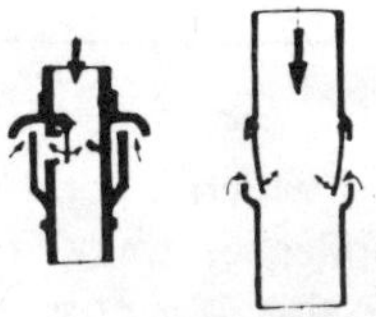	Safe principle; may have reliability problems
(b) Tee	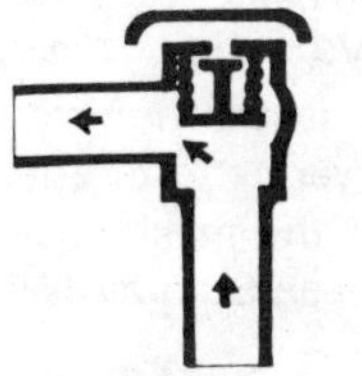	Safe principle; may have reliability problems
Reduced pressure backflow preventer	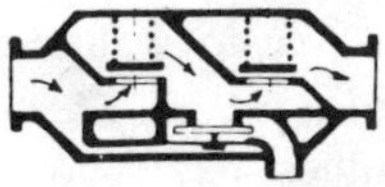	Elaborate and costly system, generally needing regular maintenance to ensure continued proper function.
Non-return valves		Simple and effective. Performance and reliability primarily depends on design and quality of components and manufacture.

Protection Systems

Theoretically it is imperative to implement those systems which prevent water returns due to these pressure disturbances. These systems should be automatic.. They can be purely hydraulic, air-hydraulic or mechanical. The hydraulic systems are theoretically the most reliable but are often expensive or difficult to install. The mechanical systems are subject to doubts regarding function and longevity. Principle types of protection are summarized in Table I.

In the presence of all these complex phenomena and conscious of the necessity to organize a water distribution of quality and to protect it against the risks of pollution, numerous countries have brought about legislation which defines the measures to be taken (see Table II).

TABLE II – STATE OF THE REGULATIONS IN FORCE IN DIFFERENT COUNTRIES

Country	Regulations with Tests, Agreements and/or Standards	Control at the Manufacturer's Site	Control at the Consumer or in the Trade
GERMANY	X	–	–
GREAT BRITAIN	X	–	X
BELGIUM	X	–	X
DENMARK	X	X	X
SPAIN	–	–	–
FRANCE	X	X	X
HOLLAND	X	X	X
ITALY	–	–	–
SWEDEN	X	X	X
SWITZERLAND	X	–	–

The principle consists of defining the rules which a good general installation has to go by and the criteria of quality of the protection appliances to be incorporated, in order that the whole mains escapes the danger of pollution. To do this, it is possible to use, individually or in combination, certain of the different hydraulic, air-hydraulic or mechanical systems examined later.

Most industrial countries have chosen to install non-return valves (NRV), taking into account the security-cost compromise and applying very strict design and control rules which considerably reduce the possible hazards connected with the mechanical character of the design. A European standard is now being prepared by the European Standards Committee (CEN).

Non-Return Valve Families

An NRV is an automatic appliance intended to prevent a fluid changing its flow direction.

Numerous classifications exist. Here we use those of the European Standards Committee (CEN) which distinguishes between four families, based on the direction of the displacement of the closing system with regard to that of the flow of water.

Note that there are other NRV families but these are not for use in the sanitary field, notably the ball valve, particularly recommended for loaded liquids.

Family 1 : linear displacement, not parallel

- perpendicular displacement
  valve with straight flap
- oblique displacement
  valve with oblique flap

Fig 4 Ball valve DN 800, PN 10 as a pump pressure valve with an electric starting motor.

Family 2 : angular displacement

- lift valve
- butterfly valve

Fig 5 Ball valve DN 1000, PN 16 as a pipe rupture safety device, with falling weight, oil pressure braking cylinder and hydraulic unit.

Family 3 : with deformable closing system

- membrane valve

Fig 6 Ball valve DN 500, PN 25 with hydraulic drive and bypass.

Family 4 : with linear parallel displacement

- "guided" valves

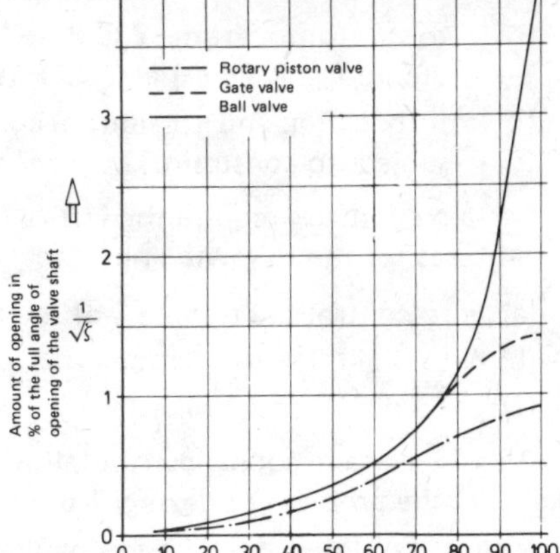

Fig 7 Coefficient of loss of various valves as a function of the amount of opening.

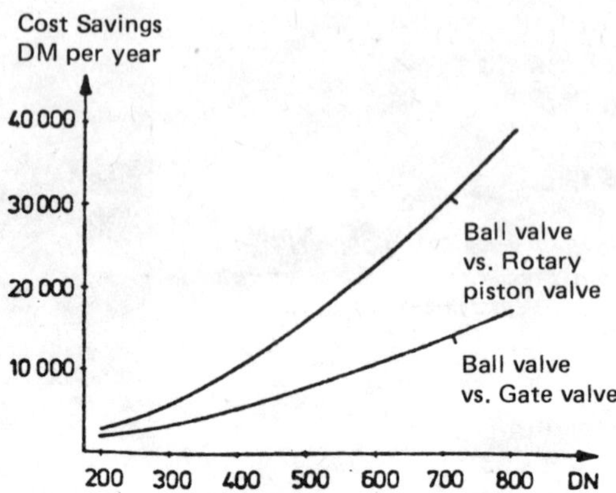

Fig 8 Annual savings in energy costs by the use of ball valves instead of other types of valves.
(Burkett)

Test Specifications on Non-Return Valves

These should embrace the following points:

mechanical characteristics of the casing
tightness
head loss
reliability

In the field of sanitary protection, different countries have enforced test specifications and methods for valves smaller than 2 in. (50 mm). We will examine here this category only and will mention the minimum and maximum requirements for each of the characteristics listed above:

Mechanical Characteristic of the Casing

The resistance is defined:

(i) to pressure: from 230 lb/in^2 (16 bar) in water at ambient temperature, to 360 lb/in^2 (25 bar) at 203°F (95°C) (duration of test 2–5 minutes).

(ii) to buckling and torsion: according to the sizes and the countries, the casings are subject to constraint, in order to determine their resistance.

Certain countries require the NRVs to be of dismountable design with regard to the connection, with bosses for the following purposes:

(a) to control the tightness of the NRV
(b) to drain the installation
(c) To disinfect the pipes.

Also in certain countries, the alloys are specified according to the nature of the water (for example, the problem of dezincification).

It must be noted that the mechanical holding tests carried out on the casing should not modify the hydraulic characteristics of the NRV.

Tightness

(i) *Low Pressure Tightness:*
All specifications prescribe that an NRV must be tight under a pressure of 1¼ inch (30 mm) water column, applied during a total time of 3 to 10 minutes.

(ii) *High Pressure Tightness:*
The valve must be tight under a pressure rising from 0–230 lb/in^2 (0–16 bar), applied during a total time of 3 to 10 minutes.

Head Loss (ΔP)

We know that the head loss is a loss of pressure induced by any plumbing appliance mounted on a pipe. The design of the appliance can reduce the head loss to a minimum. In the case of an NRV, the survey of the hydraulic profile of the internal parts determines the head loss.

It is usually expressed in feet or metres water column at a given flow. (It can also be expressed by a coefficient of ΔP, k, square function of the average speed of flow on twice the acceleration of the weight). The specifications usually prescribe, for a given nominal diameter, the achievement of a minimum flow at different values of ΔP ranging from 1.6 to 32 ft (0.5 to 10 m) water column.

Reliability

Endurance tests are very important for an NRV which, in so far as being a mechanical system, is often tainted with unfavourable prejudice in comparison with purely hydraulic systems. Without modification of its principal characteristics (tightness, ΔP), the appliance can undergo up to 100 000 cycles of operation at a pressure varying from 90–230 lb/in^2 (6–16 bar) (according to the country), one half in cold water and the other half in hot water (temperatures varying from 150 to 200°F (65 to 95°C) according to the country). This corresponds on average to a minimum longevity of 10 years.

Opening Pressure

In principle, present legislation prescribes that an NRV must be able to function in all positions, which means in this case the use of a release spring. This spring must conform to the specifications which impose a limit of opening pressure. At present, the opening pressure must always be positive and below 3.28 feet (1 m) water column, according to the country. This is very important. Experience shows that, being open for 80% of its time, an NRV functions with very weak flow rates. Under these conditions it is of prime necessity that the opening is complete, as vibrations occur which can generate strong noise nuisances, as well as there being problems of premature wear.

Valve Types Compared

At present only the linear parallel displacement valve described as "guided" seems to comply as, whatever its position, the design of this type alone gives the following results:

(i) perfect tightness at weak pressure, mainly ensured by the quality of the guide system (difficult to obtain in the valve family with angular displacement, lift or butterfly valves.)

(ii) a high pressure tightness, mainly ensured by the compression of one face against the seat and a solid closing system, which is not always the case with the family of membrane valves.

(iii) minimum head loss, the linear parallel displacement preventing a considerable deformation of the fluid flow, contrary to that given by valve families producing non-parallel linear displacement.

Role and Interest of the Regulations

Regulations can only exist if they are defined and imposed by the public authorities and if the application is controlled by the proper bodies.

Regulations depend on two aspects:

(i) *The localization of the protection features:*
It defines the list of appliances or installations in which a form of protection must be installed, in which part of the installation and what type of protection must be used.

Most European countries have prescribed their own regulations by replying to two questions: "What is to be protected?" and "In what manner?" The result is that, in certain cases, the choice of appliance is left to the decision of the installer.

(ii) *The definition of the qualitative criteria of the protection equipment:*
From the moment when the fitter has chosen a type of appliance, it is necessary for him to know that the appliance corresponds with a qualitative specification which allows it to be used in the considered case.

Conformity to Criteria

In all the countries where regulations exist, these criteria have been the object of tests, where the operating mode is defined and where the positive result is confirmed by the issue of an Approval No. and the right to affix a distinctive sign on the body of the valve.

The control of the appliance is either carried out by the body which has prescribed the regulations or by an authorized independent laboratory, whose results are confirmed by the recommending body.

It is desirable that the definition of the specification is the result of co-operation between the legislators, the manufacturers and the installers but it must not be forgotten that the opinion of the legislator is predominant.

Control of appliances when being designed and manufactured is insufficient if the complete installation itself is not also submitted to the control of the authorities. In the case of the whole installation having to be joined with an exterior source, it is the authorization of the branch which is the ultimate sanction and constitutes its certificate of conformity.

Finally, it is necessary to watch that the quality of the installation does not deteriorate with time. This is the purpose of the necessary periodic controls, carried out by the responsible authorities and which particularly apply to the appliances subjected to greater approval requests. This is the reason why NRVs are equipped with bosses, in order that controls can be carried out easily. In some cases the regulations even state quite simply that the appliances are to be exchanged every five years. Hence the importance of the valves being of standard dimensions and also being easily dismountable.

Equally it is ultimately necessary that regulations and specifications should extend beyond the national standards, so that the rules applicable to different countries are rendered identical for all. In the NRV field, the specialist commissions of the European Standards Committee are studying this problem, with the participation of the different national plumbing syndicates. It is desired that their studies result in rationalization, reduction of costs and the elimination of protectionism, which some national regulatory bodies could presently be accused of.

Water Services

Increasing use is being made of stainless steel tubing for domestic water supplies and heating and plumbing systems. Particular advantages offered by stainless steel tubes are:

(a) The appearance of finished pipework is aesthetically pleasing. Both the satin and the polished finishes are attractive, and the thin walls of capillary fittings give pipeline and fittings a neat, continuous appearance.

(b) No maintenance is required after installation; the satin finish stays satin-looking and the polished finish stays polished.

(c) The price is comparatively stable and does not fluctuate like copper. Over the past few years the price of stainless steel tubing has been slightly below that of comparable copper tube.

(d) The corrosion resistance of stainless steel is better than copper in areas of cupro-solvency and it is not prone to pitting corrosion by water.

(e) The mechanical properties are good. The strength is high which means that it is less prone to damage in service. Elongation is also high which gives it good bending properties — as a comparison, it takes the same amount of effort to bend 0.6 inch (15 mm) stainless tube as a 0.9 inch (22 mm) copper tube. It can be sawn easily with a hacksaw or roller cutter.

(f) Copper or other non-ferrous pipe and fittings are subject to pilfering on site as they have a high scrap value. Stainless steel does not have a high secondhand value.

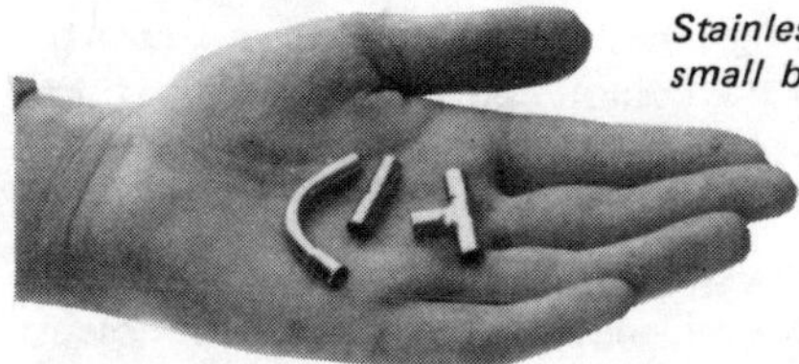

Stainless steel mini-fittings used to join small bore stainless steel tubing are now readily available.

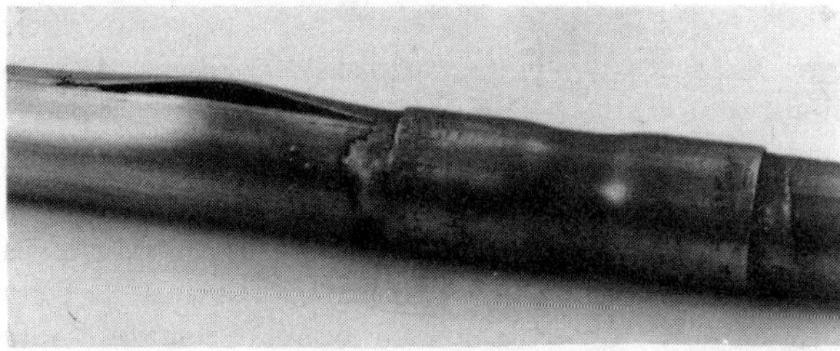

Thin-walled stainless steel tube to BS4127 joined by a stainless steel capillary fitting using Lancashire Lo-melt soft solder, after a pressure test. At a pressure of 7000 lb/in^2, the tubing burst, leaving the fitting and joint intact.

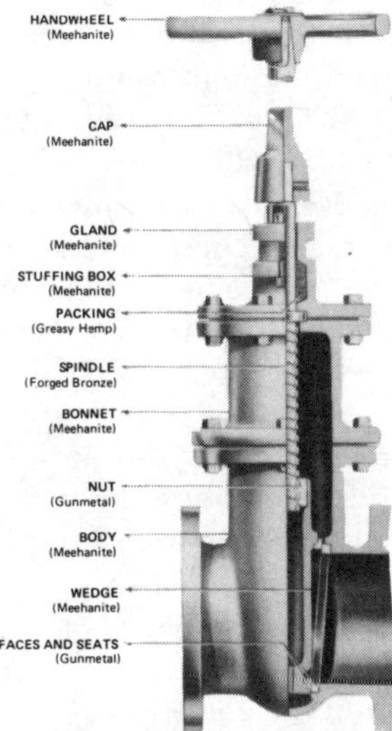

Glenfield gate valve for waterworks services. (Neptune Glenfield Ltd)

The inherent disadvantage of stainless steel tubing employed as pressure pipes has been the difficulty of making leak-free joints (*eg* stainless steel tubing used on high pressure aircraft hydraulic systems is invariably welded). However, this has now largely been overcome by the availability of suitable fittings, solders and fluxes making plumbed joints quite practicable and several jointing techniques are now well established. These fall into two major categories:

(i) techniques requiring heat

(ii) joints which can be made at room temperature.

Joints Using Heat

Techniques using heat to join are welding, soft soldering or silver soldering. Welding is not used in the majority of plumbing installations although it is an established technique for the chemical, food and cryogenic industries. Soldering implies the use of capillary fittings where the inside diameter of the fitting socket is just a few thous greater than the outside diameter of the tubing. Capillary attraction pulls the solder into this gap. To make successful joints in stainless steel it is essential to recognize two important properties of the metal:

(a) Stainless steel has a low thermal conductivity – about 1/30th of that of copper.

(b) The property which gives the metal its stainlessness – this is because a hard oxide coating is formed within seconds of a nascent surface being presented to an oxygen-bearing atmosphere. The coating steadily increases in thickness until a stable protection is achieved.

For successful soft soldering, the outside end of the tube and the inside of the fitting socket must be abraded with emergy cloth to remove the oxide skin. The prepared surfaces should then be painted with solder paste and the joint assembled.

The secret of soldering stainless steel well is not to hurry, to use a small flame and to make sure that the back of the joint (away from the plumber) gets enough heat. Using a metal reflector behind the joint can often be a help or alternatively the use of the cyclone burner produces a flame which will curl around to the back of the fitting. When the solder begins to flow a frying sound comes from the joint. The joint should be completed by end-feeding with solder wire.

Many solder paints have been made and marketed over the years. One with a five-year shelf life contains phosphoric acid base flux and also powder of the tin-lead alloy which has the lowest melting point possible. Phosphoric acid fluxes are recommended as they only become aggressive and eat up the tenacious chromium oxide when heated; they will not continue to attack the stainless steel when cold.

For silver soldering, it is necessary to use an aggressive chloride-based flux to eat the oxide and make a good joint. It is important to remember that this flux must be removed from the joint area (inside and outside) within 24 hours of making the joint. Usually swilling with water for an hour or so will remove all trace.

There are many suppliers who market silver solder and brazing rod, together with flux, especially for joining stainless steel. The majority of these are entirely satisfactory providing the suppliers' instructions are carefully followed, and there must be the same care in heating the joint as with soft soldering.

Capillary fittings are at their most efficient when the gap between tube and socket is uniform. Such uniformity is achieved with a design in which the tube is held centrally by a specific deformation of the socket at three points.

Joints made Cold

Low conductivity has undoubtedly been the cause of a number of leaking joints made by soldering and although soldering techniques have been much improved it is still useful to explore cold techniques. These might be used, for instance, where naked flames cannot be applied, on sites with fire risks or pipelines fitted with a plastic anti-burst liner.

Three basic techniques are cold adhesive bonding (now approved for cold water supplies), the use of compression fittings, and the use of screw-thread fittings. (The last named is only used for thick-walled pipe and as it is not applicable to the thin-walled tubing used in plumbing, will not be discussed further).

Experimentation has been carried out by the adhesive makers, by stainless steel capillary fitting makers and by the British Steel Corporation, to determine the best adhesive to use and to devise a method of applying it. Of the many adhesives tested, the dimethacrylate esters, which are anaerobic adhesives, have proved to be the most satisfactory.

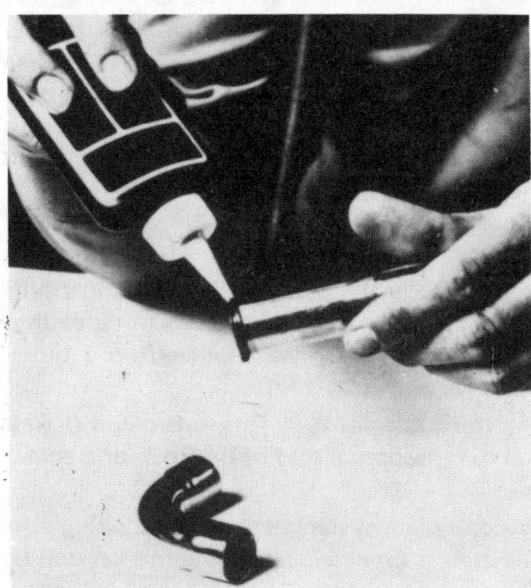

Applying a ring of cold adhesive (Loctite 638) to make a stainless steel joint. After curing for 2–4 hours at room temperature such a joint will withstand a pressure of 4500 lb/in².

These adhesives cure in the absence of air, that is, when they enter the confined capillary gap they start to harden. They will cure by themselves, but in order to make a joint cure to handling strength in two minutes it is necessary to use an accelerator which can be applied from an aerosol spray.

The bonding technique entails cleaning the tube ends with coarse emergy cloth, spraying with accelerator and allowing the latter to dry (30 seconds). A ring of the viscous adhesive is then applied to the leading edges of both the tube and the fittings. The tube is then pushed into the fitting socket and left without movement for two minutes for handling strength to be achieved. When cured, a pull-out strength of 1.1 tons is obtained for a 0.6 inch (15 mm) tube. Pipework jointed by these adhesives will withstand internal pressures of 4500 lb/in^2 (310 bar).

This technique has now been approved by the Thames Water Authority for use with potable water and by the majority of the Regional Water Authorities in the UK. The maximum temperature for which its use with water and aqueous chemicals is recommended is 140°F (60°C).

Now that mini-bore tubing is available down to ¼ inch (6 mm) o.d., together with a complete range of low cost capillary fittings, this technique (and soft soldering for applications above 150°F (60°C) will increase the use of stainless steel instrumentation, for medical gases and for various water applications.

Stainless steel compression fittings employ the principle of tightening a nut which causes a ferrule or olive to be compressed tightly on to the tube by forcing it down a cone on to the tube end. The shape of the ferrule is the subject of many patents.

Most compression fittings have two or three ferrules per fitting. Some designs dig into the tube wall — the so-called bite-ring fittings — and some are just compressed tightly on to the tube. Joints are quickly and easily made with compression fittings; with most designs just two spanners, turned in opposition, are needed. A particular model employs only one screw thread to apply the forces necessary for compressing the two ferrules and the joint can be broken and remade a number of times. These fittings are intended for use with thin-walled tube to BS4127 and this point should be checked by plumbers contemplating the purchase of stainless steel compression fittings, as some designs only work on thick-walled tube.

For the handling of some liquids such as milk, milk products and beer, it is essential to have demountable joints in order to clean stainless steel pipework. Such joints must be free from cracks and crevices which might become breeding places for bacteria. There are a number of designs of fittings available incorporating PTFE washers and neoprene O-rings for these applications. For other chemicals, cone-seated or flat-seated joints with gaskets are in common use.

Summary

The cause of leaks in stainless steel plumbed joints in the past can be mainly attributed to lack of knowledge of the differences between stainless steel and copper. It is worth repeating that the low thermal conductivity of the metal must be recognized. A big flame from a blow-lamp is not enough; time must be allowed for the heat to soak all round a joint.

The other important difference is seen in the choice of flux. Chloride-based fluxes are a hazard as they are corrosive on stainless steel. Often hygroscopic traces of hydrochloric acid can form and cause pitting.

The availability of stainless plumbing as a complete system acts as a stabilizing influence on the cost of plumbing. The growth of the stainless steel domestic plumbing market will be affected by three main factors:

(a) The price of copper tubing and fittings, which may well rise again in the not too distant future;

(b) The price of stainless steel compression fittings, which is expected to fall;

(c) The availability of an adhesive which will withstand boiling water for long periods without losing its strength. When this is available, the use of stainless steel tubing in central heating systems will be even more widely accepted.

Nuclear Valve Glands

THE VERY highest integrity is an obvious requirement of valve glands in nuclear plant, demanding satisfactory performance in the following areas:

(i) *Mechanical* – deformation, resistance and recovery.
(ii) *Dynamic* – behaviour under spindle movement, under pressure and at elevated temperatures.
(iii) *Corrosion* – elimination of valve steam corrosion.

Main fluids to be handled are demineralized water, saturated steam (BWR plants) and borated water (PWR plants); but with possibly 1 500 pieces of equipment in each division of a nuclear power plant, the multiplicity of problems involved can be immense.

The gland material now generally considered the most suitable in application is nuclear purity expanded graphite (plus corrosion inhibition, if specifically called for). This is readily formed in mechanically sound rings from tape to provide the necessary resilience and deformability to behave as an efficient seal. To accommodate extrusion of the packing, expanded graphite rings are normally combined with plaited rings at the top and bottom of the gland – a logical choice here for high temperature working being pure graphite/asbestos braided packing which can also incorporate a corrosion inhibitor and a suitable proportion of anodes material to act as a sacrificial anode. An alternating arrangement of graphite fibre and expanded graphite rings does not produce as satisfactory a seal.

The braided packing rings serve to eliminate the risk of extrusion of the expanded graphite where radial play at the bottom of the box exceeds 0.5 mm and 0.3 mm around the gland follower.

This packing arrangement, after extensive laboratory testing by EDF, has been used successfully in French nuclear power plants for many years and is rapidly being extended through all primary circuits.

Should it be necessary to avoid asbestos products, rendering even wet spun, dust free product unacceptable, a graphite fibre packing can be employed for anti-extrusion purposes. This would however introduce problems of fragility and of potential corrosion risks (the stem alloys having to be chosen with extreme care).

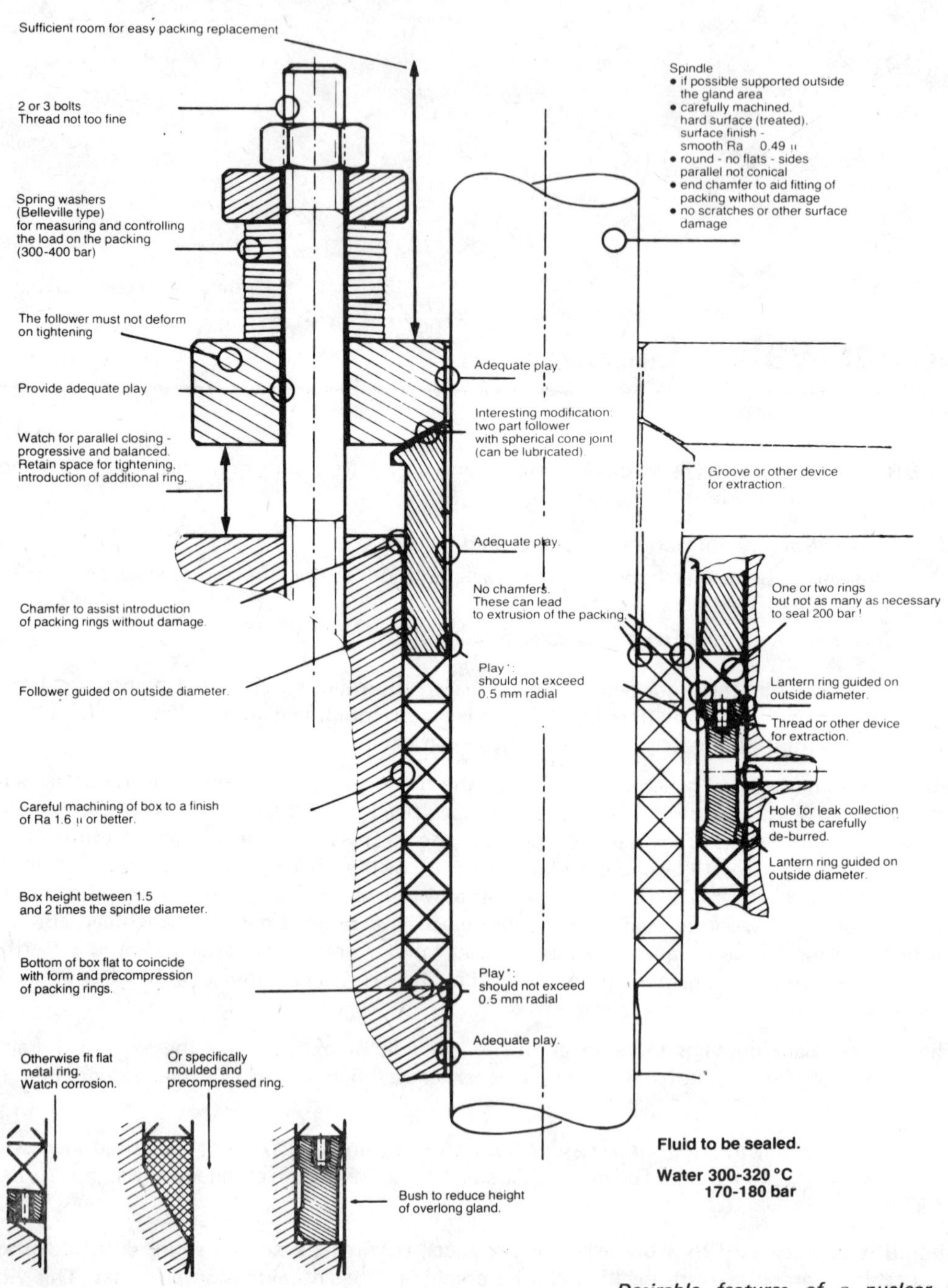

Desirable features of a nuclear valve gland. (Electyicity de France).

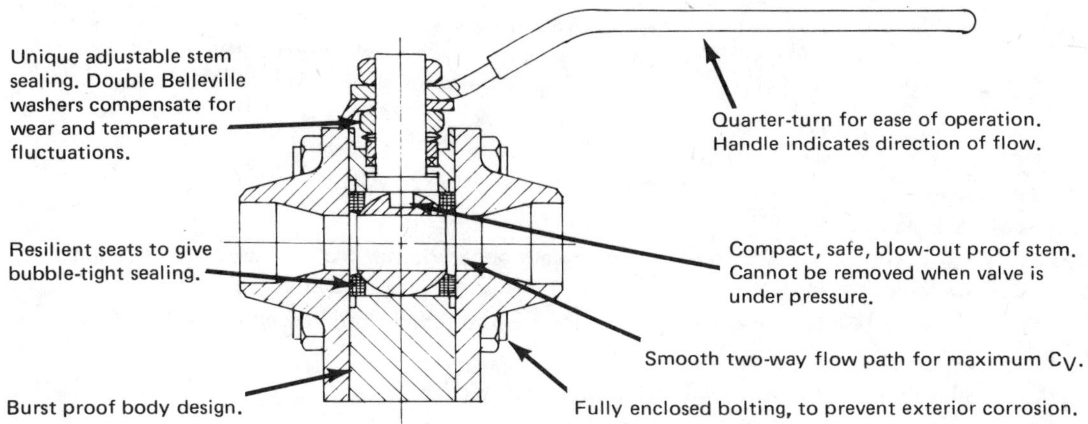

Example of ball valve design for nuclear industry.
(Canadian Worcester Controls)

Gland Dimensions

The dimensional relationship between stem, box and packing rings is of prime importance. Interference between ring and stem with play between ring and box is to be avoided as leading to high stem torque and poor sealing. It is preferable to begin with a tight fit to the box and a small clearance to the stem (for example 0.1 mm).

Surface finish is important, particularly on the valve stem to realize minimum packing wear and low operating torque. Recommended values are R_a = 0.4 μm for the stem and R_a = 1.6 μm for the gland surface.

Live Loaded Packing

The maintenance of a leak free seal is directly dependent on the maintenance of an adequate loading on the gland packing. Spring discs (Belleville type washers) can compensate for loss of loading due to:-

- relaxation of the packing (very slight for expanded graphite, of the order of 4% at 350 bar).
- wear
- differential expansions
- temperature variations

The introduction of spring discs also assists the precision with which gland loadings can be determined (*ie* by height reduction of disc). The cost of suitable spring discs is small in relation to the advantages they bring.

Gland Geometry

The depth of a valve gland should be 1.5 to 2 times the stem diameter. A greater depth serves no purpose. Many glands are too deep. Should the number of packing rings exceed 6–7, transmission of gland loading becomes very uneven and stem torque increases disproportionately. Tests made by the EDF showed that an increased number of rings/depth of gland could result in increased leakage.

Recommendations for stem diameter/ring sizes are:-

Stem diameter (mm)	Ring square section (mm)
10	4
20	6
30	8
40	8 – 10
50	10
60	12

Causes of Leakage

The initial cause of leakage developing is not always apparent, particularly as one fault can lead to another. Experience has indicated that in order of seriousness, likely causes are:-

(i) The use of braided packings that lose volume too readily and harden in use.

(ii) Damage to packing rings during fitting.

(iii) Bad meeting of ring ends where cut rings are used.

(iv) Incorrect disposition of ring joints, forming leak path.

(v) Insufficient gland loading.

(vi) Poor support to stem.

(vii) Poor ability of packing to withstand thermal shock.

(viii) Reduction in gland loading due to packing relaxation, packing wear/volume loss.

(ix) Incorrect dimensional tolerances between stem/packing/box.

(x) Gland too deep.

(xi) Stem surface finish of low order.

(xii) Play at bottom and top of box too great.

(xiii) Corrosion of stem and abrasion of packing during stem operation.

(xiv) Too many rings above lantern.

Corrective Actions

The following are points to observe, not necessarily in order of significance. Any one is important in achieving satisfactory gland performance.

(i) Correct gland design.

(ii) Entry to facilitate fitting of packing rings.

(iii) Optimum gland depth.

(iv) Correct surface finishes.

(v) Adequate capacity for gland loading/adjustment.

(vi) Spring discs to compensate for wear.

(vii) Good stem support.

(viii) Use of expanded graphite sealing rings particularly recommended (expanded graphite is permanent, resilient, has better relaxation, maintains its volume and withstands thermal shock).

(ix) Correct dimensional tolerances between packings and gland.

(x) Correct fitting and loading of gland before service operation .

See also chapter on *Seals and Packings.*

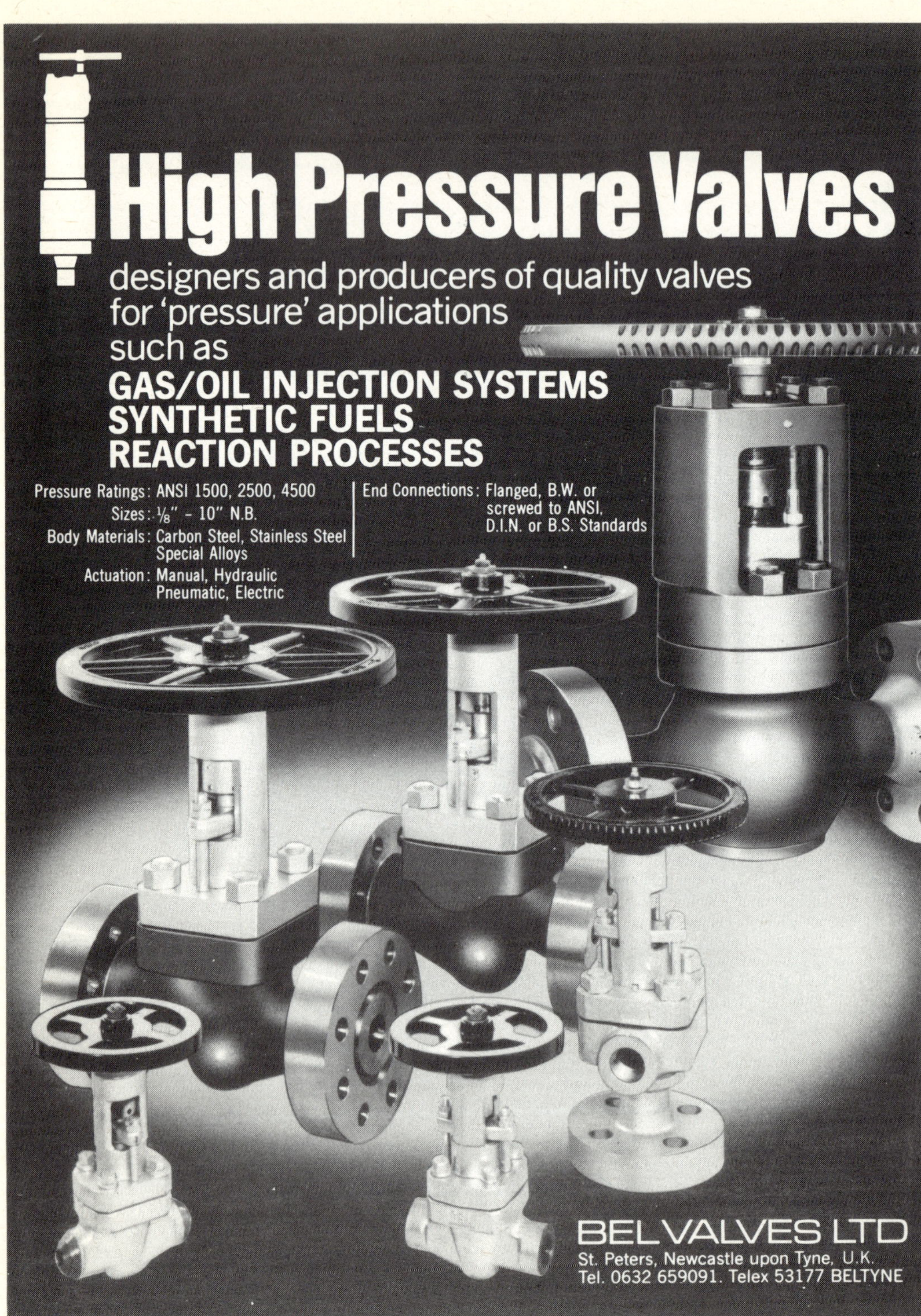
High Pressure Valves
designers and producers of quality valves
for 'pressure' applications
such as
GAS/OIL INJECTION SYSTEMS
SYNTHETIC FUELS
REACTION PROCESSES
Pressure Ratings: ANSI 1500, 2500, 4500
Sizes: 1/8" - 10" N.B.
Body Materials: Carbon Steel, Stainless Steel
Special Alloys
Actuation: Manual, Hydraulic
Pneumatic, Electric
End Connections: Flanged, B.W. or
screwed to ANSI,
D.I.N. or B.S. Standards
BELVALVES LTD
St. Peters, Newcastle upon Tyne, U.K.
Tel. 0632 659091. Telex 53177 BELTYNE

Special Valves

Cryogenic and Low Temperature Valves

Valve types used for handling liquid oxygen, nitrogen, methane, natural gas, ammonia, *etc* at cryogenic temperatures may be of ball, gate, globe, butterfly or check type, with an increasing preference for ball valves. All are special designs, normally identified by having extended bonnets to position the valve stem seals away from the cold source. This extension can also serve the purpose of providing an insulation space between the pipeline and the lever or handwheel operating the valve.

Valves for cryogenic services may range in size from 1/8 in up to 40 in (3 mm to 1 m), with pressure ratings from ultra-high vacuum up to 10 000 lb/in^2 (700 bar), capable of working down to −425°F (−254°C).

Construction may be in brass, bronze or stainless steel, rendered antistatic either by a graphite gland assembly or external bonding. On ball valves specially designed seats are required to minimize the effect of differential expansion by reducing seal volume to a minimum. Seals are normally made in PTFE.

Valve stem seals must be of a type not requiring lubrication (lubricants would freeze; also they could not be tolerated in oxygen systems). Seals used are normally graphite, TFE or PTFE. Pressure is normally equalized between the valve body and cavity within the extended bonnet. All valves for cryogenic services need to be cleaned, degreased and finally assembled in a clean room.

Low Temperature Valves

Valves designed for low temperature services but not cryogenic services may follow a similar form, but not necessarily the same extended bonnet. Valve bodies may also be in carbon steel instead of stainless steel. Suitable carbon steels are available for services down to −100°F (−73°C).

High Pressure Valves

High pressure valves fall into specific categories. All valves used on hydraulic circuits, for example, are high pressure types, typically designed and constructed to accommodate working pressures of 2 000 lb/in^2 (140 bar), or higher in other specialized systems (*eg* aircraft hydraulics).

In fluid handling, certain processes require the use of high pressure valves – the most widely known being ammonia synthesis, the oxy process, and processes for the production of urea and methanol as well as polymerization processes for the production of low density polyethylene. Apart from these, high pressure valves are also required in hydrogenation processes – *eg* coal liquefication or gasification.

The demands on the different process equipment vary depending on the process, operating pressure, operating temperature, and corrosivity of the media. Handling aggressive coal slurry, for example, which contains up to 30% pulverized coal at 10–30 microns particulate size can lead to enormous erosion problems in valves. Erosion can also be a problem in other types of high pressure valves handling clean fluids, resulting from high localized fluid velocities. Flow passage design, as well as material selection, is thus important in high pressure valves. Such valves are, therefore, normally individual designs, not modifications of standard valve types with stronger components and greater wall thicknesses.

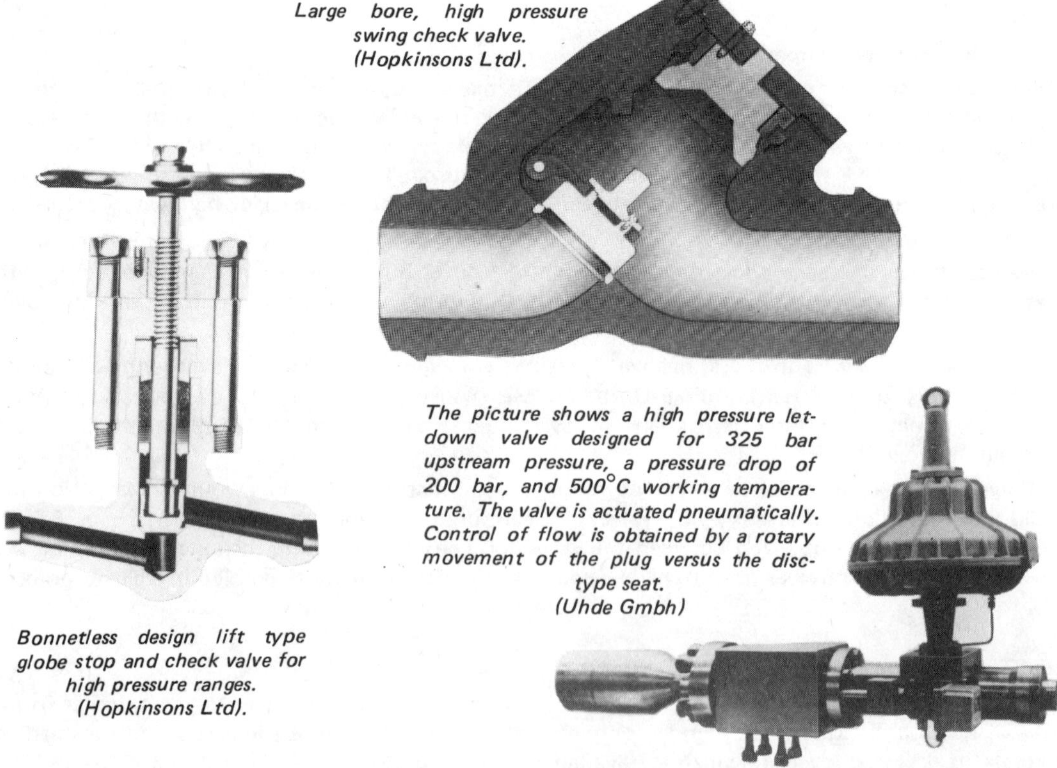

Large bore, high pressure swing check valve. (Hopkinsons Ltd).

The picture shows a high pressure let-down valve designed for 325 bar upstream pressure, a pressure drop of 200 bar, and 500°C working temperature. The valve is actuated pneumatically. Control of flow is obtained by a rotary movement of the plug versus the disc-type seat. (Uhde Gmbh)

Bonnetless design lift type globe stop and check valve for high pressure ranges. (Hopkinsons Ltd).

Special materials for seats and plugs may also be required, such as silicon nitride, to cope with severe conditions of abrasion and temperature. Special demand may also be placed on valve stem seals, particularly as operating temperature may restrict material choice.

The ultimate test of a high pressure valve is, however, the same as any other type of valve. It should be capable of performing its function reliably, without leakage, and have an acceptable life. It is only the parameters that are more arduous.

Steam Control Valves

STEAM GENERATORS produce steam at a pressure, temperature and volume not generally acceptable to a consumer. As a consequence it is necessary to apply steam conversion (*ie* change one, two or all three parameters) in the steam distribution systems of power plants for public safety, heating power plants, industrial power plants and in process industries.

Conventional practice is to reduce steam pressure in a reducing valve, and reduce temperature further down the line with desuperheaters operating on a variety of principles, *eg*

(i) direct water injection into the low pressure steam line

(ii) applying saturated steam to the hot steam

(iii) steam cooling by a separate steam cooler

The two processes of pressure reduction and desuperheating can also be accomplished either simultaneously or sequentially in a *steam converter valve.*

Fig 1 shows a steam converting valve in the heat flow diagram of a power plant.

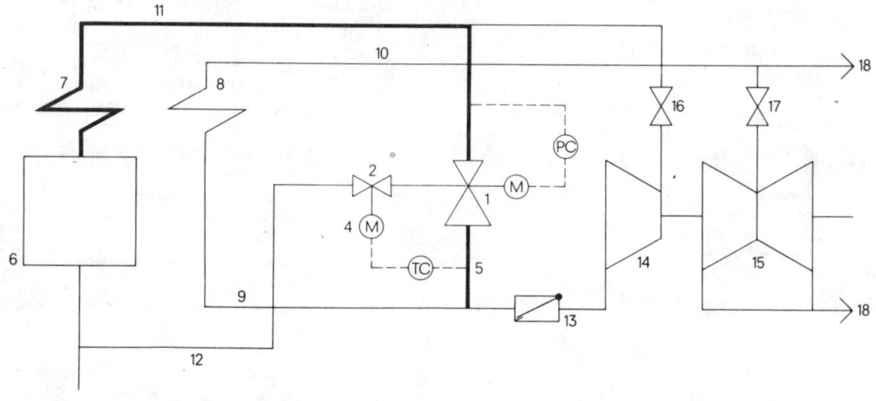

Fig 1

1 Steam converting valve
2 Spray water valve
3 Pressure controller
4 Temperature controller
5 Temperature sensor
6 Boiler
7 Superheater
8 Reheater
9 Cold reheat line
10 Hot reheat line
11 Main steam line
12 Cooling water line
13 Nonreturn valve
14 HP turbine
15 IP turbine
16 HP stop/control valve
17 IP stop/control valve
18 To LP system

Pipe desuperheater with spray unit.

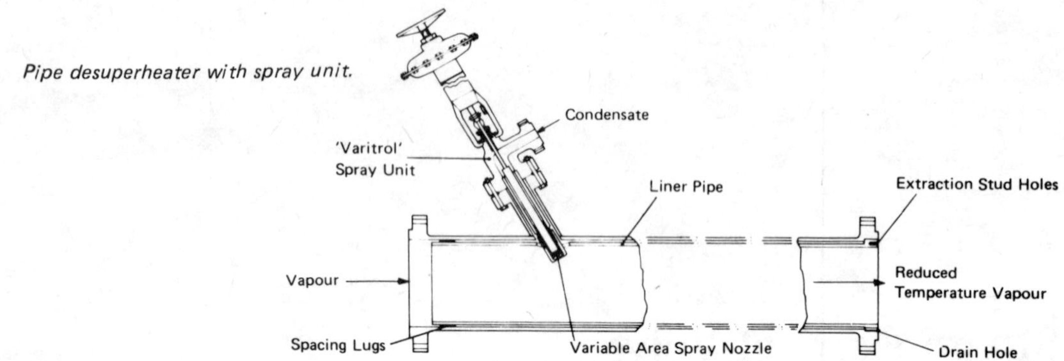

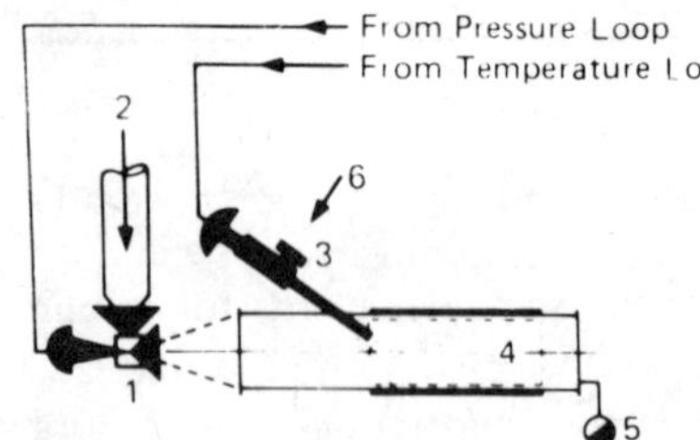

1. Introl 'Machtrol' Valve.
2. Superheated Steam.
3. 'Varitrol' Spray Unit.
4. Introl Pipe Desuperheater and Liner.
5. Drain Valve.
6. Condensate.

Reducing and desuperheating station for variations in steam demand.

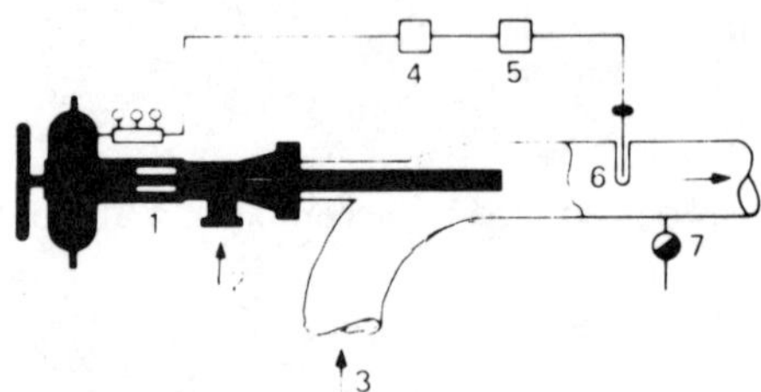

1. 'Varitrol' Spray Unit.
2. Condensate.
3. Superheated Steam.
4. Temperature Controller.
5. Temperature Transmitter (Pneumatic Output).
6. Thermocouple Fe/Co and Pocket.
7. Drain Valve.

Desuperheating station sited on pipework bend.

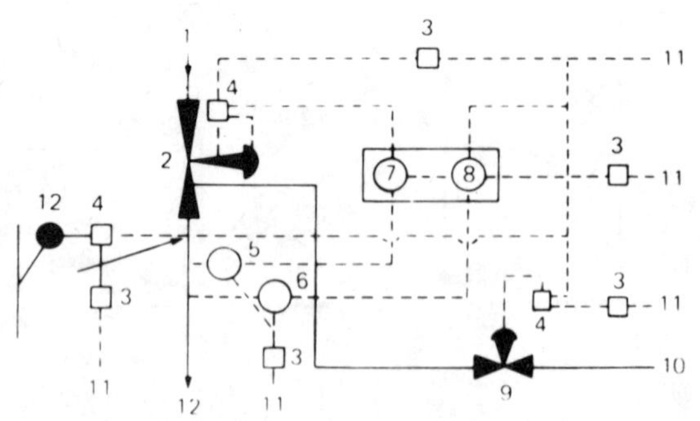

1. Main Steam Supply.
2. 'Capitrol I' (In line) Steam Reducing Desuperheating Unit – Introl Pneumatic Actuator.
3. Airset – Filter/Regulator.
4. Positioner (Split Range Operation).
5. Pressure Transmitter.
6. Temperature Transmitter.
7. Pressure Indicating/Controller – Auto Manual Facility.
8. Temperature Indicating/Controller – Auto Manual Facility.
9. 'Classic 10' Control Valve – Introl Pneumatic Actuator.
10. Cooling Water/Condensate Supply.
11. Air Supply.
12. Pipe Desuperheater with Varitrol Spray Unit.
13. Reduced/Desuperheated Steam.

Desuperheating installation with pipe desuperheater unit supplementing capacity of a combined pressure reducing desuperheating unit.
(All examples by Kent Process Controls Ltd).

In the case of a load step change of the turbine, caused by load rejection or turbine trip, the steam converting valve bypasses the main steam totally or partially into the cold reheat line. During this process, the steam is expanded and desuperheated simultaneously to match the steam conditions in the cold reheat line. The opening of the steam converting valve is controlled according to the temperature on the discharge side of the steam converting valve.

Converter Valve Design

An example of a steam converter valve design is shown in Fig 2. The body (20) and the cover (8) of low-pressure valves are sealed by means of a bolted joint and those of high-pressure valves by means of a self-sealing joint.

Depending on the design, the valve seat (22) is either integral with the body with a facing of hard-wearing material or a screwed in ring, hard faced on the seat, which is seam welded for nominal pressures above PN 40.

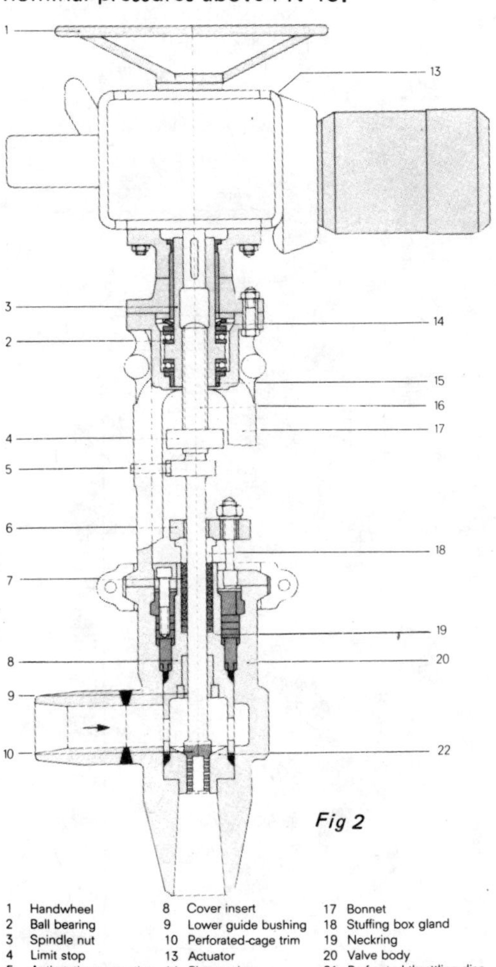

Fig 2

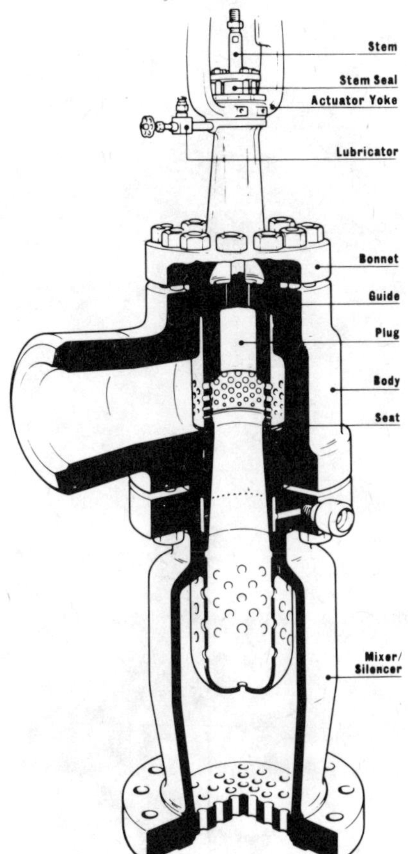

Blakeborough steam reducing/desuperheating unit.

The perforated-cage trim (10) is arranged at the lower end of the valve spindle (16). The valve stem and trim are moved axially by the actuator. This movement, together with the specified characteristic according to which the valve area is opened (equal percentage or linear) determines the flow characteristics of the valve together with the pressure drop across it.

The valve spindle is top guided by the guide bush (15) and sealed by means of gland packings (7) held by the gland and gland bush.

The valve bonnet (17) is mounted on the valve cover (8) or, alternatively, directly on the body flange for valves with self-sealing joints. The valve bonnet carries the actuator which provides the axial thrust for moving the valve stem and trim.

The travel indicator arranged above the gland also serves to prevent the valve spindle and trim from rotating (5).

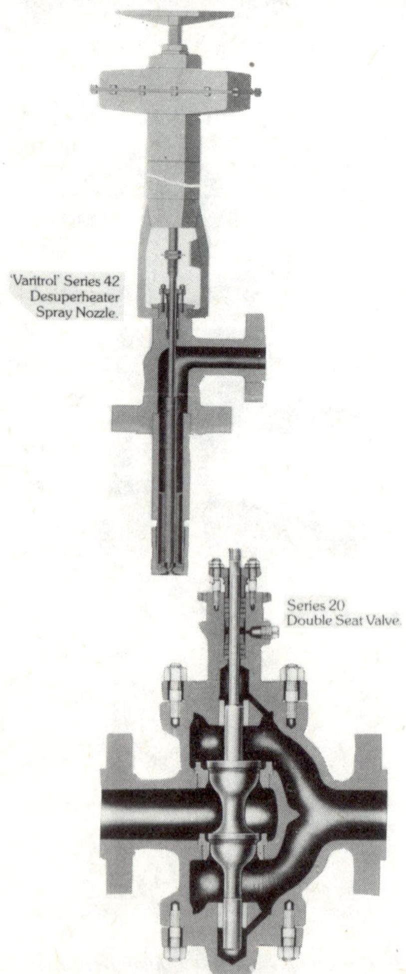

Introl steam control valves (Kent Process Controls Ltd).

Fig 3

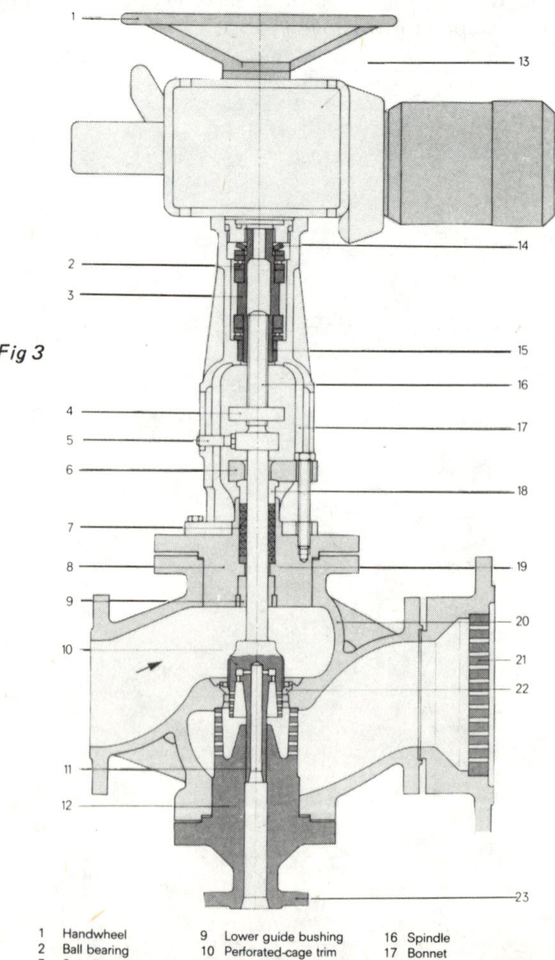

1 Handwheel
2 Ball bearing
3 Spindle nut
4 Limit stop
5 Antirotation protection
6 Packing flange
7 Gland
8 Cover insert
9 Lower guide bushing
10 Perforated-cage trim
11 Telescopic pipe
12 Throttling insert with cooling water connection
13 Actuator
14 Plate spring
15 Upper guide bush
16 Spindle
17 Bonnet
18 Stuffing box gland
19 Neckring
20 Valve body
21 Perforated throttling disc
22 Valve seat
23 Cooling water flange

A plate spring (14) placed above the upper ball bearing ensures smooth contact when the valve trim is moved into the fully closed position and also compensates for axial thermal stress.

The operating nut (3) is fitted in the top of the bonnet (17). Here, the torque applied by the actuator (13) is transformed into thrust by an acme thread.

Valve Operation (Fig 3)

The steam sealing action is obtained by lapped sealing faces arranged on the perforated-cage trim (10) and the seat ring (22). The spray water supply is shut off in a similar manner by the separate spray water valve. When the opening signals of the pressure and temperature controllers occur, both valves operate without any time delay.

The steam passes through the throttling holes of the perforated-cage trim into its inside and mixes there with the spray water. The spray water passes from the spray water valve into the steam via the spray water feed connection, the telescopic pipe (11) and the ring duct, as well as the injection holes of the perforated-cage trim (10).

Once the steam is mixed with the water, the steam/water mixture flows past the pressure-reducing and silencing internal throttling cone (12) and throttling disc (21) into the outgoing pipe. Up to five throttling discs can be fitted according to the pressure drop and the noise requirements.

The telescopic pipe (11) slides in the guide pipe, following the operating position of the valve. Suitable material combinations and temperature-dependent tolerances assure smooth operation. The configuration and size of the holes in the cage trim can be computed so as to obtain any desired characteristic. With equal-percentage characteristic, it is normal standard to obtain a mass flow range of 1:50.

STEAM CONTROL VALVES*

Temperature Controlling Equipment		Combined Steam Converting and Safety Valve
Superheated Steam Cooler	Nozzle Injectors	
Circuit 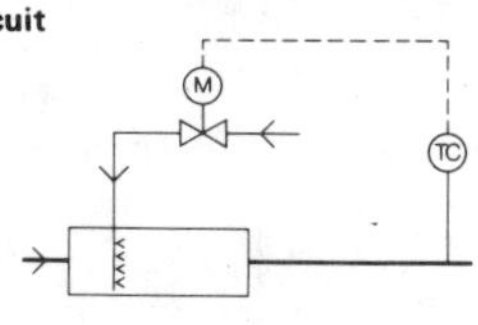	**Circuit** 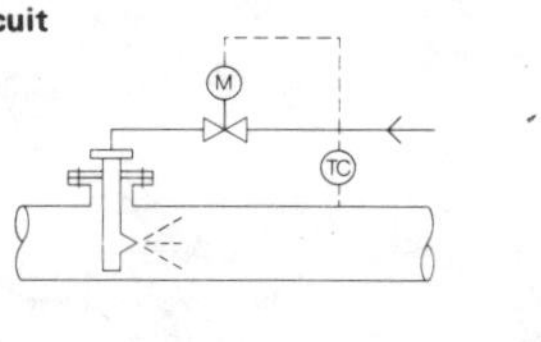	**Circuit** 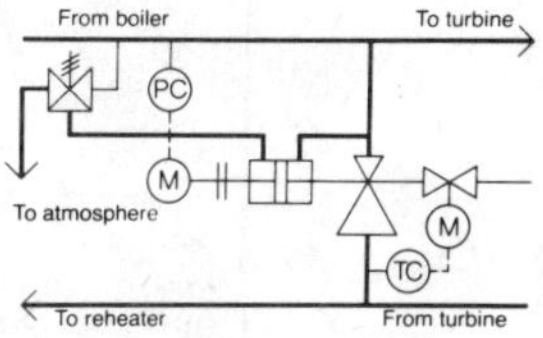
Application: For special temperature-control tasks in industrial power plants.	**Application:** For special temperature control tasks in industrial power plants.	**Application:** For high pressure (HP) bypass stations during start-up and bypass operation, primarily in public utility power plants.
Typical Design: With drawn pipes, welded, flanged (to DIN, ANSI, etc).	**Typical Design:** Nozzle arrangement as required. Division of the spray water among several immersion pipes possible depending upon purpose.	**Typical Design:** Forged, welded connections (to DIN, ANSI, etc). Type 500: inlet at side Type 600: inlet from below
Remarks: Low pressure loss, no additional atomizing steam. Low noise level. No moving parts – simple maintenance. Fitted in all positions. Sizes: DN 125 to DN 1600 (mm).	**Remarks:** High-duty nozzles for easy installation in every pipe line. No moving parts, simple maintenance. DN > 50 (mm).	**Remarks:** In conjunction with the auxiliary control system it functions as a safety valve. No external control media are used, only the existing live steam. With an electric pressure control system the valve functions as a normal control valve.

*Siemens

cont...

STEAM CONTROL VALVES* (Contd).

Feedwater Control Valves	Conventional Control Valves	Level Control Valves
Circuit	Circuit	Circuit
Application: For controlling all or part of the feed-water, also available in a special version as a boiler filling valve.	**Application:** In plants for pressure control of oil, water, steam, etc. As auxiliary control valve, drain valve of start-up flash tank.	**Application:** On high and low pressure (HP and LP) feed-water tanks, condensers etc.
Typical Design: Cast, forged, straight-through valves, angle valves, welded connections, flanges (to DIN, ANSI, etc).	**Typical Design:** Cast, forged, welded pipe construction, straight-through valves, angle valves, welded connections, flanges (to DIN, ANSI, etc).	**Typical Design:** Cast, forged, straight-through valves, angle valves, welded connections, flanges (to DIN, ANSI, etc). Valve parameters (DN, PN) as required.
Remarks: Versions available for all capacities en-countered in practice. DN 50 to DN 500 (mm)	**Remarks:** Typical sizes and pressure ranges. From DN 15 to DN 1500 (mm) From PN 10 to PN 640 (bar) (145 to 9300 lb/in^2)	**Remarks:** For plant-specific problems with evaporating media (cavitation, etc).

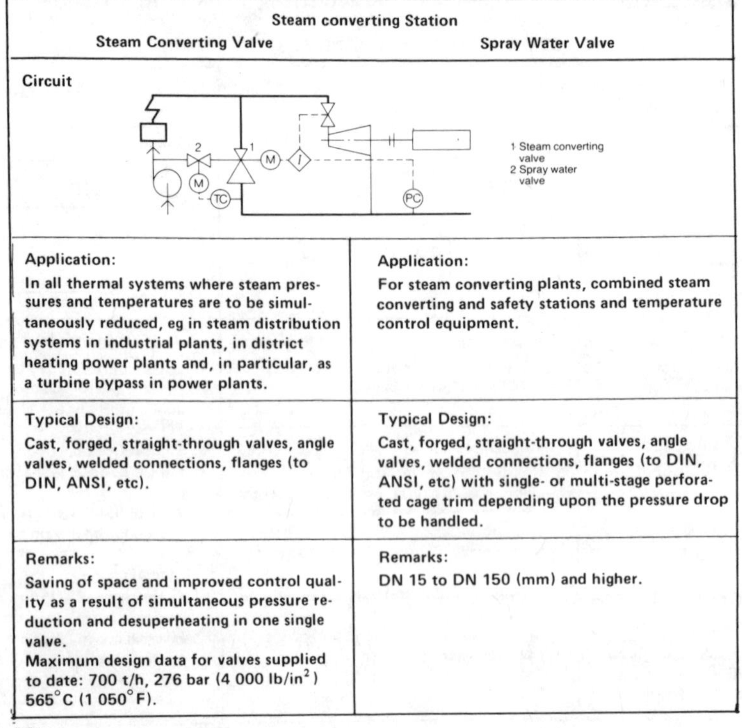

Steam converting Station Steam Converting Valve	Spray Water Valve
Circuit 1 Steam converting valve 2 Spray water valve	
Application: In all thermal systems where steam pres-sures and temperatures are to be simul-taneously reduced, eg in steam distribution systems in industrial plants, in district heating power plants and, in particular, as a turbine bypass in power plants.	**Application:** For steam converting plants, combined steam converting and safety stations and temperature control equipment.
Typical Design: Cast, forged, straight-through valves, angle valves, welded connections, flanges (to DIN, ANSI, etc).	**Typical Design:** Cast, forged, straight-through valves, angle valves, welded connections, flanges (to DIN, ANSI, etc) with single- or multi-stage perfora-ted cage trim depending upon the pressure drop to be handled.
Remarks: Saving of space and improved control qual-ity as a result of simultaneous pressure re-duction and desuperheating in one single valve. Maximum design data for valves supplied to date: 700 t/h, 276 bar (4 000 lb/in^2) 565°C (1 050°F).	**Remarks:** DN 15 to DN 150 (mm) and higher.

*Siemens

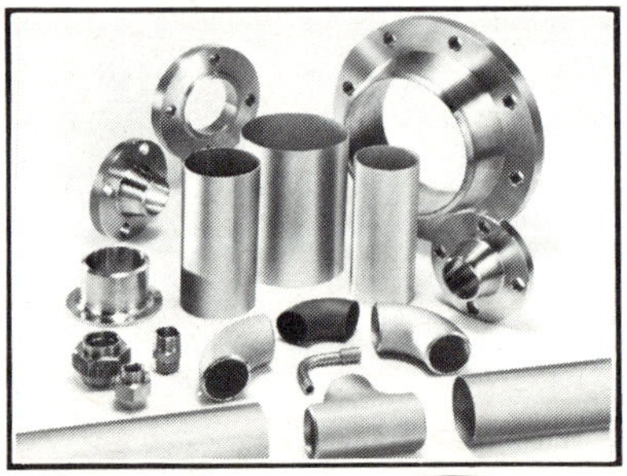

Hygienic Pipelines

STAINLESS STEEL thin walled pipes and stainless steel fittings are produced to various standards for use in hygienic applications. Some standards are

BS1864 – Stainless steel milk pipes and fittings using the recessed O-ring joint.

BS3581 – Stainless steel cone joint pipe fittings.

American 3A – dimensionally similar to BS3581 with metal-to-metal cone type joint.

IDF (International Dairy Federation) – lighter in construction than BS1864 and employing a specially shaped rubber joint to give a flush crevice-free seal.

The most common material used for hygienic pumps, valves and pipes, is 316 stainless steel or 18/10/3 stainless, also known as BS316 S16, with equivalent specifications as follows:-

United States – AISI type 316	France – Z.8CND
Sweden – 832SK and RRNJ44	Germany – V4A Supra

Corrosion of stainless steel pipelines can readily occur, however, if sterilizing agents based on halogens are allowed to remain in contact with the metal for extended periods. See also chapter on *Corrosion of Stainless Steels.*

Glass Tubing

Whilst plastic tubing meets much of the demand for non-toxic pipework, glass tubing may be preferred, or even become essential, for sterile services. Glass is attacked by only a few reagents, which include hydrofluoric and hot concentrated phosphoric acids (both of which produce serious corrosion), superheated water and alkaline solutions.

Cold alkaline solutions attack glasses very slowly, but as the temperature increases the rate of attack rises rapidly. Attack also increased with increasing alkalinity. Attack by super-heated water is seldom serious enough to prevent satisfactory service life from glass tubes, although the rate of attack increases with the temperature and alkalinity of the water. Borosilicate glasses and silica glasses are more resistant to most forms of chemical attack and would thus normally be used where corrosion may be a problem.

See also chapter on *GRP Pipes.*

Thermoplastic pipes also have their attractions because of their chemical inertness and a structure which does not harbour bacteria. Not all such materials are hygienic in the sense that they are free from tainting the product, even PTFE not being ideal for handling foodstuffs. All such materials, too, suffer from relatively low maximum service temperatures, which can make certain types unsuitable for sterilization via cleaning in place. Certain food products are, however, successfully handled by elastomeric pipes (hoses) or rubber lined,pipes.

Vacuum Services

THE MAIN types of valves used for vacuum services are the diaphragm valve and ball valve. Typical forms of diaphragm vacuum valves are shown in Fig 1.

The factor which limits the pressure at which a diaphragm valve can be used is that a large area of elastomer (usually nitrile rubber) is exposed to the process. At low pressures the molecules trapped on the surface of the elastomer are given off, limiting the ultimate pressure which can be obtained. Although materials such as Viton and PTFE, which have lower outgassing rates, can be used, it is more usual to employ different types of valve for low pressure applications.

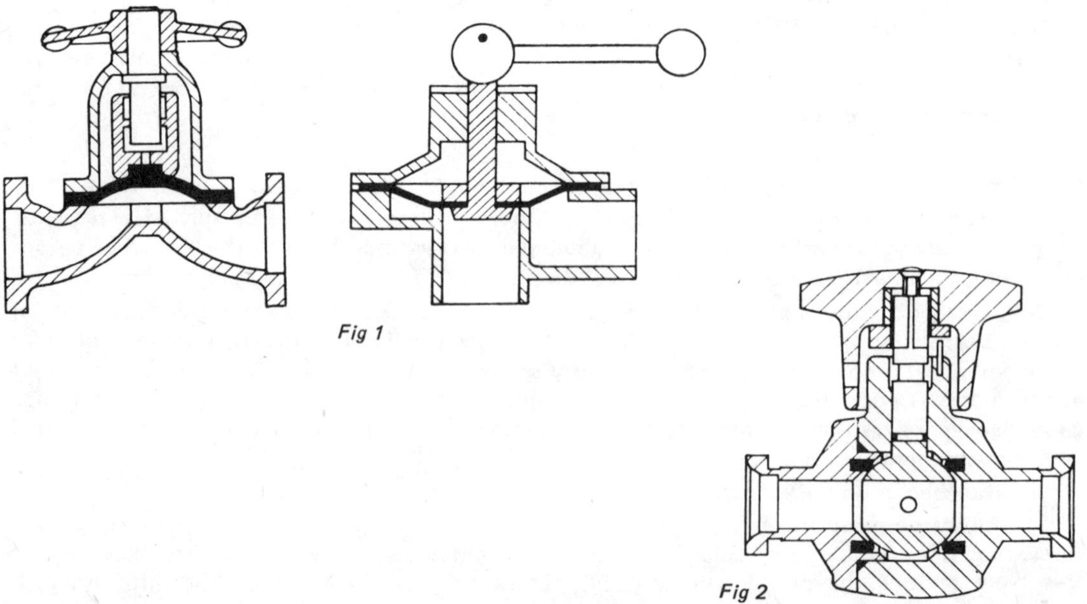

Fig 1

Fig 2

An alternative type, commonly used in pipelines, is the ball valve with a specially moulded seal – Fig 2. This presents a clean bore but the type is susceptible to damage if the fluid is dirty.

In medium to high applications where automatic or remote control is required, magnetically operated valves can be used. These valves have nitrile rubber washer seals which are kept open under spring load. Air admittance valves which automatically open when switched off are also available. See Table I.

As a general rule, reinforced diaphragms are used in valves for vacuum services.

TABLE I – VALVES FOR VACUUM SERVICES

Vacuum	Valve Type	Remarks
Rough to Medium	Diaphragm	Outgassing from elastomeric diaphragm limits the ultimate pressure which can be obtained in the system.
	Globe	With metallic bellows bonnet seal
	Ball	Generally more suitable than other types
Medium to High	Ball	More precisely machined than for medium-rough service
	Plate	May be preferred to ball or diaphragm valve for services down to 10^{-7} torr
	Quarter-turn swing valve	As pipeline valve, or incorporated in a pumping stack
	Gate	Higher conductance than quarter-turn swing valve
	Baffle valves	Baffles associated with an isolating valve for isolating a working vapour diffusion pump when the system is let up to atmospheric pressure
Very High	Right-angle plate valve Quarter-swing valve Ball valve Gate valve Baffle valve	With non-elastomeric seals
Ultra-High	High conductance type	Special, designs and constructions

SECTION 8

Materials

Thermoplastic Pipe Materials

PLASTIC PIPES in a variety of thermoplastic materials have largely replaced traditional pipe materials in a number of situations (*eg* domestic plumbing), as well as becoming attractive low-cost alternatives in others. Their main advantages are freedom from corrosion and good to outstanding resistance to chemical attack, characteristic low resistance to flow because of their smooth bore and general freedom from deposit build-up. Flexible plastic pipes have the further advantage of being particularly easy to cut, fit and joint. The main limitations of thermoplastic pipes are their limited working temperature above ambient and their higher thermal expansion/contraction rates. Rigid plastic pipes, too, can prove difficult to joint satisfactorily using simple techniques.

Plastic pipe production follows National and International standards – see Tables I and II.

TABLE I – BRITISH AND METRIC PIPE CLASSES

BS3505			
Class	Pressure Rating lb/in^2	bar	Equivalent*
B	85	6	NP 6
C	130	9	NP 9
D	175	12	NP 12
E	215	15	NP 15

*For metric sizes

DIN 8061/2		
Class	Pressure Rating lb/in^2	bar
4	58	4
6	89	6
10	145	10
16	230	16

TABLE II – INTERNATIONAL STANDARDS

Country	Standard	Pipe Sizes Covered
America	ANSI/ASTM D1248	Up to 48 in (1200 mm)
British	BS1972 and BS3284	Up to 12 in
Finland	SFS 2336/7	Up to 1600 mm metric
Germany	DIN 8074/5	Up to 1200 mm metric
Italy	UNI 7611/7615	Up to 1200 mm metric
International	ISO–R–161	Up to 1200 mm metric

Polyvinyl Chloride (PVC)

Polyvinyl chloride (PVC) can be produced in a variety of forms with properties ranging from soft, flexible tubes to hand sizes solids in unplasticized PVC. Also the normal softening point (circa 160°F (70°C)) can be raised to about 212°F (100°C) by suitable compounding. The basic material can be transparent or opaque, making it capable of being rendered in a wide range of colours by the addition of dyes or pigments.

Relatively limited use is made of flexible PVC tubes. PVC *hose,* however, is now widely used as a water discharge hose for land drainage, building and industrial sites, *etc*. This is normally of composite construction, comprising a rigid PVC spiral fused with a translucent plastic wall, the assembly being bonded and curved in a continuous heating process. Pressure rating of such hose is about ten times that of flexible PVC tubes and crush resistance about four times greater. Normal recommended operating range for pressure work is –5°F to +70°F (–20°C to +20°C).

UPVC (Unplasticized Polyvinyl Chloride)

Rigid pipes are produced in UPVC and are widely used for cold water services, including underground drainage. A number of manufacturers also produce matching UPVC fittings; and also UPVC/clay pipe connectors to enable new UPVC systems to be connected to existing (or new) clay systems.

Jointing large bore 'Polyorc' pipes to BS3505 with the 'Anger' joint to BS4346:Part II. (IMI Yorkshire Imperial Plastics Ltd).

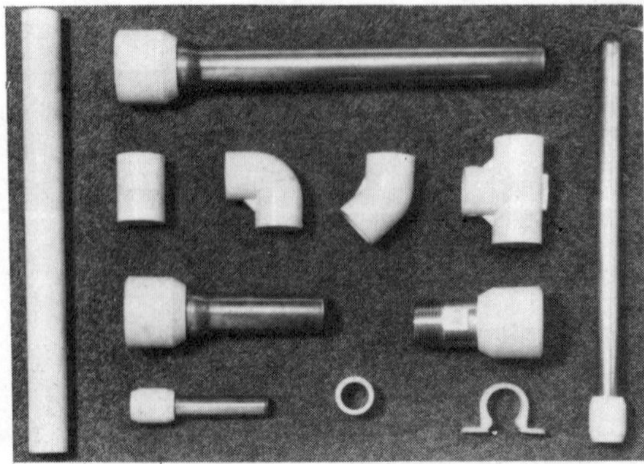

CPVC fittings for hot and cold water systems. (Hunter Genoa)

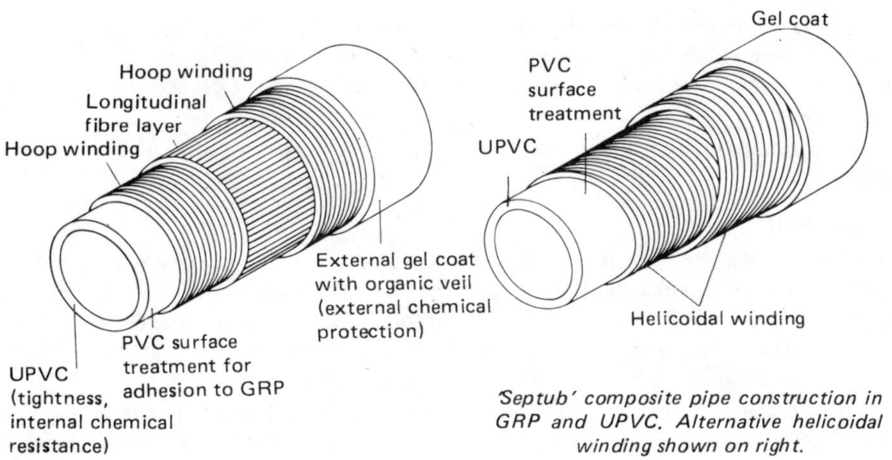

'Septub' composite pipe construction in GRP and UPVC. Alternative helicoidal winding shown on right.

CPVC (Chlorinated Polyvinyl Chloride)

Chlorinated PVC is a rigid PVC material with one more chlorine atom than the PVC molecule, giving it a higher elevated temperature strength. CPVC, in fact, is suitable for both cold and hot water services having a pressure/temperature rating of 406 lb/in^2 (28 bar) at 68°F (20°C) and 87 lb/in^2 (6 bar) at 180°F (82°C).

PVC, UPVC and CPVC are among the few plastic materials for which solvent cements are available, simplifying jointing (although the two rigid materials are also used with mechanical pipe joints).

Polybutylene (PB)

Polybutylene pipes are also suitable for both cold and hot water services, PB being a flexible material complementing CPVC for such services as well as being used on its own (*eg* for long runs where the pipe may be buried in the floor screed). PB cannot be solvent welded and so must be used with mechanical couplings.

Nylon

Nylon tubing is semi-rigid with the tensile strength of various grades of nylon ranging from about 6 000 lb/in^2 (420 bar) to 12 000 lb/in^2 (840 bar). It is generally available only in small diameter sizes where it finds application in fluid pressure lines with working pressures ranging up to about 1 250 lb/in^2 (90 bar). Maximum service temperature for homogeneous nylon pressure tubing is about 60°C (140°F), at which level the normal pressure rating quoted for 20°C is degraded by up to 40%.

The strength (and thus pressure rating) of nylon tubes can be enhanced by glass fibre reinforcement, or braided reinforcement. Braided nylon tubes are produced with both medium and high pressure ratings, the latter providing fatigue-free pressure hose directly competitive in performance with conventional reinforced hose in small bore sizes.

Polyethylene (PE)

Polyethylene — or polythene as it is commonly called — is available in two main grades; normal or low density (LD) or high strength. The latter is a denser polymer with roughly twice the tensile strength of LD polythene, and higher softening and melting points.

Polythene has extremely good chemical resistance and the material is non-toxic, making it suitable for handling foodstuffs, *etc*, as well as drinking water. A particular application is for underground water services, being very economic to lay. Jointing is (almost) invariably by metal compression fittings.

Particular limitations of polythene are its relatively low pressure rating and high coefficient of thermal expansion.

Polypropylene (PP)

Propylene is one of the lightest thermoplastics (density circa 0.9 g/cm^3) with a high softening point (180°C), good mechanical properties and excellent chemical resistance (almost comparable with that of polyethylene). It is superior to polyethylene for resistance to detergents, and superior to nylon for resistance to acids. Working temperature range is also high for a thermoplastic material (–20°C to +500°C).

Normally the material is translucent with a relatively poor resistance to ultra-voilet light. It can be light stabilized by special treatment or the addition of pigments.. Grey or beige-grey is a typical colour for polypropylene pipes.

Polypropylene pipes are more suitable than most other thermoplastic pipes for combined temperature/pressure applications.

ABS

Acrylonitrile butadiene styrene (ABS) copolymers and blends are formulated in a wide variety of grades with hardnesses ranging from about 30 Shore to 100 Shore. The harder, rigid types are used for pipes, normally modified by rubber to give high impact strength. Working temperature range is wide (–40°C to 50°C) and the material is a direct alternative to PVC where greater resistance to mechanical change is required. ABS, however, has a lower chemical resistance than PVC and thus a more limited range of applications. Like PVC, ABS can be jointed with solvent cements.

Polyvinylidene Fluoride (PVDF)

Polyvinylidene fluoride (PVDF) has superior temperature and pressure resistant characteristics to all other thermoplastics, together with excellent chemical resistance. Working temperature range is –40°C to +140°C. It retains high impact resistance, even at low temperatures.

The main limitation of pipes in this material is the difficulty of obtaining satisfactory joints. Methods normally recommended are heat fusion and welding.

Support of Plastic Piping

Because of their high coefficients of thermal expansion adequate support is needed via clips or hinges (or even channels) to avoid excessive sagging at service temperatures above ambient (*eg* when carrying wam fluids). Conversely, if the pipes carry cold fluids appreciable contraction may take place. It is therefore necessary to allow for expansion/contraction characteristics and to provide expansion loops and/or expansion joints on long runs. At the same time if there is movement present, pipe clips or supports should not be so tight that all movement is restricted, otherwise there is the possibility of pipes failing at such points.

Recommended support centres for various types and sizes of plastic pipe are given in Table III.

Frictional Losses

All plastic piping is smooth bore and frictional losses are similar for most materials. Frictional losses with polythene and PVC pipes can be estimated from Fig 1.

TABLE III – RECOMMENDED SUPPORT CENTRES FOR PLASTIC PIPES

	UPVC		ABS		Polyprop.		PVDF			
Temperature (°C)	20°	50°	20°	50°	20°	50°		20°	80°	140°
Diameter (inches)	m	m	m	m	m	m				
3/8	0.6	0.30	0.7	0.6			20 mm	0.95	0.75	0.60
1/2	0.8	0.40	0.9	0.7	0.60	0.55	25 mm	1.00	0.85	0.70
3/4	0.8	0.40	1.0	0.7	0.70	0.65	32 mm	1.10	0.90	0.75
1	0.9	0.45	1.0	0.8	0.76	0.70	40 mm	1.25	1.00	0.80
1¼	1.0	0.50	1.1	0.9	0.87	0.80	50 mm	1.40	1.15	0.95
1½	1.1	0.55	1.2	1.0	0.90	0.85	63 mm	1.50	1.20	1.00
2	1.2	0.60	1.3	1.0	1.00	0.95				
3	1.5	0.75	1.6	1.3	1.20	1.10	75 mm	1.65	1.30	1.10
4	1.7	0.85	1.9	1.5	1.40	1.25	90 mm	1.80	1.45	1.20
6	2.1	1.05	2.1	1.8	1.60	1.50	110 mm	2.00	1.60	1.30
8	2.5	1.25	2.4	1.9	1.70					

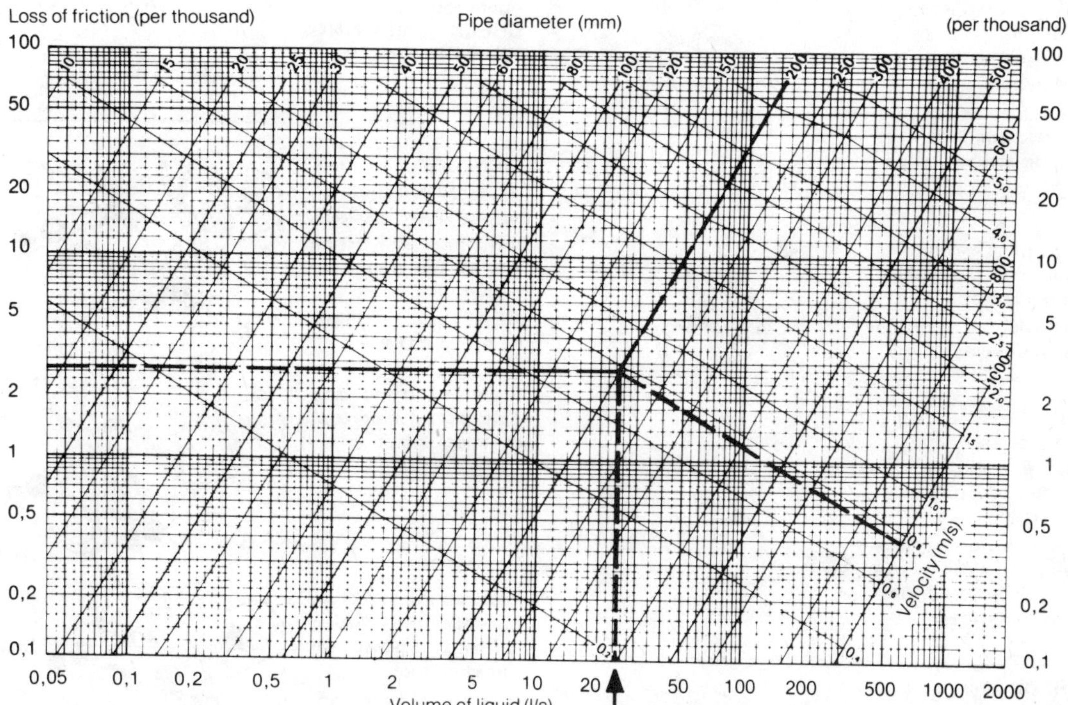

Friction diagram for PE and PVC pipes calculated according to Colebrook.
For diameters up to 200 mm, k = 0.01 mm,
and for diameters exceeding this, k = 0.05 mm.
Water temperature + 10° Cent.

Example: Flow 25 l/sec through 200 mm dia polythene fibre. Connect flow valve to reach diameter diagonal. Read across horizontally to determine loss = 2.8 per thousand. Follow diagonal down to read flow velocity = 0.75 m/sec.

Fig 1 Frictional loss for polythene and PVC pipes.
(Wilk & Hoeglund (UK) Ltd)

Fusion Jointing

The following procedures are typical of those recommended for the fusion jointing of polyethylene pipes for water distribution, sewage, industrial effluents, *etc* and gas distribution. (Details relate specifically to Stewart & Lloyds Plastics polythene system).

Socket Fusion by Hand

Equipment Required

Heating tool (electrical 110 volt)* incorporating heat control and temperature indicator and extension handle.

Depth gauge.

Re-rounding clamp.

Fittings holder (for socket and reducers).

Pipe cutting tool or five toothed wood saw.

Cleaning material — rag or paper — lint free.

Sharp knife.

PREPARATION

CHECK both pipe and fitting are of same material.

Cut pipe end square and deburr.

Clean pipe end using a clean cloth (lint free) or disposable paper towel.

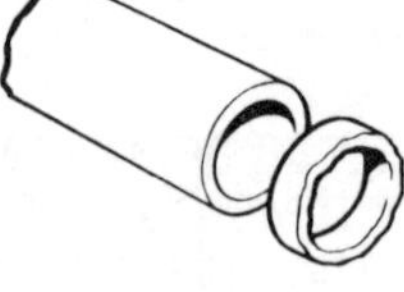

Position re-rounding tool on pipe using depth gauge.

CHECK fusion surface is clean.

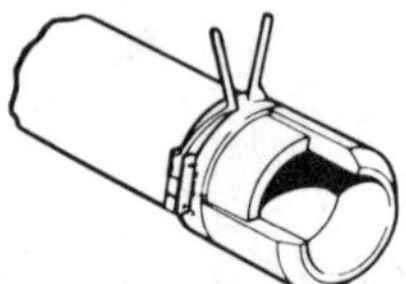

Secure fitting in holder (do not over-tighten).

Clean fusion surface using a clean cloth (lint free) or disposable paper towel.

CHECK fusion tool surfaces are undamaged.

CHECK fusion tool surfaces are clean.

CHECK fusion tool temperature is in the green zone.

(Face temperature 275 ± 15°C).

OPERATION

Before commencing operation

CHECK HEATING TIME REQUIRED.

Size mm	Tool Surface Temp. °C	Heating Time secs.
20	All Sizes 275 ±15°C	5
25		7
32		10
63		20
90		30
125		45

Position pipe and fitting on fusion tool.

Apply firm pressure until pipe and fitting are at full depth on fusion tool.

HEATING TIME STARTS NOW.

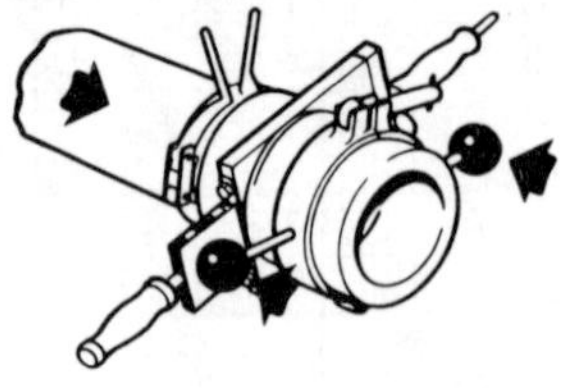

Separate fitting and pipe from fusion tool by 'snap' action.

Push pipe and fitting together until fitting is hard against re-rounding tool.

Hold using firm, constant pressure for two minutes minimum.

Remove re-rounding tool and fitting clamp.

INSPECT JOINT FOR EVENNESS OF MELT PATTERN.

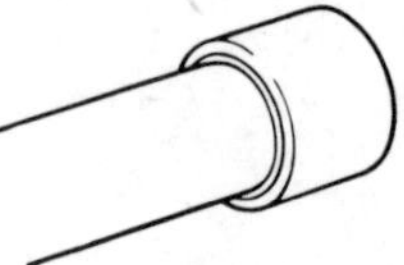

*Gas heated tools are available but are not recommended due to the less accurate temperature control.

Socket Fusion Jointing by Machine

A socket fusion machine should be used wherever possible on 90 mm and 125 mm pipe and fittings

PREPARATION

CHECK both pipe and fitting are of same material.

Close machine.

Place fitting in its clamp.

Butt end against pipe clamp.

Close fitting clamp and secure.

Align pipe clamp and fitting bores.
If necessary, adjust the fittings clamp position.

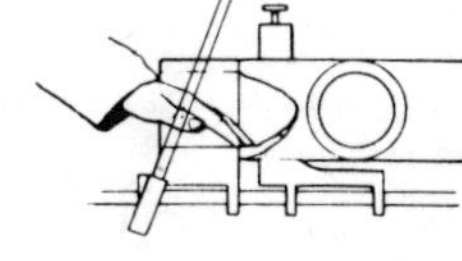

Clean both fusion surfaces using clean cloth or paper (lint free).

Lightly scrape away tarry deposits.

Position pipe in clamp using depth gauge.

CHECK pipe end is aligned with socket.

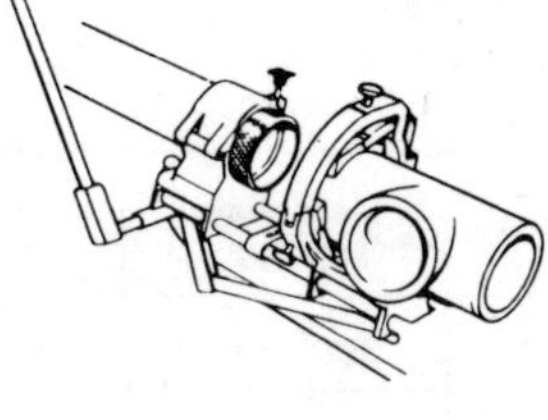

CHECK fusion tool surfaces are undamaged.

CHECK fusion tool surfaces are clean.

CHECK fusion tool temprature is in the green zone or light flashes.

(Face temperature 275 ± 15°C)

OPERATION

Before commencing operation
CHECK HEATING TIME REQUIRED.

Size mm	Heating time secs.
90	30
125	45

Place fusion tool in slot provided.

Operate lever applying firm pressure until pipe and fitting are to full depth on fusion tool.

HEATING TIME STARTS NOW.

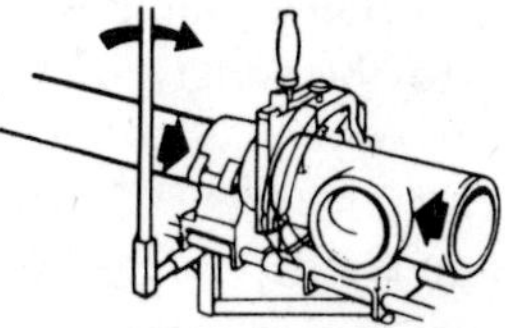

Operate lever to separate fitting and pipe from fusion tool by 'snap action'.

Remove the fusion tool.

Operate lever to push pipe and fitting together until fitting is hard against pipe clamp.

Hold using firm, constant pressure for two minutes minimum.

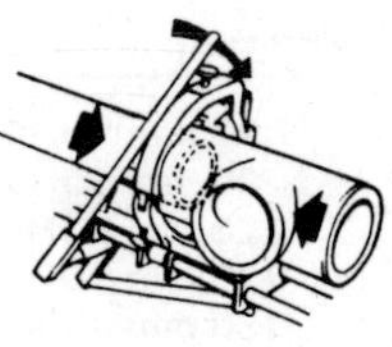

Remove the clamps.

INSPECT JOINT FOR EVENNESS OF MELT PATTERN.

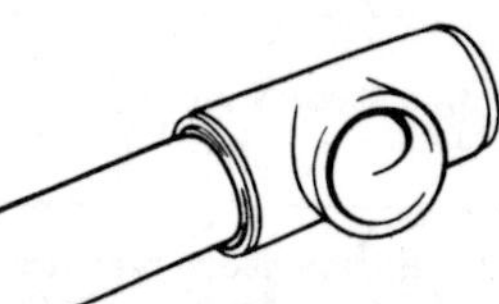

Butt Fusion Jointing

Equipment Required

Butt welding clamps.

Separate or integral hydraulic unit.

Heating plate (110 volt electric) complete with heating control and temperature indicator. (Face temperature 205 ± 8°C).

Planning tool

Pipe supports.

Data plate.

Butt Fusion Conditions

Due to the variety of butt fusion welding machines now available a single set of operating conditions which is correct for all machines cannot be set down. The following parameters are, however, applicable to all machines:-

Fusion temperature – 205 ± 8°C.

Heating/fusion pressure – 1.5 kg/cm^2 of the pipe end cross-sectional area.

Heat soak time – three minutes maximum for all sizes.

Cooling time – dependant on wall thickness, but not less than ten minutes in clamps at maintained fusion pressure followed by at least ten minutes out of clamps.

Machine Operating Conditions

The Table shows times, pressures and temperatures for butt fusion machines in common use on pipe sizes 90, 125 and 180 mm of SDR11 and SDR17 ratings.

Operation	**Machine**	**Fusion Equipment Ltd BF1**	**Fusion Equipment Ltd BF2**	**Hardwick HT180**	**Griesheim Midiplast**	**Haxey Mk II**
Trimming Pressure		**300 lb/in^2**	**300 lb/in^2**	**300 lb/in^2**	**20 bar**	**0.5 bar**
Heating and Fusion Pressure	90 mm SDR11	90 lb/in^2	135 lb/in^2	100 lb/in^2	4 bar	1.0 bar
	90 mm SDR17	60 lb/in^2	90 lb/in^2	65 lb/in^2	2.6 bar	0.7 bar
	125 mm SDR11	180 lb/in^2	270 lb/in^2	200 lb/in^2	7 bar	1.5 bar
	125 mm SDR17	120 lb/in^2	175 lb/in^2	130 lb/in^2	4.5 bar	1.0 bar
	180 mm SDR11		590 lb/in^2	390 lb/in^2	14 bar	3.1 bar
	180 mm SDR17		380 lb/in^2	250 lb/in^2	9 bar	2.0 bar
Heat Soak Time at Zero Pressure		Three minutes maximum all machines				
Cooling Times		Ten minutes in clamps at maintained fusion pressure				
		Ten minutes minimum out of clamps				

Note: Fusion gauge pressure should be the fusion pressure given in Table above, added to drag pressure on gauge when moving pipe length to be welded.

PREPARATION

Check machine planer and heating plate are in working order.

CHECK both pipes are of same material and SDR rating.

Clean pipe ends, inside and outside, with a clean cloth. Lightly scrape away tarry deposits.

Fully open carriage and position planer in machine.

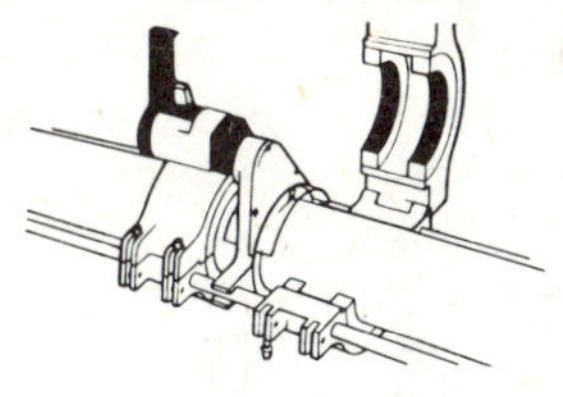

Place pipes in clamps with pipe markings in line (if possible) and ends touching planer.

Align and level pipes using pipe supports. Tighten pipe clamps.

Close carriage and hold pipe ends lightly against planer until continuous shavings are produced from each end.

Continue planing while reducing pressure to zero to avoid steps on pipe ends.

Remove planer.

Remove shavings from machine and inside pipes.

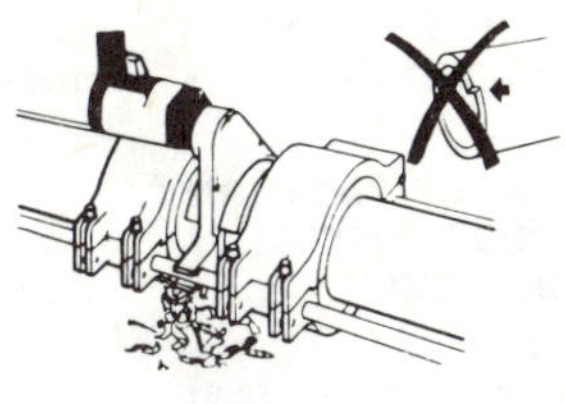

Do not touch pipe ends.

CHECK pipe ends are completely planed.

Bring pipe ends together.

CHECK there is no visible gap and minimize mismatch. If necessary adjust pipe supports and clamps and then re-plane pipe ends.

When checks are satisfactory ensure shavings have been removed from machine and inside pipes.

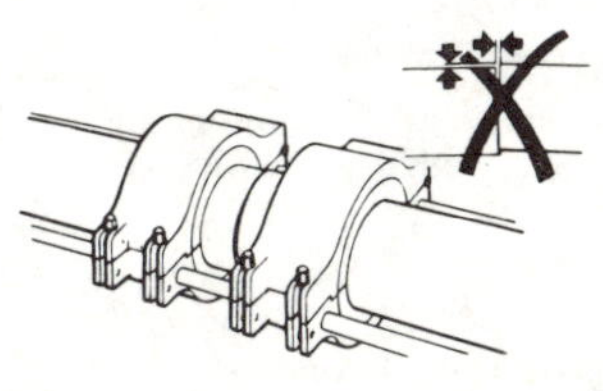

Do not touch pipe ends.

CHECK heating plate surfaces are clean and undamaged.

CHECK heating plate temperature is in the green zone or light flashes.

(Face temperature (205 ± 8°C).

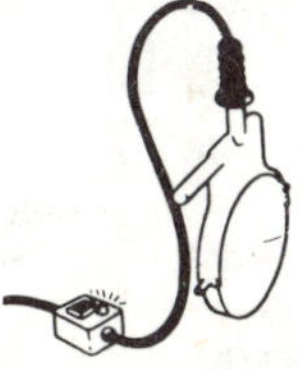

On machine's data plate note:

Heating Soak Time/Fusion Pressure/Fusion Cooling Time

OPERATION

Close carriage and note drag pressure (repeat if necessary).

Position heating plate on supports in machine.

Apply correct pressure (fusion pressure + drag pressure).

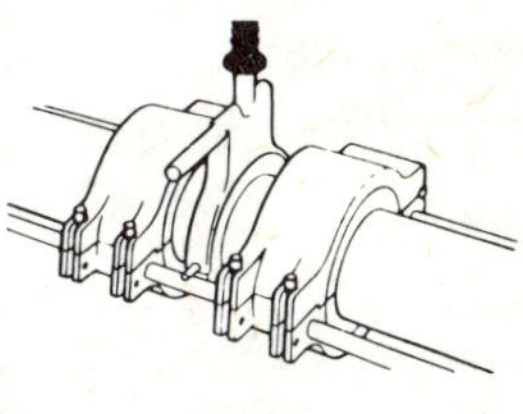

When a 2 mm bead forms completely around each pipe end, release pressure completely.

CHECK heating plate is still gripped by pipe ends.

HEATING SOAK TIME STARTS NOW.

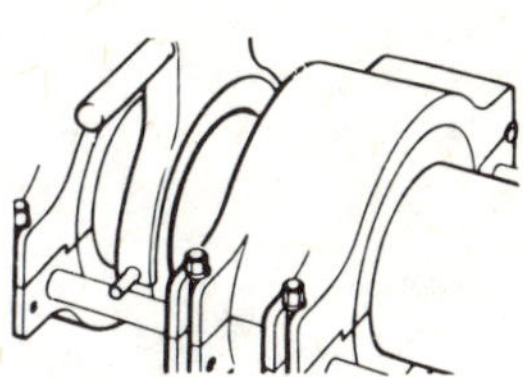

After heating soak time, open carriage.

By hand, tap heating plate from pipe end and carefully remove.

Join pipes and apply correct pressure (fusion pressure + drag pressure).

Maintain pressure for fusion cooling time.

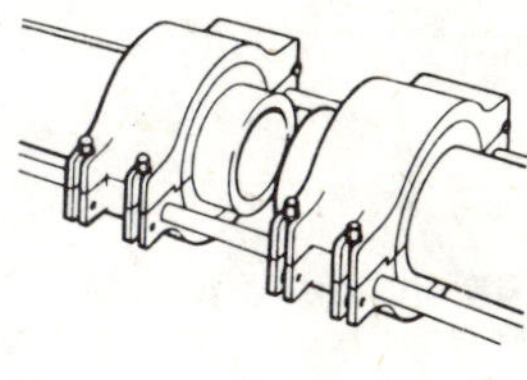

Carefully remove pipe from machine.

CHECK complete weld bead for evenness.

CHECK bead width is: 7–11 mm wide.

If checks are not satisfactory, cut out a short section containing defective butt joint.

Allow pipe to cool for ten minutes before site handling.

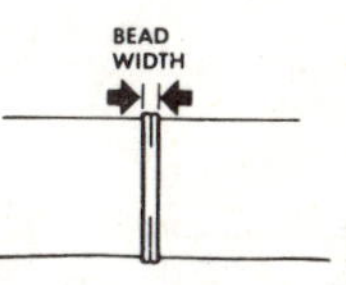

Saddle Fusion Jointing by Hand (Saddles up to 32 mm)

Equipment Required

Heating Tool (electrical 110 volt)* incorporating heat control temperature indicator and extension handle. Window clamp for pipe size being fitted with saddle.

Cleaning material – rag or paper – lint free.

Sharp knife.

PREPARATION

CHECK both pipe, fitting and fusion tool are of same material;

CHECK size of pipe.

Clean pipe using clean cloth.

Fit the correct size window clamp around the pipe.

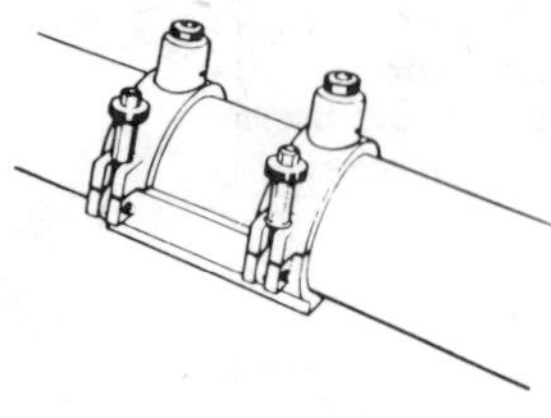

Lightly scrape the exposed area of pipe and saddle base with a square edged blade.

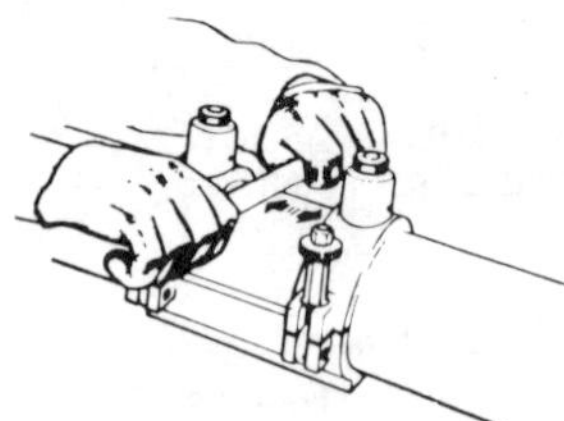

CHECK the fitting is the correct size.

CHECK the fitting fits the pipe.

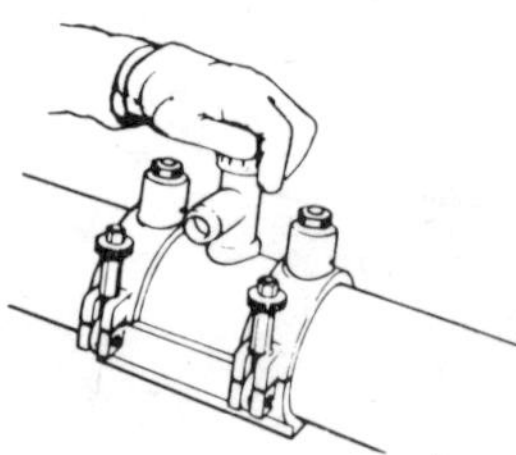

CHECK fusion tool surfaces are clean and undamaged.

CHECK the heating head is the correct size.

CHECK heating plate temperature is in the green zone or light flashes.

(Face temperatures 275 ± 15°C).

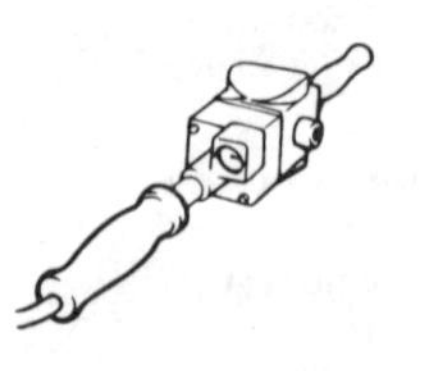

OPERATION

Before commencing operation **CHECK HEATING TIME REQUIRED.**

Size mm	Heating time secs.
All Sizes	25

Firmly press fusion tool onto pipe for a total of 25 seconds. Snap off fusion tool.

CHECK melt pattern is substantially complete.

IF NOT – STOP CHECK steps 1, 2, 3 and 4.

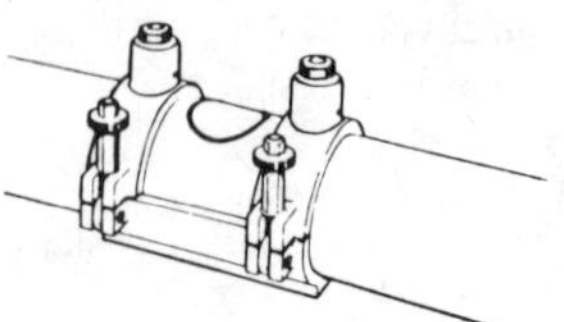

Make a new melt pattern 250 mm (10 inches) from abandoned position.

IF MELT PATTERN IS SUBSTANTIALLY COMPLETE.

Carefully replace fusion tool on melt pattern.

Position fitting on fusion tool.

Apply firm pressure onto fitting.

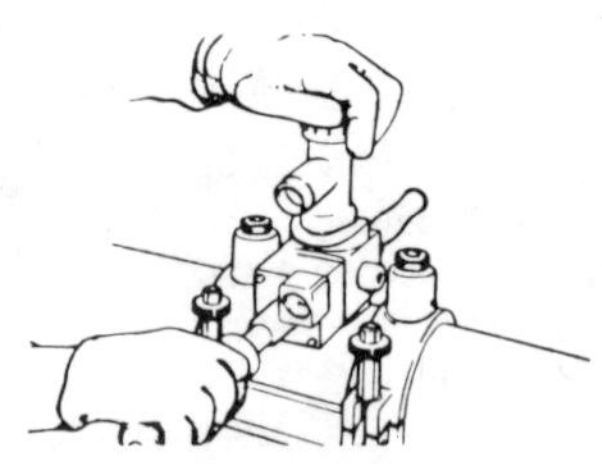

HEATING TIME STARTS NOW

When small complete melt bead forms around fitting, relax pressure.

At end of heating time, snap fitting off fusion tool.

Snap fusion tool off pipe.

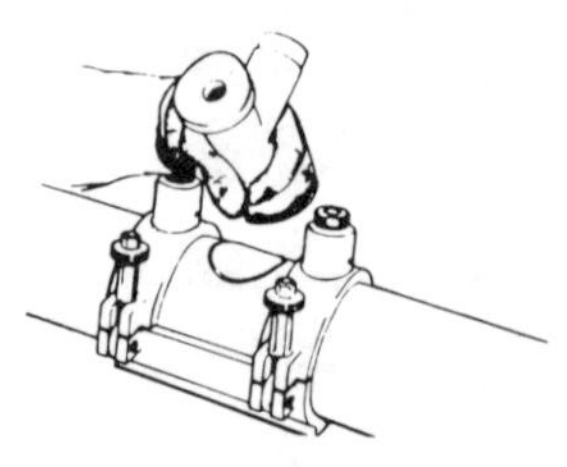

CHECK (quickly) melt patterns are complete, even and clean.

Position fitting within pipe melt pattern.

Press firmly to form a continuously double weld bead.

Hold pressure for 45 seconds.

Allow joint to cool for fifteen minutes before making service connection.

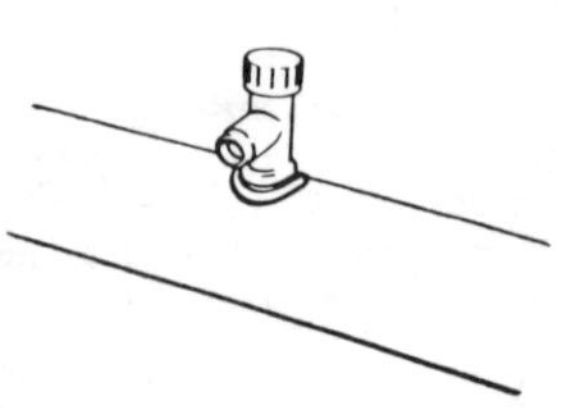

Do not re-use the fitting or reheat the same area on the pipe after it has been through the heating cycle.

*Gas heated tools are available but are not recommended due to the less accurate temperature control.

Saddle Fusion Jointing by Hand (Simple Saddle 63 mm)

PREPARATION

CHECK both pipe and fitting are of same material.

CHECK size of pipe.

Clean pipe using clean cloth.

Assemble the correct size window clamp around the pipe.

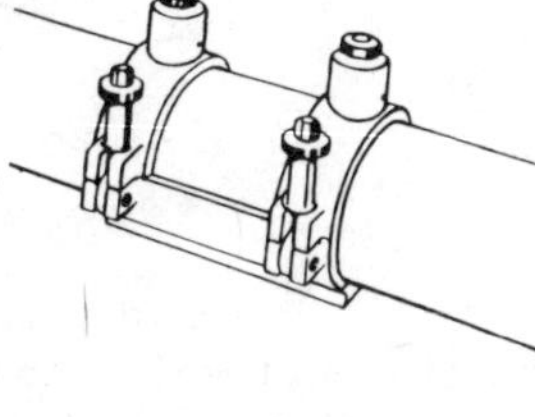

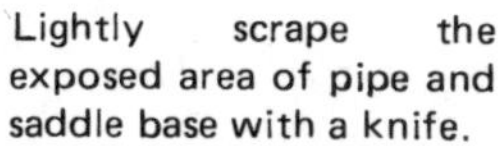

Lightly scrape the exposed area of pipe and saddle base with a knife.

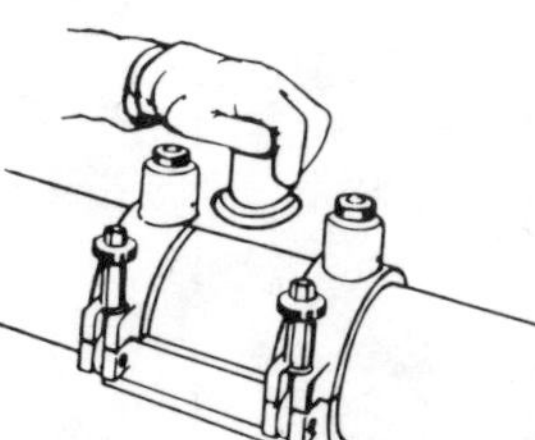

CHECK the fitting is the correct size.

CHECK the fitting fits the pipe.

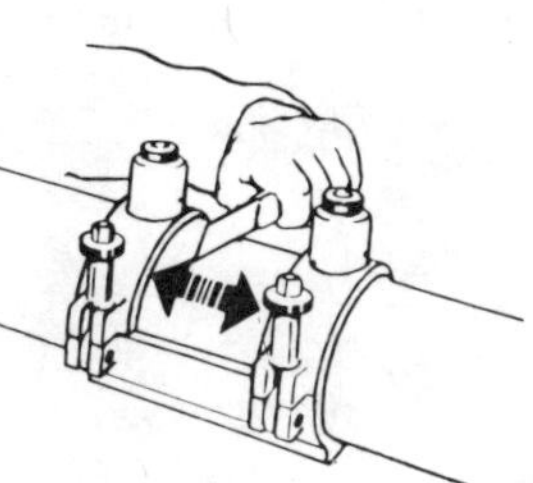

CHECK fusion tool surfaces are clean and undamaged.

CHECK the heating head is the correct size.

CHECK fusion tool temperature is in the green zone.

(Face temperature 275 ± 15°C).

OPERATION

Before commencing operation **CHECK** HEATING TIME REQUIRED.

Size mm	Tool Surface Temp. °C	Heating time secs.
All Sizes of Main	260-290	60

Place heating tool on pipe surface.

Press firmly for thirty seconds and check for bead formation round base of tool.

Snap off tool and check melt pattern substantially complete.

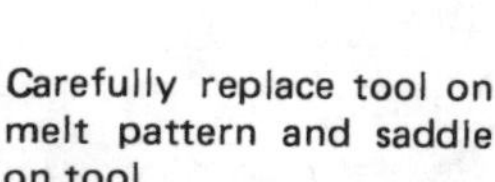

Carefully replace tool on melt pattern and saddle on tool.

HEATING TIME STARTS NOW

Press down firmly on saddle localizing pressure as required to achieve complete internal and external bead formation.

Snap tool with saddle off pipe then saddle off tool.

Check quickly for evenness of melt patterns.

Carefully position saddle within pipe melt pattern.

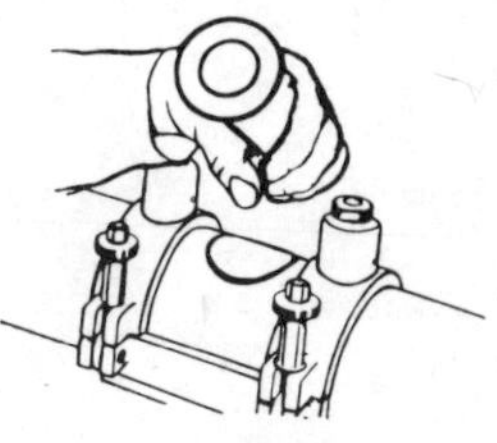

Press down firmly on saddle to form double bead internally and externally.

Hold pressure for sixty seconds.

Allow to cool – fifteen minutes.

Do not re-use the fitting or re-heat the same area on the pipe after it has been through the heating cycle.

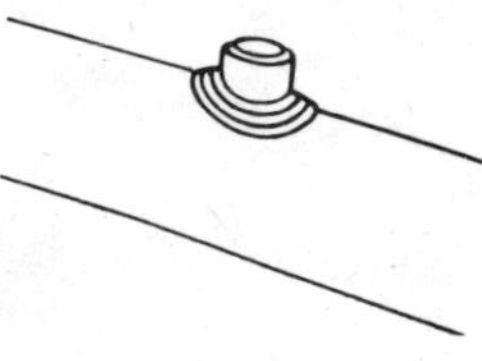

Plastic Valves (Thermoplastic Materials)

Various types of valves are now produced in all-plastic construction (with the exception of valve spindles which are normally steel, plastic-coated), although mainly restricted to smaller sizes. Typical types and sizes produced in UPVC include:-

Angle seat valves – 3/8 inch up to 3 inches.

Ball valves – 3/8 inch up to 3 inches.

Butterfly valves – up to 8 inches.

Other thermoplastics used for valve construction include polypropylene, polyethylene and polyvinyliden fluoride (VDF). PVDF is particularly attractive since it can withstand a range of temperatures and pressures greater than has hitherto been possible with thermoplastic materials. Temperature/pressure maxima for PVDF pipes and valves are shown in Fig 2.

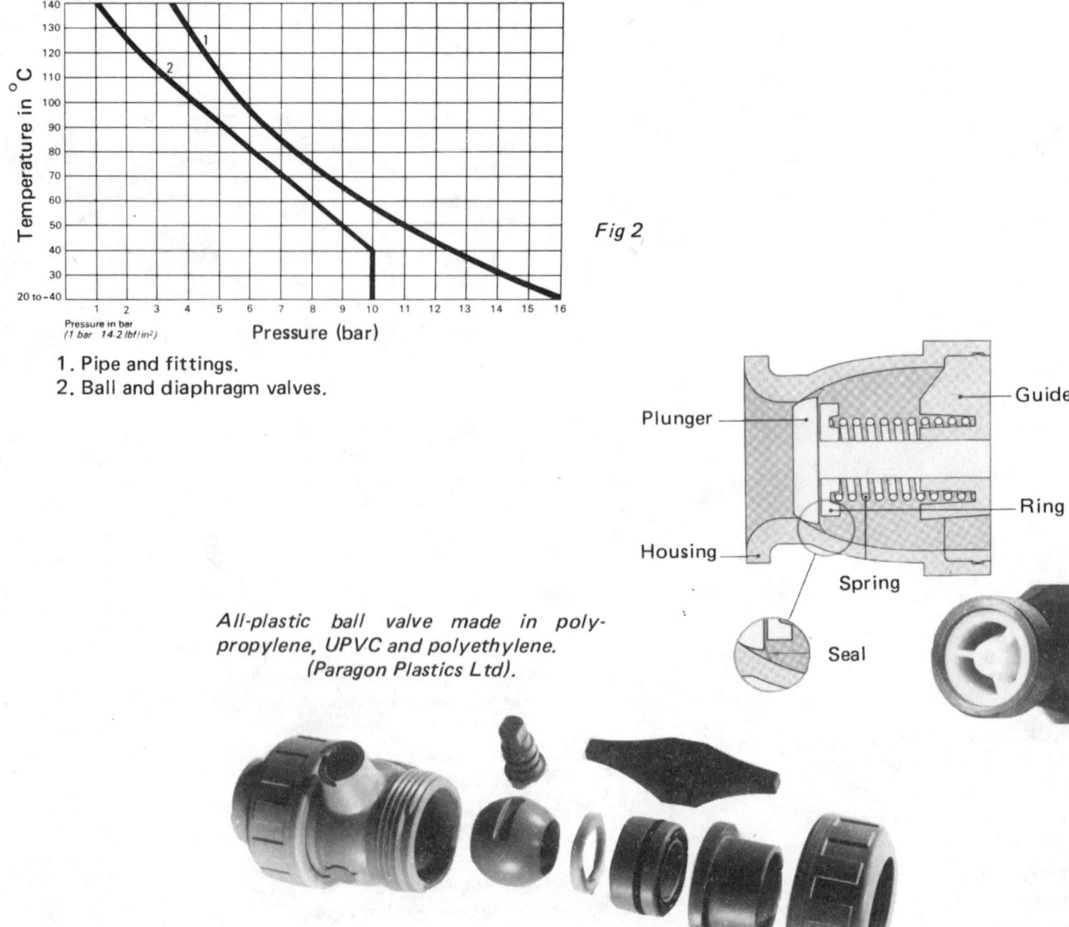

Fig 2

1. Pipe and fittings.
2. Ball and diaphragm valves.

All-plastic ball valve made in polypropylene, UPVC and polyethylene. (Paragon Plastics Ltd).

GRP Pipes

GLASS REINFORCED PLASTIC (GRP) pipes offers the advantage of light weight, high strength/weight ratio and very good resistance to corrosion. Initially production was limited to smaller diameters, with a major disadvantage of high cost. It was not until large diameter GRP pipes (above 24 inch (600 mm) size) were introduced around 1970 that this pipe construction made any significant impact on other than the petrochemical and specialized industries as an alternative to steel pipes. Today GRP pipes are regarded as a cost-effective alternative in the water, sewage and drainage industries, amongst others.

There are two distinct types of GRP pipes as such – those based on conventional glass fibre construction and those including an inert filler, usually sand, in the pipe wall. These are sufficiently different to be referred to by a different name – RPM (reinforced matrix pipes). A particular advantage offered by RPM pipes is a constant stiffness for any given pressure rating.

Materials

Resins used include polyester, epoxy, furane and vinyl ester. Glass fibre reinforcement may be chopped strand, rovings or cloth, or a combination of types. Glass tape may also be incorporated to improve horizontal strength. Aramid fibre is potentially an alternative to glass fibre as a reinforcement material, capable of producing a lighter pipe for the same strength, but has very limited application for pipe construction because of its higher cost.

Sand, where used as a filter in RPM pipes, may account for up to 50% of the total volume, and is employed to build up wall thickness.

Production Methods

Production methods for GRP pipes embrace the following techniques

(i) hand lay-up (conventional GRP moulding)

(ii) filament winding

(iii) pressure/vacuum moulding

(iv) centrifugal spinning

Filament winding is normally preferred for both GRP and RPM pipes, particularly in the larger sizes, employing a steel mandrel. This process also makes it practical to wind a socket integral with the pipe for ease of jointing.

Sizes and Classes

Literally any size (or section shape) can be produced in GRP or RPM pipes, but the majority of production is for standard class sizes in a range of pressure ratings – *eg* 6, 10, 12.5 and 16 bar (87, 145, 181 and 232 lb/in^2) in the UK. Such pipes are normally made to a standard stiffness class.

Buried Stress

A particular difference between a GRP or RPM pipe compared with other standard materials (*eg* steel or concrete) is that it deflects under vertical load with diametrical distortion. In buried applications it thus relies on the support available from embedment to withstand vertical loads. At the same time as the diameter deflects under vertical load, passive soil pressure is induced as the sides move outwards against the embedment.

This means that it is desirable to ensure that buried pipes are laid under conditions which ensure minimal diametric deflections – typically not more than 2–5%. Recommended types of pipe embedment are summarized in Fig 1. Large GRP and RPM pipes laid in unstable soils may have to be supported on piers.

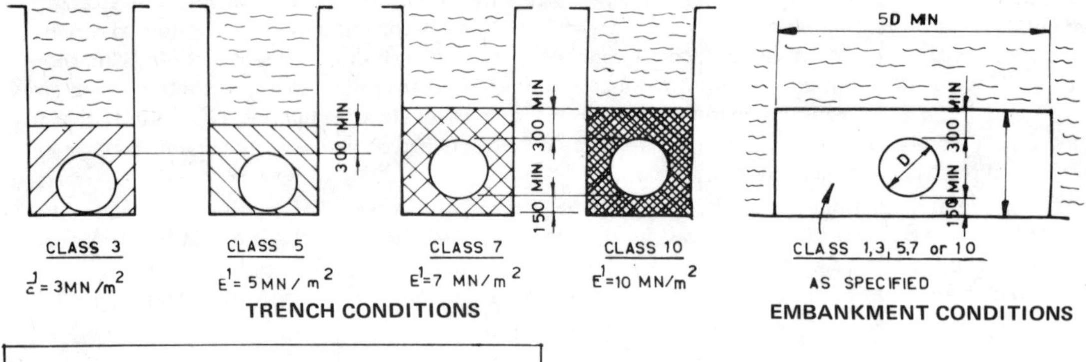

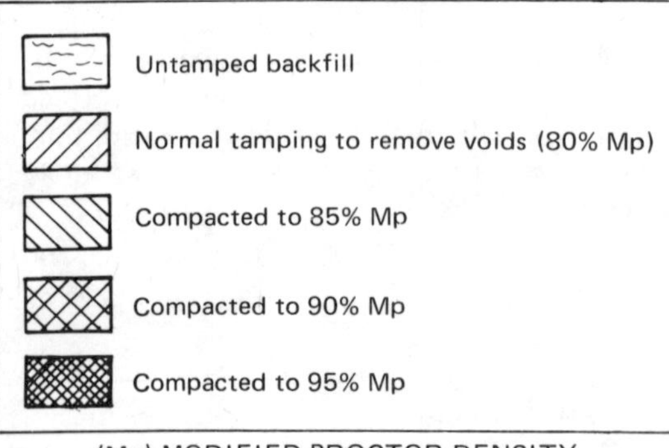

(Mp) MODIFIED PROCTOR DENSITY

Fig 1 Recommended types of GRP and RPM pipe embedment.
(Stanton & Staveley)

Pipe Joints

Joints on RPM pipes are commonly of push-in type. A typical push-in joint comprises an integral socket and moulded spigot combination, incorporating an O-ring seal. Such joints can be pressure tight up to the order of 600 lb/in^2 (40 bar) when properly assembled.

Fittings may be fabricated in RPM or steel. The former is generally limited in pressure applications.

Johnston 'Armaflo' GRP pipe being withdrawn from mould after curing.

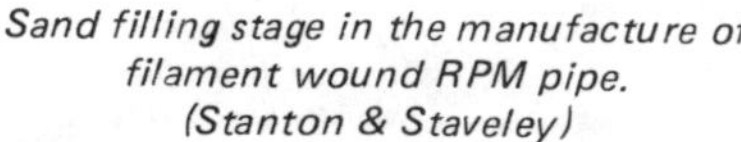

Sand filling stage in the manufacture of filament wound RPM pipe. (Stanton & Staveley)

Cost Effectiveness

RPM pipes can be considered to be competitive with other pipeline materials for both pressure and non-pressure applications when considered on a total installation cost basis. RPM pipe is not necessarily the cheapest material to buy, but, in the majority of cases, the laying costs can be significantly reduced to overcome any differentials in material costs. Savings in laying costs come from a number of sources, *eg:*

(i) A range of pressure classes to match the pipe for the design requirements, unlike some other materials which have too high a performance for a given application and incur material cost penalties as a result.

(ii) Lower transport costs because of their lighter weight. The farther away from

(ii) Lower transport costs because of their lighter weight. The farther away from manufacturing plant and the greater the distance to transport on site the bigger the savings.

(iii) Light weight results in lower duty equipment being required on site and also allows for more rapid movement of pipe during handling and lowering into trench, *etc.*

(iv) Slim profile reduces excavation, the amount of surplus excavated material and backfill costs.

(v) Long lengths result in faster mainlaying operations; this, coupled with light weight can mean that laying to fill can be achieved much more rapidly than with other materials.

(vi) Quick and reliable jointing results in considerable savings when compared with welded steel jointing costs.

Pipe Cutting and Bending

Pipe Cutting and Threading

Large diameter pipes are normally cut by mechanical saws or oxygen (flame) cutters in the manufacturer's shop. Special cutting machines are used for accurate production of profiles for pipe branchwork, saddles and square cuts.

Pipe cutting machines and tools for general or on-site use range in size from powered tools capable of handling the largest sizes of pipes down to hand-operated cutters. The actual cutters may be circular saws (or cutting discs), reciprocating saws, wheels or knife edges. Hacksaws may also be used for hand cutting smaller pipes.

Circular saw cutters are usually robust bench- or stand-mounted machines, but differ appreciably in detail design. Some employ milling cutter wheels, others conventional circular saw blades. The majority encircle the pipe to be cut in a self-centring vice, but some are designed to travel around the pipe as it makes the cut.

Reciprocating saw cutters range from simple machines with minimal guidance to guillotine saws where the tool is clamped to the pipe and the horizontal sawblade is fully guided to ensure a square cut.

Machining large pipes.
(Aiton)

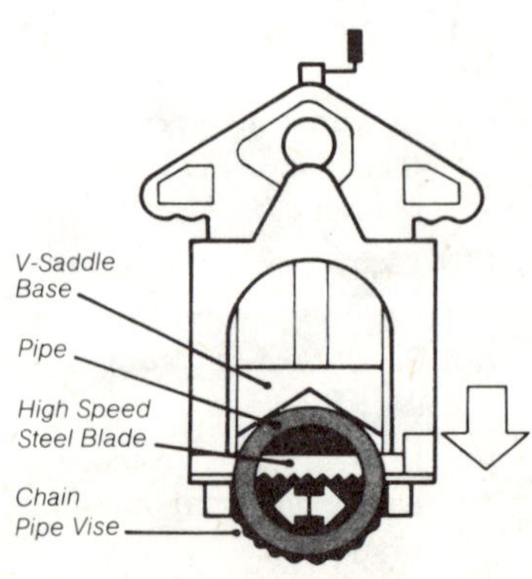

Guillotine pipe saw.
(E. H. Wachs Co).

Trav-L-Cutter portable pipe cutter.
(E. H. Wachs Co.)

Cutter for PVC pipes.
(E. Pass & Co).

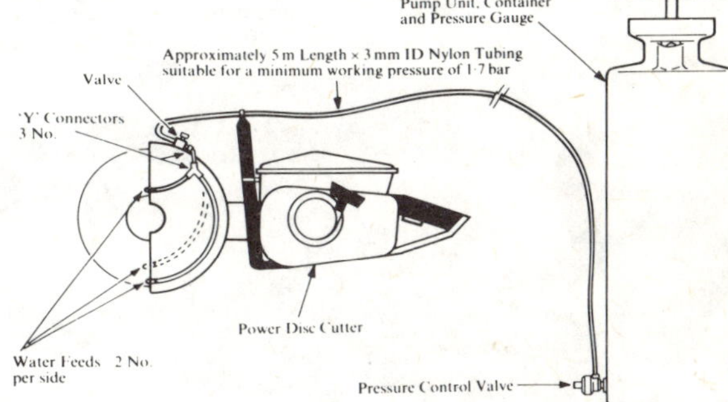

Disc cutter for wet cutting asbestos-cement pipes.
(Dentite, UK)

RA pipe cutting machine
(George Fisher Sales Ltd)

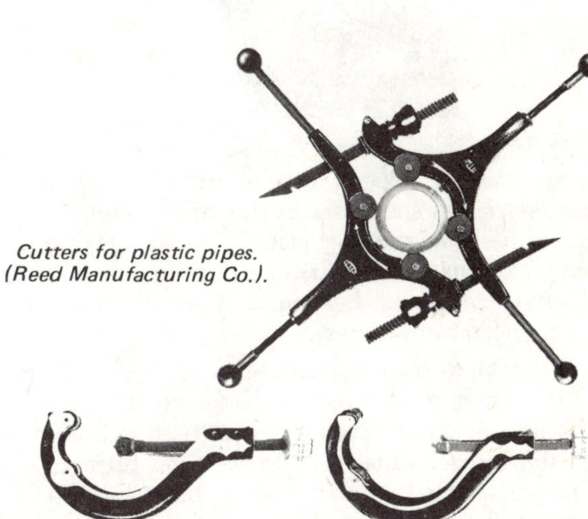

Cutters for plastic pipes.
(Reed Manufacturing Co.).

Wheel or 'knife-edge' cutters are designed to encircle the pipe and hold it concentrically as the cutter is rotated around the pipe. On larger, power-operated machines of this type the actual cutters are symmetrically positioned on a ring which is rotated by power. Each cutter is held in a tool box which automatically increases the depth of cut with rotation of the carrier ring and is profiled to remove equal amounts from the cut simultaneously.

Manually operated cutters normally employ cutting wheels rather than 'knife-edges', with a similar working principle (*ie* the tool is rotated around the pipe to produce the cut). The number of wheels employed may range from one to four. Cutter advance is by rotation of the handle (*ie* is not automatic). The other major difference is that the cutter does not have to be rotated continuously but can be 'rocked' backwards and forwards to make the cut if it incorporates three or four cutting wheels. This is a distinct advantage for close field work as only a relatively small clearance is then needed around the pipe being cut.

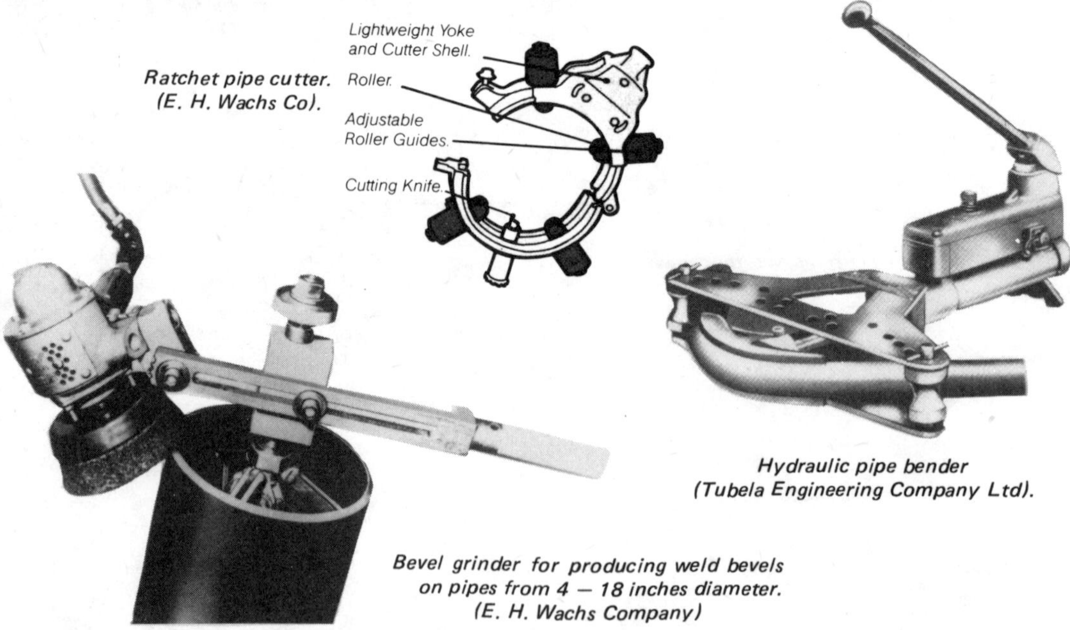

Ratchet pipe cutter. (E. H. Wachs Co).

Hydraulic pipe bender (Tubela Engineering Company Ltd).

Bevel grinder for producing weld bevels on pipes from 4 – 18 inches diameter. (E. H. Wachs Company)

Pipe Bends

Large diameter pipes are bent in the manufacturers' pipe bending shops using various types of bending tables and furnaces (for hot bending). Three main types of bends are used: plain bending, crease bending and corrugated bending. The former two may require the pipes to be filled with sand. Corrugated bends are produced by first corrugating the pipe in the straight on a corrugating machine with local heating, and are then bent empty with each corrugation heated separately (*eg* by portable furnaces).

Plain pipes from ½ in to 12 in (12 mm–300 mm) may be bent cold on hydraulically operated cold bending machines, with bends up to 180° possible on radii depending on the pipe diameter and wall thickness. Plain pipes up to 36 in (900 mm) diameter or more can be bent hot, depending on the furnace capacity. Temperatures up to 2 000°F (1 1100°C) may be used and heating time depends on the pipe size, type of bend and pipe material. For bending the heated pipe is pegged to the bending table at one end and the other end pulled by a winch.

The bending operator, who works to a curved template conforming accurately to the shape of the bend taken directly from a floor drawing, controls the rate of bending and the contour by using a coolant on the exterior of the pipe to control the heat spread. Water is normally used for low carbon steels, but is not recommended for alloy steels. Alternatively, stop pegs are used for control instead of a coolant; these are removable pegs inserted into holes in the bending table as required during the pulling operation.

After bending and allowing to cool, the pipe is emptied of sand, dressed and examined, and then checked on a surface table against a full-scale drawing. In all instances of hot bending, a great deal depends upon the skill of the pipe bender; movement may take place during cooling which must be predicted and allowed for, and the whole procedure of hot bending a large pipe calls for many years of accumulated experience for which there is no substitute. With thinner pipes, subsequent dressing with a flatter may be necessary to take care of slight rippling while the possible ovality of a pipe caused in bending is always under careful control. Sometimes, a resetting may be necessary to correct error, for which gas-fired portable furnaces using premixed gas and air supplies are used.

Tube Bending

Pipe bending by hand is only practicable in the smaller sizes of tubes, and then is not always satisfactory. On larger sizes, or with tubing with thin walls, it is difficult to prevent local collapse of the inner wall unless a pipe bender or filler is used. Provided the tube material is reasonably ductile, all such bends are made cold.

The general rule for minimum radius of bend is that this should not be smaller than three times the outside diameter of the tube. A more generous figure is to be preferred – particularly in the case of the smaller sizes which are usually hand bent. Bend radii larger than the minimum values should always be used as far as possible, since these produce less frictional loss and are less liable to result in deformation of the pipe section through wrinkling or stretching, or introducing ovality. The latter is a common fault, even with pipe benders, unless extreme care is taken, and can materially reduce the working strength of the tube at the bend.

The four basic methods of machine bending are:-

(i) *Press bending* – particularly adapted to the bending of heavy gauge wall tubing up to six times the tube o.d. radii with included bend angles of 120 degrees maximum. It can also be used with wing dies to achieve a minimum bend radius of three times the tube o.d.

(ii) *Roll bending* – also suitable for heavy gauge wall tubing and capable of achieving a satisfactory bend radius down to six times the tube o.d. The bend angle is unlimited since by using three power driven rolls the tube can be fed continuously through the machine to produce complete coils. Four-roll bending machines are capable of producing true arcs right to the extreme end of the tube.

(iii) *Stationary die* – a simple and popular method of bending smaller diameter sizes (usually up to about 5/8 in (16 mm) o.d.). The tube is simply wrapped or 'whipped' around a grooved bending die. Both circular and non-circular bends can be produced, also bends in two planes (using special dies). This method is, therefore, extremely versatile.

(iv) *Revolving die* – in this case the die is rotated whilst the bending shoe remains stationary. The particular advantage of this method is that the tubing can be entirely confined internally and externally at the point of bend, thus minimizing the risk of distortion. The revolving die machine is particularly suited to handling thin-walled tubing.

Proprietary pipe benders are usually based on one or other of these methods. All aim at producing the bends down to a specified minimum radius with minimum distortion of the tube material. Many, it will be appreciated, involve a 'wiping' or rubbing action over the outer surface of the tube and in such cases lubrication is important. Light mineral oil is a satisfactory lubricant for bending steel tubes. Some designs of pipe benders are designed to compensate for 'thinning' effects on the outside wall of the formed bend so that the distribution of material over the cross section remains unaltered. Others may produce appreciable thinning.

Instead of pipe benders, filler may be used to support the inside of pipes and tubes for manual manipulation, or even be used with pipe benders to prevent distortion. The best type of filler for the purpose is a low melting point alloy which can be removed by gently heating after the bend is completed. The use of low melting point metallic fillers is not, however, generally recommended for bending hydraulic tubes as it is difficult to remove completely all traces of the metal. The only effective way of ensuring complete removal is usually to blow through the pipes with steam. Sand is not generally used as a filler as it is difficult to ensure its complete removal after forming, unless elaborate pressure-cleaning methods are used.

The quality of the bend produced, whether manipulated by hand or machine, is very much dependent on the operator. Jerky or irregular actions may produce kinks or wrinkles. Wrinkling may also occur on the inside of the bend due to the compression of the material in this region, unless the machine compensates for this by applying tension to the inner radius.

Thickness of Bends

The minimum thickness (tb) of a straight pipe from which a pipe bent to a radius in accordance with Table I is to be made shall be determined from equation (i) or equation (ii), except where it can be demonstrated that the use of a thickness less than tb would not reduce the thickness below tf at any point after bending.

For pipes 219.1 mm outside diameter and below, and for pipes above 219.1 mm outside diameter bent to the radii specified in the Table, column 2,

$$tb = 1.125\ tf \qquad \text{(i)}$$

For pipes above 219.1 mm outside diameter, where tf is 32 mm or more, bent to the radii specified in the Table, column 3,

$$tb = 1.1\ tf \qquad \text{(ii)}$$

The value of tb is the minimum thickness and provision shall be made for minus tolerances. Manufacturing considerations may make it necessary for pipes thicker than this minimum to be used.

Radii of Bends

Pipes complying with the requirements of BS1387 and BS3601 shall not be bent to radii less than those given in Table I. Other pipes of a thickness determined by BS formulas shall not be bent to radii less than those given in Table I unless

(a) it can be demonstrated that the use of this thickness will not reduce the thickness at any point after bending to below tf, and,

TABLE I – MINIMUM BENDING RADII FOR PIPES OF THICKNESS DETERMINED BY BS FORMULAS

Outside diameter	Radii measured to centre line of pipe	
	tb = 1.125 tf all thicknesses	tb = 1.1 tf tb = 35 mm or above
mm	mm	mm
26.9	65	
33.7	75	
42.4	100	
48.3	115	
60.3	150	
76.1	190	
88.9	230	
101.6	265	
114.3	305	
139.7	380	
168.3	460	
193.7	630	
219.1	710	
244.5	810	1140
273.0	1020	1270
323.9	1220	1520
355.6	1500	1780
406.4	1730	2030
457.0	2030	2280

(b) where the design temperature of the piping is higher than 430°C in the case of carbon steels or higher than 525°C in the case of alloy steels and the radius is less than three times the inside diameter, it can be additionally demonstrated that the thickness at the intrados of the bend is not less than that resulting from the following

$$ti \geqslant tf \; \frac{2R - r}{2R - 2r}$$

where

ti is the thickness at intrados (mm);
R is the radius of the bend (mm);
r is the mean radius of the pipe (mm).

In general it will be necessary to increase the thickness above that determined by BS formulas in order to meet the aforementioned requirements.

There is a minimum thickness for each size of pipe, depending on bending procedure, below which the allowance for thinning will be exceeded, and in such cases the radius given in Table I shall be increased where necessary to ensure that the thickness is not below tf at any point after bending.

Corrosion of Stainless Steel

STEELS CONTAINING more than 11% chromium – generally known as stainless steels – exhibit 'passive' characteristics in that a corrosive ambience a very thin oxide film is formed on the surface limiting further attack (although not necessarily providing total resistance to all corrosive conditions). Where corrosion does subsequently occur it may be of one or more distinct types.

General Corrosion

General corrosion is characterized by uniform attack over a large surface area exposed to the corrosive medium and in the case of stainless steels it implies that passivity has broken down. If the maximum rate of attack is less than 0.004 in (0.1 mm) per year, repassivation can take place after every attack and the steel is considered to be virtually resistant to the corrosive medium. In some cases up to 0.02 in (0.5 mm) per year can be tolerated whereas above 0.04 in (1.0 mm) per year the steel is considered unusable.

Stainless steels usually suffer from general corrosion in strong acidic and alkaline media. They are insensitive to general corrosion in neutral salt solutions. The resistance to oxidizing acids such as nitric increases with the chromium content of the steel (Fig 1). Resistance to reducing acids such as sulphuric is improved by molybdenum and in certain cases copper (Fig 2).

Intergranular/Intercyrstaline Corrosion

This type of corrosion implies that the steel is dissolved or cracked along the grain boundaries (Fig 3). For this type of attack to take place it is necessary to have an acidic aqueous medium of pH less than 4 and for the zones alongside the grain boundaries to be depleted of chromium. When stainless steels (generally austenitic) are heated or cooled slowly through the temperature range 1020°F to 1560°F (550°C to 850°C), chromium carbides are precipitated at the grain boundaries. This leads to an area adjacent to the grain boundaries that is highly depleted in chromium, and which is vulnerable to selective corrosion attack.

To prevent this type of attack, the carbon content can be lowered to 0.03% maximum, or in a higher carbon steel, titanium or niobium are added to tie up or stabilize the surplus carbon. The steels are then insensitive to intergranular/intercrystaline corrosive attack.

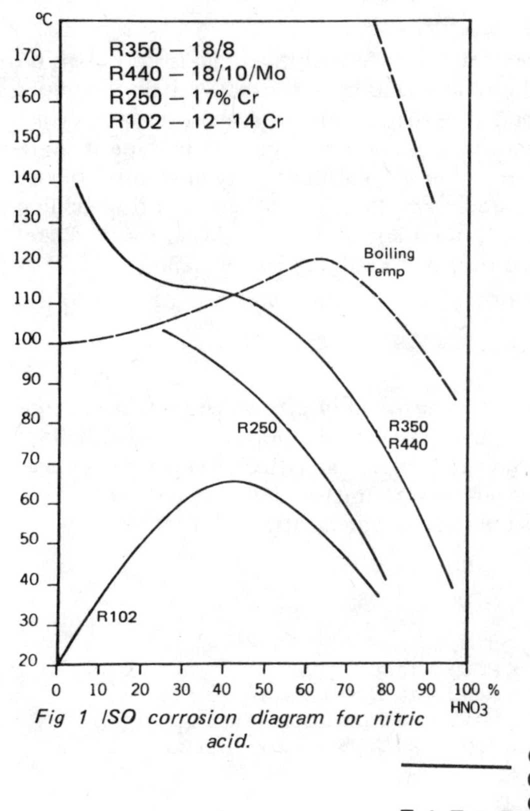

Fig 1 ISO corrosion diagram for nitric acid.

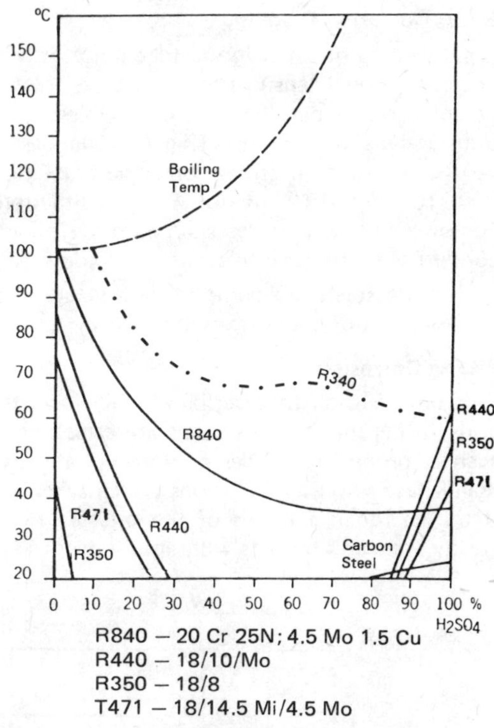

Fig 2 ISO corrosion diagram for sulphuric acid.

——— Corrosion rate 0.1 mm/year

– · – · – Corrosion rate 0.5 mm/year

- - - - - - - Boiling temp of acid

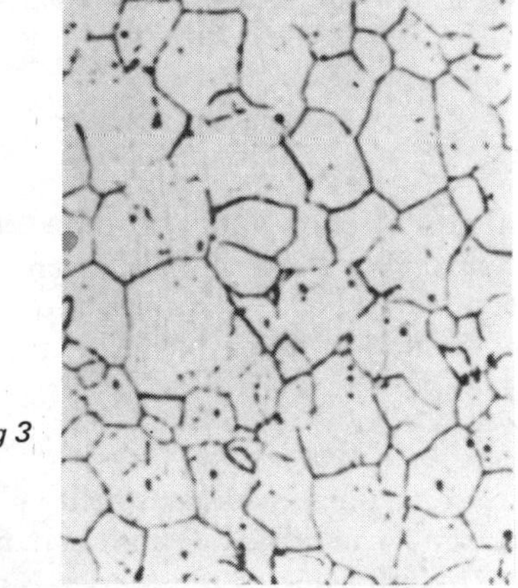

Fig 3

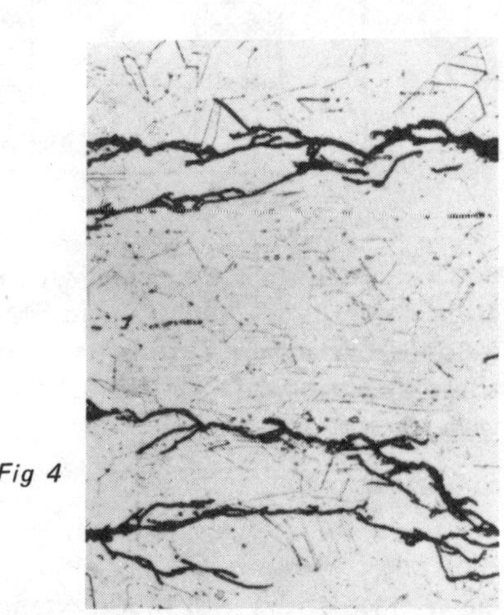

Fig 4

Stress Corrosion Cracking

For this type of corrosion to take place, it is necessary for the surface of the metal to be in tension or have residual tensile stresses left over from cold forming and to be in contact with certain types of corrosive media. The cracking which initiates and propagates rapidly is characterized by its trans-granular appearance (Fig 4). The basic austenitic steels such as type 304 or type 316 are very susceptible to stress corrosion cracking in neutral chloride solutions at temperatures higher than 108°F (70°C); in hot water containing 100 parts per million of chlorides, and in boiling concentrated hydroxides such as those of potassium, lithium and sodium. With increasing nickel content the resistance to stress corrosion cracking improves but is still not entirely eliminated.

Ferritic steels are completely insensitive to chloride stress corrosion and this has led to the development of easily workable steels.

Pitting Corrosion

This term is used to described highly localized attack concentrated in pits which grow predominantly in depth. Stainless steels are attacked usually in the presence of chlorides but other halides such as bromides, iodides or thiosulphates can also give rise to pitting. Attack frequently starts at manganese sulphide inclusions which when dissolved give rise to microscopic pits on the surface. If the corroded surfaces of the pits are not completely repassivated, attack will enter into the steady state and the pits will rapidly deepen.

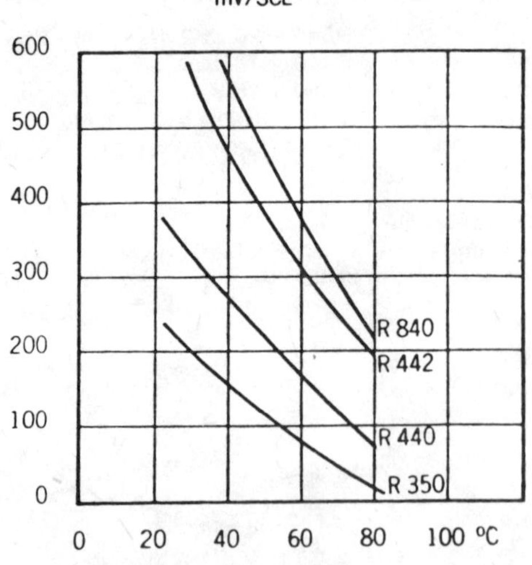

R350 – 18/8 – (T304)
R440 – 18/10/Mo (T316)
R442 – 18/10/Mo Low manganese
R840 – 20 Cr 25 Ni 4.5 Mo 1.5 Cu

Fig 5 Pitting potential – temperature for R350, R440, R442 and R840 in 0.1N NaCl

Resistance to pitting increases with increasing contents of chromium and molybdenum and decreasing contents of sulphur (Fig 5). Resistance to pitting is drastically improved if the manganese content is lowered from its normal level of between 1% and 2% to less than 0.3%. Low manganese steels possess almost as good resistance to pitting as much higher alloys and more expensive steels.

Crevice Corrosion

This occurs in fine crevices and under deposits and coatings in a medium which will normally induce pitting corrosion. The attack mechanism is very similar to pitting when it commences, but at an early stage grows into more extended corrosion. The influence of the crevice is to generate a highly acidic solution within the crevice and make the medium far more corrosive. Crevice corrosion can be slowed down in the same way as pitting by increasing the contents of chromium and molybdenum and by lowering the content of manganese to less than 0.3%.

Erosion-Corrosion

Most corrosion resistant metals receive their protection from the formation of a 'passive' surface coating of an initial corrosion product. They then remain resistant to corrosion as long as this passive surface is not disrupted. An obvious source of disruption is erosion by abrasive particles (*eg* present in fluids carried in pipes). Another source of corrosion developing in pipes carrying water is high fluid velocities, particularly when air is entrained in the water.

Copper alloy pipes in particular are susceptible to this form of erosion-corrosion, the severity of which is normally directly related to the flow velocity. A general guide to resistance to erosion-corrosion in this respect is given below:-

Metal	**Resistance to Erosion-Corrosion at Flow Velocity of**			
	3 ft/sec (1 m/sec)	**4–7 ft/sec (1–2 m/sec)**	**8–15 ft/sec (2–5 m/sec)**	**over 15 ft/sec (over 5 m/sec)**
Copper	fair	poor	very poor	very poor
Red brass	good	poor	very poor	very poor
Aluminium brass	excellent	excellent	good	fair
Phosphor bronze	excellent	excellent	good	fair
Tin bronze (over 5% tin)	excellent	excellent	good	fair
Tin bronze (under 5% tin)	good	fair	poor	poor
Silicon bronze	good	fair	poor	very poor
Manganese bronze	good	good	good	good
Aluminium bronze	good	good	good	good
70/30 copper/nickel	excellent	excellent	good	fair to good
90/10 copper/nickel	excellent	excellent	excellent	excellent

From the aforementioned it will be seen that copper pipes, for example, have only a limited resistance to erosion-corrosion at flow velocities up to 3 ft/sec (1 m/sec) and are highly likely to suffer marked corrosion at higher flow velocities. Aluminium bronze or phosphor bronze pipes, on the other hand would appear unlikely to suffer from erosion-corrosion effects at flow velocities less than 15 ft/sec (5 m/sec).

See also chapter on *Valve Spindle Corrosion.*

Valve Spindle Corrosion

CORROSION OF valve spindles on water services is a common occurrence in many applications, aggravated by the fact that once corrosion has started gland packings are damaged and leaks develop as the surface is attacked. Replacement of the gland packing may only provide temporary relief; it may be necessary to change the valve spindle as well. Surprisingly, too, replacement of the valve spindle with a more exotic stainless steel will not necessarily provide an answer, but merely reduce the corrosion rate.

A further problem is that corrosion may even make its appearance during stocking of the valve, after hydraulic test; but more especially in service on valves remaining unoperated for long periods.

The basic problem is one of a wetted packing in contact with a stainless steel under conditions providing differential aeration and exclusion of oxygen from the surface of the steel under the packing. As a result the steel becomes active rather than passive in this region establishing an electrolytic corrosion cell with the actual steel area becoming the anode. This can quickly give rise to 'pinhole' corrosion, enlarging to cavities in the surface of the stem. The rate of corrosion is markedly affected by the amount of chlorine and oxygen in the water, and also by the quality of the gland packings.

The answer to the problem is best found in the choice of gland packing rather than stem material. Certain packing materials have proved markedly better than others in this respect, *eg* expanded graphite may be preferred to a braided packing, but is not necessarily a complete answer in all applications. Even pure expanded graphite is not necessarily free from promoting electrolytic action.

First attempts to overcome the unexpected limitations shown by asbestos/graphite packings replaced the flake graphite lubricant by mica, and then commercial grade asbestos with chemically pure asbestos. Finally 99.99% pure expanded graphite replaced asbestos valve packings in critical services, but still electrolytic corrosion of valve spindles persisted. It was found that this could be overcome by raising the quality from standard 13% chrome stainless steel to a chrome nickel alloy, but at considerable extra cost. The introduction of a corrosion inhibitor and sacrificial anode into expanded graphite enables even 13% chrome steel spindles to be used without corrosion occurring. The success of such treatment does, however, depend on uniform dispersion of the corrosion inhibitor or anode material through the packing at the material manufacturing stage.

In general 13% chrome steels are to be avoided in the presence of a packing which does not contain a suitable corrosion inhibitor or a sacrificial anode. A classification of steels in order of their resistance to valve stem fitting is:-

	USA	UK	France	Germany	Sweden
(i)	403–410	(56A) 410S21	212C13	4006	2302
(ii)	420	(56D) 420S21	230C13	4028	2303/2304
(iii)	416	(56AM) 416S21	213CF13	4005	2380
(iv)	304	(58E) 203S15	260N18-09	4301	2332/2333

Corrosion test data would indicate the superiority of nuclear purity expanded graphite for valve stem seals, particularly if compounded with a corrosion inhibitor or sacrificial anode. Such a resistance may eliminate corrosion regardless of the contact grade of stainless steel (including 13% chrome).

See also chapters on *Seals and Packings, Nuclear Valve Glands* and *Corrosion of Stainless Steel.*

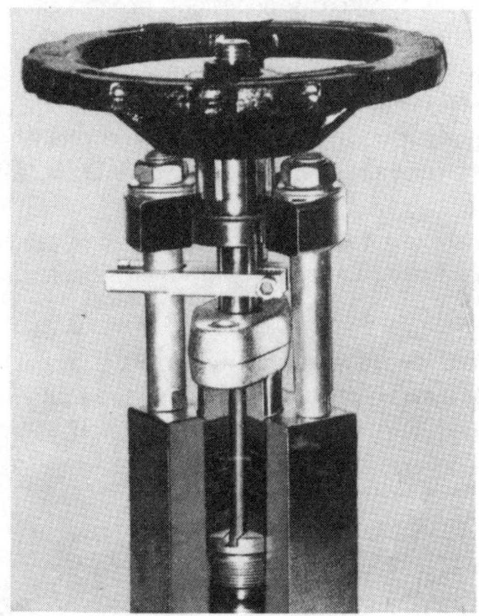

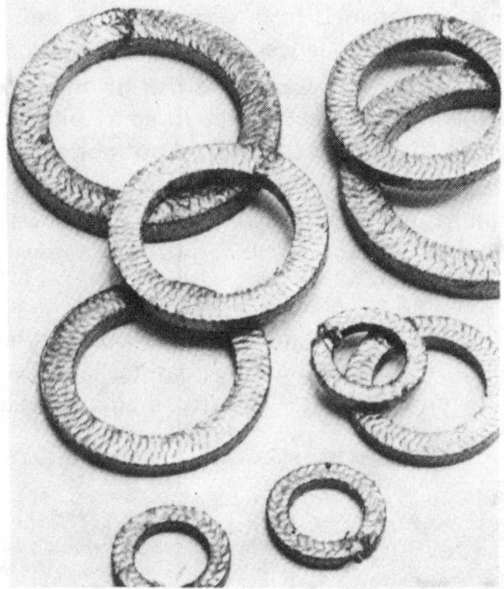

Valve stem seal using moulded rings of LATTYTEX® 117 – a superplaited packing based on chemically pure dust-free graphite asbestos yarn produced by a wet process enabling a 40/60 graphite/asbestos mix to be achieved. This packing is further reinforced with inconel wire, incorporates a sacrificial anode and corrosion inhibitor to provide superior stem protection.

Protective Coatings

SPECIALLY DEVELOPED plastic coatings have greatly increased the range of basic materials which can be used for pipelines and equipment in a wide range of services and process plant. The main benefits obtained from such coatings are protection against corrosion and chemical attack, increased abrasive resistance and non-toxicity, together with cost savings.

Plastic coatings also lessen the risk of impact damage, reduce erosion and in certain instances, improve the flow characteristics of some piping systems. Coating valves, pipes and pipelines can allow metals to be used that would not normally resist, for example, aggressive chemicals, and this can result in substantial cost savings.

Preferably, the selection of the coating materials should take place at the design stage of a project so that the best possible, most economical way of producing the component is selected.

General Coatings

Coatings for general use include PVC, polyethylene, polypropylene and nylon. PVC in particular can be used to produce bearing coatings and is widely used to protect both valves and pipelines in systems handling seawater, gravity drains, service water and sanitary sewers, as well as potable

Vylastic RS60 coating on salt water carrying pipework in the engine room of MS Winter Wave.

water in the water treatment industry. Suitable formulated PVC powders have good impact resistance and high tear strength, making them particularly suitable for pipework. Such coatings allow no bacterial growth – a particular advantage in salt water systems.

Nylon coatings are also suitable for water services, yielding a safe non-toxic finish which aids water flow and eliminates build-up of bacterial growth. Nylon coatings exhibit good abrasion resistance, low water absorption and are also resistant to corrosion from petroleum residue, alkalis, and most solvents.

Polyethylene is a preferred coating for pipework, process vessels, tanks, *etc* where protection from acids is required. Reducer vessels employed in oil refineries have been successfully finished in polyethylene for fifteen years to give protection from the corrosive effects of hydrocarbon solutions.

Valve Linings

Similar plastic coatings may be used as valve linings, although here some of the more exotic plastics may be employed as a preferred alternative to rubber or glass linings for valves handling corrosive and hazardous liquids. Plastic materials used include:-

(i) Polypropylene (PP)
Polypropylene is an excellent general purpose lining with particular applications for water treatment, effluent lines, especially hot effluents from dyestuffs, chemical processing, plating fluids, steelworks picking lines, food and drinking water.

(ii) Polyvinylidenefluoride (PVDF)
Polyvinylidenefluoride is a tough well established thermoplastic material. It is resistant to most inorganic acids and bases. Highly recommended on services such as sodium hypochlorite, aliphatic and aromatic hydrocarbons as well as wet and dry chlorine. It is safe for use with foods, has good temperature and general resistance, especially for use on halogens, halogenated solvents and alcohol services.

(iii) Ethylene Tetrafluoroethylene (ETFE)
Ethylene Tetrafluoroethylene has an outstanding balance of properties and chemical resistance. A very tough fluoropolymer with high resistance to abrasion. Used extensively in the broad chemical field but especially for the fine chemical, petrochemical and pharmaceutical industries. Ideal for process lines, filtration and effluents. It resists strong acids, bases, solvents and is satisfactory for foods. It has superior resistance to environmental stress cracking.

(iv) Ethylene Chlorotrifluoroethylene (Halar®)
A versatile hard material applied to all body sizes and types, including angle bodies and check valves. It is a tough, durable and continuous lining 0.010–0.025 inches (0.25–0.64 mm) thick. Resistant to many industrial chemicals throughout the processing field. Excellent for strong mineral acids such as sulphuric, oxidizing acids, and bases. Temperature resistance generally to 200°F (95°C) and in many cases up to 250°F (120°C). Its inherent property of good impact resistance makes this valve coating an attractive proposition for the mining and heavy chemical industries.

(v) Polytetrafluoroethylene (PTFE)

PTFE is a melt processable perfluoro polymer with the inert properties of TFE, additionally exhibiting excellent physical properties at high temperatures. PTFE is inert to strong mineral, oxidizing and inorganic acids and is resistant to bases, halogens, metal salt solutions, organic acids and anhydrides. Aromatic and aliphatic hydrocarbons, alcohols, aldehydes, ketones, ethers, amines and esters.

It is suitable for services from −40° F to +350° F (−40° C to +175° C). PTFE has good mechanical properties at these elevated temperatures and in many cases improves with continuous exposure. A robust lining offering superior corrosion resistance over an exceptionally wide temperature range.

Vylastic PVC plastisol has been used to coat this 78 inch diameter butterfly valve recently installed at the C.E.G.B's Oldbury Power Station near Bristol.

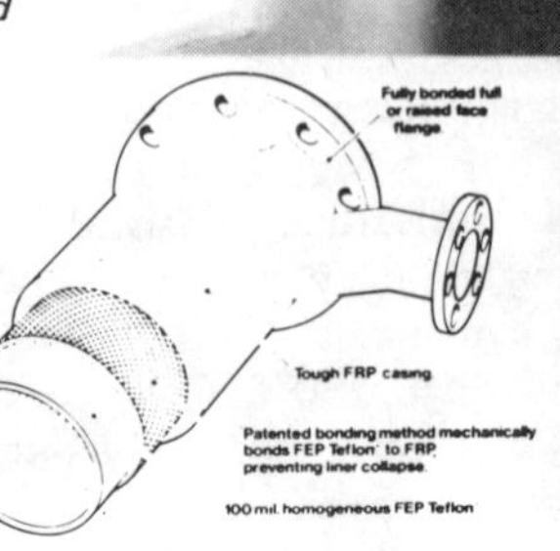

A section of Maxar® FEP piping system, showing the external GRP casing bonded to the internal Teflon FEP liner.

Specialist Coatings

Pipelines which are to go on land, underground or underwater and which may be required to resist attack from moisture, bacteria, fungi, weak acids and alkalis are often coated in epoxy.

Epoxy coatings have established themselves as one of the prime means of protecting plant and pipelines from corrosion because of their excellent chemical abrasion resistance for industrial metalwork. Pipework coated in epoxy is unlikely to suffer abrasion damage on site even if roughly handled, and in the event of damage occurring small areas can be repaired on site using a cold applied epoxy material. The coatings can be applied to either the internal or external surfaces of the pipework, or both. Flange faces and bolt holes can be left uncoated and flanges machined flat if necessary.

Fusion bonded epoxy coatings can be used over a wide temperature range −40°F to +240°F (−40°C to +116°C), which has led to their use in the extreme climates of the Middle East and North Sea.

They are also strongly recommended for use in coating oil pipelines in the North Sea, or other similar environments requiring good chemical resistance.

EXAMPLES OF PLASTIC COATINGS AND APPLICATIONS

	Coating	Normal Thickness Range	Working Temperature Range in Air		Chemical Resistance			Impact Resistance	Coating Applied By	Applications
			Minimum	Maximum	Dilute Acids	Dilute Alkalis	Solvents			
GENERAL COATINGS	Vylastic PVC RS60	0.06–0.20 in 1.5–5.0 mm	−4°F −20°C	140°F 60°C	E	F	N.R.	E	Dipping	Heavy industrial coating, chemical resistance, good salt water resistance
	Vyflex PVC NT80S	0.01–0.03 in 250–750 μm	−4°F −20°C	140°F 60°C	G	F	N.R.	G	Dipping in fluidized bed	Use with potable water, non-toxic.*
	LD polyethylene	0.012–0.036 in 300–900 μm	−76°F −60°C	140°F 60°C	F	F	N.R.	G	Dipping in fluidized bed	Excellent electrical and inorganic chemical resistance.*
	HD polyethylene	0.012–0.024 in 300–600 μm	−58°F −50°C	140°F 60°C	G	G	N.R.	G	Dipping in fluidized bed	Acid and chemical resistance.
SPECIALIST COATINGS	Epoxies	0.01–0.028 in 250–700 μm	−40°F −40°C	194°F 90°C	G	G	G	V.G.	Fluidized bed, electrostatic spray	Chemical and abrasion resistance for industrial work.*
	PTFE (general)	0.0006–0.012 in 15–300 μm	−328°F −200°C	570°F 300°C	G	G	G	G	Spraying	Surface engineering properties, high dielectric strength dry lubrication.
	PTFCE	0.008–0.01 in 200–250 μm	−112°F −80°C	330°F 165°C	E	E	N.R.	V.G.	Spraying	Electrical insulation and high temperature, chemical resistance.
	Polypropylene	0.012–0.026 in 300–650 μm one coat system 0.024–0.09 in 600–2000 μm two coat system	−22°F −30°C	212°F 100°C	G	G	G	G	Dipping in fluidized bed or spraying	One coat system – decorative textured finish. Two coat system – good impact, abrasion and chemical resistance.
	EVA	0.01–0.012 in 250–300 μm	−76°F −60°C	140°F 60°C	G	G	G	F	Dipping in fluidized bed	Pipe fittings for the water and waste treatment industry, excellent weathering resistance, non-toxic.*
LININGS	FEP	0.002–0.088 in 50–2200 μm	−410°F −240°C	500°F 260°C	E	E	E	E	Loose linings or adhesives	Developed for harsh aggressive chemicals the most inert man-made material.
	PVDF	0.04–6 in 1–150 mm	−76°F −60°C	250°F 120°C	E	E	E	E	Loose linings or adhesives	High temperature chemical resistance.

*Non-toxic grades are usually white

Key:
E – Excellent
F – Fair
G – Good
N.R. – Not recommended
V.G. – Very good

Fluoroplastics

Fluoroplastics have seen widespread use over the last decade on pipes and valves, but they are still regarded as the coating for the more specialist applications.

PTFE, the most commonly used fluoroplastic, may be used at extremes of temperature, is totally unaffected by most chemicals and offers corrosion resistance when applied to certain metals where the substrate has been treated, thereby counteracting the porosity of the coating.

PTFE based coatings are also used on filters where their non-wetting characteristics simplify the separation of water from, for example, petroleum. Low friction, non-stick finishes for actuator valves and linings for tanks are amongst their other applications.

Aggressive Chemicals

PTFCE (polytrifluoromonochloroethylene) coatings are used in contact with very aggressive chemicals. The coatings provide a completely non-porous surface which offers a high degree of chemical resistance even at temperatures up to 330°F (165°C).

These chemical properties, combined with excellent electrical insulation, make PTFCE coatings of particular use on chemical instrumentation such as thermocouples, bursting discs and for the internal coating of flowmeter tubes, as well as for valves, seals and fittings.

PTFCE can be used in cryogenic applications and is recommended for use under the severely corrosive conditions found in marine environments. In the food industry, it resists the concentrated acids commonly used and remains non-toxic and non-tainting.

Linings

FEP (fluorinatedethylenepolypropylene) linings are chosen for use where a non-porous coating is required for vessels in a highly aggressive chemical environment.

The linings are available in sheet and film form with or without a fibre glass backing and are amongst the most inert man-made material ever constructed for use in highly corrosive conditions. FEP linings show inertness to virtually all known chemicals except molten alkali metals, fluorine and strong fluorinated agents.

Seamless extruded FEP is available for lining pipes up to 4 ft (1.21 m) diameter and lengths up to 40 ft (12.19 m).

Other linings available from plastic coatings include PVDF (polyvinylidenefluoride) used instead of FEP coating where greater thickness is required for extra corrosion protection. Typical applications again include the coating of tanks and pipework for chemical process and petrochemical industries and in the nuclear industry.

Methods of Application

Methods of application vary, depending on the size of the component to be coated, the material used and the characteristics of the component itself.

PVC coatings are normally applied by dipping. Some PVC formulations and polyethylene powders are applied by the fluidized bed coating of the metal component. Larger components (*eg* large pipes) can be coated by spraying techniques.

Larger vessels can also be coated by electrostatic spraying. The possible size of the component is determined by the ability to cure it in the oven before the coating, but otherwise the entire outside of vessels and pipelines can be sprayed and in the majority of cases internal application is also possible.

Coatings commonly sprayed include nylon, polypropylene, epoxy, polyester and fluoroplastics (including PVDF).

Polythene Sleeving

The additional protection most commonly used for spun iron pipes buried in aggressive soils in the UK and in Europe is polyethylene sleeving applied over the normal works-applied coating immediately before the pipe is laid in the trench. Studies on this material were first made in the USA in 1952 and reported after eleven years experience in 1963. The first operational pipeline in Britain to use polyethylene sleeving was laid in Kent in 1963 in soils of resistivity 1000–2500 ohm.cm and since then very many gas and water lines laid in the UK have been given this additional protection. A recent account of the use of this protection by the British Gas Industry quotes the results of a survey of 18 sites of protected mains ranging in age from three to thirteen years in which only three examinations showed any pitting.

The most important features in the application of polyethylene sleeving are that the pipe surface must be clean (it does not matter if it is wet) and that the sleeving must be wrapped tightly around the barrel of the pipe. Fitting the sleeving snugly around the joint requires special attention. It is now recommended that the ends of the sleeving applied to the barrel should be secured by a turn of petrolatum-impregnated tape which is then carried on over the whole of the joint contour to mate with the polyethylene sleeving applied to the adjacent barrel. The tape is then given mechanical protection by a further short length of polyethylene sleeving secured by adhesive tape, string or more petrolatum-impregnated tape.

Polyethylene sleeving is an attractive and economical method of protection, readily applied on site thus obviating the risk of transit damage. Its efficiency has been confirmed by extensive field experience and laboratory tests and as a result it is now used throughout the world.

Protective Treatments

Protective treatments and coatings used for ferrous metal pipes range from rust-inhibiting coatings to paints and hot enamels and plastic coatings on the one hand, and cold-applied tapes on the other. Rust inhibiting coatings do not provide long-term protection and are only suitable as temporary primers prior to coating with a spearhead barrier coating (*eg* epoxy, polyurethane or vinyl paints). Zinc dust coatings (normally containing 90% zinc dust in the dry film) combine the properties of a conventional coating with cathodic protection and may be used with or without top coats of compatible paints.

The problem with most paint-type coatings is that surface preparation is critical. Epoxy coatings are the most vulnerable in this respect, with coal tar epoxy or coal tar more tolerant, and hot enamels better still.

Pipewraps

Various specialized tapes are produced for use as pipeline coatings consisting essentially of a barrier compound coated into a fabric reinforcement (tape). Such tapes may also incorporate an additional outerwrap for mechanical protection. Some surface preparation (and priming) is normally necessary before the application of such tapes, but, once properly applied, they should be free from debonding under cathodic protection which can occur with paint type coatings. A further feature is that such tapes have high electrical resistance, giving excellent characteristics when cathodic protection is employed.

Pipe wraps fall into two broad categories — those of a flexible nature designed for cold application and those plasticized with a high melting point compound (usually bitumen) which has to be softened by heating to facilitate wrapping and promote bonding both to the pipe and at the tape overlaps.

Cold-applied tape wraps are commonly based on petrolatum and are particularly suitable for the wrapping of small and medium size pipes as well as over joints, couplings and valves. For the protection of buried or submarine pipelines, formulations based on rubber bitumen are used for cold applied tapes; and coal tar or bitumen for hot applied tapes. All types are applied over a compatible primer, the choice of which is also dictated by the pipe surface condition which can be classified as:

(A) *Wet* – Free beads of moisture on the surface of a non-porous body, or wetness on a porous body.

(B) *Damp* – Water on a surface but not enough to give droplets.

(C) *Dry* – Total absence of free moisture.

(1) *Very Clean* – (a) A continuous coat of well adhered primer; (b) Fresh weld; (c) Fresh shot or sand blasted surface.

(2) *Clean* – A surface free from all loose rust, scale, primer or coating, mud *etc.*

(3) *Dirty* – A surface having heavy and loose rust or scale, mud, dirt, poorly adhered primer or coating, *etc.*

Table I is typical of the relationship between choice of primer and surface condition.

TABLE I – TAPEWRAP PRIMERS AND SURFACE CONDITION

Primer	**For Petrolatum Tapes**			**Bituminous Tapes**	**Coal Tar Tapes**
COMPOSITION	petrolatum paste	water-displacing petrolatum paste	petrolatum solution	bitumen solution	synthetic resin
MINIMUM SURFACE CONDITION	B2	A2	B2	C1	C1

SECTION 9

Buyers' Guide

Sub-section (a)

TRADES NAMES INDEX

AIR AUTOMATION – Air control equipment – Joucomatic Controls Ltd.
ALLENAIR – Solenoid valves – Helipebs Controls Ltd.
AMETAL – Corrosion resistant copper alloy – Tour & Andersson Ltd.
APEX – Air relief valves (clean water) – Glenfield & Kennedy Ltd.
ATICOMATIC – Solenoid valves – Helipebs Controls Ltd.
AUTOFLOW – Electric positioner control device – Worcester Controls (UK) Ltd.
AX 90, AX 99 – Pneumatic actuators – Gachot SA

BALLOSTAR° – Ball valve – Rich. Klinger Aktiengesellschaft
BALL VALVE II – Ball valves – Masoneilan Ltd.
BEV–L–GRINDER – Pipe grinder – E. H. Wachs Co.
BLAKEBOROUGH – Valves – J. Blakeborough & Sons Ltd.

CAMFLEX II – Rotary valves – Masoneilan Ltd.
CASCADE – Control valve trim – J. Blakeborough & Sons Ltd.
COCKBURNS – Butterfly valves – Hindle Cockburns Ltd.
CORROFLON – Flexible hose – Aflex Hose Ltd.
CORROFLON – PTFE lined flexible hose – Standard Hose Ltd.
CRYOMISER – Cryogenic valve – Worcester Controls (UK) Ltd.

DIALAFLOW – Electric control device – Worcester Controls (UK) Ltd.
DISCOCOMPACT, DISCOVANNE – Butterfly valves – Gachot SA.
DISCO–O–SEAL – Butterfly valve – Worcester Controls (UK) Ltd.
DUNLOPIPE – Corrosion protected steel pipe – Dunlop Ltd, Dunlopipe Division
DUO–CHEK – Wafer style check valves – TRW Mission Ltd.
DYNA–CON – Quick release hose couplings – IMI Dyna-Quip Ltd.
DYNA–QUIP – Ball type shut-off valves, gas, steam, oxygen valves, quick release hose couplings, blow guns – IMI Dyna-Quip Ltd.

ECON–O–MEE – Brass valve – Worcester Controls (UK) Ltd.
ECON–O–MISER – Three piece valve – Worcester Controls (UK) Ltd.

ECON–O–MITE – Brass valve – Worcester Controls (UK) Ltd.
EKO – Brass drain bibcock – Enolgas Bonomi sas
END PREP LATHE – Portable pipe lathe – E. H. Wachs Co.
ENERGYMISER – Steam valve – Worcester Controls (UK) Ltd.
EPEX – Air relief valves (sewage/effluent systems) – Glenfield & Kennedy Ltd.
EURO 16 – Resilient seal gate valve – Eurovalve Ltd (Selling agent Stanton & Staveley Ltd).
EURO 16 – Valves – Pont A. Mousson SA.
EUROPAM – Valves – Pont A. Mousson SA.
EUROSTOP – Valves – Pont A. Mousson SA.
EXPRESS – Ductile iron pipes – Pont A. Mousson SA.

FIRE-FLOW – Fire safe butterfly valve – Mark Control International
F & R VALVES – Valves, rotary, slide and pinch – Bush & Wilton Valves Ltd.
FLEXMASTER – Portable, mechanical, penstock and valve actuators, flexible shaft drive machines for grinding, sanding, polishing, milling and filing rotary burs – Flexible Drives (Gilmans) Ltd.
FLEXO–RING – Tanker valves – J. Blakeborough & Sons Ltd.
FLOW MATE – Pneumatic actuators – Worcester Controls (UK) Ltd.
FLOWSEAL – High performance butterfly valve – Mark Control International
FLUSH BOTTOM – Tank valve – Yarway Europa BV.
FURMANITE – Under pressure leak sealing service – Furmanite International Ltd.

GASTOP – Plug gas valves – Enoglas Bonomi sas
GESTRA® – Valves, control equipment, process engineering, steam traps, automatic drain equipment – Gustav F. Gerdts GmbH & Co KG.
GUILLOTINE SAW – Portable pipe cutting – E. H. Wachs Co.
GUN–PAKT – Expansion joints – Yarway Europa BV.
GYROVANE – Control device – Worcester Controls (UK) Ltd.

HINDLE – Valves – Hindle Cockburns Ltd.
HI–VENT – Air valves – J. Blakeborough & Sons Ltd.
HYDRA – Metal hoses, expansion joints, metal bellows, pipe hangers – Witzenmann GmbH Metallschlauch- Fabrik Pforzheim.

ISI – Ball valve – Tour & Andersson Ltd

JET–SET – Flexible shaft operated self-priming, submersible pump – Flexible Drives (Gilmans) Ltd.
JOUCOMATIC – Control valves and control equipment, Unilog logic equipment, Trinorm air cylinders – Joucomatic Controls Ltd.

KLINGER° – Piston valves (types KVN, KVI, KVD) – Rich. Klinger Aktiengesellschaft
KLINGERFLON° – PTFE products – Rich. Klinger Aktiengesellschaft
KLINGERIT° – CAF gasket sheets – Rich. Klinger Aktiengesellschaft
KTM – Ball valves – KTM (UK)

LARNER-JOHNSON – Streamline valves – J. Blakeborough & Sons Ltd.
LATTYFLON – PTFE packings for valve stem glands – Latty International
LATTYGRAF EMG – Expanded graphite joints for valve bonnet and body seals – Latty International
LATTYGRAF E and E/1 – Expanded graphite packings for valve stem glands – Latty International
LATTYTEX – Plaited packings for valve stem glands – Latty International
LOGOS – Brass ball valve (reduced bore) – Enolgas Bonomi sas

MARKAIR – Valve actuator – pneumatic – Mark Control International
MARKTRON – Valve actuator – electric – Mark Control International
MARPAC – Ball valve – Mark Control International
MICROPAK – Small flow application valves – Masoneilan Ltd
MIL – Steam traps, control valves, bellows seal valves, strainers – BVMI Ltd.
MINITORK II – Butterfly valves – Masoneilan Ltd.
MISTCOOL – Liquid spraying equipment – IMI Norgren Enots Ltd.
MITI 1 – Pneumatic actuator – Worcester Controls (UK) Ltd.

NELDISC® – Butterfly valves – Neles OY
NORBRO – Valve actuator – Worcester Controls (UK) Ltd.
NORMIFLO – Water flow regulator – Tour & Andersson Ltd.

OCEAN – Check, diaphragm, hydraulically operated and non-return valves – Ocean BV
OLYMPIAN – Compressed air processing system – IMI Norgren Enots Ltd.
OMEGA – Brass ball valve (full bore) – Enolgas Bonomi sas
ONDASTOP – Membrane check valve – AMIS SpA Apparecchi e Macchine Idrauliche Speciali

PAR/LARC – Automatic pump recirculating valve – Yarway Europa BV
PNEUTROL – Flow control valves – Helipebs Controls Ltd.
POLYFILL – PTFE based substance – Worcester Controls (UK) Ltd.

Q–BALL® – Ball valve with low noise trim – Neles OY

RADIETT – Twin entry radiator valve – Tour & Andersson Ltd.
RADIFIX – Radiator valve manifold assembly – Tour & Andersson Ltd.
RAY – Pressure snubbers – IMI Norgren Enots Ltd. assoc. co. IMI Shipston Ltd.
RECOIL – Non-slam check valves – Glenfield & Kennedy Ltd.
ROTORK 'A' RANGE – Electric valve actuator – Rotork Controls Ltd.
ROTORK 'H' RANGE – Hydraulic valve actuator – Rotork Controls Ltd.
ROTORK 'P' RANGE – Pneumatic valve actuator – Rotork Controls Ltd.

SAMPLE MASTER – Sampling valve – Yarway Europa BV.
SECURO – Plug gas valve – Enolgas Bonomi sas
SIGMA F – Globe valves (pneumatically actuated) – Masoneilan Ltd.
SKATOSKALO – Flexible shaft drive machines and accessories for boiler descaling and tube cleaning – Flexible Drives (Gilmans) Ltd.
STANDARD – Ductile iron pipes – Pont A. Mousson SA.
STEAM FORM – Steam conditioning valve – Yarway Europa BV.
STEM–BALL® – Ball valves – Neles OY
SYNCROPAK – Integral starter and control package for electric actuator – Rotork Controls Ltd.

TEMPLOW – Desuperheater valve – Yarway Europa BV
TRAPAID – Steam trap failure indicator – Furmanite International Ltd.
TRAV–L–CUTTER – Portable pipe cutting and bevelling – E. H. Wachs Co.
TREVITEST – Safety valve testing service – Furmanite International Ltd.
TRIDUCT – Ductile iron pipes – Pont A. Mousson SA
TURBO–CASCADE – High pressure fluid control – Yarway Europa BV

ULTRAIRE – High efficiency compressed air filters – IMI Norgren Enots Ltd.
ULTRASEAL – Ball valves – Hindle Cockburns Ltd.

V3, V9, V16, V22 – Ball valves – Gachot SA
VALLEY VALVES – Valves, rotary, slide, butterfly diverter, and pinch – Bush & Wilton Valves Ltd.
VALSTOP – Non-return valve – Enolgas Bonomi sas
VANOX – Gate valves – Gachot SA
VARIPOS – Positioner control device – Worcester Controls (UK) Ltd.
VERSA – Air and solenoid valves – Helipebs Controls Ltd.

WEY – Valves, penstocks – Sistag Maschinenfabrik Sidler Stalder AG
WEY – Slide valves, penstocks – The Reiss Engineering Co Ltd.

Sub-section(b)

CLASSIFIED INDEX OF MANUFACTURERS PRODUCING VALVES, PIPES, PIPELINES AND/OR RELATED EQUIPMENT

BACKFLOW PREVENTERS
SOCLA

BALL COCKS
Enolgas Bonomi sas
Rich.Klinger Aktiengesellschaft
Starmet SRL

BLOW GUNS
IMI Dyna-Quip Ltd.

BOLT TENSIONERS FOR VALVE MOUNTING
Uhde GmbH, Werk Hagan

CASTINGS
Eurovalve Ltd (Selling agent Stanton & Staveley Ltd).
Fagersta AB.
Fundiciones Caetano SA
Haapakosken Tehdas OY

CATHODIC PROTECTION
Moody Tottrup International (Holdings) Ltd.

CHAINWHEELS
Costruzioni Meccaniche Lupi SNC

COCKS
Enolgas Bonomi sas
Rich.Klinger Aktiengelsellschaft
Pont A.Mousson SA

COLLARS – See PIPE COUPLINGS
Fagersta AB

CONTROL ROOM INSTALLATION ELECTRIC/ HYDRAULIC
Schrader Bellows
Superfos Hydraulic A/S

CONTROL ROOM INSTALLATION HYDRAULIC
Superfos Hydraulic A/S

CONTROL ROOM INSTRUMENTATION, PNEUMATIC
Helipebs Controls Ltd.
IMI Norgren Enots Ltd
Joucomatic Controls Ltd.

COUPLINGS – See PIPE COUPLINGS
Fagersta AB.
Tour & Andersson Ltd.

COUPLINGS, QUICK RELEASE HOSE
IMI Dyna-Quip Ltd.

DESUPERHEATER
A. Schneider

DIAPHRAGMS
Fluorocarbon Co Ltd.

FIELD INSTRUMENTATION, PNEUMATIC
IMI Norgren Enots Ltd

FILTERS
B.V.M.I. Ltd.
Ducroux
Gachot SA
IMI Norgren Enots Ltd.
Joucomatic Controls Ltd
Ets Klein
Naegelen SA
Schrader Bellows

FITTINGS, STAINLESS STEEL
Fagersta AB
Frank Stacey

FLANGE LEAK REPAIR UNDER PRESSURE
Furmanite International Ltd.

FLANGED GASKETS
The Victaulic Co Ltd.

FLANGES
Fagersta AB
Haapakosken Tehdas OY
Mannesmann Demag Hüttentechnik
Uhde GmbH, Werk Hagan

FLANGES, STAINLESS STEEL
Frank Stacey

FLOW CONTROLLER
Tour & Andersson Ltd.

FLOW METERS
Pont A.Mousson SA

GASKETS, BONNET AND BODY
Latty International

GAUGES, TEMPERATURE, PRESSURE, REFRIGERANT, DIFFERENTIAL
Eriks-Allied Polymer Ltd.

HANDSTOPS
The Reiss Engineering Co Ltd.

HANDWHEELS
Costruzioni Meccaniche Lupi SNC
Glenfield & Kennedy Ltd.

HOSES
Aflex Hose Ltd.
Eriks-Allied Polymer Ltd.
Schrader Bellows
Standard Hose Ltd.
Witzenmann GmbH Metallschlauch-Fabrik Pforzheim

HYDRANTS
J. Blakeborough & Sons Ltd.

IDENTIFICATION TAPES (PIPELINE)
Ian Campbell & Co Ltd.

IDENTIFICATION TAPES (GAS MEDICAL)
Ian Campbell & Co Ltd.

IDENTIFICATION TRIANGLES (PIPELINE)
Ian Campbell & Co Ltd.

IN SITU MACHINING
Furmanite International Ltd.

INSPECTION FOR PIPES, PIPELINES AND EQUIPMENT
Moody Tottrup International (Holdings) Ltd.

INSULATION, THERMAL
Rubber-Astic & Co Ltd.

JETTING EQUIPMENT
Eriks-Allied Polymer Ltd.

JOINTING COMPOUNDS
Fundiciones Caetano SA
Rich.Klinger Aktiengesellschaft

JOINTS
Mannesmann Demag Hüttentechnik

JOINTS, EXPANSION
Naegelen SA
SOCLA
Witzenmann GmbH Metallschlauch-Fabrik Pforzheim
Yarway Europa BV

JOINTS, FLEXIBLE
Aflex Hose Ltd
Eriks-Allied Polymer Ltd.
Eurovalve Ltd. (Selling agent Stanton & Staveley Ltd)

Fundiciones Caetano SA
Rubber-Astic & Co Ltd.
Standard Hose Ltd.
Witzenmann GmbH Metallschlauch-Fabrik Pforzheim

JOINTS, HYGIENIC
Fluorocarbon Co Ltd

LINE BLINDS
Hindle Cockburns Ltd.

LUBRICATORS, AIR LINE
IMI Norgren Enots Ltd.

MANIFOLDS
Ducroux
Fundiciones Caetano SA
Schrader Bellows

METAL BELLOWS FOR GLANDLESS SEALING OF VALVES
Witzenmann GmbH Metallschlauch-Fabrik Pforzheim

NOISE REDUCTION EQUIPMENT
Eriks-Allied Polymer Ltd.
IMI Norgren Enots Ltd.

NOZZLE CHECK VALVES
Mannesmann Demag Hüttentechnik

PACKINGS, STEM GLAND
Latty International

PENSTOCKS
B.V.M.I. Ltd.
Glenfield & Kennedy Ltd.
Haapakosken Tehdas OY
The Reiss Engineering Co Ltd.
Sistag Maschinen Fabrik Sidler Stalder AG

PIPE CLAMPS
Fagersta AB

PIPE CLIPS
Fagersta AB

PIPE CUTTERS
E. H. Wachs Co

PIPE FITTINGS, PLASTIC
Schrader Bellows
Stewarts & Lloyds Plastics

PIPE HANGERS ETC.
Fagersta AB
Fluorocarbon Co Ltd.
Witzenmann GmbH Metallschlauch-Fabrik Pforzheim

PIPE JOINTS
Eurovalve Ltd (Selling agent Stanton & Staveley Ltd)
Fundiciones Caetano SA
Naegelen SA
Rubber-Astic & Co Ltd.
Witzenmann GmbH Metallschlauch-Fabrik Pforzheim

PIPE MATERIALS, FERROUS
Eurovalve Ltd (Selling agent Stanton & Staveley Ltd)
Fundiciones Caetano SA
Haapakosken Tehdas OY

PIPE MATERIALS, NON-FERROUS
Eurovalve Ltd (Selling agent Stanton & Staveley Ltd)
Perrin GmbH

PIPE MATERIALS, PLASTIC
Eurovalve Ltd (Selling agent Stanton & Staveley Ltd)
Fluorocarbon Co Ltd

PIPE PACKS/SUPPORTS
Corak Ltd

PIPE TAPS
Starmet SRL

PIPE, WELDED STAINLESS STEEL
Frank Stacey

PIPELINE DESUPERHEATERS
J. Blakeborough & Sons Ltd.

PIPELINE LEAK REPAIR UNDER PRESSURE
Furmanite International Ltd.

PIPELINES, CHEMICAL
R. Blackett Charlton & Co Ltd.
Fluorocarbon Co Ltd.
Naegelen SA

PIPES, ASBESTOS CEMENT
Fundiciones Caetano SA

PIPES, COATED
Fluorocarbon Co Ltd
Haapakosken Tehdas OY

PIPES, CONCRETE
Eurovalve Ltd (Selling agent Stanton & Staveley Ltd)

PIPES, COPPER
Schrader Bellows

PIPES, DUCTILE IRON
Pont A. Mousson SA

PIPES, FLEXIBLE
Aflex Hose Ltd
Eriks-Allied Polymer Ltd
Naegelen SA
Standard Hose Ltd
Witzenmann GmbH Metallschlauch-Fabrik Pforzheim

PIPES, GRP
Dunlop Ltd, Dunlopipe Division
Eurovalve Ltd (Selling agent Stanton & Staveley Ltd)

PIPES, INSULATED
Pont A.Mousson SA

PIPES, IRON AND STEEL
Dunlop Ltd, Dunlopipe Division
Eurovalve Ltd (Selling agent Stanton & Staveley Ltd)
Fundiciones Caetano SA
Haapakosken Tehdas OY
Pont A.Mousson SA

PIPES, PLASTIC
Aflex Hose Ltd
Fluorocarbon Co Ltd
Joucomatic Controls Ltd
Schrader Bellows
Stewarts & Lloyds Plastics

PIPES, POLYETHELYNE
Stewarts & Lloyds Plastics

PIPES, PTFE LINED
Naegelen SA

PIPES, SEAMLESS STAINLESS STEEL
Frank Stacey

PIPES, SMALL BORE
Stewarts & Lloyds Plastics

PIPES, STAINLESS STEEL
Fagersta AB
Frank Stacey

PIPES, THERMOPLASTIC
Fluorocarbon Co Ltd
Stewarts & Lloyds Plastics

PIPES, STEEL – CORROSION PROTECTED
Dunlop Ltd, Dunlopipe Division

PORTABLE PIPE GRINDER
E. H. Wachs Co.

PORTABLE PIPE LATHE
E. H. Wachs Co.

PORTABLE WELD CROWN GRINDER
E. H. Wachs Co.

POSITIONERS
Arca-Regler GmbH
Gachot SA

PROCUREMENT FOR PIPES, PIPELINES AND EQUIPMENT
Moody Tottrup International (Holdings) Ltd.

QUALITY ASSURANCE FOR PIPES, PIPELINES AND EQUIPMENT
Moody Tottrup International (Holdings) Ltd.

QUICK RELEASE COUPLINGS
Mannesmann Demag Hüttentechnik

RADIOGRAPHIC EQUIPMENT
Moody Tottrup International (Holdings) Ltd.

REGULATORS
Gustav F. Gerdts GmbH & Co KG
IMI Norgren Enots Ltd.
Schrader Bellows

REPAIR OF EROSION AND CORROSION DAMAGE
Furmanite International Ltd.

RUBBER JOINT SEALING RINGS
The Victaulic Co Ltd.

SAFETY VALVE TESTING ON-LINE
Furmanite International Ltd

SEALS
Eriks-Allied Polymer Ltd
Rich. Klinger Aktiengesellschaft
Latty International
Rubber-Astic & Co Ltd
The Victaulic Co Ltd

SEALS, PRESSURE TEST
A. Schneider

SEALS, STEM GLAND
Latty International

SIGHT FLOWS
Naegelen SA

SILENCERS
IMI Norgren Enots Ltd.
Joucomatic Controls Ltd
A. Schneider
Schrader Bellows

SLEEVES
Haapakosken Tehdas OY

SNUBBERS
IMI Norgren Enots Assoc. Co. IMI Shipston Ltd.

SPLIT-DISC CHECK VALVES
Gustav F. Gerdts GmbH & Co KG

STAINERS
Caen SA

STAINLESS STEEL FITTINGS
Meca-Inox

STEAM TRAP FAILURE INDICATORS
Furmanite International Ltd.

STOP LOGS
The Reiss Engineering Co Ltd.

STOPPING-OFF EQUIPMENT
Furmanite International Ltd.

STRAINERS
Naegelen SA

SWING CHECK VALVES
Gustav F. Gerdts GmbH & Co KG

SWITCHES, LIMIT
Gachot SA

TESTING EQUIPMENT
Uhde GmbH, Werk Hagan

TESTING EQUIPMENT, NON-DESTRUCTIVE
Moody Tottrup International (Holdings) Ltd.

TRAPS
Gustav F. Gerdts GmbH & Co KG

TRAPS, STEAM
B.V.M.I. Ltd.
Eriks-Allied Polymer Ltd.
Gustav F. Gerdts GmbH & Co KG

TUBE, SEAMLESS STAINLESS STEEL
Frank Stacey

TUBE, WELDED STAINLESS STEEL
Frank Stacey

TUBES, NON-FERROUS
Starmet SRL

TUBES, STAINLESS STEEL
Fagersta AB

ULTRASONIC EQUIPMENT (FOR HIRE)
Moody Tottrup International (Holdings) Ltd.

UNDER PRESSURE LEAK SEALING
Furmanite International Ltd.

UNDER PRESSURE SAFETY VALVE TESTING
Furmanite International Ltd.

UNIONS

Fagersta AB
Meca-Inox
Perrin GmbH

VACUUM BREAKERS

B.V.M.I. Ltd.
Costruzioni Meccaniche Lupi SNC
Gustav F. Gerdts GmbH & Co KG
Glenfield & Kennedy Ltd
Schrader Bellows
SOCLA

VALVE ACCESSORIES

Glenfield & Kennedy Ltd
Haapakosken Tehdas OY
Helipebs Controls Ltd
IMI Norgren Enots Ltd
Masoneilan Ltd
The Reiss Engineering Co Ltd
Schrader Bellows
Superfos Hydraulic A/S

VALVE ACTUATORS, AUTOMATED

J. Blakeborough & Sons Ltd
Helipebs Controls Ltd
Joucomatic Controls Ltd
Mark Control International
Masoneilan Ltd.
Merobel SA
Orbit Valve Ltd.
Rotork Controls Ltd.
Schrader Bellows
Tour & Andersson Ltd.

VALVE ACTUATORS, COMPUTER-CONTROLLED

J. Blakeborough & Sons Ltd.
Rotork Controls Ltd.
Schrader Bellows
Tour & Andersson Ltd.

VALVE ACTUATORS, ELECTRIC

Arca-Regler GmbH
J. Blakeborough & Sons Ltd.
Flexible Drives (Gilmans) Ltd
Gachot SA
Haapakosken Tehdas OY
Helipebs Controls Ltd
Joucomatic Controls Ltd
Mark Control International
Merobel SA
Rotork Controls Ltd
Schrader Bellows
A. Schneider
Tour & Andersson Ltd
E. H. Wachs Co.
Worcester Controls (UK) Ltd.

VALVE ACTUATORS, ELECTRO-PNEUMATIC

Gachot SA
General Torque (UK) Ltd.
Helipebs Controls Ltd.
Joucomatic Controls Ltd.
Masoneilan Ltd.
Rotork Controls Ltd.
Schrader Bellows
A. Schneider

VALVE ACTUATORS, HYDRAULIC

Fundiciones Caetano SA
General Torque (UK) Ltd.
Haapakosken Tehdas OY
Helipebs Controls Ltd
Merobel SA
Rotork Controls Ltd.
A. Schneider
Superfos Hydraulic A/S
E. H. Wachs Co.
Walton Engineering Co Ltd.

VALVE ACTUATORS, LINEAR

J. Blakeborough & Sons Ltd.
Helipebs Controls Ltd.
Joucomatic Controls Ltd.
Masoneilan Ltd.
Orbit Valve Ltd.
Rotork Controls Ltd.
Schrader Bellows
Tour & Andersson Ltd.
Walton Engineering Co Ltd.

VALVE ACTUATORS, MANUAL

Fundiciones Caetano SA
Haapakosken Tehdas OY
Mark Control International
Merobel SA
Schrader Bellows
Tour & Andersson Ltd.

VALVE ACTUATORS FOR NUCLEAR DUTIES

Rotork Controls Ltd.

VALVE ACTUATORS, PNEUMATIC

Arca-Regler GmbH
J. Blakeborough & Sons Ltd.

EFFEBI SpA
Eriks-Allied Polymer Ltd.
Gachot SA
Haapakosken Tehdas OY
Helipebs Controls Ltd.
Joucomatic Controls Ltd.
Ets Klein
KTM (UK)
Mark Control Internatinal
Masoneilan Ltd.
Merobel SA
Mono Valve Engineering Ltd.
Neles OY
Orbit Valve Ltd.
Perrin GmbH
Rotork Controls Ltd
Schrader Bellows
A. Schneider
Tour & Andersson Ltd
E. H. Wachs Co
Walton Engineering Co Ltd.
Worcester Controls (UK) Ltd.

VALVE ACTUATORS, PORTABLE
Flexible Drives (Gilmans) Ltd
E. H. Wachs Co.

VALVE ACTUATORS, ROTARY
Flexible Drives (Gilmans) Ltd.
Helipebs Controls Ltd.
Masoneilan Ltd.
Rotork Controls Ltd.
Schrader Bellows
A. Schneider
Superfos Hydraulic A/S
Tour & Andersson Ltd.

VALVE, JACKETS
Ducroux
Gachot SA
Masoneilan Ltd.

VALVE LEAK REPAIR UNDER PRESSURE
Furmanite International Ltd.

VALVE, PACKINGS
Eirks-Allied Polymer Ltd.
Furmanite International Ltd.
Masoneilan Ltd.

VALVE POSITIONERS – See VALVE ACTUATORS
Arca-Regler GmbH
Masoneilan Ltd.
Neles OY
Rotork Controls Ltd.

VALVE TEST BENCH
Uhde GmbH, Werk Hagan

VALVES AND FITTINGS
Mannesmann Demag Hüttentechnik

VALVES, AIR RELIEF
J. Blakeborough & Sons Ltd.

VALVES, AUTOMATIC CONTROL
Arca-Regler GmbH
R. Blackett Charlton & Co Ltd.
J. Blakeborough & Sons Ltd.
B.V.M.I. Ltd.
General Torque (UK) Ltd.
Gustav F. Gerdts GmbH & Co KG
Glenfield & Kennedy Ltd.
Joucomatic Controls Ltd
Magisco Valves Ltd
Masoneilan Ltd.
Neles OY
Perrin GmbH
Schrader Bellows
Tour & Andersson Ltd
Uhde GmbH, Werk Hagan
Walton Engineering Co Ltd.
Charles Winn (Valves) Ltd.
Worcester Controls (UK) Ltd.

VALVES, AUTOMATIC TEMPERATURE CONTROL
J. Blakeborough & Sons Ltd.
B.V.M.I. Ltd.
Tour & Andersson Ltd.
Walton Engineering Co Ltd.

VALVES, BALANCING
Tour & Andersson Ltd.

VALVES, BALL
EFFEBI, SpA
Enolgas Bonomi sas
Eriks-Allied Polymer Ltd
Gachot SA
General Torque (UK) Ltd.

Hindle Cockburns Ltd.
IMI Dyna-Quip Ltd.
IMI Norgren Enots Ltd.
Joucomatic Controls Ltd.
Rich.Klinger Aktiengesellschaft
KTM (UK)
Mark Control International
Masoneilan Ltd
Meca-Inox
Naegelen SA
Neles OY
Orbit Valve Ltd.
Perrin GmbH
Pont A. Mousson SA
Schrader Bellows
SOCLA
Starmet SRL
Tour & Andersson Ltd
Worcester Controls (UK) Ltd.

VALVES, BELLOWS SEAL
B.V.M.I. Ltd.

VALVES, BELLOWS SEALED
Ets Klein

VALVES, BLOCK AND BLEED
General Torque (UK) Ltd.
Hindle Cockburns Ltd.
KTM (UK)
Mark Control International
Neles OY
Orbit Valve Ltd.
Perrin GmbH

VALVES, BLOWDOWN
Gustav F. Gerdts GmbH & Co KG
Masoneilan Ltd.
Neles OY
Orbit Valve Ltd.

VALVES, BUTTERFLY
Arca-Regler GmbH
J. Blakeborough & Sons Ltd.
Bush & Wilton Valves Ltd.
B.V.M.I. Ltd.
Caen, SA
Eriks-Allied Polymer Ltd.
Gachot SA
General Torque (UK) Ltd.
Glenfield & Kennedy Ltd.
Hindle Cockburns Ltd.
Joucomatic Controls Ltd.
Mark Control International
Masoneilan Ltd.
Mono Valve Engineering Ltd.
Naegelen SA
Neles OY
Pont A. Mousson SA
The Reiss Engineering Co Ltd.
Sistag Maschinenfabrik Sidler Stalder AG
Tour & Andersson Ltd.
Charles Winn (Valves) Ltd.
Worcester Controls (UK) Ltd.

VALVES, CHECK
Amis SpA Apparechi e Macchine Idrauliche Speciali
B.E.L. Valves
R. Blackett Charlton & Co Ltd.
J. Blakeborough & Sons Ltd.
B.V.M.I. Ltd.
Caen SA
Ducroux
EFFEBI SpA
Enolgas Bonomi sas
Eriks-Allied Polymer Ltd.
Gachot SA
General Torque (UK) Ltd.
Gustav F. Gerdts GmbH & Co KG
Glenfield & Kennedy Ltd
Haapakosken Tehdas OY
Helipebs Controls Ltd.
Hindle Cockburns Ltd.
Joucomatic Controls Ltd.
Ets Klein
Meca-Inox
Naegelen SA
Ocean BV
Schrader Bellows
A. Schneider
Sistag Maschinenfabrik Sidler Stalder AG
SOCLA
Starmet SRL
Tour & Andersson Ltd.
TRW Mission Ltd.
Uhde GmbH, Werk Hagan

VALVES, CONTROL
A. Schneider

VALVES, CORROSION RESISTANT
B.E.L. Valves

VALVES, CRYOGENIC
Arca-Regler GmbH
R. Blackett Charlton & Co Ltd.
J. Blakeborough & Sons Ltd.
B.V.M.I. Ltd.
Caen, SA
Ducroux
Gachot SA
General Torque (UK) Ltd.
Helipebs Controls Ltd.
Hindle Cockburns Ltd.
KTM (UK)
Masoneilan Ltd.
Neles OY
TRW Mission Ltd
Charles Winn (Valves) Ltd.
Worcester Controls (UK) Ltd.

VALVES, DIAPHRAGM
Arca-Regler GmbH
R. Blackett Charlton & Co Ltd.
Fundiciones Caetano SA
Helipebs Controls Ltd.
IMI Norgren Enots Ltd.
Joucomatic Controls Ltd.
Masoneilan Ltd.
Merobel SA
Ocean BV
Schrader Bellows
Charles Winn (Valves) Ltd.

VALVES, DIVERTER
Bush & Wilton Valves Ltd.
Glenfield & Kennedy Ltd.
Helipebs Controls Ltd.
Joucomatic Controls Ltd.
Masoneilan Ltd.
Perrin GmbH
The Reiss Engineering Co Ltd.
Starmet SRL
Tour & Andersson Ltd.
Walton Engineering Co Ltd.
Charles Winn (Valves) Ltd.

VALVES FOR ELECTRIC TRANSFORMER
Caen, SA

VALVES, ELECTRICALLY OPERATED
R. Blackett Charlton & Co Ltd.
J. Blakeborough & Sons Ltd.
B.V.M.I. Ltd.
General Torque (UK) Ltd.
Glenfield & Kennedy Ltd.
Haapakosken Tehdas Oy.
Helipebs Controls Ltd.
Joucomatic Controls Ltd.
Ets Klein
Magisco Valves Ltd
Mark Control International
Neles OY
The Reiss Engineering Co Ltd.
Schrader Bellows
A. Schneider
Sistag Maschinenfabrik Sidler Stalder AG
Tour & Andersson Ltd.
Walton Engineering Co Ltd.
Charles Winn (Valves) Ltd.

VALVES, EMERGENCY STOP
A. Schneider

VALVES, FIRE HYDRANT
Eurovalve Ltd (Selling agent Stanton & Staveley Ltd).

VALVES, FIRE RESISTANT
Helipebs Controls Ltd.
KTM (UK)
Mark Control International
Neles OY
Orbit Valve Ltd.
Perrin GmbH
Charles Winn (Valves) Ltd.

VALVES, FLAPPER
B.E.L. Valves
The Reiss Engineering Co Ltd.

VALVES, FLOAT
J. Blakeborough & Sons Ltd.
Ducroux
Glenfield & Kennedy Ltd.
Hindle Cockburns Ltd.
Magisco Valves Ltd
Sistag Maschinenfabrik Sidler Stalder AG
Starmet SRL

VALVES, FLOW
J. Blakeborough & Sons Ltd.
Glenfield & Kennedy Ltd.
Helipebs Controls Ltd.
Magisco Valves Ltd
Schrader Bellows
Tour & Andersson Ltd.
Charles Winn (Valves) Ltd.

VALVES, FLOW CONTROL
Worcester Controls (UK) Ltd.

VALVES, FOOT
J. Blakeborough & Sons Ltd.
Ducroux
EFFEBI SpA
Enolgas Bonomi sas
Eriks-Applied Polymer Ltd.
Glenfield & Kennedy Ltd.
Haapakosken Tehdas OY
Helipebs Controls Ltd.
Joucomatic Controls Ltd.
Schrader Bellows
SOCLA
Starmet SRL

VALVES, GAS/AIR
Arca-Regler GmbH
B.E.L. Valves
R. Blackett Charlton & Co Ltd.
J. Blakeborough & Sons Ltd.
B.V.M.I. Ltd.
Caen, SA
EFFEBI SpA
Enolgas Bonomi sas
Eriks-Allied Polymer Ltd
Fundiciones Caetano SA
Gachot SA
General Torque (UK) Ltd.
Gustav F. Gerdts GmbH & Co KG
Glenfield & Kennedy Ltd.
Helipebs Controls Ltd.
IMI Dyna-Quip Ltd.
IMI Norgren Enots Ltd.
Joucomatic Controls Ltd
Ets Kelin
Magisco Valves Ltd
Merobel SA
Schrader Bellows
Sistag Maschinenfabrik Sidler Stalder AG
Starmet SRL
Tour & Andersson Ltd
TRW Mission Ltd.

VALVES, GATE
R. Blackett Charlton & Co Ltd
J. Blakeborough & Sons Ltd.
B.V.M.I. Ltd.
Ducroux
EFFEBI SpA
Eriks-Allied Polymer Ltd.
Eurovalve Ltd (Selling agent Stanton & Staveley Ltd).
Gachot SA
General Torque (UK) Ltd.
Glenfield & Kennedy Ltd.
Haapakosken Tehdas OY
Hindle Cockburns Ltd.
Mark Control International
Merobel SA
Pont A.Mousson SA
The Reiss Engineering Co Ltd
A. Schneider
Sistag Maschinenfabrik Sidler Stalder AG
Starmet SRL
Tour & Andersson Ltd
Uhde GmbH, Werk Hagan

VALVES, GEAR OPERATED
B.E.L. Valves
R. Blackett Charlton & Co Ltd.
J. Blakeborough & Sons Ltd.
B.V.M.I. Ltd.
Eriks-Allied Polymer Ltd.
Gachot SA
General Torque (UK) Ltd.
Glenfield & Kennedy Ltd.
Haapakosken Tehdas OY
Hindle Cockburns Ltd.
Mark Control International
Mono Valve Engineering Ltd.
Neles OY
The Reiss Engineering Co Ltd.
A. Schneider
Sistar Maschinenfabrik Sidler Stalder AG
Uhde GmbH, Werk Hagan

VALVES, GLOBE
Arca-Regler GmbH
B.E.L. Valves
R. Blackett Charlton & Co Ltd.
J. Blakeborough & Sons Ltd.
B.V.M.I. Ltd.
Ducroux
Eriks-Allied Polymer Ltd.
Gachot S.A.
General Torque (UK) Ltd.
Helipebs Controls Ltd.
Hindle Cockburns Ltd.
Joucomatic Controls Ltd.
Ets Kelin
Rich.Klinger Aktiengesellschaft
Mark Control International

VALVES, NUCLEAR SERVICE

R. Blackett Charlton & Co Ltd.
J. Blakeborough & Sons Ltd.
B.V.M.I. Ltd.
Caen, SA
Ducroux
Gachot SA
General Torque (UK) Ltd.
Helipebs Controls Ltd.
Neles OY
Perrin GmbH
A. Schneider

VALVES, OXYGEN SERVICE

Arca-Regler GmbH
R. Blackett Chalrton & Co Ltd.
J. Blakeborough & Sons Ltd.
Caen, SA
General Torque (UK) Ltd.
Helipebs Controls Ltd.
Hindle Cockburns Ltd.
IMI Dyna-Quip Ltd.
Monteilan Ltd.
Merobel SA
Neles OY
Orbit Valve Ltd
TRW Mission Ltd.
Charles Winn (Valves) Ltd.

VALVES, PACKLESS (WITH BELLOW)

Caen, SA

VALVES, PENSTOCK/SLUICE

J. Blakeborough & Sons Ltd.
B.V.M.I. Ltd.
General Torque (UK) Ltd.
Glenfield & Kennedy Ltd.
Haapakosken Tehdas OY
The Reiss Engineering Co Ltd.
Sistag Maschinenfabrik Sidler Stalder AG

VALVES, PINCH

Bush & Wilton Valves Ltd.
Magisco Valves Ltd

VALVES FOR PIPELINES

R. Blackett Charlton & Co Ltd.
J. Blakeborough & Sons Ltd.
Eurovalve Ltd. (Selling agent Stanton & Staveley Ltd).
General Torque (UK) Ltd.
Gustav F. Gerdts GmbH & Co KG
Glenfield & Kennedy Ltd.
Haapakosken Tehdas OY
Helipebs Controls Ltd.
IMI Dyna-Quip Ltd.
IMI Norgren Enots Ltd.
Joucomatic Controls Ltd.
Magisco Valves Ltd
Mark Control International
Masoneilan Ltd.
Orbit Valve Ltd.
Schrader Bellows
TRW Mission Ltd.
Charles Winn (Valves) Ltd.

VALVES, PISTON

Arca-Regler GmbH
B.E.L. Valves
Glenfield & Kennedy Ltd.
Joucomatic Controls Ltd.
Rich.Klinger Aktiengesellschaft
A. Schneider

VALVES, PLASTIC

Gachot SA
SOCLA

VALVES, PLASTIC LINED

Masoneilan Ltd.

VALVES, PLUG

Arca-Regler GmbH
Enolgas Bonomi sas
Gachot SA
General Torque (UK) Ltd.
IMI Dyna-Quip Ltd.
Masoneilan Ltd.
A. Schneider

VALVES, PNEUMATIC

Arca-Regler GmbH
B.E.L. Valves
Ducroux
Gachot SA
General Torque (UK) Ltd.
Glenfield & Kennedy Ltd.
Helipebs Controls Ltd.
IMI Norgren Enots Ltd.
Joucomatic Controls Ltd.
Ets Klein
Masoneilan Ltd.
Perrin GmbH
The Reiss Engineering Co Ltd.
Schrader Bellows

VALVES, VACUUM

R. Blackett Charlton & Co Ltd.
B.V.M.I. Ltd.
General Torque (UK) Ltd.
Glenfield & Kennedy Ltd.
Helipebs Controls Ltd.
IMI Dyna-Quip Ltd.
Schrader Bellows

VALVES, VENTURI

General Torque (UK) Ltd.
Ets Klein
Meca-Inox

WELDED PIPES, FERROUS

R. Blackett Charlton & Co Ltd.
Fagersta AB
Pont A.Mousson SA
Moody Tottrup International (Holdings) Ltd.

X-RADIOGRAPHY EQUIPMENT

General Torque (UK) Ltd.

X-RADIOGRAPHY EQUIPMENT (FOR HIRE)

Moody Tottrup International (Holdings) Ltd.

Sub-section(c)

ALPHABETICAL LIST OF MANUFACTURERS WITH ADDRESSES, TELEPHONE NUMBERS, TELEGRAM ADDRESSES OF HEAD OFFICE, WORKS AND BRANCHES

AFLEX HOSE LTD., Owlerings Mill, Brighouse, W.Yorkshire, HD6 1EJ.
Telephone: (0484) 712311. Telex: 51654

A M I S SpA Apparecchi e Macchine Idrauliche Speciali, Via Borgomasino 71/73, 10149 Torino, Italy.
Telephone: (011) 7399696. Telex: 213517 AMIS I. Grams: Idroamis Torino

ARCA – REGLER GMBH, Kempener Str. 18, P.O.Box 20, D-4154 Tönisvorst 2
Telephone: 2156/7021-24. Telex: 853 599. Grams: ARCA Tönisvorst 2

B.E.L. VALVES, St. Peters, Newcastle-upon-Tyne NE6 1BS.
Telephone: 0632 659091. Telex: 53177

R. BLACKETT CHARLTON & COMPANY LIMITED, White Street, Walker, Newcastle-upon-Tyne, Tyne & Wear, NE6 3QH.
Telephone: 0632 625361. Telex: 53437

J. BLAKEBOROUGH & SONS LIMITED, P.O.Box 11, Brighouse, West Yorkshire, HD6, 1NH.
Telephone: 0484 715511. Telex: 51202

BUSH & WILTON VALVES LIMITED, Golden Valley Lane, Bitton, Bristol BS15 6LF.
Telephone: Bitton (027588) 2131

BVMI LTD., Shaw Road, Bushbury, Wolverhampton WV10 9NN.
Telephone: 0902 20496. Telex: 338171

CAEN, S.A., Jon Arrospide, 11-Bilbao (14), Espana.
Telephone: 94/4350675, 94/4473996

IAN CAMPBELL & CO. LTD., Britannia Works, West Road, London E15.
Telephone: 01 472 6018 or 01 470 1223. Telex: 896691 TLXIR-G-CAMPAK

CORAK LIMITED, 1 Dunraven Street, London W1Y 3FG.
Telephone: 01 408 1677. Telex: 8953652
Management Office:
Corakges S.A., 8 Avenue Calas, 1206 Geneva, Switzerland.

COSTRUZIONI MECCANICHE LUPI S.N.C.
Via Cancelliera, 27-00040 Cecchina Di Ariccia, Rome, Italy.
Telephone: 06 9315233. Telex: 614191 CMLVPI

DUCROUX,
12 bis rue du général Leclerc, Ronchin, France
Telephone: 20 53 43 55. Telex: 130206

DUNLOP LIMITED, DUNLOPIPE DIVISION,
Holbrook Lane, Coventry, CV6 4AA.
Telephone: 0203 88733. Telex: 31677. Grams: Sound Coventry

EFFEBI SpA, Via Industriale No.4, I-25060-Polaveno (Bs)
Telephone: 030 84151/2. Telex: 301289 EFFEBI. Grams: EFFEBI Polaveno

ENOLGAS BONOMI sas, via Europa 229-25062, Concesio (Bs), Italy
Telephone: 030 2751361. Telex: 300555

ERIKS-ALLIED POLYMER LIMITED, Manchester Industrial Centre, Water Street, Manchester M3 4JU
Telephone: 061 832 6784. Telex: 66216

EUROVALVE LIMITED (Selling Agent Stanton & Staveley Ltd), P.O.Box 72, Nr. Nottingham, NG10 5AA.
Telephone: 0602 322121. Telex: 37671 S&SSG. Grams: Stanstaves, Nottm. Telex.
Works:
Eurovalve Limited, P.O.Box 12, Ilkeston, Derbyshire, DE7 5RT.
Telephone: 0602 302054

FAGERSTA AB., Box 501, S-77301 Fagersta, Sweden.
Telephone: 010 46223-45000. Telex: 7525

FLEXIBLE DRIVES (GILMANS) LTD., Skatoskalo Works, Kineton Road Industrial Estate, Southam, Leamington Spa, Warwickshire CV33 ODS.
Telephone: 092 681 3818. Telex: 31451 fdg cel. Grams: Skatoskalo Southam, Leamington Spa.

FLUOROCARBON COMPANY LIMITED, Caxton Hill, Hertford, Herts, SG13 7NH
Telephone: Hertford 50731. Telex: 81435

FUNDICIONES CAETANO S.A., Apartado 952, Sevilla, Spain.
Telephone: 390260. Telex: 72565

FURMANITE INTERNATIONAL LIMITED, Furman House, Shap Road, Kendal, Cumbria LA9 6RO
Telephone: 0539 29009. Telex: 65262

GACHOT S.A., BP 14, 26 bis, avenue de Paris, 95230 Soisy s/s Montmorency, France.
Telephone: (3) 989.90.11. Telex: 698 671F

GENERAL TORQUE (U.K.) LTD., Suite No.1, Broadway Chambers, Broadway North, Pitsea, Basildon, Essex SS16 4JB.
Telephone: 0268 558545. Telex: 995905

GUSTAV F. GERDTS, GMBH & Co KG, Postfach 105549, Hemmstrasse 130, D-2800 Bremen 1.
Telephone: 0421 3 50 31. Telex: 244945 gfg d. Grams: gestra bremen

GLENFIELD & KENNEDY LIMITED, P.O.Box 3, Kilmarnock, Scotland KA1 3XH.
Telephone: 0563 21150. Telex: 77258 Gand.K.

HAAPAKOSKEN TEHDAS OY, 77520 Haapakoski, Finland.
Telephone: (9)58-48102. Telex: 55712

HELIPEBS CONTROLS LTD., Premier Works, Sisson Road, Gloucester GL2 ORE
Telephone: 0452 423201. Telex: 43364

HINDLE COCKBURNS LIMITED, Victoria Road, Leeds LS11 5UG.
Telephone: 443741. Telex: 55257

IMI DYNA-QUIP LIMITED, Shipston-on-Stour, Warwickshire.
Telephone: 0608 61676. Telex: 83208. Grams: Philsym, Shipston

IMI NORGREN ENOTS LIMITED, Shipston-on-Stour, Warwickshire.
Telephone: 0608 61676. Telex: 83208. Grams: Philsym, Shipston

JOUCOMATIC CONTROLS LIMITED, Air Automation House, Navigation Street, Wolverhampton.
Telephone: 0902 59924. Telex: 337372

Ets KLEIN, B.P.No.2, 39190 Cousance, France.
Telephone: (84) 85.90.44. Telex: 360 125

RICH.KLINGER AKTIENGESELLSCHAFT, Am Kanal 8–10, A-2352 Gumpoldskirchen, Austria, P.O.Box 19
Telephone: 02252/62406-0. Telex: 14430 z klgtg. Grams: Klingerit Gumpoldskirchen
Branches:
Rich.Klinger Ges.m.b.H., Rich.Klinger-Strasse, D-6270 Idstein, P.O.Box 1370, German Federal Republic
Richard Klinger Limited, Sidcup, Kent DA14 5AG
Richard Klinger Inc., 2350 Campbell Road, Sidney, Ohio 45365, P.O.B.105, U.S.A.
Klinger de Mexico, S.A., Calzada Olimpica 1518, S.R., Guadalajara, Jal. P.O.B.4-063, Mexico.
Richard Klinger S.A.A.C.I.Y.F., Cangallo 315-5 Piso Of.520, Buenos Aires, P.O.B.3994, Argentina
Richard Klinger Indústria e Comércio Ltda, Rua Vergueiro 1833, 10 Andar, 0100 Sao Paulo, P.O.B.30403, Brazil.

Richard Klinger Ltd., Brakpan, 1540 Transvall, P.O.B.217, South Africa.
Richard Klinger Pty.Ltd., 138–146 Browns Road, Noble Park, Victoria 3174, P.O.B.225, Australia.

KTM (U.K.), Suite No.1, Broadway Chambers, Broadway, North Pitsea, Basildon, Essex SS16 4JB
Telephone: 0268 558545. Telex: 995905

LATTY INTERNATIONAL, 82 Rue St. Lazare, Paris, France.
Telephone: 874 1044. Telex: 290145
Branches:
Latty International Limited, 50 Bracken Lane, Retford, Nottinghamshire DN22 7EX.
Telephone: 0777 704352. Telex: 312242 Mid Tlx g

MAGISCO VALVES LIMITED, Ashton House, 67A Compton Road, Wolverhampton WV3 9QZ.
Telephone: 0902 20028. Telex: CHAMCOM Wolves 338490.

MANNESMANN DEMAG HUTTENTECHNIK, Subdivision Meer, Ohlerkirchweg 66, D-4050 Mönchengladbach 1
Telephone: (2161) 350-1. Telex: 852 525

MARK CONTROLS INTERNATIONAL, Foundry Lane, Horsham, Sussex RH13 5TL
Telephone: Horsham 69727. Telex: 87660

MASONEILAN LTD., Controls House, Riverside Way, Uxbridge, Middlesex UB8 2BF
Telephone: 0895 58161. Telex: 935174

MECA-INOX, 42 Rue de Montigny, B.P.77, 95101 Argenteuil, France.
Telephone: (3) 982.40.90. Telex: 698862
Works:
32 Route des Routis, 60990 Le Coudray, St. Germer
Telephone: (4) 481.62.20. Telex: 140 318

MEROBEL S.A., 58 rue Stendhap 75020, Paris, France.
Telephone: 33.1.797.9350. Telex: 260717 ref 659

MONO VALVE ENGINEERING LIMITED, 17 Chalford Industrial Estate, Stroud, Glos.
Telephone: 0453 883866. Telex: 43587

MOODY-TOTTRUP INTERNATIONAL LIMITED, Oakfield House, Perrymount Road, Haywards Heath, West Sussex, RH16 3BP
Telephone: 0444 456441. Grams: Moody Hayheath. Telex: 877337

NAEGELEN S.A., 8 rue de l'Ill, B.P. 29 Brunstatt, 68200 Mulhouse, France.
Telephone: (89) 06.12.22. Telex: 881677 F

NELES OY, Levytie 6, P.O.Box 4, SF-00811 Helsinki 81, Finland.
Telephone: International+358-0-75841. Telex: 121211 neles sf

OCEAN B.V., P.O.Box 16, NL-6950 AA Dieren, The Netherlands.
Telephone: 08330-19004. Telex: 35365. Grams: Ocean-Dieren

ORBIT VALVE LIMITED, Orbit House, Millington Road, Hayes, Middlesex UB3 4AZ
Telephone: 01 561 8049. Telex: 938171

PERRIN GMBH, Siemensstrasse 1, 6369 Nidderau-1, W.Germany
Telephone: (0) 6187 3021-23. Telex: 04-184874

PONT. A. MOUSSON S.A., 4X 54017 Nancy Cedex, France.
Telephone: (8) 396 81 21. Telex: Pamsa 850 003 F.

THE REISS ENGINEERING CO.LTD., 2 Dalston Gardens, Stanmore, Middlesex.
Telephone: 01 204 7155. Telex: 935400. Grams: Reisengine
Works:
Ivatt Way, Westwood, Peterborough, Northants.
Telephone: 0733 266866

ROTORK CONTROLS LIMITED, Rotork House, Brassmill Lane, Bath, Avon BA1 3JQ
Telephone: 0225 28451. Telex: 44823

RUBBER-ASTIC & COMPANY LIMITED, Old Park Road, Wednesbury, West Midlands WS10 9LR
Telephone: 021 556 4271. Telex: 338057

A. SCHNEIDER GMBH CO KG, Friderikastrasse 148, 4630 Bochum 1
Works:
Hohensteinstasse 52, 4630 Bochum, Wattenscheid
Telephone: 02327/3758. Telex: 820410. Grams: Bomafa

SCHRADER BELLOWS, Walkmill Lane, Bridgtown, Cannock, Staffordshire
Telephone: Cannock 2644. Telex: 336150 Schrad G. Grams: Airvalve Cannock

SISTAG MASCHINENFABRIK SIDLER STALDER AG, CH-6274 Eschenbach/LU
Telephone: 041/89.24.44. Telex: 78189 sista ch. Grams: Sistag
S O C L A, B.P.300 71107 Chalon Sur Saone
Telephone: (85) 46.30.34. Telex: 800 259
FRANK STACEY, Upper Brook Street, Walsall, WS2 9PD.
Telephone: 0922 644333. Telex: 338291

STANDARD HOSE LTD, Owler Ings Mill, Owler Ings Road, Brighouse, Yorkshire HD6 1EJ.
Telephone: Brighouse 2311-2-3. Telex: 51654.

STARMET S.R.L., 14 Via Domenichino, 20149 Milano, Italy.
Telephone: 48.33.55/43.58.29. Telex: 333485 Strmet I. Grams: Starmet
STEWARTS & LLOYDS PLASTICS, St. Peters Road, Huntingdon, Cambridgeshire PE18 7DJ
Telephone: 0480 52121. Telex: 32221 VICSLP G
SUPERFOS HYDRAULICS A/S, Aaderupvej 41, 4700 Naestved, Denmark.
Telephone: 45 3 724225. Telex: 46231
TOUR & ANDERSSON LTD., 149 Lower Luton Road, Harpenden, Herts AL5 5EQ.
Telephone: 05827 67991. Telex: 826465
TRW MISSION LIMITED, Berkeley Square House, Berkeley Square, London W1X 6JE
Telephone: 01 491 3322. Telex: 21379
Works:
Alexander Road, Cregagh, Belfast BT6 9HJ, Northern Ireland
Telephone: 0232 51771. Telex: 747256
UHDE GMBH, WERK HAGEN, Buschmühlenstrasse 20, D-5800 Hagen 1, Fed.Rep. of Germany
Telephone: 0231 6921. Telex: 0823798 uhde d
THE VICTAULIC COMPANY LIMITED, St. Peters Road, Huntingdon, Cambridgeshire PE18 7DJ
Telephone: 0480 52121. Telex: 32221 Vicslpg
E. H. WACHS COMPANY, 100 Shepard St, Wheeling, Illinois 60090, U.S.A.
Telephone: (312) 537 8800. Telex: 283483
WALTON ENGINEERING CO.LTD., 50 Pall Mall, London SW1
Telephone: 01 839 2300/1597. Telex: 24372
Works:
Gauges (St. Albans) Ltd., 17-19 Sutton Road, St.Albans,Herts.
Telephone: St. Albans 58539. Telex: 24372
CHARLES WINN (VALVES) LIMITED, 70 Warwick Street, Birmingham B12 ONL
Telephone: 021 772 6981. Telex: 338100. Grams: Winn B'Ham, Telex.
WITZENMANN GMBH Metallschlauch-Fabrik Pforzheim, P.O.Box 1280, D-7530 Pforzheim
Telephone: 07231/581-1. Telex: 7 83 828-0. Grams: Metallschlauch
WORCESTER CONTROLS (UK) LTD., Burrell Road, Haywards Heath, Sussex RH16 1TL
Telephone: 0444 414133. Telex: 87189
YARWAY EUROPA BV, Rechtzaad 17, P.O.Box 7900, AB Roosendaal, The Netherlands.
Telephone: 01650 50600. Telex: 78146.

INDEX

A

B

C

D

E

F

G

H

I

J

K

L

Q

R

S

T

Index to Advertisers